Surface Science

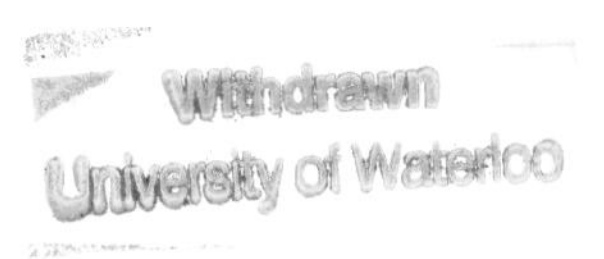

Advanced Texts in Physics

This program of advanced texts covers a broad spectrum of topics which are of current and emerging interest in physics. Each book provides a comprehensive and yet accessible introduction to a field at the forefront of modern research. As such, these texts are intended for senior undergraduate and graduate students at the MS and PhD level; however, research scientists seeking an introduction to particular areas of physics will also benefit from the titles in this collection.

Springer
Berlin
Heidelberg
New York
Hong Kong
London
Milan
Paris
Tokyo

K. Oura V. G. Lifshits A. A. Saranin
A.V. Zotov M. Katayama

Surface Science

An Introduction

With 372 Figures and 16 Tables

Springer

Professor K. Oura
Professor M. Katayama
Department of Electronic Engineering
Faculty of Engineering
Osaka University
Suita, Yamada-oka 2-1
Osaka 565-0871 Japan

Professor V.G. Lifshits
Professor A.A. Saranin
Institute of Automation
and Control Processes
5 Radio Street
690041 Vladivostok
Russia

Professor A.V. Zotov
Vladivostok State University
of Economics and Service
690600 Vladivostok
Russia

Library of Congress Cataloging-in-Publication Data: Surface science: an introduction/ Oura ... [et al.]. p. cm. – (Advanced texts in physics, ISSN 1439-2674) Includes bibliographical references and index. ISBN 3-540-00545-5 (acid-free paper) 1. Surfaces (Physics) I. Oura, K. (Kenjiro), 1941– II. Series. QC 173.4.S94S96425 2003 530.4'27–dc21 2003042812

ISSN 1439-2674

ISBN 3-540-00545-5 Springer-Verlag Berlin Heidelberg New York

Springer-Verlag Berlin Heidelberg New York
a member of BertelsmannSpringer Science+Business Media GmbH

http://www.springer.de

Printed in Germany

Data conversion: Frank Herweg, Leutershausen
Cover design: *design & production* GmbH, Heidelberg

Printed on acid-free paper SPIN 10768993 57/3141/tr 5 4 3 2 1 0

Preface

Surface Science – An Introduction is designed as a textbook for undergraduate and graduate students in engineering and physical sciences who want to get a general overview of surface science. It also provides necessary background information for researchers just starting out in the field. The book covers all the most important aspects of modern surface science with all subjects being presented in a concise and clear form accessible to a beginner.

It includes:

- Experimental background on ultra-high-vacuum technology;
- An overview of the most widely used analytical techniques;
- The basics of two-dimensional crystallography;
- Information on atomic structure of well-defined clean and adsorbate-covered crystal surfaces;
- Consideration of a variety of surface phenomena and properties;
- Application of surface science to thin film growth and nanostructure formation.

The whole is rounded off with:

- Numerous vivid examples from classical and very recent original works;
- Problems and practice exercises which aim to develop a deeper awareness of the subject;
- Principal references for further reading.

Osaka and Vladivostok
January 2003

K. Oura
V.G. Lifshits
A.A. Saranin
A.V. Zotov
M. Katayama

Contents

1. Introduction 1

2. Basics of Two-Dimensional Crystallography 3
2.1 Two-Dimensional Lattices 3
2.1.1 Lattice, Basis, and Crystal Structure (3D Case) 3
2.1.2 Concept of a 2D Lattice 3
2.1.3 2D Bravais Lattices 4
2.2 Miller Indices for Crystal Planes 6
2.2.1 Definition of Miller Indices 6
2.2.2 Low-Miller-Index Planes of Some Important Crystals 8
2.2.3 High-Miller-Index Stepped Surfaces 8
2.3 Indices of Directions 11
2.4 Notation for Surface Structures 11
2.4.1 Matrix Notation 11
2.4.2 Wood's Notation 12
2.4.3 Some Examples 13
2.5 2D Reciprocal Lattice 14
2.6 Brillouin Zone 16
Problems 17
Further Reading 18

3. Experimental Background 19
3.1 Why Ultra-High Vacuum? 19
3.2 Vacuum Concepts 20
3.3 Ultra-High-Vacuum Technology 23
3.3.1 UHV Materials 23
3.3.2 UHV Pumping System 24
3.3.3 UHV Hardware 32
3.4 Preparation of Atomically Clean Surfaces 35
3.4.1 Cleavage 35
3.4.2 Heating 37
3.4.3 Chemical Treatment 37
3.4.4 Ion Sputtering and Annealing 37
3.5 UHV Deposition Technology 38

3.5.1 Deposition Concepts 38
3.5.2 Deposition Sources 39
3.5.3 Deposition Monitors 43
3.5.4 Exposure to Gases 44
Problems 45
Further Reading 46

4. Surface Analysis I. Diffraction Methods 47
4.1 Low-Energy Electron Diffraction (LEED) 47
4.1.1 Ewald Construction in LEED Conditions 47
4.1.2 LEED Experimental Set-Up 49
4.1.3 Interpretation of a LEED Pattern 51
4.2 Reflection High-Energy Electron Diffraction (RHEED) 59
4.2.1 Ewald Construction in RHEED Conditions 59
4.2.2 RHEED Set-Up 60
4.2.3 RHEED Analysis 61
4.3 Grazing Incidence X-Ray Diffraction (GIXRD) 66
4.3.1 Refraction of X-Rays at Grazing Incidence 66
4.3.2 Ewald Construction in GIXRD Conditions and Basics of the Kinematic Approximation 67
4.3.3 GIXRD Experimental Set-Up 69
4.3.4 Structural Analysis by GIXRD 69
4.4 Other Diffraction Techniques 73
4.4.1 Transmission Electron Diffraction (TED) 73
4.4.2 Atom Scattering 73
4.4.3 Photoelectron Diffraction (PED) and Auger Electron Diffraction (AED) 74
Problems 74
Further Reading 75

5. Surface Analysis II. Electron Spectroscopy Methods 77
5.1 General Remarks 77
5.1.1 Surface Specificity 77
5.1.2 Spectrum of Secondary Electrons 77
5.1.3 Electron Energy Analyzers 78
5.2 Auger Electron Spectroscopy 82
5.2.1 Physical Principles 82
5.2.2 AES Experimental Set-Up 84
5.2.3 AES Analysis 86
5.3 Electron Energy Loss Spectroscopy 89
5.3.1 Core Level Electron Energy Loss Spectroscopy 90
5.3.2 Electron Energy Loss Spectroscopy 92
5.3.3 High-Resolution Electron Energy Loss Spectroscopy 95
5.4 Photoelectron Spectroscopy 98
5.4.1 Photoelectric Effect 98

5.4.2 PES Experimental Set-Up ... 99
5.4.3 PES Analysis ... 101
Problems ... 107
Further Reading ... 108

6. Surface Analysis III. Probing Surfaces with Ions ... 109
6.1 General Principles ... 109
6.1.1 Classical Binary Collisions ... 109
6.1.2 Scattering Cross-Section ... 113
6.1.3 Shadowing and Blocking ... 114
6.1.4 Channeling ... 117
6.1.5 Sputtering ... 118
6.1.6 Ion-Induced Electronic Processes ... 120
6.2 Low-Energy Ion Scattering Spectroscopy ... 123
6.2.1 General Remarks: Merits and Problems ... 123
6.2.2 Alkali Ion Scattering and Time-of-Flight Techniques ... 124
6.2.3 Quantitative Structural Analysis in Impact-Collision Geometry ... 126
6.3 Rutherford Backscattering and Medium-Energy Ion Scattering Spectroscopy ... 129
6.3.1 General Remarks ... 129
6.3.2 Surface Peak ... 131
6.3.3 Thin Film Analysis ... 136
6.4 Elastic Recoil Detection Analysis ... 137
6.5 Secondary Ion Mass Spectroscopy ... 138
Problems ... 142
Further Reading ... 143

7. Surface Analysis IV. Microscopy ... 145
7.1 Field Emission Microscopy ... 145
7.2 Field Ion Microscopy ... 147
7.3 Transmission Electron Microscopy ... 149
7.4 Reflection Electron Microscopy ... 152
7.5 Low-Energy Electron Microscopy ... 154
7.6 Scanning Electron Microscopy ... 156
7.7 Scanning Tunneling Microscopy ... 159
7.8 Atomic Force Microscopy ... 164
Problems ... 168
Further Reading ... 168

8. Atomic Structure of Clean Surfaces ... 171
8.1 Relaxation and Reconstruction ... 171
8.2 Relaxed Surfaces of Metals ... 173
8.2.1 Al(110) ... 173
8.2.2 Fe(211) ... 174

8.3 Reconstructed Surfaces of Metals ... 176
8.3.1 Pt(100) ... 176
8.3.2 Pt(110) ... 178
8.3.3 W(100) ... 178
8.4 Graphite Surface ... 179
8.5 Surfaces of Elemental Semiconductors ... 180
8.5.1 Si(100) ... 181
8.5.2 Si(111) ... 183
8.5.3 Ge(111) ... 187
8.6 Surfaces of III-V Compound Semiconductors ... 188
8.6.1 GaAs(110) ... 188
8.6.2 GaAs(111) and GaAs($\bar{1}\bar{1}\bar{1}$) ... 189
Problems ... 192
Further Reading ... 194

9. Atomic Structure of Surfaces with Adsorbates ... 195
9.1 Surface Phases in Submonolayer Adsorbate/Substrate Systems ... 195
9.2 Surface Phase Composition ... 196
9.2.1 Coverage of Adsorbate ... 197
9.2.2 Coverage of Substrate Atoms ... 199
9.2.3 Experimental Determination of Composition ... 200
9.3 Formation Phase Diagram ... 205
9.4 Metal Surfaces with Adsorbates ... 210
9.4.1 Family of $\sqrt{3}\times\sqrt{3}$ Structures on (111) fcc Metal Surfaces ... 210
9.4.2 Ni(110)2×1-CO ... 213
9.4.3 $n\times1$ Structures in Pb/Cu(110), Bi/Cu(110), Li/Cu(110), and S/Ni(110) Systems ... 214
9.5 Semiconductor Surfaces with Adsorbates ... 217
9.5.1 Family of $\sqrt{3}\times\sqrt{3}$ Structures on Si(111) and Ge(111) ... 217
9.5.2 2×1, 1×1, and 3×1 Phases in the H/Si(100) System ... 224
Problems ... 226
Further Reading ... 227

10. Structural Defects at Surfaces ... 229
10.1 General Consideration Using the TSK Model ... 229
10.1.1 Point Defects ... 229
10.1.2 Steps, Singular and Vicinal Surfaces, Facets ... 232
10.2 Selected Realistic Examples ... 237
10.2.1 Adatoms ... 237
10.2.2 Vacancies ... 239
10.2.3 Anti-Site Defects ... 243
10.2.4 Dislocations ... 245
10.2.5 Domain Boundaries ... 247

10.2.6 Steps 252
10.2.7 Facetting 257
Problems 260
Further Reading 260

11. Electronic Structure of Surfaces 261
11.1 Basics of Density Functional Theory 261
11.2 Jellium Model 263
11.3 Surface States 266
11.4 Electronic Structure of Selected Surfaces 272
11.4.1 Si(111)2×1 272
11.4.2 Si(111)7×7 273
11.4.3 Si(111)1×1-As 275
11.4.4 Si(111)$\sqrt{3}\times\sqrt{3}$-In 277
11.5 Surface Conductivity 277
11.6 Work Function 282
11.6.1 Work Function of Metals 282
11.6.2 Work Function of Semiconductors 286
11.6.3 Work Function Measurements 286
Problems 293
Further Reading 293

12. Elementary Processes at Surfaces I. Adsorption and Desorption 295
12.1 Adsorption Kinetics 295
12.1.1 Coverage Dependence 296
12.1.2 Temperature Dependence 301
12.1.3 Angular and Kinetic Energy Dependence 305
12.2 Thermal Desorption 305
12.2.1 Desorption Kinetics 305
12.2.2 Thermal Desorption Spectroscopy 308
12.3 Adsorption Isotherms 315
12.4 Non-Thermal Desorption 319
Problems 322
Further Reading 323

13. Elementary Processes at Surfaces II. Surface Diffusion 325
13.1 Basic Equations 325
13.1.1 Random-Walk Motion 325
13.1.2 Fick's Laws 327
13.2 Tracer and Chemical Diffusion 330
13.3 Intrinsic and Mass Transfer Diffusion 331
13.4 Anisotropy of Surface Diffusion 333
13.5 Atomistic Mechanisms of Surface Diffusion 335
13.5.1 Hopping Mechanism 336

13.5.2 Atomic Exchange Mechanism 338
13.5.3 Tunneling Mechanism 338
13.5.4 Vacancy Mechanism 340
13.6 Surface Diffusion of Clusters 341
13.7 Surface Diffusion and Phase Formation 345
13.8 Surface Electromigration 348
13.9 Experimental Study of Surface Diffusion 349
13.9.1 Direct Observation of Diffusing Atoms 349
13.9.2 Profile Evolution Method 350
13.9.3 Capillarity Techniques 353
13.9.4 Island Growth Technique 354
Problems 356
Further Reading 356

14. Growth of Thin Films 357
14.1 Growth Modes 357
14.2 Nucleation and Growth of Islands 359
14.2.1 Island Number Density 359
14.2.2 Island Shape 365
14.2.3 Island Size Distribution 368
14.2.4 Vacancy Islands 374
14.3 Kinetic Effects in Homoepitaxial Growth 374
14.4 Strain Effects in Heteroepitaxy 377
14.5 Thin Film Growth Techniques 379
14.5.1 Molecular Beam Epitaxy 379
14.5.2 Solid Phase Epitaxy 381
14.5.3 Chemical Beam Epitaxy 382
14.6 Surfactant-Mediated Growth 383
Problems 386
Further Reading 387

15. Atomic Manipulations and Nanostructure Formation 389
15.1 Nano-Size and Low-Dimensional Objects 389
15.2 Atomic Manipulation with STM 393
15.2.1 Lateral Atomic Displacement 394
15.2.2 Atom Extraction 399
15.2.3 Atom Deposition 403
15.3 Self-Organization of Nanostructures 404
15.4 Fullerenes and Carbon Nanotubes 408
Further Reading 415

References 417

Index 433

1. Introduction

The inception of modern surface science, as well as the appearance of the term "surface science" in common use, dates back to the early 1960s, although there were many previous studies of surface phenomena, and many basic theoretical concepts had already been developed (for details see, for example, [1.1, 1.2]). The breakthrough in the field resulted from a combination of factors, including progress in vacuum technology, the development of surface analytical techniques and the appearance of high-speed digital computers. Since then, surface science has undergone great development, which still continues. The milestones in the progress of surface science are reflected in the collections of review articles compiled from the anniversary volumes of the journal *Surface Science* and entitled *Surface Science: The First Thirty Years* (1994) [1.3] and *Frontiers in Surface and Interface Science* (2002) [1.4].

The main distinctive feature of modern surface science is that it deals with crystal surfaces, which are well-defined from the viewpoint of their structure and composition. In other words, these surfaces are either clean adsorbate-free on the atomic level, or, in the case of adsorbate-covered surfaces, contain adsorbate species added intentionally in amounts controlled also on the atomic level. Thus, surface science experiments are typically conducted in ultra-high vacuum, as it is the only possible environment where such surfaces can be prepared and maintained. The main results of modern surface science, therefore, refer to the solid–vacuum and solid–vapor interfaces. When speaking about solid surfaces, one conventionally means the topmost few atomic layers of the crystal, i.e., the region ~10 Å thick, whose atomic arrangement and electronic structures differ substantially from those in the crystal bulk. Note, however, that the surface might affect the properties of the much deeper bulk layers (for example, the near-surface space-charge layer in semiconductors might extend for microns).

The present book covers all the most important aspects of modern surface science with all subjects being presented in a concise and clear form acceptable for a beginner. In particular, the contents of the book includes the discussion of the following items: Chapter 2 gives a brief overview of surface crystallography, in other words, it provides the reader with the nomenclature accepted in surface science to describe the structure of crystal surfaces. Chapter 3 can be considered as a kind of excursion to the surface science experimental laboratory, where one learns about ultra-high-vacuum systems, their operation and main components, as well as about some general procedures used for sample preparation. Chapters 4–7 are devoted to the experimental methods of surface analysis, including, diffraction techniques, electron spectroscopy, ion-scattering spectroscopy and microscopy, respectively. Within each chapter a set of particular methods are considered. The scheme of the presentation of the technique is as follows: First, the physical phenomena, which constitute the basis of the technique, are introduced. Then, the experimental set-up is described. Finally, information which can be gained by the technique is specified. The purpose of Chap. 8 is to give an impression of what perfect atomically clean surfaces of a crystal constitute on the atomic scale. Chapter 9 illustrates the structure of clean crystal surfaces, on which submonolayer films (i.e., those with thickness below one atomic monolayer) have been adsorbed. Here, the formation of two-dimensional ordered surface phases, phase transitions and phase diagrams are discussed. The defect-free crystal is an apparent idealization. In practice, the crystal surface always contains a certain number of defects (for example, adatoms, vacancies, steps, etc.), which are discussed in Chap. 10. Among other defects, atomic steps are given especial attention, as their discussion involves such important topics as surface stability, surface morphology and equilibrium crystal shape. Surface electronic structure and properties are addressed in Chap. 11. Chapters 12 and 13 are devoted to elementary atomic processes on the surface, namely, adsorption, desorption and surface diffusion. In Chap. 14, the surface phenomena involved in the growth of thin films (with thickness exceeding the monolayer range) and their effect on the growth mode, as well as on the structure and morphology of the grown films, are discussed. Chapter 15 reflects very recent trends in surface science and deals with so-called nanostructures fabricated using atomic manipulations and self-organization processes.

2. Basics of Two-Dimensional Crystallography

In this chapter, the nomenclature used to describe surface structures is developed. An understanding of this nomenclature is very important, as it will be used continuously throughout all the following chapters of the textbook. It should be noted, however, that here only the formal definitions and concepts are given.

2.1 Two-Dimensional Lattices

2.1.1 Lattice, Basis, and Crystal Structure (3D Case)

Recall, first, the main concepts accepted in bulk crystallography. The structure of an ideal crystal is described conventionally in terms of a *lattice*. For the bulk three-dimensional (3D) crystal, the lattice is defined by three *fundamental translation vectors* $\boldsymbol{a}$, $\boldsymbol{b}$, $\boldsymbol{c}$ such that the atomic arrangement of the crystal looks absolutely the same when viewed from point $\boldsymbol{r}$ as when viewed from the point

$$\boldsymbol{r}' = \boldsymbol{r} + n\boldsymbol{a} + m\boldsymbol{b} + k\boldsymbol{c} \,, \tag{2.1}$$

where n, m, k are integers ($0, \pm 1, \pm 2, \ldots$). Hence, a lattice can be visualized as a set of points $\boldsymbol{r}'$ that fit (2.1). A lattice is a geometrical abstraction. *The crystal structure* (a physical object) is formed when an atom or a group of atoms called *a basis* is attached to every lattice point, with every basis being identical in composition, arrangement and orientation (Fig. 2.1). Thus, it can be written

$$lattice + basis = crystal\ \ structure. \tag{2.2}$$

2.1.2 Concept of a 2D Lattice

As for crystalline surfaces and interfaces, these are essentially two-dimensional (2D) objects. Although the surface (or interface) region is, in principle, a 3D entity having a certain thickness, all symmetry properties of the surface are two-dimensional, i.e., the surface structure is periodic only in two directions.

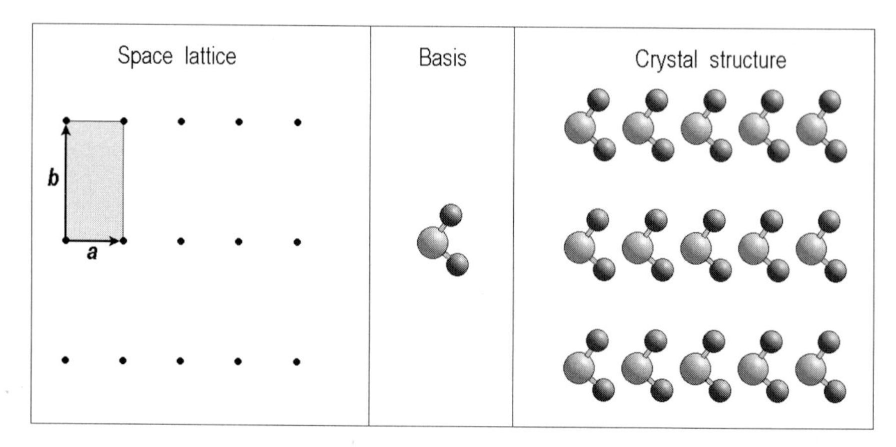

Fig. 2.1. The crystal structure is formed by the addition of the basis to every point of the space lattice

Thus, surface crystallography is two-dimensional and for describing the surface lattice only two translation vectors $\boldsymbol{a}$ and $\boldsymbol{b}$ are used and (2.1) is rewritten as

$$\boldsymbol{r}' = \boldsymbol{r} + n\boldsymbol{a} + m\boldsymbol{b} \,. \tag{2.3}$$

Sometimes a surface lattice is called a *net*. The parallelogram with sides $\boldsymbol{a}$ and $\boldsymbol{b}$ is called a *unit cell* or a *mesh*. The unit cell with the minimum area is called a *primitive cell*.

Another possible type of primitive cell is represented by a *Wigner–Seitz cell*, which is constructed according to the following procedure (see Fig. 2.2):

- The chosen arbitrary lattice point is connected by straight lines with all neighboring lattice points.
- Through the midpoints of these lines, perpendicular lines are drawn (planes in the 3D case).
- The smallest area (volume in the 3D case) enclosed in this way comprises the Wigner–Seitz primitive cell.

To describe completely the structure of a given surface, one should define its 2D lattice and its basis (i.e., the arrangement of the surface atoms within a unit cell). This is similar to the concept introduced by logic (2.2) for the bulk crystal. Moreover, one may notice that the schematic example shown in Fig. 2.1 actually corresponds to the 2D case.

2.1.3 2D Bravais Lattices

All the great variety of surface lattices are organized into five main types, called two-dimensional *Bravais lattices* (recall that there are 14 Bravais lat-

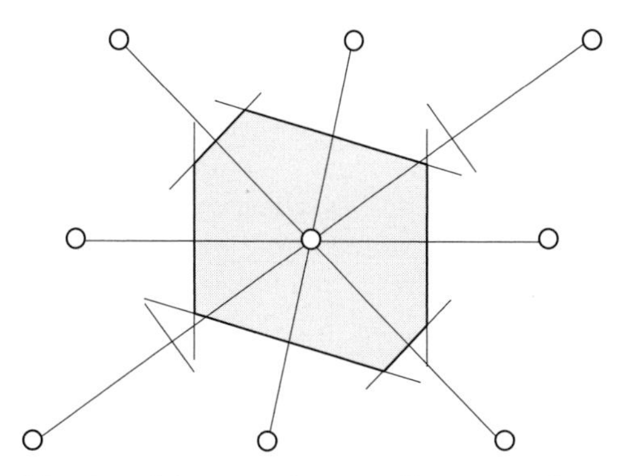

Fig. 2.2. Schematic diagram illustrating the construction of the Wigner–Seitz primitive cell

tices in three dimensions). The 2D Bravais lattices are shown in Fig. 2.3. They are

- oblique lattice $|\boldsymbol{a}| \neq |\boldsymbol{b}|, \quad \gamma \neq 90^\circ,$
- rectangular lattice $|\boldsymbol{a}| \neq |\boldsymbol{b}|, \quad \gamma = 90^\circ,$
- centered rectangular lattice $|\boldsymbol{a}| \neq |\boldsymbol{b}|, \quad \gamma = 90^\circ,$
- square lattice $|\boldsymbol{a}| = |\boldsymbol{b}|, \quad \gamma = 90^\circ,$
- hexagonal lattice $|\boldsymbol{a}| = |\boldsymbol{b}|, \quad \gamma = 120^\circ.$

Note that in Fig. 2.3 two types of unit cell are shown for the centered rectangular lattice. The primitive unit cell is non-rectangular, while the rectangular cell is non-primitive. Nevertheless, in practice, one often uses it for convenience of description.

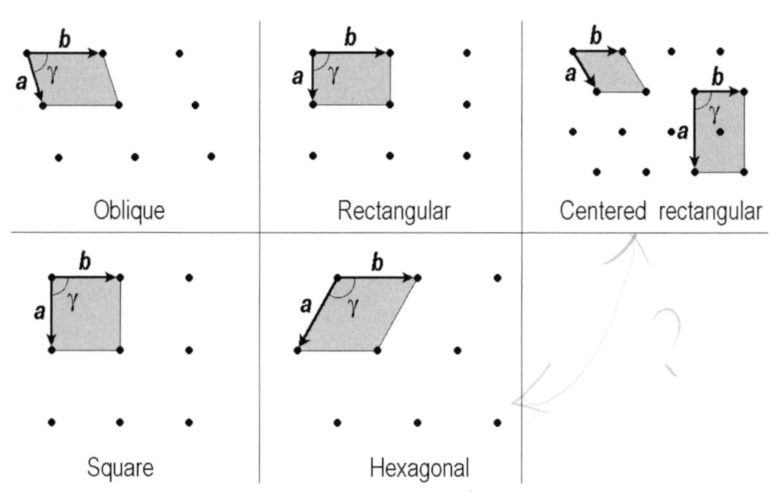

Fig. 2.3. Five two-dimensional Bravais lattices. Translation vectors $\boldsymbol{a}$ and $\boldsymbol{b}$ are shown, unit cells (meshes) are hatched

2.2 Miller Indices for Crystal Planes

2.2.1 Definition of Miller Indices

Before describing a particular surface structure one should specify first what the plane of the crystal is under consideration. The orientation of the plane is denoted by Miller indices which are determined in the following way:

- First, the intercepts of the plane on the axes are found in terms of the lattice constants a, b, c, which may be those of a primitive or non-primitive unit cell.
- Then, the reciprocals of the obtained numbers are taken.
- Finally, they are reduced to three integers having the same ratio, usually the smallest three integers.

The result enclosed in parentheses (hkl) constitutes the *Miller index* of the crystal plane.

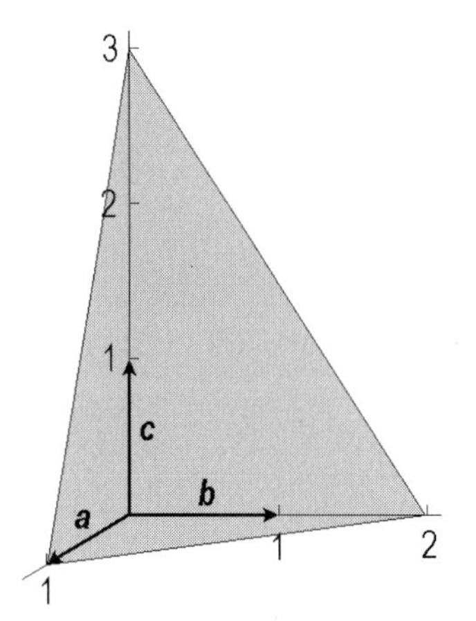

Fig. 2.4. This plane intercepts the $\boldsymbol{a}$, $\boldsymbol{b}$, $\boldsymbol{c}$ axes at $1a$, $2b$, $3c$. Thus, the Miller indices of the plane are (632) (see text)

For example, if the intercepts of the plane are

$$1, 2, 3$$

(Fig. 2.4), then the reciprocals are

$$1, \frac{1}{2}, \frac{1}{3}$$

and the smallest three integers that have the same ratio are

$$6, 3, 2,$$

i.e., the index of this plane is (632). If the plane is parallel to the axis, it is accepted that the intercept is at infinity and the corresponding Miller index is zero. If the plane intercepts the axis on the negative side of the origin, the corresponding index is negative. To indicate this, a minus is placed above the index: $(\bar{h}kl)$. As an example, Fig. 2.5 illustrates the Miller indices of some important planes in a cubic crystal. The set of planes that are equivalent by symmetry are denoted by braces around the Miller indices (for example, $\{100\}$ for the cube faces).

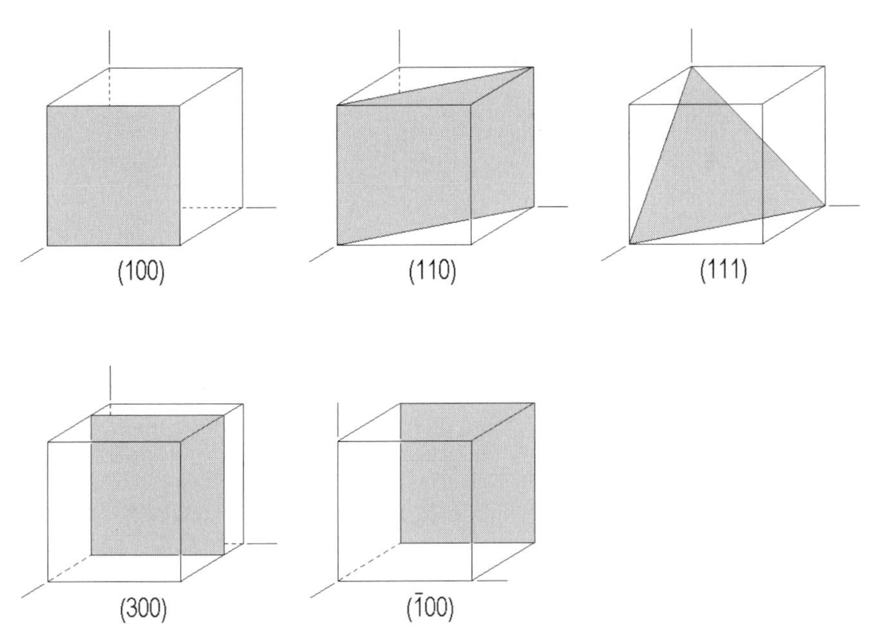

Fig. 2.5. Miller indices of some important planes in a cubic crystal

For the description of the hcp (hexagonal close-packed) lattice, four axes are conventionally introduced, three of equal length a in the basal plane inclined at 120° to each other and one axis of length c normal to this plane. Hence, four-index notation is employed for planes in the hcp crystal. One can see that the three-index notation (hkl) corresponds to the four-index notation $(h, k, -h-k, l)$, i.e., the three-index notation is obtained from the four-index one by simply omitting the third index. Note that the necessary requirement to use this rule is that, in the three-index notation, the axes in the basal plane make an angle of 120°. If the axes are chosen so that they make an angle of 60°, then the corresponding four-index notation is $(h, k-h, -k, l)$.

2.2.2 Low-Miller-Index Planes of Some Important Crystals

Figures 2.6, 2.7, 2.8, and 2.9 show the atomic arrangement of the principal low-index planes of fcc (face-centered cubic), bcc (body-centered cubic), hcp (hexagonal closed-packed) and diamond crystals, respectively. The topmost atoms are shown by white circles, the atoms in the deeper layers by gray circles: the deeper the layer, the darker the circles. The layer numbers, counted from the topmost one, are indicated. The surface unit cells are outlined.

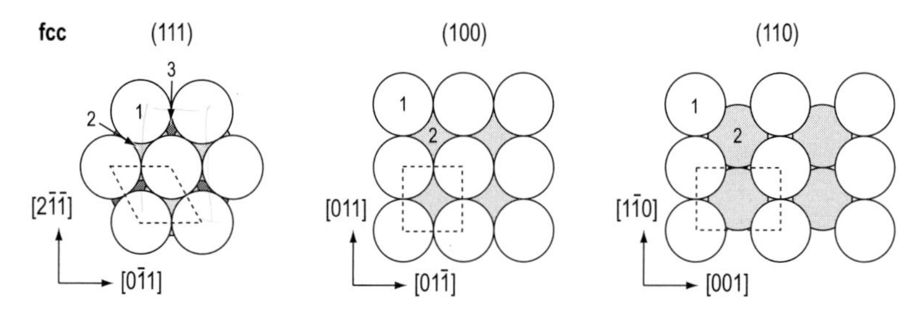

Fig. 2.6. Main low-index planes of a fcc (face-centered cubic) crystal

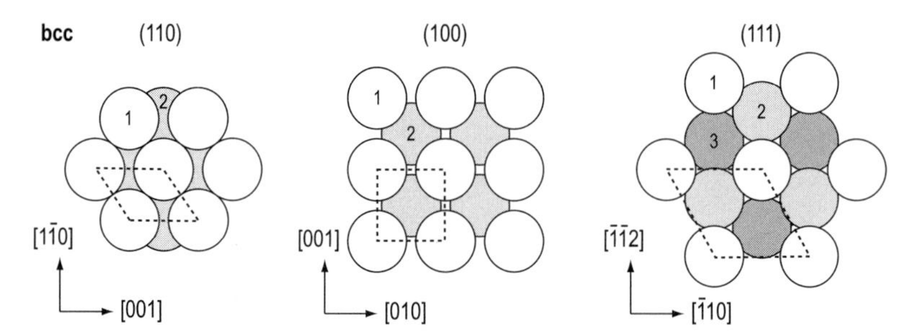

Fig. 2.7. Main low-index planes of a bcc (body-centered cubic) crystal

2.2.3 High-Miller-Index Stepped Surfaces

If the crystal surface is misoriented from the low-index plane by a small angle, it can be described by the combination of three parameters: *tilt angle*, *tilt azimuth*, and *tilt zone*. The tilt zone specifies the axis, around which the rotation from the basal low-index plane to the tilted plane is conducted, the azimuth specifies the direction of the rotation, and the tilt angle specifies the angle of rotation (see Fig. 2.10).

On the atomic scale, such a surface, called *a stepped* or *vicinal surface*, is composed conventionally of terraces separated by steps of monatomic height,

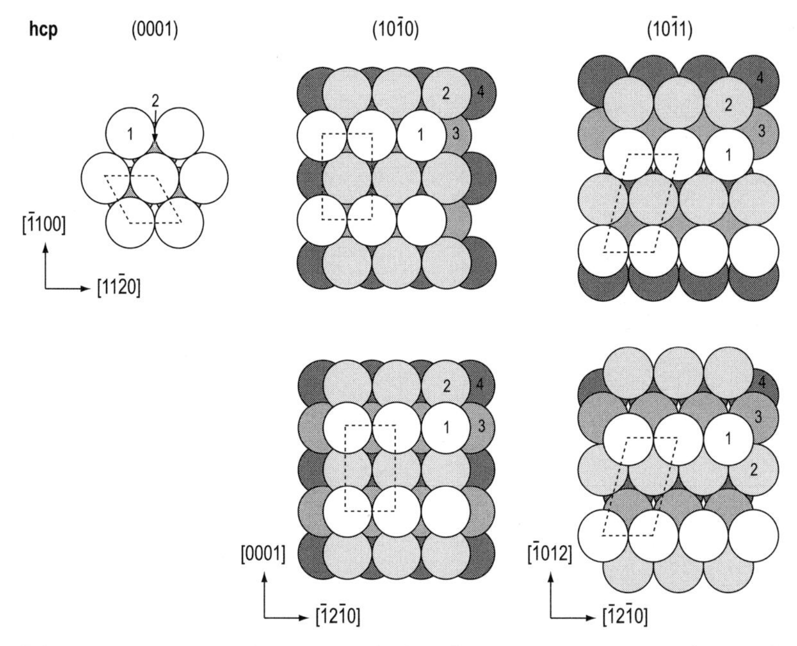

Fig. 2.8. Main low-index planes of a hcp (hexagonal close-packed) crystal. For the (10$\bar{1}$0) and (10$\bar{1}$1) planes, two possible types of surface termination are shown

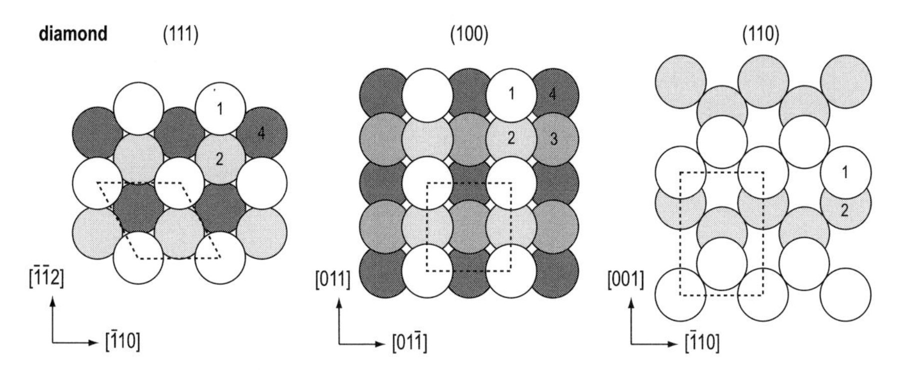

Fig. 2.9. Main low-index planes of a diamond crystal

which may also have kinks in them (Fig. 2.11). Although this surface can be designated by its corresponding Miller indices, for example, (755) for Fig. 2.11a, this notation does not indicate, at a glance, the geometric structure of the surface. A more vivid notation, devised by Lang, Joyner, and Somorjai [2.1], gives the structure in the form of

$$n(h_\mathrm{t}k_\mathrm{t}l_\mathrm{t}) \times (h_\mathrm{s}k_\mathrm{s}l_\mathrm{s}) \, .$$

Here $(h_\mathrm{t}k_\mathrm{t}l_\mathrm{t})$ and $(h_\mathrm{s}k_\mathrm{s}l_\mathrm{s})$ are the Miller indices of the terrace plane and step plane, respectively, and n gives the number of atomic rows in the terrace

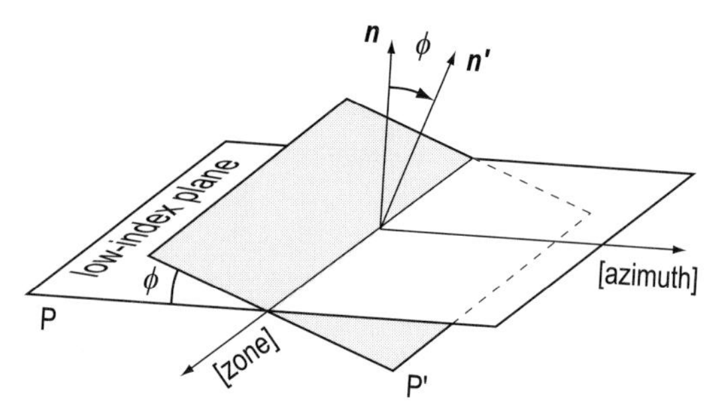

Fig. 2.10. The misoriented plane P' is obtained from the low-index plane P by rotation around the [zone] axis towards the [azimuth] direction by the tilt angle ϕ. The vectors $\boldsymbol{n}$ and $\boldsymbol{n}'$ denote the normals to the planes P and P', respectively

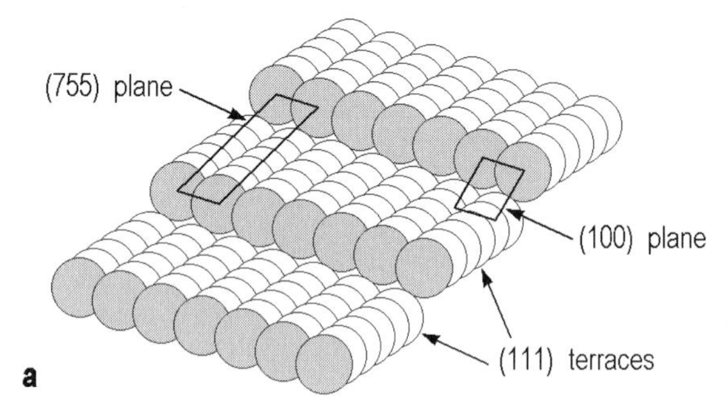

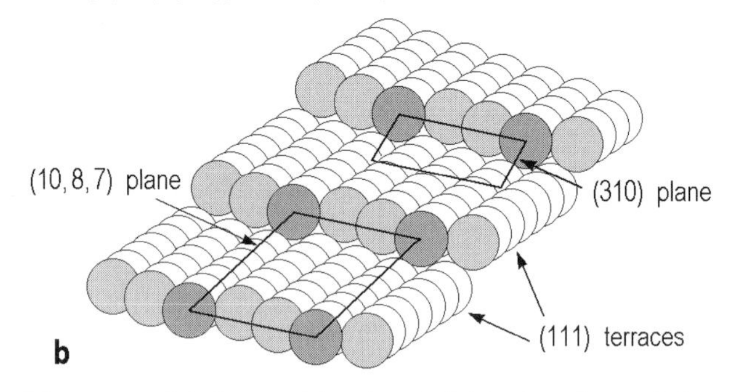

Fig. 2.11. (**a**) Stepped (755) and (**b**) kinked (10 8 7) fcc crystal faces

parallel to the step edge. Thus, the (755) surface of the fcc crystal is denoted 6(111)×(100), as it consists of (111) terraces, 6 atoms wide, separated by monatomic steps of (100) orientation (Fig. 2.11a). A stepped surface with steps that are themselves high-Miller-index surfaces is called a *kinked surface.*

The fcc (10 8 7), i.e., 7(111)×(310) surface (Fig. 2.11b) furnishes an example of a kinked surface.

2.3 Indices of Directions

To specify a certain direction in a crystal or in its surface, the direction vector is expressed by indices in square brackets, $[hkl]$. The values h, k, l are the set of the smallest integers that have the ratio of the components of a vector, referred to the axes of the crystal. Thus, the directions of the axes are designated as [100], [010] and [001]. The negative component of the direction vector is indicated by a minus placed above the index, $[\bar{h}kl]$. The full set of equivalent directions is denoted as $\langle hkl \rangle$. In cubic crystals, the direction $[hkl]$ is normal to the plane (hkl) with the same indices; however, this does not hold generally for other crystal types.

2.4 Notation for Surface Structures

The structure of the surface layer is not necessarily the same as that of the underlying bulk planes even for clean surfaces (i.e., adsorbate-free). The term *superstructure* is used conventionally to outline the specific structure of the top atomic layer (or a few layers). The notation used to describe a superstructure ties its 2D lattice to that of the underlying substrate plane. This is done conventionally in one of two ways.

2.4.1 Matrix Notation

The notation proposed by Park and Madden [2.2] resides in the determination of the matrix which establishes a relation between the basic translation vectors of the surface under consideration and those of the ideal substrate plane. That is, if $\boldsymbol{a}_\mathrm{s}$, $\boldsymbol{b}_\mathrm{s}$ and $\boldsymbol{a}$, $\boldsymbol{b}$ are the basic translation vectors of the superstructure and substrate planes, respectively, than they can be linked by the equations

$$\begin{aligned} \boldsymbol{a}_\mathrm{s} &= G_{11}\boldsymbol{a} + G_{12}\boldsymbol{b} \\ \boldsymbol{b}_\mathrm{s} &= G_{21}\boldsymbol{a} + G_{22}\boldsymbol{b} \end{aligned} \tag{2.4}$$

and the superstructure is specified by the matrix:

$$G = \begin{pmatrix} G_{11} & G_{12} \\ G_{21} & G_{22} \end{pmatrix} . \tag{2.5}$$

The values of the matrix elements G_{ij} determine whether the structure of the surface is *commensurate* or *incommensurate* with respect to the substrate.

Commensurability means that a rational relationship between the vectors $\boldsymbol{a}_\mathrm{s}$, $\boldsymbol{b}_\mathrm{s}$ and $\boldsymbol{a}$, $\boldsymbol{b}$ can be established. If there is no rational relationship between the unit vectors of the surface superstructure and the substrate, the superstructure is incommensurate. In other words, the incommensurate superstructure is registered in-plane incoherently with the underlying substrate lattice.

2.4.2 Wood's Notation

A more vivid but less versatile notation for surface superstructures was proposed by Wood [2.3]. In this notation, the ratio of the lengths of the basic translation vectors of the superstructure and those of the substrate plane is specified. In addition, one indicates the angle of rotation (if any) which makes the unit mesh of the surface to be aligned with the basic translation vectors of the substrate. That is, if on a certain substrate surface $X(hkl)$ a superstructure is formed with the basic translation vectors of

$$|\boldsymbol{a}_\mathrm{s}| = m|\boldsymbol{a}|, \quad |\boldsymbol{a}_\mathrm{s}| = n|\boldsymbol{b}| \tag{2.6}$$

and with the rotation angle of $\varphi°$, then this surface structure is labeled as

$$X(hkl)m \times n\text{-}R\,\varphi° \,. \tag{2.7}$$

If the unit mesh of the superstructure is aligned along the axes of the substrate net, i.e., $\varphi = 0$, than the notation does not specify this zero angle (for example, Si(111)7×7). A possible centering is expressed by the character c (for example, Si(100)c(4×2)). If the superstructure is induced by the adsorbate, this adsorbate is specified by its chemical symbol at the end of the notation (for example, Si(111)4×1-In). Sometimes, the number of adsorbate atoms per unit cell is indicated (for example, Si(111)$\sqrt{3}\times\sqrt{3}$-R30°-3Bi).

Note that Wood's notation is applicable only for the cases where the included angles of the superstructure and substrate meshes are the same. This requirement is matched when both meshes have the same Bravais lattice or when one is rectangular and the other is square. But, in general, it does not provide an adequate description for mixed symmetry meshes, in which case the matrix notation should be used. Sometimes, however, the Wood's-type notation is used in the literature even for the cases where strictly speaking it is not applicable. In these instances, the notation is included in quotes, which indicates that it does not provide an exactly true relationship. The clean Si(110) surface can serve as an example. The structure of this surface is properly described in matrix notation as Si(110)$\begin{pmatrix} 2 & 2 \\ 17 & 1 \end{pmatrix}$. However, it is often labeled as Si(110)"2 × 16" with quotes indicating that the unit mesh of the "2×16" structure is non-rectangular in contrast to the rectangular (1×1) mesh of the ideal Si(110) plane.

2.4.3 Some Examples

In conclusion of this section, let us consider some simple examples of using the above nomenclature. Figures 2.12 and 2.13 display several superlattices on hexagonal and square lattices, respectively. The 2D lattice of the substrate is shown by black dots and the lattice of the superstructure is shown by empty circles. Note that we deal here with lattices only, not structures (see Sect. 2.1).

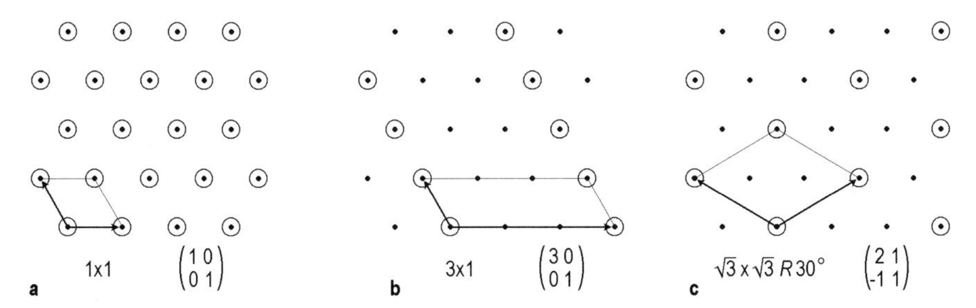

Fig. 2.12. Wood's and matrix notation for some superlattices on a hexagonal 2D lattice

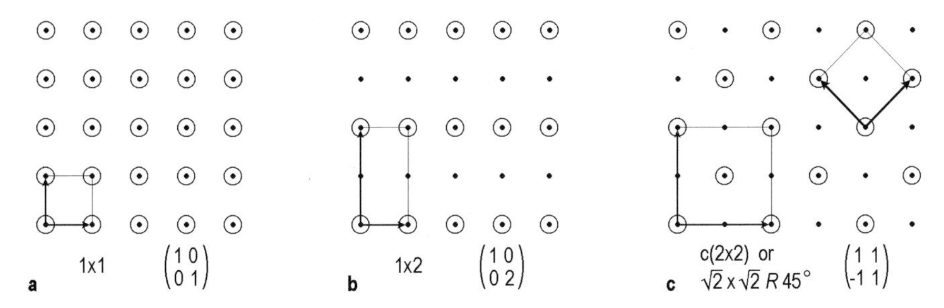

Fig. 2.13. Wood's and matrix notation for some superlattices on a square 2D lattice

When the superstructure has a unit cell of the same size as the substrate unit cell and parallel to it, i.e., the two lattices coincide (Fig. 2.12a and Fig. 2.13a), then the superstructure is designated as

$$1 \times 1 \text{ or } \begin{pmatrix} 1 & 0 \\ 0 & 1 \end{pmatrix} .$$

If the unit cell of the superstructure is three times as long as the substrate unit cell along one major crystallographic axis and has the same length along the other (Fig. 2.12b), the superstructure is designated as

$$3 \times 1 \quad \text{or} \quad \begin{pmatrix} 3 & 0 \\ 0 & 1 \end{pmatrix} .$$

The qualitatively similar case with the 1×2 or $\begin{pmatrix} 1 & 0 \\ 0 & 2 \end{pmatrix}$ superstructure is shown in Fig. 2.13b.

Figure 2.12c represents the $\sqrt{3}\times\sqrt{3}$-$R30°$ superlattice, in which the basic vectors are $\sqrt{3}$ times as long as those of the substrate, and the superstructure unit cell is rotated by 30° with respect to the substrate unit cell. The superstructure can be designated in matrix notation as $\begin{pmatrix} 2 & 1 \\ -1 & 1 \end{pmatrix}$.

The superlattice shown in Fig. 2.13c can be designated in one of three possible ways. First, it can be denoted as c(2×2), since it may be viewed as a (2×2) surface lattice with an extra lattice point in its center. If one considers the primitive unit cell, the superstructure can be designated as

$$\sqrt{2} \times \sqrt{2}\text{-}R45° \quad \text{or} \quad \begin{pmatrix} 1 & 1 \\ -1 & 1 \end{pmatrix} .$$

2.5 2D Reciprocal Lattice

The concept of the reciprocal lattice is very useful when one deals with structural investigations by means of diffraction techniques. This point will be addressed in Chap. 4 devoted to diffraction methods of surface analysis. Here only the basic definitions are introduced.

The *2D reciprocal lattice* is a set of points whose coordinates are given by the vectors

$$\boldsymbol{G}_{hk} = h\boldsymbol{a}^* + k\boldsymbol{b}^* , \tag{2.8}$$

where h, k are integers (0, ±1, ±2, ...) and the primitive translation vectors, $\boldsymbol{a}^*$ and $\boldsymbol{b}^*$, are related to the primitive translation vectors of the real-space lattice, $\boldsymbol{a}$ and $\boldsymbol{b}$, as

$$\boldsymbol{a}^* = 2\pi \cdot \frac{\boldsymbol{b} \times \boldsymbol{n}}{|\boldsymbol{a} \times \boldsymbol{b}|} , \quad \boldsymbol{b}^* = 2\pi \cdot \frac{\boldsymbol{n} \times \boldsymbol{a}}{|\boldsymbol{a} \times \boldsymbol{b}|} , \tag{2.9}$$

where $\boldsymbol{n}$ is a unit vector normal to the surface.

From (2.9), one can easily distinguish the following properties of the vectors $\boldsymbol{a}^*$, $\boldsymbol{b}^*$:

- The vectors $\boldsymbol{a}^*$, $\boldsymbol{b}^*$ lie in the same surface plane as the real-space vectors $\boldsymbol{a}$, $\boldsymbol{b}$.
- The vector $\boldsymbol{a}^*$ is perpendicular to vector $\boldsymbol{b}$; $\boldsymbol{b}^*$ is perpendicular to vector $\boldsymbol{a}$.

- The lengths of vectors $\boldsymbol{a}^*$, $\boldsymbol{b}^*$ are

$$|\boldsymbol{a}^*| = \frac{2\pi}{a \cdot \sin \measuredangle(\boldsymbol{a}, \boldsymbol{b})},$$
$$|\boldsymbol{b}^*| = \frac{2\pi}{b \cdot \sin \measuredangle(\boldsymbol{a}, \boldsymbol{b})}.$$

Note that while the real-space vectors $\boldsymbol{a}$, $\boldsymbol{b}$ have the dimensions of [length], the reciprocal lattice vectors $\boldsymbol{a}^*$, $\boldsymbol{b}^*$ have the dimensions of [1/length].

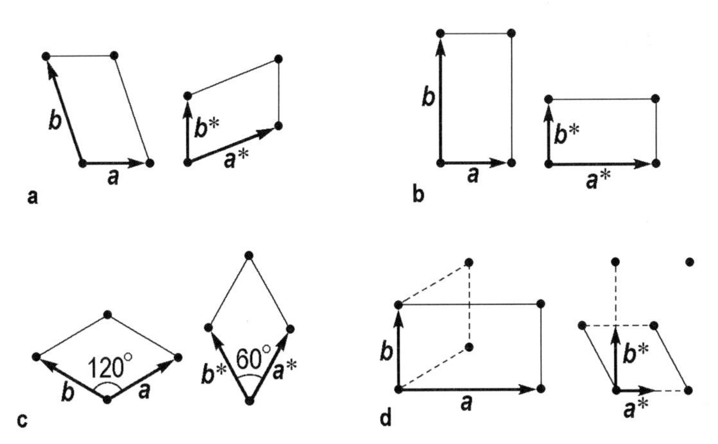

Fig. 2.14. Translation vectors and unit meshes of the real-space and corresponding reciprocal 2D Bravais lattices: (**a**) oblique lattice; (**b**) rectangular lattice (the square lattice is essentially the same with $|\boldsymbol{a}| = |\boldsymbol{b}|$); (**c**) hexagonal lattice, (**d**) centered rectangular lattice

Figure 2.14 shows the reciprocal lattices for the real-space 2D Bravais lattices. Here only the square lattice is omitted as it can be considered as a simple specific case of a rectangular lattice with $|\boldsymbol{a}| = |\boldsymbol{b}|$. With reference to Fig. 2.14 two general features can be seen:

- Each pair consisting of a real-space lattice and the corresponding reciprocal lattice belong to the same type of Bravais lattice, i.e., if the real-space lattice is, say, hexagonal, than the reciprocal lattice is also hexagonal; if the real-space lattice is centered rectangular, the corresponding reciprocal lattice is also centered rectangular, etc.
- The angle between the reciprocal unit vectors $\measuredangle(\boldsymbol{a}^*, \boldsymbol{b}^*)$ is related to that between the real-space unit vectors $(\boldsymbol{a}, \boldsymbol{b})$ by

$$\measuredangle(\boldsymbol{a}^*, \boldsymbol{b}^*) = 180° - \measuredangle(\boldsymbol{a}, \boldsymbol{b}). \tag{2.10}$$

Thus, for rectangular and square lattices these angles are the same, 90°, but for the real-space and reciprocal hexagonal lattices the angle is 120° and 60°, respectively.

2.6 Brillouin Zone

A Wigner–Seitz primitive cell (see Fig. 2.2) in the reciprocal lattice is referred to as the first *Brillouin zone*. The concept of the Brillouin zone is of prime importance for the analysis of the electronic energy-band structure of crystals. As an example, Figures 2.15, 2.16, and 2.17 show the 2D Brillouin zones of the main planes of the fcc, bcc and hcp crystals, respectively, in relation to the respective bulk Brillouin zones. Symmetry points and directions are indicated using BSW (Bouckaert–Smoluchowski–Wigner) notation [2.4].

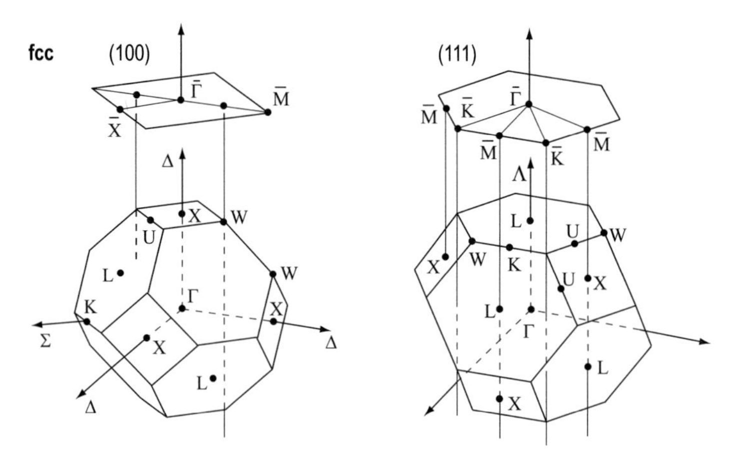

Fig. 2.15. Relation between the 2D Brillouin zones of the (100) and (111) planes of the fcc crystal and the bulk Brillouin zone. Note that the reciprocal lattice of the fcc lattice is the bcc lattice

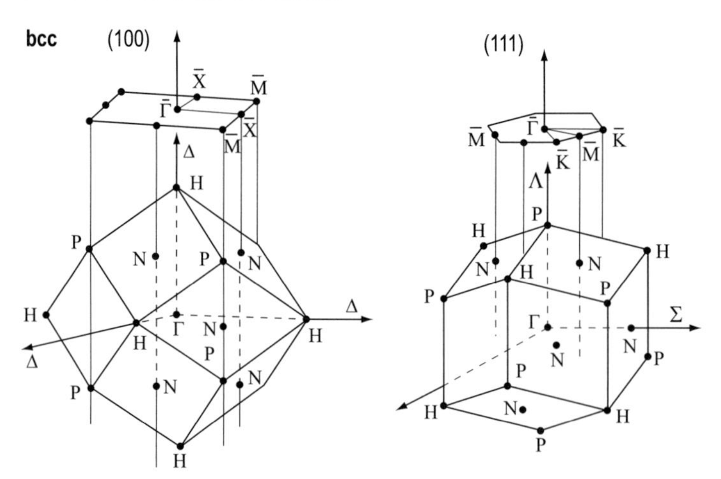

Fig. 2.16. Relation between the 2D Brillouin zones of the (100) and (111) planes of the bcc crystal and the bulk Brillouin zone. Note that the reciprocal lattice of the bcc lattice is the fcc lattice

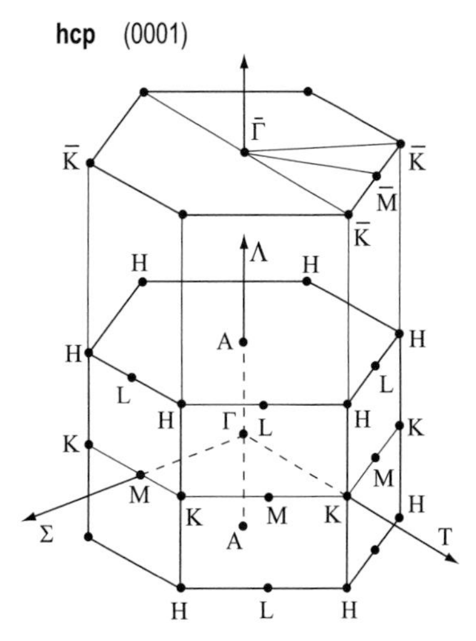

Fig. 2.17. Relation between the 2D Brillouin zone of the (0001) plane of the hcp crystal and the bulk Brillouin zone

Problems

2.1 Do the points shown in the following figure form a 2D lattice? If "yes", show its primitive translation vectors. If "no", provide the arguments.

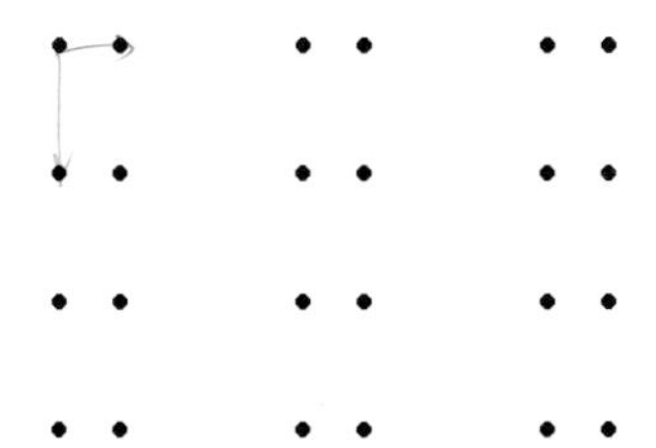

2.2 Show the (133), $(33\bar{1})$, and (113) planes in the simple cubic crystal. Are they equivalent planes?

2.3 Specify the type of the 2D Bravais lattice of the (111) plane of the fcc (face-centered cubic) crystal. What is the period of this 2D lattice, if the edge of the fcc cell is a?

2.4 Prove that the matrix notation for the $\sqrt{3}\times\sqrt{3}$–$R30°$ superstructure on the 2D hexagonal lattice is $\begin{pmatrix} 2 & 1 \\ -1 & 1 \end{pmatrix}$, when the basic translation vectors make an angle of 120°. How will the matrix notation change, if one chooses basic vectors which make an angle of 60°?

2.5 Nickel adsorption on the Si(111) surface (hexagonal lattice) induces a commensurate superstructure $\sqrt{7}\times\sqrt{7}\text{-}R\varphi^\circ$. Find the value of the angle φ° and construct the 2D lattice of this superstructure superposed on the 1×1 lattice. How is this superstructure designated in matrix notation?

Further Reading

1. C. Kittel: *Introduction to Solid State Physics*, 7th edn. (John Wiley, New York 1996) Chapter 1
2. G.A. Somorjai: *Introduction to Surface Chemistry and Catalysis* (John Wiley, New York 1994) Chapter 2

3. Experimental Background

The purpose of this chapter is to familiarize the reader with the experimental background of surface science. First, the necessity for usage of UHV is justified. Then, vacuum concepts are briefly considered before introducing UHV technology. Finally, the main experimental techniques for the preparation of atomically clean surfaces and the deposition of materials are discussed.

3.1 Why Ultra-High Vacuum?

The characterization of a solid surface on an atomic level implies unambiguously that the surface composition remains essentially unchanged over the duration of an experiment. This means that the rate of arrival of reaction species from the gas environment should be low, or, in other words, the experiments should be conducted in vacuum. The concept of vacuum is normally understood in terms of *molecular density*, *mean free path* and the *time constant to form a monolayer*. According to the kinetic theory of gases, the flux I of molecules impinging on the surface from the environment is given by the expression

$$I = \frac{p}{\sqrt{2\pi m k_B T}} . \tag{3.1}$$

Here p is the pressure, m is the mass of the molecule, k_B is Boltzmann's constant, and T is the temperature. Then, one can easily obtain

$$n = \frac{p}{k_B T}, \quad \text{molecular density,} \tag{3.2}$$

$$\lambda = \frac{f}{n\sigma^2}, \quad \text{mean free path,} \tag{3.3}$$

$$\tau = \frac{n_0}{I} = \frac{n_0\sqrt{2\pi m k_B T}}{p}, \quad \text{time constant to form a monolayer ,} \tag{3.4}$$

where σ^2 is the molecular cross-section, and n_0 is the number of atoms in a monolayer.

Table 3.1 illustrates how these values vary with the pressure. What can be learned from the table concerning the vacuum requirements for surface science? Let us accept as a reasonable criterion that the number of atoms

Table 3.1. Molecular density n, arrival rate I, mean free path λ, and the time constant to form a monolayer τ for nitrogen molecules at room temperature ($T \approx 293$ K). The sticking coefficient is assumed to be unity. The density of one monatomic layer is defined to be $n_0 = 10^{15}$ cm^{-2} (which is in fact close to the real values for solid surfaces)

Pressure, p, Torr	Molecular density, n, cm^{-3}	Arrival rate, I, cm^{-2}s^{-1}	Mean free path, λ	Monolayer arrival time, τ
760	2×10^{19}	3×10^{23}	700 Å	3 ns
1	3×10^{16}	4×10^{20}	50 µm	2 µs
10^{-3}	3×10^{13}	4×10^{17}	5 cm	2 ms
10^{-6}	3×10^{10}	4×10^{14}	50 m	2 s
10^{-9}	3×10^{7}	4×10^{11}	50 km	1 hour

or molecules adsorbed to the surface from the gas phase in about one hour (the time usually needed to perform an experiment) should not exceed a few per cent of a monolayer (the usual accuracy of current surface science techniques). One can see that a vacuum of the order of 10^{-10} Torr or better is required to fit this criterion. Though our estimation refers to the worst case with a sticking coefficient of unity, many surfaces of interest do match these conditions.

Some complementary notes are as follows. Even in ultra-high vacuum [1] the molecular density is still rather high and there are still lots of gas molecules around. On the other hand, their mean free path is much greater than the typical dimensions of the vacuum chamber, which means that the gas molecule will collide with the chamber walls or elements of its inner construction many times before it meets another gas molecule.

As a final remark, it should be noted that several units are used for measuring vacuum pressures. The most common are Torr (or millimeters of mercury, mmHg), Pascal (SI unit, 1 Pa = 1 N/m^2) and millibar (1 mbar = 100 Pa). Table 3.2 shows the relationship between these units.

3.2 Vacuum Concepts

To introduce briefly the basic vacuum concepts, consider a very simple model of a pumping system with a vacuum chamber connected to a pump by a cylindrical tube (Fig. 3.1). During pumping, the gas flows from the chamber to the pump in response to the pressure difference between the ends of the tube. Passing through the tube, a small volume of gas in the inlet will expand to become a larger volume at the outlet. *The throughput* Q is defined as:

[1] Vacuum better than 10^{-9} Torr is defined conventionally as an ultra-high vacuum (UHV)

Table 3.2. Conversion table for the most common pressure units

To convert	Into	Multiply
mbar	Pa	100
mbar	Torr	0.75
Pa	mbar	0.01
Pa	Torr	7.5×10^{-3}
Torr	mbar	1.33
Torr	Pa	133

$$Q = p\frac{\mathrm{d}V}{\mathrm{d}t} \qquad [\mathrm{Torr\ l\ s^{-1}}] \tag{3.5}$$

and *the pumping speed* S as:

$$S = \frac{Q}{p} \qquad [\mathrm{l\ s^{-1}}]\,. \tag{3.6}$$

Pumping speeds in a vacuum system are always limited by the finite conductance of the tubes through which the gas is pumped. In analogy to the concept of conductance in electricity, *the conductance* C of a vacuum element, say, a tube, characterizes the ease of gas flow through this element:

$$C = \frac{Q}{\Delta p} \qquad [\mathrm{l\ s^{-1}}]\,, \tag{3.7}$$

where Δp is the pressure difference between the inlet and outlet of the vacuum element.

In further analogy with the electrical case (Kirchhoff's laws),

- two tubes in parallel have a conductance: $C_{\mathrm{total}} = C_1 + C_2$;
- two tubes in series have a conductance: $\dfrac{1}{C_{\mathrm{total}}} = \dfrac{1}{C_1} + \dfrac{1}{C_2}$.

Useful expressions are, for the conductance of an *aperture* of area A:

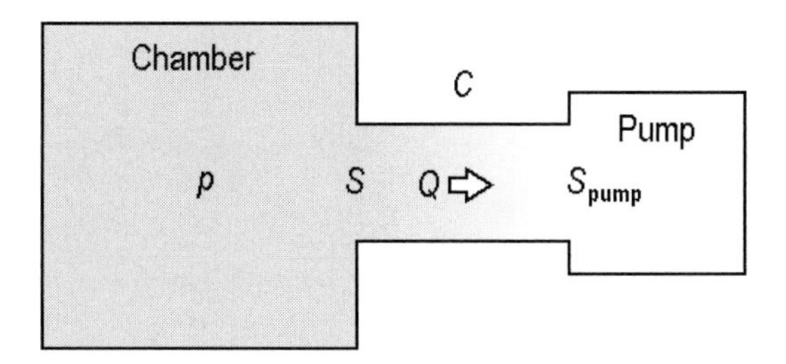

Fig. 3.1. Vacuum chamber and pump connected by a tube

$$C_{\text{aperture}} = \sqrt{\frac{k_{\text{B}}T}{2\pi m}}\,A \tag{3.8}$$

and for the conductance of a *cylindrical tube* of diameter D and length L:

$$C_{\text{tube}} = \frac{\dfrac{D^3}{6L}\sqrt{\dfrac{2\pi k_{\text{B}}T}{m}}}{1+\dfrac{4D}{3L}}\ . \tag{3.9}$$

Both equations are valid under conditions of *molecular flow* (i.e., when the mean free path of the gas molecules exceeds the geometric dimensions of the vacuum element; for conventional vacuum systems this holds for $p < 10^{-3}$ Torr).

It is clear that to have higher conductance, one should use short and wide tubes. Straight tubes are preferable, since, for example, the conductance of a tube bent through 90° is about half that of the straight one.

For a pump connected to a tube with a conductance C (Fig. 3.1), the effective pumping speed at the tube inlet is:

$$\frac{1}{S} = \frac{1}{S_{\text{pump}}} + \frac{1}{C}\ , \tag{3.10}$$

where S_{pump} is the pumping speed of the isolated pump.

The balance equation, describing the change of the amount of the gas in the chamber, is referred conventionally as *the pumping equation* which is:

$$-V\,\frac{\mathrm{d}p}{\mathrm{d}t} = Sp - Q_{\text{T}}\ , \tag{3.11}$$

where V is the chamber volume, p is the pressure, the product Sp gives the gas amount removed by the pump and Q_{T} is the total gas load. In a steady state, the system reaches its *base pressure* given by:

$$p_{\text{ base}} = \frac{Q_{\text{T}}}{S}\ . \tag{3.12}$$

The gas load Q_{T} contains two main components: first, *leaks* due to atmospheric gases passing from outside the chamber (*real leaks*) or due to gases trapped somewhere in the vacuum system and slowly releasing (*virtual leaks*); and second, *degassing*, i.e., the desorption of the gases from the internal surfaces of the vacuum system. Once exposed to air, the surfaces of the inner walls and components within the chamber become covered by a skin of molecules of water, nitrogen, oxygen and other atmospheric gases. When the system is pumped down, desorption of the adsorbed gases hampers the achievment of the desired vacuum. The *bakeout* of the system serves to accelerate gas desorption and, hence, to decrease greatly the surface coverage of the gases. As a result, upon subsequent cooling to room temperature, the desorption rate becomes reduced and lower pressures can be obtained.

3.3 Ultra-High-Vacuum Technology

Conducting experiments in ultra-high vacuum necessitates special instrumentation. Any UHV system for surface science experiments integrates a chamber (or several chambers), a pumping stack, valves, process equipment (such as load lock facilities, sample manipulators, heaters, and evaporators), service components (such as systems of power supply and control), and facilities for surface analysis. As an example, Fig. 3.2 shows a photograph of UHV system designed for surface studies by means of ion scattering spectroscopy.

3.3.1 UHV Materials

The main requirements for materials used for the manufacture of UHV systems are low vapor pressure and the ability to endure bakeout. It is also desirable that UHV materials be non-magnetic. This demand stems from the fact that many tools for surface analysis employ beams of low-energy electrons.

The most common material for vacuum chambers and associated components is *304 stainless steel*. This steel is a suitable vacuum material due to its low gas permeability, resistance to corrosion, and ability to take a high polish. Other commonly used metals are *copper*, *aluminum*, and refractory metals like *tantalum*, *tungsten*, and *molibdenum*. The latter metals are used for the construction of evaporators and sample holders. The need to avoid the influence of magnetic fields requires the use of μ-*metal* for magnetic shields.

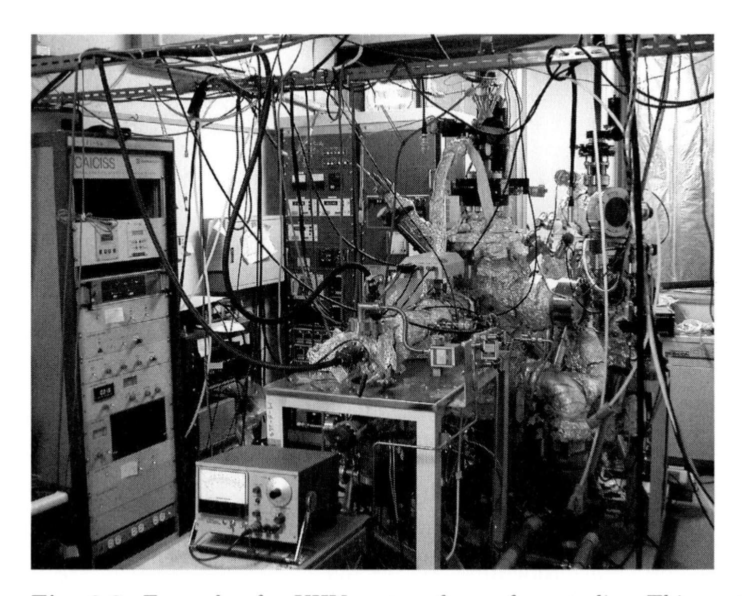

Fig. 3.2. Example of a UHV system for surface studies. This particular system employs ion scattering spectroscopy for surface analysis

Glass is also a common material used in UHV, as many types of glass have low gas permeability (except for helium) and good vacuum characteristics. However, glass is more susceptible to damage. For electrical isolation inside the UHV chamber various types of *ceramics* are employed.

The majority of plastics and rubbers have poor vacuum characteristics and are not compatible with UHV. They are useful only in forelines. However, there are several types of high-temperature plastics which can be used in UHV, for example, *teflon* (for insulating electrical wires) and *viton* and *silicone* (for sealing gaskets). They withstand high temperatures up to 200–250°C for short periods. Nevertheless, silicone tends to harden throughout prolonged high-temperature treatment, sticking to the metal surfaces and making the sealing gasket difficult to replace.

Though the maximum temperature to which 304 stainless steel components could be baked is as high as 450°C, essentially a lower baking temperature is used in practice. Typically, that is about 200°C with the duration of heating being 12–24 hours. However, some UHV systems may contain materials such as piezoceramics or high-temperature plastics, which are very sensitive to the exact bakeout temperature, say, between 150 and 250°C.

Finally, it should be noted that many common materials are not recommended for use in UHV systems and one should be very cautious when using arbitrarily chosen materials. For instance, zinc and cadmium-plated steels should never be used inside the chamber due to their outgassing properties. Even cadmium-plated screws can cause certain operational troubles.

3.3.2 UHV Pumping System

The purpose of a pumping system is to obtain and maintain the vacuum in the chamber. The main components of the pumping system are pumps, gauges and valves.

Pumps. The transition from atmospheric pressure (760 Torr) to UHV (10^{-10}–10^{-11} Torr) means changing the pressure value by ~13–14 orders of magnitude, which is well beyond the pumping characteristics of any single pump. So, two or more different pumps are needed to pass this way. The most common pumps used for preliminary pumping are *rotary pumps* and *cryosorption pumps* and those to reach the UHV level are *ion pumps* and *turbomolecular pumps* with auxiliary *titanium sublimation pumps*.

Rotary Vane Pumps. (Fig. 3.3) These pumps are widely used for pumping the system from atmospheric pressure down to about 10^{-3} Torr. They are also employed as backing pumps for turbomolecular pumps. The principle of the operation of rotary vane pumps is illustrated in Fig. 3.3b. Gas enters the inlet port and becomes trapped in the volume between the rotor vanes and the stator. Upon rotation of the eccentrically mounted rotor, the gas is compressed and then expelled into atmosphere through the exhaust discharge valve. The pumps employ vacuum oil as a sealant and lubricant. To prevent

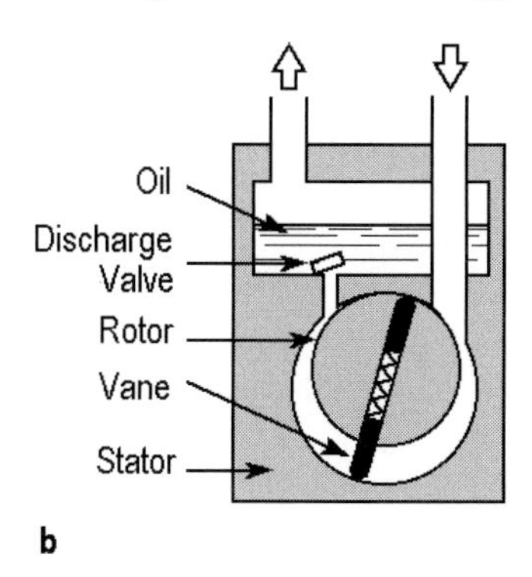

a b

Fig. 3.3. (**a**) Photograph of a rotary vane pump (Alcatel); (**b**) schematic diagram of its operation. The rotor is rotated clockwise

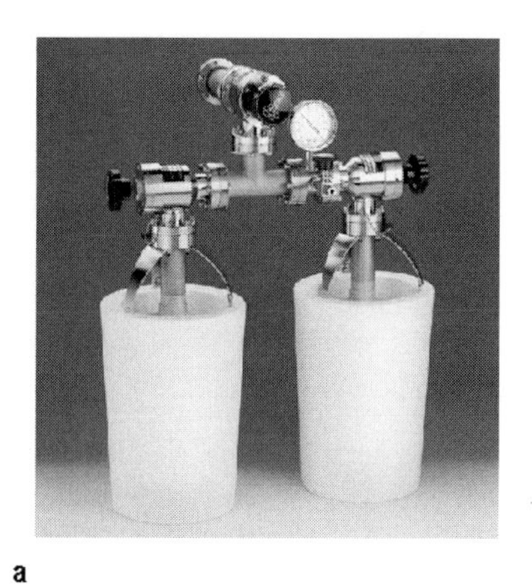

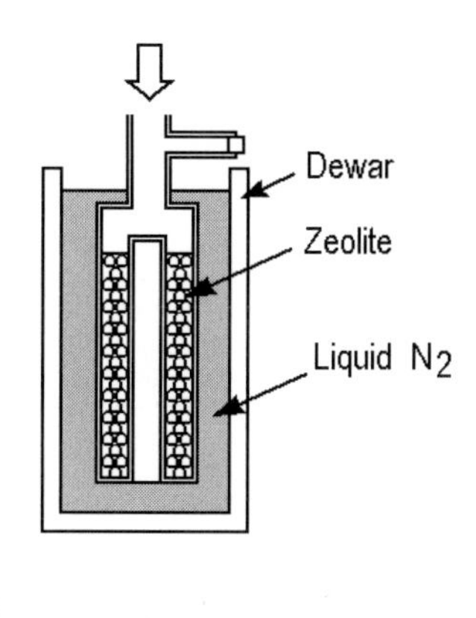

a b

Fig. 3.4. (**a**) Photograph of two cryosorption pumps mounted at the foreline (MDC Vacuum Product Corporation); (**b**) schematic diagram showing cryosorption pump design

the oil vapor from backstreaming, a liquid-nitrogen trap can be mounted at the inlet port.

Cryosorption Pumps. Their common application is as preliminary pumps in ion-pumped UHV systems, which are not often raised to atmospheric pressure. They evacuate the system from atmosphere to about 10^{-4} Torr. The cryosorption pump contains molecular sieve pellets (zeolite) in the closed tube (Fig. 3.4). The great enhancement of the sorption activity of the molecular sieve when cooling the pump walls with liquid nitrogen comprises the pumping principle. For better operation, conventionally two pumps are used in sequence. After completing the pumping cycle, the molecular sieve is regenerated just by heating; in so doing the adsorbed gases are expelled to atmosphere through a blow-off valve.

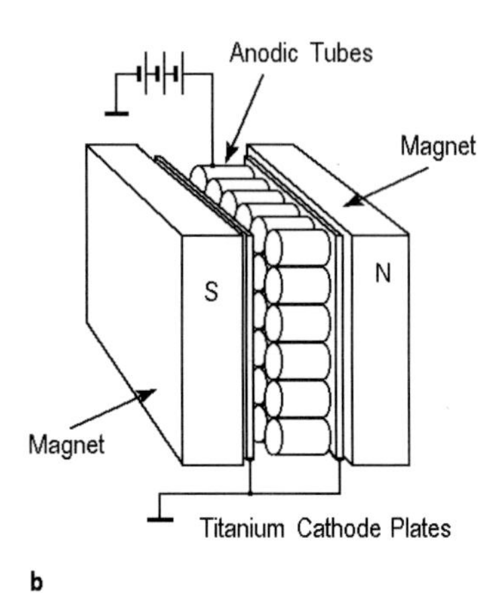

a b

Fig. 3.5. (**a**) Photograph of an ion pump (Thermionics Vacuum Products) and (**b**) schematic diagram of its design

Ion Pumps. These are the most popular UHV pumps. They start to operate at about 10^{-3} Torr and can reach the 10^{-11} Torr level. The basic configuration of an ion pump (Fig. 3.5) includes two plates made of Ti (cathode) mounted close to the open ends of the short stainless steel tubes (anode), a strong magnetic field being applied parallel to the tube's axes. The pumping action is as follows. Electrons are emitted from cathodic plates and move along helical trajectories in the anodic tubes causing ionization of the gas molecules. The ionized molecules are accelerated by the electric field and sputter titanium, when striking the cathode. The sputtered Ti coats the tubes, the cathode and pump walls. As a result of chemical reaction with the gas molecules, ion burial and neutral burial can be realized as possible pumping mechanisms. The major advantages of ion pumps are cleanliness, the ability to pump different gases and to withstand bakeout, vibration-free operation, low power consumption, long operating life, and the ability to read pressure (from the value of the pump current).

Turbomolecular Pumps. Operating in the 10^{-4} to 10^{-10} Torr range, a turbomolecular pump resembles a jet engine: a stack of rotors with multiple blades with angled leading edges are rotated at very high speed (50,000–100,000 rpm) and sweep the gas molecules in the direction of the exhaust connected to the foreline (Fig. 3.6). The compression ratio may be as high as 10^8 for N_2, but much lower ($\sim 10^3$) for H_2 and He. The turbomolecular pump is backed conventionally by a rotary pump or sometimes by combination with an additional small turbo pump and rotary pump (when ultimate UHV is required). Turbomolecular pumps are clean and reliable, but due to induced vibrations they are not suitable in systems with precise positioning, like those for microscopy or surface microanalysis.

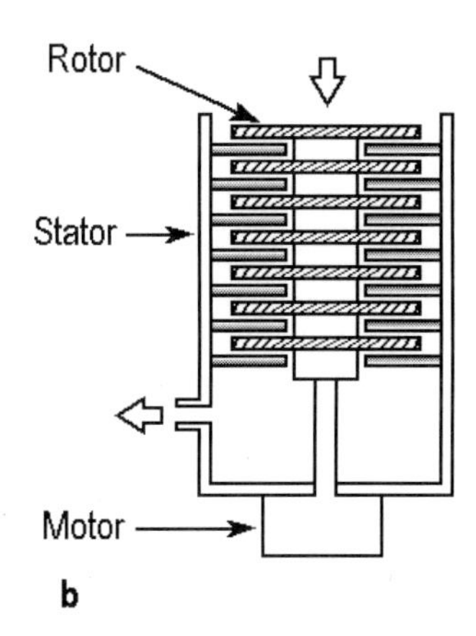

a b

Fig. 3.6. (**a**) Photograph of a turbomolecular pump (Osaka Vac Ltd.) and (**b**) schematic diagram of its design

Titanium Sublimation Pumps. These pumps are used intermittently in addition to the main UHV pumps, say, ion pumps. Titanium is sublimated from the Ti-covered filament heated by passing an electrical current through it. Active gases, for example, N_2, O_2, H_2, react with the fresh Ti film to form non-volatile compounds. The cooling of the Ti-adsorbed surfaces by liquid nitrogen enhances the pumping efficacy.

Gauges. As with pumps, there is no universal gauge which can operate in the whole range from atmospheric pressure down to UHV. Depending on the pressure, different instruments are used. *Mechanical gauges* usually operate in the pressure range from atmosphere to 1–10 Torr. They are usually employed in the places where the major concern is whether a vacuum exists, rather than its accurate measurement. Thermocouple and Pirani gauges measure a vacuum in the range of 10–10^{-3} Torr, and ionization gauges cover the range lower than 10^{-4} Torr.

Thermocouple and Pirani Gauges. A general characteristic of these gauges is to determine the pressure from the energy that is transported by gases from a hot filament to a cold surface (generally the outer walls of the gauge tube). The thermocouple gauge contains a heated filament and a bimetallic thermocouple junction for the measurement of filament temperature as a function of gas pressure (Fig. 3.7b). The filament is fed from a constant current supply. At higher pressure, more molecules hit the filament and remove more heat energy, causing the thermocouple voltage to decrease.

In a Pirani gauge (Fig. 3.7c) proposed by M. Pirani in 1906, the pressure is determined from the change in the resistance of the filament, often platinum. The resistance is measured using a Wheatstone bridge scheme, where the

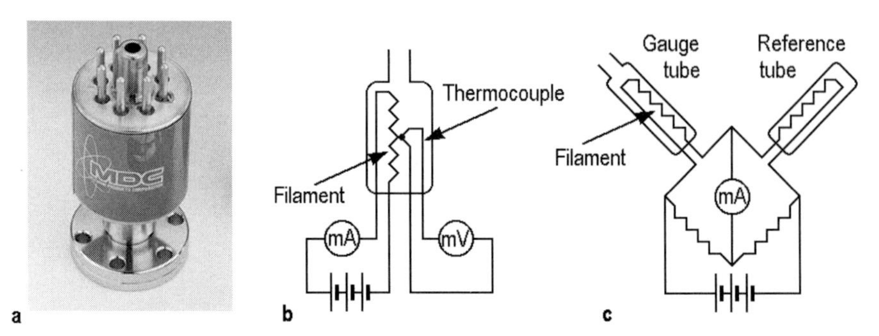

Fig. 3.7. (**a**) Photograph of a Pirani gauge (MDC Vacuum Product Corporation). A thermocouple gauge has a similar appearance. Schematic diagrams of the operational principle of (**b**) thermocouple gauge and (**c**) Pirani gauge. Both gauges measure system gas heat conduction which are dependent on gas composition and are effective in the vacuum range from 10 to 10^{-3} Torr

reference filament is immersed in a permanently evacuated chamber, while the measurement filament is exposed to the system gas. Both filaments are heated by a constant current through the bridge. The reference filament temperature is usually maintained between 100°C and 200°C. When the pressure in the system volume is equal to that of the reference filament, the current between the two arms of the bridge circuit is zero. The higher the pressure in the system volume, the lower the resistance of the measurement filament and, consequently, the higher the electrical current between the arms of the bridge. Pirani gauges have roughly the same pressure measurement range as the thermocouple gauges and are used extensively in foreline monitoring.

Ionization Gauges. The basic gauge for the measurement of pressures lower than 10^{-3} Torr is the ionization gauge. The operation of all ionization gauges is based on the ionization of molecules of a system gas. As the ionization rate and, hence, the ion current are directly dependent on the gas pressure, the pressure can be determined. There are different types of ionization gauges: hot-filament gauges, called ion gauges (Bayard–Alpert and Schulz–Phelps) and cold-cathode gauges (Penning, inverted magnetron). The former use thermionic emission of electrons from a hot filament, while the latter use electrons induced by glow discharge or plasma. In the *Bayard–Alpert gauge* (Fig. 3.8b), the electrons from the filament are accelerated towards the grid which is at a potential of about +150 V with respect to the filament. These electrons ionize atoms and molecules, which are then attracted towards the fine-wire grounded collector situated at the center of the gauge. Finally, the collector current is converted to a pressure indication. The low-pressure limit of the gauge arises mainly from the excitation of x-rays by electrons bombarding the grid. Ion gauge measurements are seriously affected by the gas composition. They are typically calibrated for nitrogen.

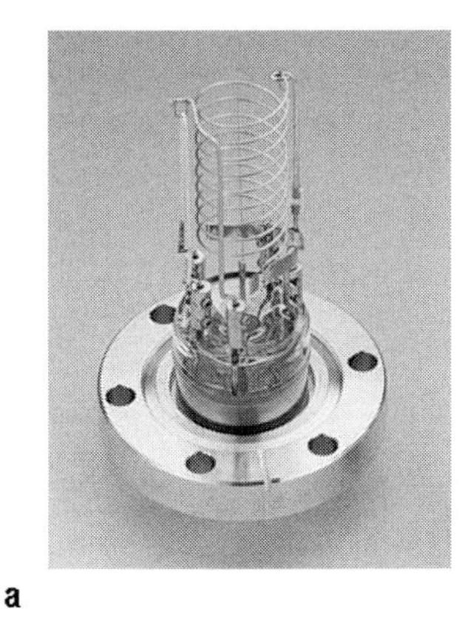

a

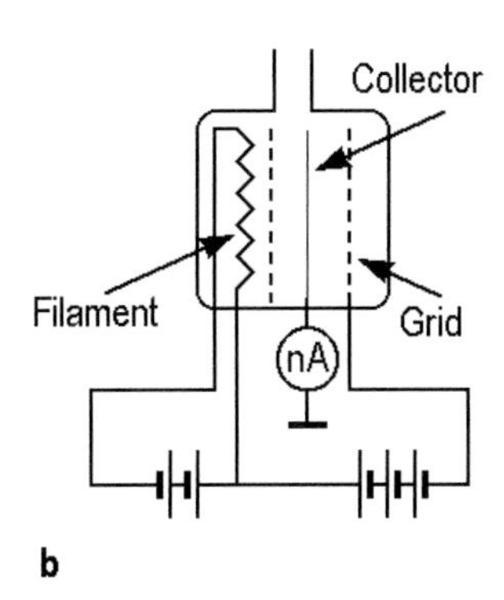

b

Fig. 3.8. (**a**) Photograph of a Bayard–Alpert gauge (MDC Vacuum Product Corporation) and (**b**) schematic diagrams of its operational principle. It works below 10^{-3} Torr and has a lower limit down to 10^{-11} Torr

Residual Gas Composition. Knowledge of the gas composition in a vacuum system is highly desirable in many cases. This can be learned with a compact quadrupole mass spectrometer known as a *residual gas analyzer* (*RGA*). Most of the RGAs are used as nothing more than leak detectors and are very useful as a maintenance tool for the vacuum system.

Typically a residual gas atmosphere of the properly baked vacuum system consists mainly of a mixture of H_2, CO, CO_2 and a small amount of H_2O, and it is useful to record this mass spectrum as a "standard." If you come across any problem with your vacuum system, you can take a look at the current spectrum to see if it differs from the "standard," especially in the appearance of new peaks. For example, the presence of water peak at mass 18 indicates improper bakeout of the system as illustrated in Fig 3.9, while an air leak shows up with an oxygen peak at mass 32. If you have an air leak, you can tune the RGA for helium and start probing your system with helium gas from outside until you find and fix the leak.

Valves. Operation of the pumping system implies that sometimes one needs to isolate one section of a vacuum system from another. Valves are used for this purpose. The main component of any valve is a "flapper," i.e., some movable part that seats against the valve body and seals the valve exit from its entrance (Fig. 3.10 and Fig. 3.11). The common sealing materials for UHV valves are viton and copper. However, when using viton, one should bear in mind that the viton gasket can withstand baking to about 200°C if the valve is open, but only to about 120°C if the valve is closed. In contrast, while using a stainless steel knife edge cutting into soft copper (metal-to-metal seal), no precautions of this kind are required. The actuating mechanism should also be sealed against atmosphere. To reach this goal, bellows are used and these are welded or sealed to the valve body.

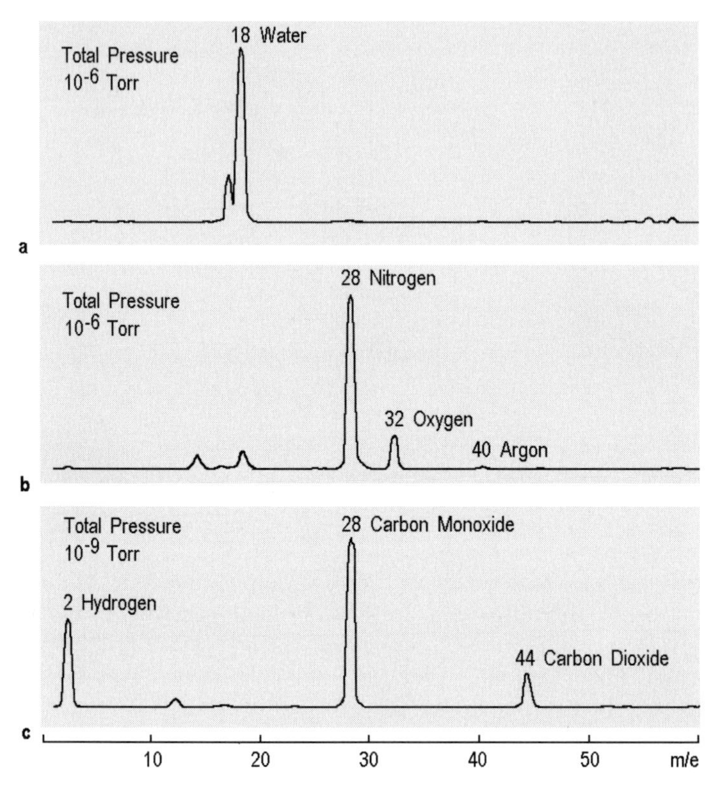

Fig. 3.9. Example of the characteristic mass spectrum for (**a**) just pumped down system (total pressure 10^{-6} Torr), (**b**) system with an air leak (total pressure 10^{-6} Torr), and (**c**) properly baked-out system (total pressure 10^{-9} Torr). Note that absolute peak heights are different for each spectrum

Example of a Pumping System. The pumps, valves, and gauges connected by tubes to the chamber (or chambers) and to each other comprise the pumping system. Figure 3.12 shows an example of a routine pumping system. Here two chambers are used. The larger one is the main chamber for experiments and analysis, and the smaller one is the load-lock chamber for changing the samples. Four pumps are in operation: a turbomolecular pump backed with a rotary vane pump, an ion pump and an auxiliary Ti-sublimation pump. The turbomolecular and rotary pumps are employed to pump the system from atmospheric pressure down to the UHV range. The ion pump serves to maintain the UHV conditions during the experiment. As the ion pump is vibration-free, it allows the usage of microprobe techniques for analysis. For pressure measurements, the ion gauge is installed in the analysis chamber with a Pirani gauge at the outlet of the turbomolecular pump. There are several valves, each having its own specific function. The large pneumatic gate valve serves to cut off the main chamber from the tur-

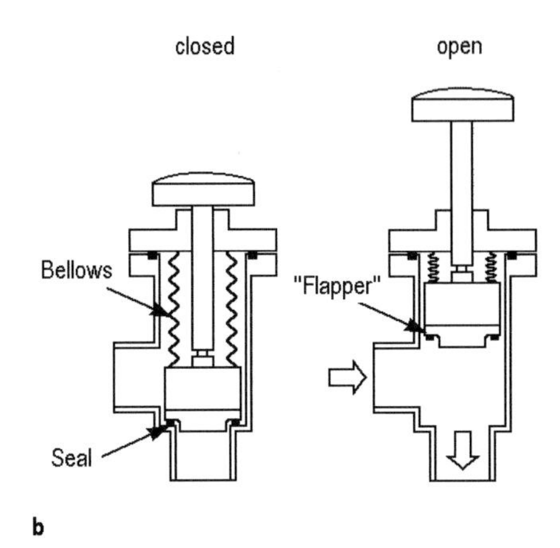

Fig. 3.10. Right-angle valve: (**a**) photograph (MDC Vacuum Product Corporation); (**b**) schematic diagram showing the valve in the closed and open positions. These valves are widely used both as foreline valves and UHV shut-off valves

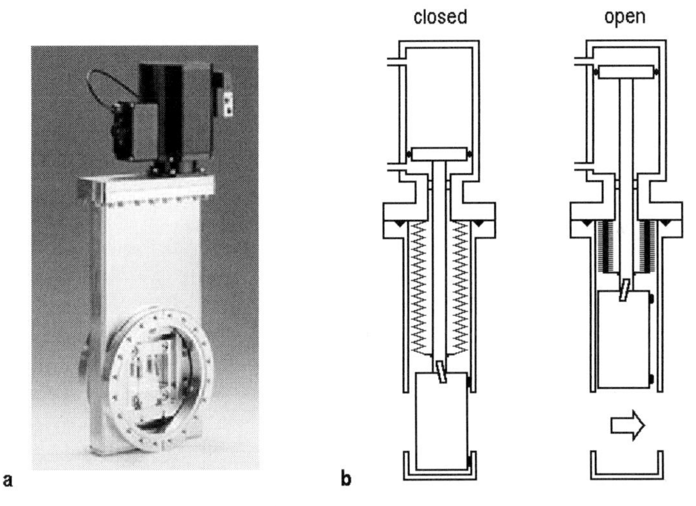

Fig. 3.11. Gate valve: (**a**) photograph (MDC Vacuum Product Corporation); (**b**) schematic diagram showing the valve in the closed and open positions. The gate valves are used when one needs to leave a wide unobstructed bore and to have a high conductance path. Their main applications are to close the chamber from UHV pumps, say, for venting and to transfer samples or radiation from one section of a system to another. The gate valve shown is operated by compressed air, but manual gate valves are also in common use

bomolecular pump. The smaller manual gate valve is for closing and opening the channel between the analysis chamber and the load-lock chamber. Two venting valves allow the analysis chamber and the load-lock chamber to be

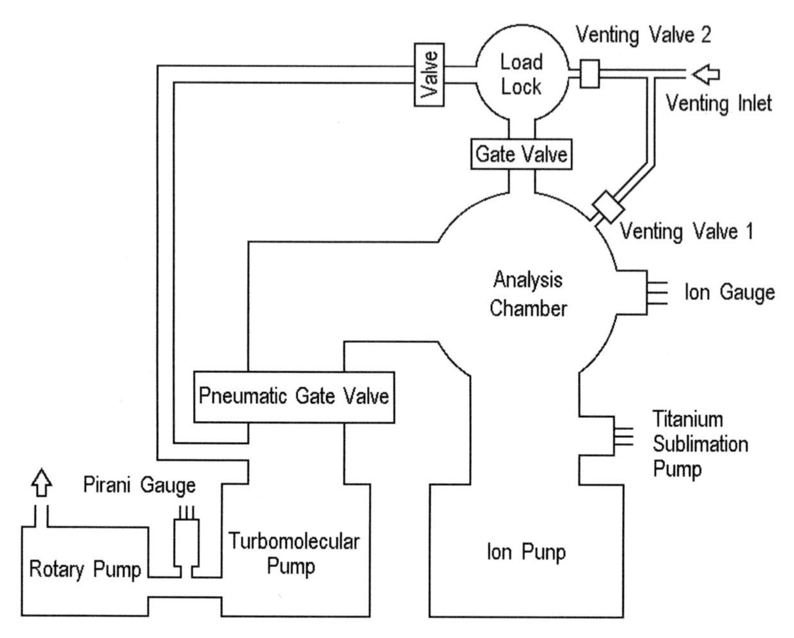

Fig. 3.12. Schematic diagram of a routine UHV pumping system

exposed to atmosphere independently. One more valve is used to control the connection of the load-lock chamber to a turbomolecular pump.

3.3.3 UHV Hardware

Chamber and Flanges. The analysis chamber is usually a central part of the vacuum system, precisely where the UHV experiments are conducted. It is typically made of stainless steel and has a number of ports with different sizes to attach the other chambers (preparation or load-lock), analytical instruments, deposition systems, sample manipulators, windows and so on. Usually its design is application-specific. Each port has a flange also made of stainless steel. Metal seal flanges employ a soft metal gasket, usually copper, compressed between two identical flanges as shown in Fig. 3.13. Standard copper gaskets are made from high-purity and oxygen-free copper and conventionally are for one-time use only. These flanges are usually referred to as Conflat (CF) flanges (after Varian) and are the most widely used for applications. They have a number of standard sizes in outside diameters (OD) ranging from 70 to 419 mm. The most popular sizes are listed in Table 3.3 in mm and inches. Inside diameters for bored flanges are also indicated. The CF sealing mechanism is suitable for pressures from atmosphere to 10^{-13} Torr. Flanges can be baked to 450°C and cooled to −200°C. Modular design of UHV systems and compatibility of UHV components gives the possibility of

a

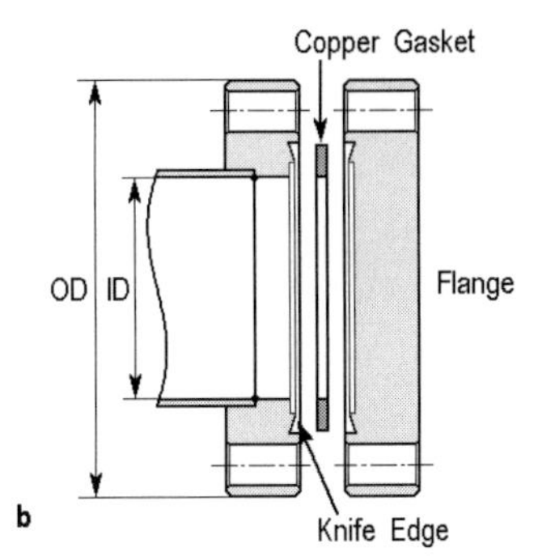

b

Fig. 3.13. (**a**) Photographs of blank and bored Conflat flanges (MDC Vacuum Product Corporation). (**b**) Cross-section of a pair of Conflat flanges used in UHV equipment. The seal is made by bolting together two identical flanges with a flat metal ring gasket between the knife edges. The knife edges press annular grooves in each side of the copper gasket, filling all voids and defects and producing a leak-tight seal

Table 3.3. The most popular sizes of Conflat flanges. Outside (OD) and inside (ID) diameters are indicated in mm and inches

OD		ID	
mm	inches	mm	inches
34	1.33	19	0.75
70	2.75	38	1.5
114	4.5	63	2.5
150	6	100	4
200	8	150	6
250	10	200	8

easy exchange of the different parts of the UHV systems and for versatile upgrading. Likewise, worn or damaged components may be replaced easily and quickly. There are a number of manufacturers who produce a large variety of UHV components on CF flanges. As an example, some of these components are shown in Fig. 3.14.

Viewports. CF flange-mounted viewports (Fig. 3.14a) are observation windows allowing the operator to view a vacuum process. Usually they are transparent for visible light and made of glass. However, for special application, UV or IR-transparent viewports are available. In the viewports, the glass is sealing to the low-expansion alloy Fe/Ni/Co, known as Kovar. Nevertheless, viewports are quite sensitive to rapid temperature changes or gradients. To reduce temperature gradients viewports are usually covered with layers of aluminum foil before bakeout. They can easily be stressed or even damaged

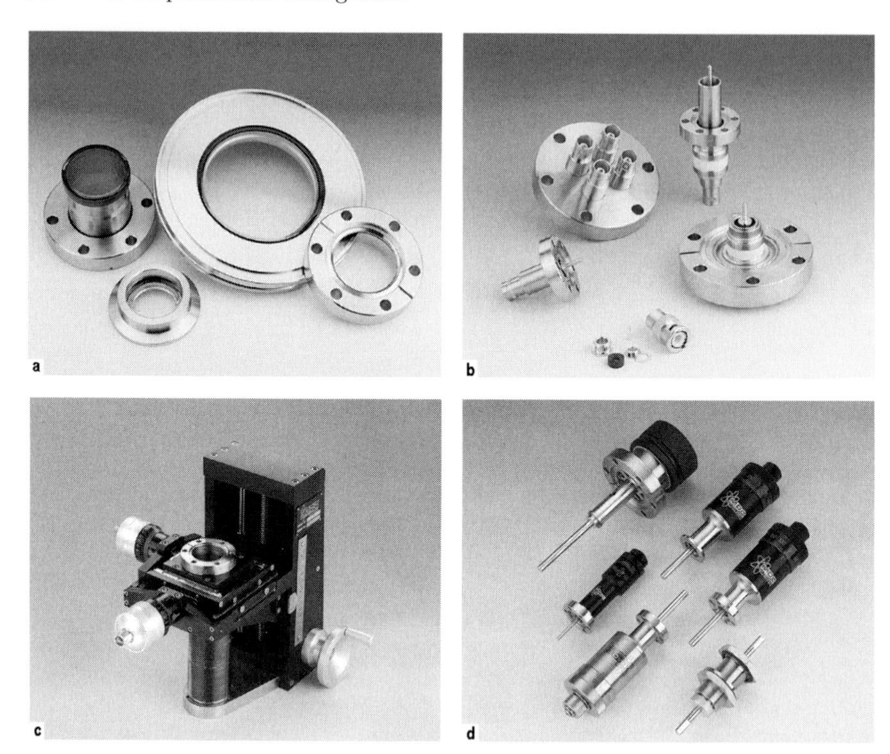

Fig. 3.14. Photographs of typical CF-mounted UHV components (MDC Vacuum Product Corporation): (**a**) viewports; (**b**) coaxial electrical feedthroughs; (**c**) XYZ stage manipulator; and (**d**) rotary motion feedthroughs

during bolt tightening, so it is strongly recommended to settle the viewport with minimum physical distortion.

Electrical Feedthroughs. These are used to transfer electrical power from the air side to the vacuum side of the system. A conventional electrical feedthrough (Fig. 3.14b) has one or many conductors that penetrate the vacuum wall. Each conductor is isolated from the flange by a metal–ceramic (often alumina) sealing. Different designs of feedthroughs cover all possible applications from low to high operating voltages and currents.

Sample Manipulators. Any experiment in a UHV chamber would hardly be possible without sample manipulators. For instance, they are used to move the sample from the load-lock chamber to the analysis chamber, detach the sample from the transporter to the XYZ manipulator, position the sample with respect to analytical techniques, and examine a specific region of the sample. All these motions are carried out while maintaining UHV integrity. Devices that move samples in three orthogonal axes are called *XYZ manipulators* (Fig. 3.14c). The flexible edge-welded bellows used in their design allows them to achieve long travel in the Z direction for particular applications. Usually high-precision XYZ manipulators are enhanced by *rotational feedthroughs* (Fig. 3.14d) to provide precise sample rotation about one or two axes. When depositing materials, rotational feedthroughs are also used to move shutters and, thus, to start and terminate material deposition onto the sample surface. *Linear drives* for long travel are generally used to transfer the sample from the load lock chamber to the sample manipulator. Relatively simple hand-operated movement of the sample inside the vacuum chamber can be done by using a *wobble stick*. Its vacuum end, which contains a tapped hole or pincers, can be used to transfer the sample from one analytical stage to another.

3.4 Preparation of Atomically Clean Surfaces

Once UHV conditions are ensured, one can start with the experiments. The first thing to be done is to prepare a clean sample surface. Before installation into the UHV chamber, the sample undergoes a set of various cleaning treatments, say, mechanical polishing, chemical etching, boiling in organic solvents, rinsing in deionized water, etc. However, all this processing constitutes rather a pretreatment stage, as the final preparation of an atomically clean surface (i.e., a surface which contains foreign species in amounts of a few per cent (and preferably less) of a monolayer) can be conducted only in situ in the UHV chamber. In most instances, it is also highly desirable that the surface should be well ordered on an atomic scale. The most common methods of the in situ cleaning are (Fig. 3.15):

- cleavage;
- heating;
- chemical processing; and
- ion sputtering.

3.4.1 Cleavage

Cleavage in UHV seems to be the most straightforward and self-explanatory way to prepare a fresh clean surface. It is applicable to such brittle materials as oxides (for example, ZnO, TiO_2, SnO_2), alkali halides (for example, NaCl, KCl), elemental semiconductors (Si and Ge) and compound semiconductors (for example, GaAs, InP, GaP). A typical cleavage set-up includes a bar of a sample material with notches cut into it and a wedge controlled mechanically, magnetically, or electrically from the outside of the chamber.

The surfaces produced by cleavage are intrinsically clean and, in the case of compound materials, generally stoichiometric. Some drawbacks should be mentioned as follows. First, cleavage is suitable for brittle materials only. Second, the cleaved surfaces are not flat but, instead, contain a high density

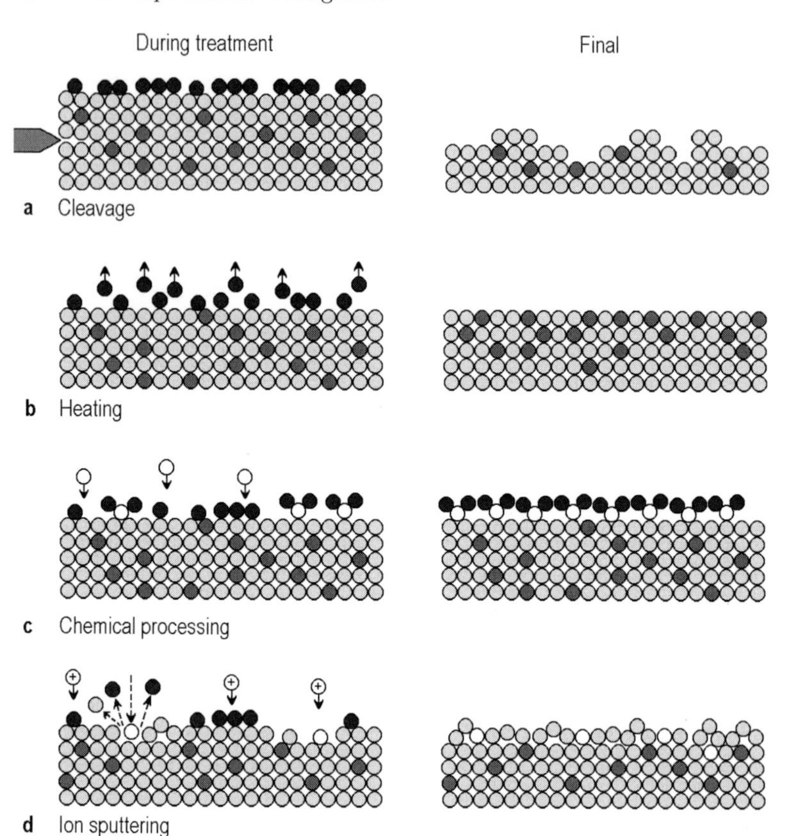

Fig. 3.15. Schematic diagram illustrating the main techniques for cleaning the samples in situ: (**a**) cleavage; (**b**) heating; (**c**) chemical reaction; and (**d**) ion bombardment. The atoms of the sample material are shown by light gray circles. The dark gray circles denote the foreign species in the sample bulk and black circles correspond to surface contaminants. Reactive gas molecules in (**c**) and impinging ions in (**d**) are shown by white circles

of steps. Moreover, the step density varies from cleave to cleave, which may result in random variation of the surface properties. Third, cleavage is only possible along certain crystallographic directions. For example, for Si and Ge, the cleavage plane is (111); for III-V semiconductors, it is (110). The surfaces with other orientations cannot be obtained by cleavage. Fourth, the cleaved surface does not necessarily posses an equilibrium structure. For example, cleaved Si(111) and Ge(111) surfaces exhibit a 2×1 structure, whereas Si(111) and Ge(111) surfaces in equilibrium display 7×7 and c(2×8) structures, respectively. On the other hand, this fact can be considered as an advantage, if one is interested in studying metastable surfaces.

3.4.2 Heating

Some of the crystal surfaces can be cleaned in situ just by heating by passing electrical current through the sample, using electron bombardment, or laser annealing. The main requirement is that the adsorbed species and/or surface oxides can be evaporated at temperatures below the melting point of the material under investigation. This holds, for example, for W and similar high-melting-point metals, as well as for Si. However, even for these materials, the heat treatment has shortcomings as follows. The annealing may result in redistribution of impurities in the sample bulk and even cause their segregation to the surface. Some impurities (for example, carbon) can form exceedingly strongly bound compounds with the substrate material during annealing and, hence, they can hardly be eliminated completely.

3.4.3 Chemical Treatment

To facilitate thermal cleaning, sometimes chemical processing of the sample surface is employed ex situ or in situ. The ex situ chemical pretreatment resides in the formation of a relatively clean protective layer which can be removed in situ at moderate temperatures. An example is the RCA procedure developed for Si by W. Kern [3.1] while working for RCA (Radio Corporation of America), hence, the name.

In situ chemical processing is generally as follows. The reactive gas is admitted into the vacuum chamber at low pressures (typically, around 10^{-6} Torr or less) and the sample is annealed in this gas environment. The gas reacts with surface impurities to form volatile or weakly bound compounds. For example, to get rid of carbon, the tungsten is annealed in oxygen at 1400–1500°C. This treatment converts C into CO, which can then be removed from the W surface by heating to about 2000°C.

3.4.4 Ion Sputtering and Annealing

Surface contaminants can be sputtered off together with the substrate top layer by bombardment of the surface by noble gas ions (usually, Ar^+). To produce an ion beam, Ar gas is admitted through a leak valve directly into the ion gun (Fig. 3.16) or into the UHV chamber as a whole. Ionization of the gas atoms proceeds via electron impact in the ionizer of the ion gun. Electrons are emitted from the cathode. The produced ions are extracted from the ionizer, accelerated to the desired energy (typically, 0.5–5.0 keV) and directed towards the sample.

Ion sputtering is a very effective cleaning technique. The side effect of ion bombardment is the degradation of the surface structure. Therefore, subsequent annealing is required to restore the surface crystallography and to remove embedded and adsorbed Ar atoms. Annealing implies the problems discussed above, but fortunately the required temperatures are usually lower

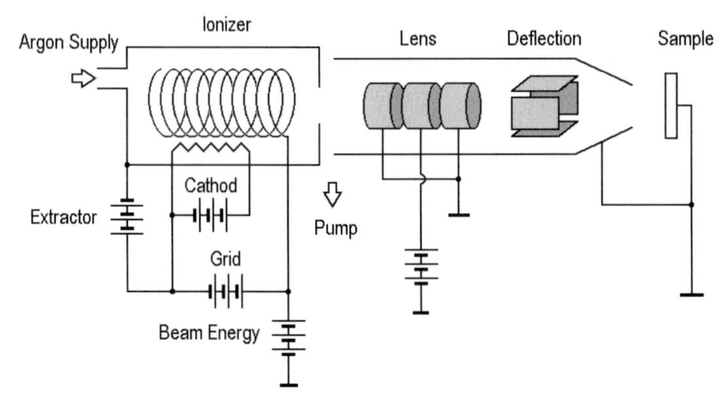

Fig. 3.16. Schematic diagram showing a typical ion sputter gun used for sample cleaning. Such a gun can also be applied for sputter depth profiling, secondary ion mass spectrometry, and ion scattering spectrometry

compared to those when only heating is used. In practice, several ion bombardment/annealing cycles are required to prepare a well-ordered atomically clean surface.

In conclusion, it is apparent that no universal cleaning procedure exists. For each material, an individual method with definite parameters (or maybe a combination of several techniques) is required. Sometimes, the procedure might be very complicated and may even include some additional techniques. For example, heating or ion bombardment of the compound materials may noticeably change their surface stoichiometry due to preferential desorption/sputtering of one component. A possible solution is to overgrow an epitaxial layer with perfect stoichiometry above the prepared surface.

3.5 UHV Deposition Technology

3.5.1 Deposition Concepts

Most UHV deposition technologies employ thermal evaporation or sublimation of materials. When a substance is heated to a relatively high temperature, some atoms or molecules have enough energy to break the chemical bonds and leave the substance. The liquid is evaporating and the solid is subliming. The flux I of atoms or molecules onto a substrate located at a distance L away from a deposition source can be calculated using (3.1)

$$I = \frac{p(T)A}{\pi L^2 \sqrt{2\pi m k_\mathrm{B} T}} , \tag{3.13}$$

where $p(T)$ is the equilibrium vapor pressure of the deposited material, A is the evaporation or sublimation area, and other parameters are the same as in

(3.1). To achieve a typical deposition rate of 0.1 to 1 ML/min, with a distance L of 10 cm and an area A of 0.5 cm^2, an equilibrium vapor pressure of about 10^{-5}–10^{-4} Torr is required. The temperature of the source in order to reach these pressures can be estimated from equilibrium vapor pressure plots like those shown in Fig. 3.17 for selected elements. Tables of temperatures for given vapor pressures are also useful. Table 3.4 furnishes an example for the vapor pressures of 10^{-8}, 10^{-6} and 10^{-4} Torr.

3.5.2 Deposition Sources

Depending on the particular experimental design, various deposition sources are employed. We shall give an overview below of the most commonly used types.

Simple Thermal Sources. Simple thermal sources constitute a group of devices that are of simple design and usually home-made. They are essentially open heaters without radiative or insulating shielding or any means of reducing thermal gradients. They are made of refractory metal foils (*boats* and *tubes*) or shaped filaments (*filaments* and *baskets*) and are heated directly by passing an electrical current through them. The main shortage of these sources is that they do not produce a quite stable deposition rate.

Filaments and Baskets. These are made of single- or multiple stranded tungsten wire. Typical filaments are bent into a "V" shape (Fig. 3.18a). A small piece of material for deposition is melted onto the wire. To increase the amount of material and, hence, the operation time, basket-shaped sources are used (Fig. 3.18b). The material deposited from the filament or basket should wet the tungsten wire, but not react with it. Otherwise one can use a ceramic crucible that sits in a wire filament coiled to fit the crucible's outside dimensions (Fig. 3.18c).

Boats and Tubes. Tungsten, tantalum, or molybdenum foil is also a suitable material for making simple thermal sources. The most popular shapes are boats (Fig. 3.18d) and tubes. Tubes may have one end open (Fig. 3.18e) or, being closed from both sides, have a small hole in the middle (Fig. 3.18f). It is apparent that tubes produce a more directed vapor stream, compared to boats.

Sublimation Sources. Some materials, like Fe and Si, have sufficient vapor pressure well below the melting point and can be sublimated. In this case, the wire or rod made of these materials is heated directly by passing the electrical current through it.

Knudsen Cells. When highly constant evaporation rates are required, Knudsen cells (or simply K cells) are used. A Knudsen cell utilizes the principle of molecular effusion (demonstrated by Knudsen as early as 1909). In the cell, the crucible (tungsten in a classic K cell or alumina, pyrolytic boron

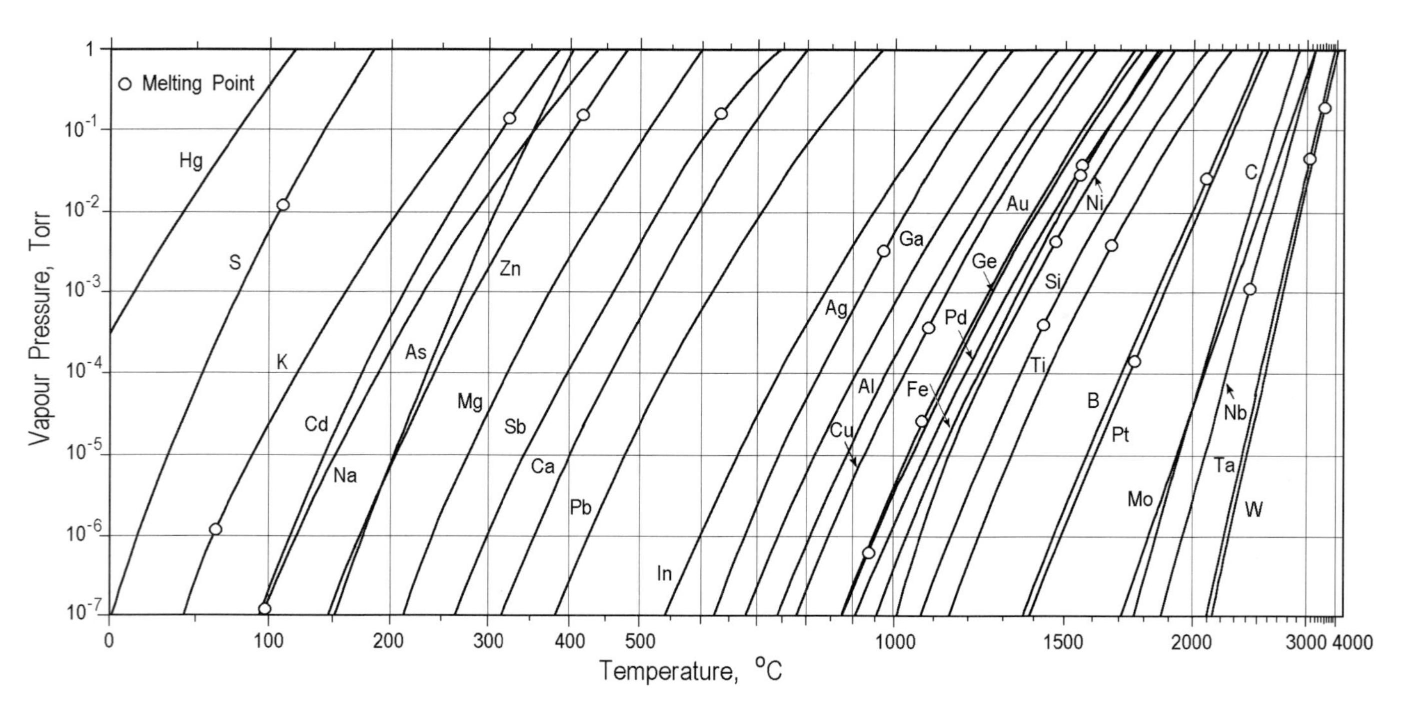

Fig. 3.17. Equilibrium vapor pressure of selected elements as a function of temperature

Table 3.4. Melting points (MP) and the temperatures for given vapor pressures of selected elements [3.2]

Material	Symbol	MP, °C	Temperature °C for a given vapor pressure Torr		
			10^{-8}	10^{-6}	10^{-4}
Aluminum	Al	660	685	812	972
Antimony	Sb	630	279	345	425
Arsenic	As	817	104	150	204
Bismuth	Bi	271	347	409	517
Boron	B	2300	1282	1467	1707
Cadmium	Cd	321	74	119	177
Carbon	C	3652	1657	1867	2137
Copper	Cu	1083	722	852	1027
Gallium	Ga	30	619	742	907
Germanium	Ge	937	812	947	1137
Gold	Au	1064	807	947	1132
Indium	In	157	488	597	742
Iron	Fe	1535	892	1032	1227
Magnesium	Mg	649	185	246	327
Mercury	Hg	−39	−72	−44	7
Molybdenum	Mo	2610	1592	1822	2117
Nickel	Ni	1455	927	1072	1262
Platinum	Pt	1772	1292	1492	1747
Potassium	K	63	21	65	123
Silicon	Si	1410	992	1147	1337
Silver	Ag	962	547	685	832
Sodium	Na	98	74	123	193
Tantalum	Ta	2996	1957	2237	2587
Tin	Sn	232	682	807	997
Titanium	Ti	1660	1062	1227	1442
Tungsten	W	3410	2117	2407	2757

nitride, or graphite in a modern cell) is surrounded by a heater and several layers of radiation shielding (Fig. 3.19). The depositing material emerges from the aperture in a cosine distribution of flux, provided the aperture is small enough. The deposition rate is extremely stable, being determined by the temperature of the furnace which is monitored by a thermocouple. To

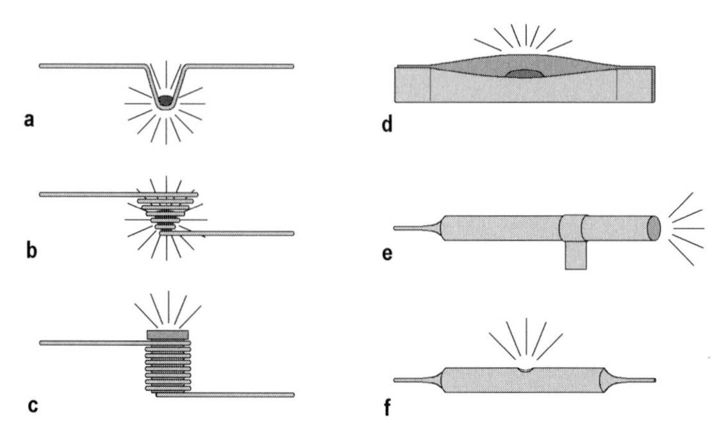

Fig. 3.18. Simple thermal evaporators: (**a**) "V" shaped filament; (**b**) basket; (**c**) basket with crucible; (**d**) boat; (**e**) open; and (**f**) closed tubes

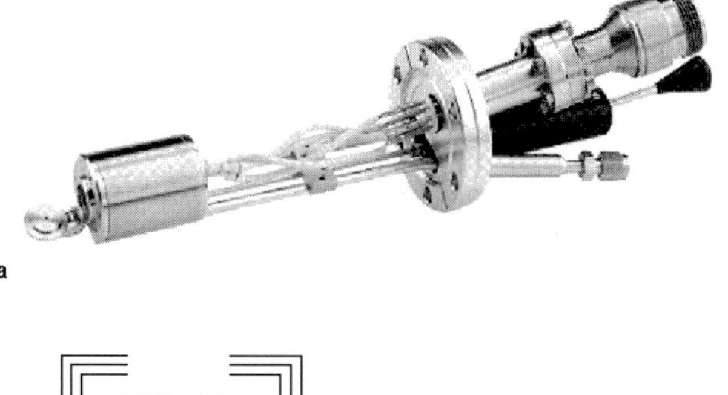

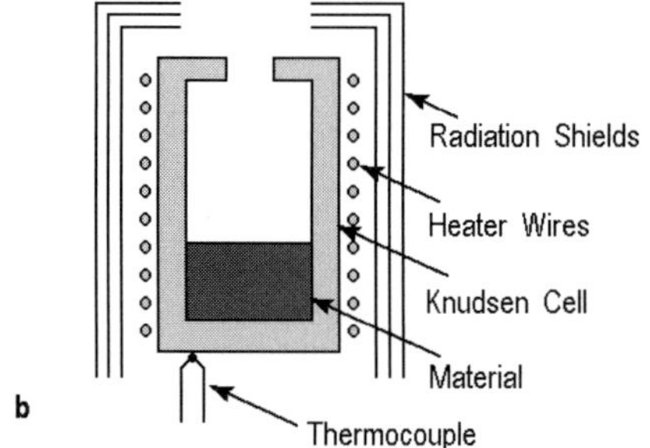

Fig. 3.19. Knudsen cell: (**a**) photograph (Maxtex, Inc.); (**b**) schematic diagram

start and stop deposition, a movable shutter located near the cell aperture is used.

Simple thermal evaporators and Knudsen cells are suited mostly for materials which have relatively high vapor pressures and relatively low melting points (for example, Al, Ag, As, Bi, Sb, Cu, Ga, In, Mg).

Electron Beam Evaporators. In electron beam evaporators, the solid material in the form of powder, granules, or lumps is placed in a hearth

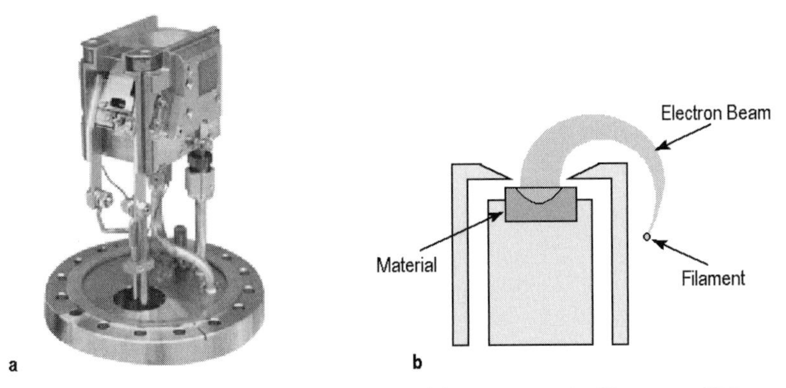

Fig. 3.20. Electron beam evaporator: (**a**) photograph (Omicron Vakuumphysik GmbH); (**b**) schematic diagram

and heated by a high-current electron beam. The electrons are generated by thermionic emission from a hot filament placed below the hearth and focused on the material surface by permanent electromagnets (Fig. 3.20). As a result of local melting, a pool of liquid forms from which material evaporates. Since the melt is surrounded by solid material of the same composition, electron beam evaporators are less prone to contamination than other types of thermal sources.

Electron beam sources find application in production-scale machines due to their high deposition rates and large material capacity. They are suitable for deposition of materials with relatively low vapor pressure, like Mo, Nb, Si, Ta, W, and Zr.

SAES Getters. Alkali metals possess properties that are very inconvenient for use as a deposition material in UHV. First, they are highly reactive with respect to the water vapor present in the air, making loading of a source very problematic. Second, they have very high vapor pressures even at bakeout temperatures. Fortunately, special types of deposition sources, called SAES getters, are commercially available for alkali (Li, K, Na, Cs) and alkaline earth (Ba, Sr) metals. In these sources, a porous material (getter) filled with metal is used. The deposition is conducted simply by heating the getter.

3.5.3 Deposition Monitors

Deposition rate and final film thickness are the most important parameters for any deposition process. A *quartz crystal thickness monitor* is the main instrument for measuring these characteristics. It consists of an AT-cut quartz crystal which vibrates at its fundamental frequency f_0 which depends on the crystal thickness d_q as

$$f_0 = N/d_\mathrm{q} \,, \tag{3.14}$$

where $N = 1.67 \cdot 10^6$ Hz mm. That frequency (usually of $\sim$ 5–10 MHz) is stable at a fixed temperature until the crystal's mass changes. If a crystal sensor is placed near the substrate during deposition, the deposited material on the crystal increases its mass by Δm and lowers its resonant frequency by Δf:

$$\Delta f = \frac{K f_0^2 \Delta m}{\rho_q N A} = \frac{K N \Delta m}{\rho_q d_q^2 A} , \tag{3.15}$$

where ρ_q is the density of the quartz (2.65 g/cm^3), A is the area of the crystal, and K is a constant, usually close to unity, which depends on the spatial distribution of the deposited material over the area A of the crystal. Note that, if only a fraction of the area of the crystal is covered by electrodes, used to excite the oscillations, only the area of the electrodes should be treated as A in (3.15), as the crystal oscillations are negligibly small outside this area.

In practice, several factors should be taken into account when using (3.15) for the determination of the deposited film thickness.

- First, the linear relation between Δf and Δm holds as long as the thickness of the deposited material is much lower than the crystal thickness.
- Second, the crystal does not receive exactly the same deposition as the substrate. The correction factor must be calculated for the system geometry.
- Third, one should ensure that the sticking coefficient of deposited atoms or molecules is indeed unity for a given substrate temperature.
- Fourth, it is important that the temperature of the sensor does not vary during the measurements, since the fundamental frequency of the quartz crystal depends on its temperature. For AT-cut quartz crystal the value of the frequency–temperature coefficient is close to zero at 30°C and remains lower than $\pm 5 \cdot 10^{-6}\,^\circ\mathrm{C}^{-1}$ in the temperature range of $\pm$30°C.

All precautions being taken, the quartz crystal monitor provides an accuracy down to fractions of a monolayer.

3.5.4 Exposure to Gases

Exposure of samples to high-purity gases is another way to produce well-characterized overlayers in UHV. The interaction of solid surfaces with gas molecules and atoms is of great scientific and practical interest. Much experimental research is conducted in this field. In experiments of this kind, the UHV chamber is backfilled with a certain gas from a container through a *leak valve*. The purpose of the leak valve is to admit small amounts of gas at a controlled leak rate. The conductance of the leak valve is very low even when fully open.

When doing experiments with gases, one should bear in mind that the state of a gas is very important, say, if gas is in the atomic or molecular state.

For example, atomic hydrogen reacts readily with the Si surface at room temperature, while molecular hydrogen demonstrates a negligible reactivity. For cracking of molecular hydrogen, the heated tungsten filament is mounted in front of the Si sample surface. The operation of some UHV devices (for example, ion pumps, ion gauges, electron guns) might lead to the excitation of the gas molecules and atoms, which also affects their reactivity.

Problems

3.1 Calculate the time required for a N_2 molecule to travel a distance equal to its mean free path in a vacuum of 10^{-10} Torr at 300 K.

3.2 A vacuum chamber of volume 50 l is backfilled with molecular hydrogen at 300 K to a pressure of 10^{-6} Torr. What is the volume occupied by the same amount of hydrogen at normal conditions (1 Atm., 300 K)?

3.3 Using the scheme of the pumping system in Fig. 3.12, consider the sequence of operations with pumps and valves needed

(a) to exchange the sample without breaking the UHV conditions in the analysis chamber;
(b) to expose the analysis chamber to atmospheric pressure;
(c) to pump down the system from atmosphere to UHV (including baking-out of the system).

For the answers, fill the following table. As a starting point, consider the state of the pumping system with UHV in the analysis chamber.

Pumps			Valves				
Ion	Turbo	Rotary	Pneum. gate	Manual gate	Vent. 1	Vent. 2	Valve 3
ON	OFF	OFF	OPEN	CLOSE	CLOSE	CLOSE	CLOSE

3.4 After deposition of Al film, the basic frequency of the quartz crystal monitor (10 MHz) is shifted by 10 Hz. Calculate the amount of deposited Al. The answer should be expressed in thickness units (Å). Assume the density of the Al film to be the same as for bulk metal Al (2.70 g/cm^3).

3.5 Show that the thickness variation $d(x)$ of a metal film deposited from a filament source is given by:

$$d(x) = \frac{d(0)}{1 + \left(\frac{x}{D}\right)^2},$$

where D is the normal distance to the source and x is the distance from the substrate to the normal between substrate and source.

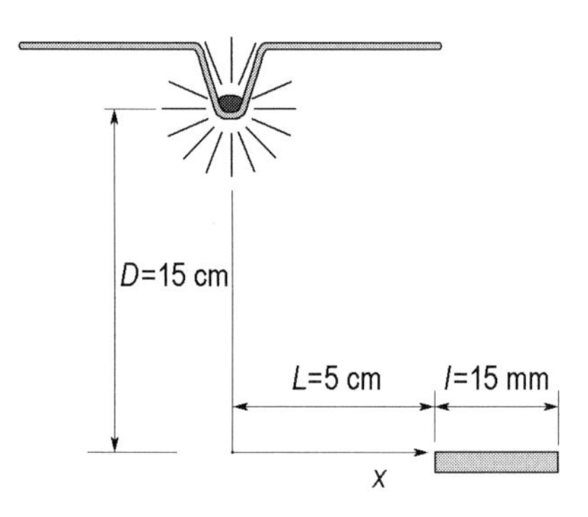

Estimate the thickness variation along the sample in %, if $D = 15$ cm, $L = 5$ cm, and the sample length l =15 mm.

Further Reading

1. S. Dushman: *Scientific Foundations of Vacuum Technique*, 2nd edn. (John Wiley, New York 1962)
2. N. Harris: *Modern Vacuum Practice* (McGraw-Hill, New York 1989)
3. J.F. O'Hanlon: *A Users Guide to Vacuum Technology*, 2nd edn. (John Wiley, New York 1990)
4. J.M. Lafferty (Ed): *Foundations of Vacuum Science & Technology* (Wiley-Interscience, New York 1998)
5. H.Lüth: *Surfaces and Interfaces of Solid Materials*, 3rd edn. (Springer, Berlin, Heidelberg 1995) Panel 1 and Chapter 2
6. A. Roth: *Vacuum Technology*, 2nd edn. (North Holland, Amsterdam 1982)

4. Surface Analysis I. Diffraction Methods

Diffraction techniques utilizing electrons or x-ray photons are widely used to characterize the structure of surfaces. The structural information is gained conventionally from the analysis of the particles/waves scattered elastically by the crystal. The intensity of the diffracted beams contains information on the atomic arrangement within a unit cell. The spatial distribution of the diffracted beams tells us about the crystal lattice. The evaluation of the crystal lattice is straightforward, as the diffraction pattern is directly related to the crystal reciprocal lattice by the condition:

$$\boldsymbol{k} - \boldsymbol{k}_0 = \boldsymbol{G}_{hkl} \; . \tag{4.1}$$

Here $\boldsymbol{k}_0$ is the incidence wave vector, $\boldsymbol{k}$ is the scattered wave vector, and $\boldsymbol{G}_{hkl}$ is the reciprocal lattice vector (Sect. 2.5). As scattering is elastic,

$$|\boldsymbol{k}| = |\boldsymbol{k}_0| \; . \tag{4.2}$$

Equations (4.1) and (4.2) express the laws of conservation of momentum and energy, respectively.

Diffraction can be represented graphically using the *Ewald construction* as follows (Fig. 4.1).

- Construct the reciprocal lattice of the crystal.
- Draw the incidence wave vector $\boldsymbol{k}_0$ with origin chosen such that $\boldsymbol{k}_0$ terminates at the reciprocal lattice point.
- Draw a sphere of radius $k = 2\pi/\lambda$ centered at the origin of $\boldsymbol{k}_0$.
- Find the reciprocal lattice points lying on the surface of the sphere and draw the scattered vectors $\boldsymbol{k}$ to these points.

One can see that the scattered wave vectors $\boldsymbol{k}$ obtained in this way satisfy (4.1) and (4.2).

4.1 Low-Energy Electron Diffraction (LEED)

4.1.1 Ewald Construction in LEED Conditions

Consider now the case of low-energy electron diffraction. The usage of low-energy electrons for surface analysis stems from the following two main reasons:

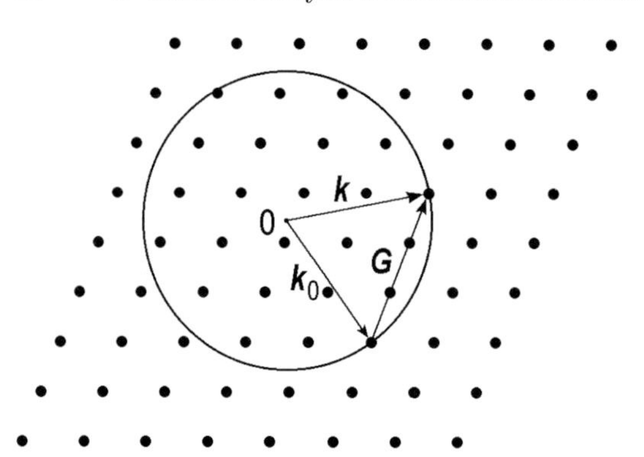

Fig. 4.1. Ewald construction for diffraction from a 3D lattice

- First, as the de Broglie wavelength of an electron is given by

$$\lambda = \frac{h}{\sqrt{2mE}} , \quad \lambda[\text{Å}] = \sqrt{\frac{150}{E(\text{eV})}} , \tag{4.3}$$

  in the typical range of energies used in LEED (30–200 eV) the electrons have a wavelength (∼1–2 Å) that satisfies the atomic diffraction condition, namely, λ is of the order of or less than the interatomic distances.
- Second, the mean free path of the low-energy electrons is very short, of the order of a few atomic layers. Thus, most elastic collisions occur in the very top layers of a sample.

Therefore, LEED provides mostly information about the 2D atomic structure of the sample surface.

For the case of diffraction on the 2D surface, the crystal periodicity in the direction normal to the surface is lacking and (4.1) becomes

$$\boldsymbol{k}^{||} - \boldsymbol{k}_0^{||} = \boldsymbol{G}_{hk} . \tag{4.4}$$

That is, the law of conservation of momentum concerns only the wave vector components parallel to the surface, namely, the scattering vector component parallel to the surface, $(\boldsymbol{k}^{||} - \boldsymbol{k}_0^{||})$, must be equal to the vector of the 2D surface reciprocal lattice, $\boldsymbol{G}_{hk}$. Note that the wave vector component normal to the surface is not conserved in this process.

The Ewald construction modified for diffraction on a 2D lattice is shown in Fig. 4.2. In contrast to the 3D reciprocal lattice points, here the *reciprocal lattice rods* perpendicular to the surface are attributed to every 2D reciprocal lattice point. (A 2D lattice can be conceived as a 3D lattice with infinite periodicity in the normal direction, $|\boldsymbol{c}| \to \infty$. This will lead to $|\boldsymbol{c}^*| \to 0$, i.e., the reciprocal lattice points along the normal direction are infinitely dense, thus forming rods.) The incidence wave vector $\boldsymbol{k}_0$ terminates at the reciprocal lattice rod. The intercepts of the rods with the Ewald sphere define the scattered wave vectors $\boldsymbol{k}$ for diffracted beams.

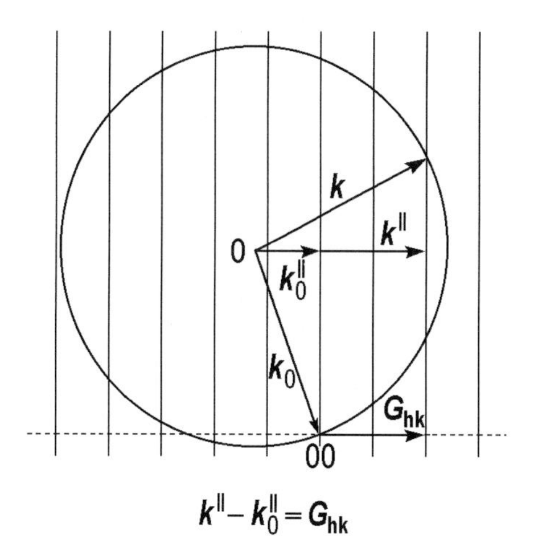

Fig. 4.2. Ewald construction for diffraction on a 2D surface lattice

4.1.2 LEED Experimental Set-Up

The standard experimental set-up for LEED allowing direct observation of the diffraction pattern is shown schematically in Fig. 4.3. The main elements are

- an *electron gun* to produce a collimated beam of low-energy electrons,
- a *sample holder* with the sample under investigation, and
- a hemispherical *fluorescent screen* with a set of *four grids* to observe the diffraction pattern of the elastically scattered electrons. Note that the sample is located at the center of curvature of the grids and screen.

The electron gun unit consists of a cathode filament with a Wehnelt cylinder followed by an electrostatic lens. The cathode is at negative potential ($-V$), while the last aperture of the lens, the sample, and the first grid are at earth potential. Thus, the electrons emitted by the cathode are accelerated to an energy of eV within the gun and then propagate and scatter from the sample in the field-free space. The second and third grids are used to reject the inelastically scattered electrons. The potential of the second and third grids is close to that of the cathode, but somewhat lower in magnitude, $-(V - \Delta V)$. The greater ΔV, the brighter the LEED pattern, but the higher its background intensity. So, the retarding voltage is adjusted to get a LEED pattern with the highest spot-to-background contrast. The fourth grid is at earth potential and screens other grids from the field of the fluorescent screen, which is biased to a high voltage of about +5 kV. Thus, the elastically scattered, diffracted electrons, after passing the retarding grids, are reaccelerated to a high energy to cause fluorescence of the screen, where the diffraction pattern is observed.

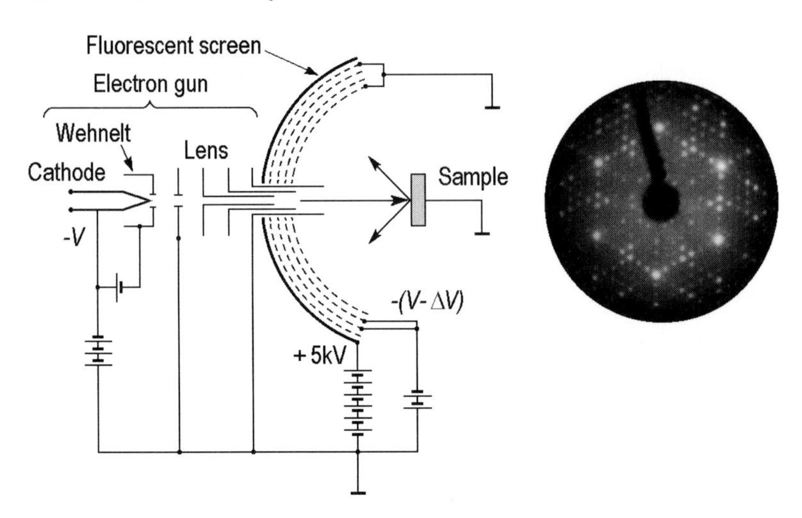

Fig. 4.3. Schematic diagram of a standard four-grid LEED set-up and LEED pattern of a Si(111)7×7 surface as seen in the reverse-view LEED arrangement

From the viewpoint of LEED pattern observation, two types of LEED systems are in common use:

- In the *normal-view arrangement*, a viewport is placed in front of the back-side of the sample and thus the LEED pattern is viewed past the sample through the grids. The sample holder with the sample partially obscures the view and, therefore, should be reasonably small.
- In the *reverse-view arrangement*, the LEED pattern is viewed through a viewport placed behind the transmission phosphorescent screen. In this case, the electron gun has to be miniaturized, while there are no specific limitations on the size and shape of the sample holder.

Comparison of the LEED set-up (Fig. 4.3) and the Ewald construction under LEED conditions (Fig. 4.2) reveals that the diffraction pattern seen at the screen is essentially a view of the surface reciprocal lattice. Just imagine that the fluorescent screen corresponds to the Ewald sphere surface and diffracted beams produce spots where reciprocal lattice rods intersect the Ewald sphere.

The observed spots are indexed in the same way as the points of the reciprocal lattice, namely, by the values of h, k in (2.8). The specular spot is accepted as the (0,0) point. For normal incidence of the primary electron beam, it is located in the center of the LEED pattern (Fig. 4.4). It is clear that the number of spots seen in a given LEED pattern depends on the size of the Ewald sphere. By increasing the electron energy, one decreases the wavelength and, consequently, increases the Ewald sphere radius. As a result, diffraction spots become closer to the specular (0,0) beam and more spots are seen.

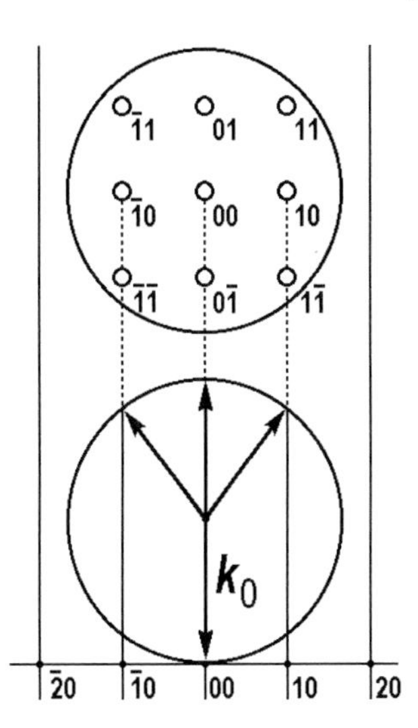

Fig. 4.4. Labelling of LEED spots at normal incidence of the primary electron beam. Spots have the same indices as the corresponding points on the 2D surface reciprocal lattice

4.1.3 Interpretation of a LEED Pattern

Consider now what kind of information concerning the surface structure can be gained from the LEED data.

Sharpness of a LEED Pattern. The very first inspection of the LEED pattern resides conventionally in the qualitative estimation of the structural perfection of the surface under investigation. The well-ordered surface exhibits a LEED pattern with bright sharp sports and low background intensity. The presence of structural defects and crystallographic imperfections results in broadening and weakening of the spots and in increasing of the background intensity. The absence of any spots in the LEED pattern is an indication of a disordered, amorphous, or finely polycrystalline, surface.

LEED Spot Geometry. The next step is the inspection of the geometrical spot positions. To illustrate what information can be gleaned from the inspection, consider some representative examples. The purpose of these examples is to illustrate the correspondence of the surface 2D lattice to the observed LEED patterns. In general, to draw a sketch of a LEED pattern for a given surface structure one should use (2.8) and (2.9) for the reciprocal lattice construction.

The 1×1 LEED pattern is seemingly the simplest case. It is obvious that a bulk-like terminated surface would exhibit such a pattern, however, this is not the only possible case. A perpendicular displacement of a surface layer from its bulk ideal position will not change the 1×1 surface periodicity, nor will a variation in composition or atomic positions within a 2D 1×1 unit cell.

Some clean metal surfaces (for example, Ni(110), Pd(111), Rh(111), W(110), etc.) furnish examples of the 1×1 structure.

In some instances, one speaks about an *apparent 1×1* LEED pattern, which refers to surfaces that display a 1×1 LEED pattern, but, in fact, do not possess long-range ordering (for example, a high-temperature Si(111)1×1 structure with randomly migrating adatoms or a highly heterogeneous Si(111) 1×1 surface after pulse-laser quenching). To indicate that it does not correspond to true 1×1 periodicity, an apparent 1×1 pattern is sometimes labeled in quotes, "1×1".

When a superstructure is formed at the surface, new spots develop in the LEED pattern. These spots are called *superspots* or *extra-spots* to distinguish them from those of the 1×1 pattern, which are called *integer-order* or *main spots*. Figure 4.5 shows sketches of LEED patterns for some typical superlattices (2×1, 2×2, $c(4 \times 2)$) formed on a substrate with a square lattice and Fig. 4.6 shows those (2×1, 2×2, $\sqrt{3}\times\sqrt{3}$) on a substrate with a hexagonal lattice. In particular, one can see that doubling the period in real space results in half-order spots (i.e., at half the distance between the main spots) and, in general, the longer the superstructure period, the more closely the extra-spots are spaced.

In analogy to (2.4) and (2.5) for real space lattices, the primitive translation vectors $\boldsymbol{a}^*_\mathrm{s}$, $\boldsymbol{b}^*_\mathrm{s}$ of the superstructure reciprocal lattice can be expressed in terms of those of the substrate reciprocal lattice $\boldsymbol{a}^*$, $\boldsymbol{b}^*$ by

$$\begin{aligned} \boldsymbol{a}^*_\mathrm{s} &= G^*_{11}\boldsymbol{a}^* + G^*_{12}\boldsymbol{b}^* \;, \\ \boldsymbol{b}^*_\mathrm{s} &= G^*_{21}\boldsymbol{a}^* + G^*_{22}\boldsymbol{b}^* \;. \end{aligned} \tag{4.5}$$

The reciprocal space matrix G^* is related to real space matrix G by

$$G^* = (G^{-1})^T \;, \tag{4.6}$$

where the components of the transpose inverse matrix $(G^{-1})^T$ are

$$\begin{aligned} G^*_{11} &= \frac{G_{22}}{\det G} \;, \quad G^*_{12} = \frac{G_{21}}{\det G} \;, \\ G^*_{21} &= \frac{G_{12}}{\det G} \;, \quad G^*_{22} = \frac{G_{11}}{\det G} \;. \end{aligned} \tag{4.7}$$

Expressions (4.5)–(4.7) can be used to predict for a given superlattice the location of the extra-spots in the LEED pattern with respect to the main spots.

It is not necessary that the entire sample surface be occupied by a single superstructure, as considered in the above examples. In many cases, the symmetry of the substrate allows the existence of several symmetrically equivalent domains. For example, two orthogonal domains are possible on a square-lattice substrate (Fig. 4.7) and two mirror domains on a substrate with a rectangular lattice. On a substrate with hexagonal symmetry, stripe-like superstructures (for example, 2×1, 3×1, 4×1, etc.) can occur in three

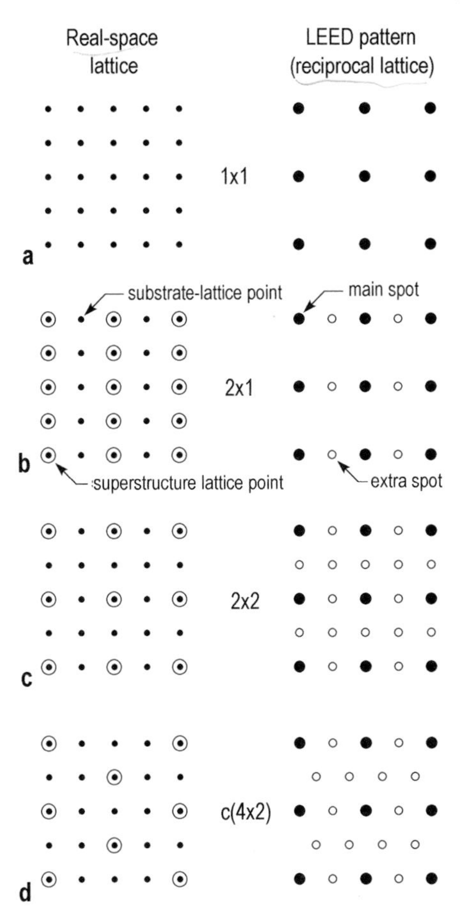

Fig. 4.5. Examples of real space superlattices on a substrate with a square lattice and corresponding LEED patterns: (**a**) 1×1; (**b**) 2×1; (**c**) 2×2; (**d**) $c(4 \times 2)$. In real space, the substrate lattice points are shown by black dots, the superstructure lattice points are shown by open circles. In LEED patterns, the main spots are shown by large black circles, the extra-spots are shown by small open circles

120°-rotated domains and when such a superstructure lacks mirror symmetry with respect to the main axis (for example, 5×2), the number of domains is already six. Coexistence of domains with quite different superstructures (for example, 2×2 and $\sqrt{3}\times\sqrt{3}$, as shown in Fig. 4.8) might also take place.

Note that scattered waves interfere to produce a diffraction pattern only when they originate from the same surface region of the size of the *coherence length*, where surface atoms can be considered to be illuminated by a simple plane wave. When the waves are scattered from surface points spaced by more than the coherence length, their intensities are added rather than their amplitudes. If the size of the superstructure domains exceeds the coherence length, the resultant LEED pattern is just a superposition of the diffraction

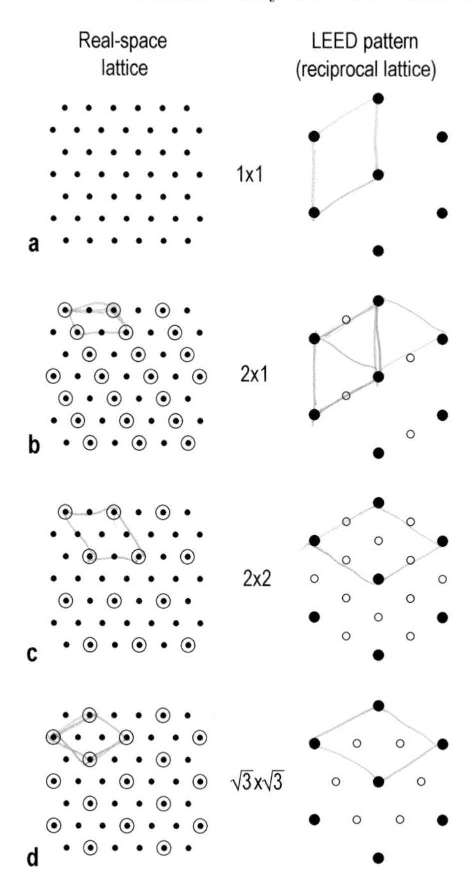

Fig. 4.6. Examples of real space superlattices on a substrate with a hexagonal lattice and corresponding LEED patterns: (**a**) 1×1; (**b**) 2×1; (**c**) 2×2; (**d**) $\sqrt{3}\times\sqrt{3}$-$R30°$. In real space, the substrate lattice points are shown by black dots, the superstructure lattice points are shown by open circles. In LEED patterns, the main spots are shown by large black circles, the extra-spots are shown by small open circles

patterns from separate domain. To interpret multidomain LEED pattern, one should first distinguish the contribution of each domain. Sometimes this is rather difficult and for some instances this is even impossible to do in unique way (for example, on the hexagonal lattice substrate the three-domain 2×1 pattern coincides with the 2×2 pattern; another example is superposition of the $\sqrt{3}\times\sqrt{3}$-$R30°$ and the three-domain 3×1 patterns, which produces a 3×3 pattern).

LEED Spot Profile. Additional information can be extracted by probing the *spot profile*, i.e., the intensity distribution across width of a spot. This information mainly concerns surface imperfections, as any deviation from an ideal 2D periodicity disrupts sharp spot profile. For example, reducing

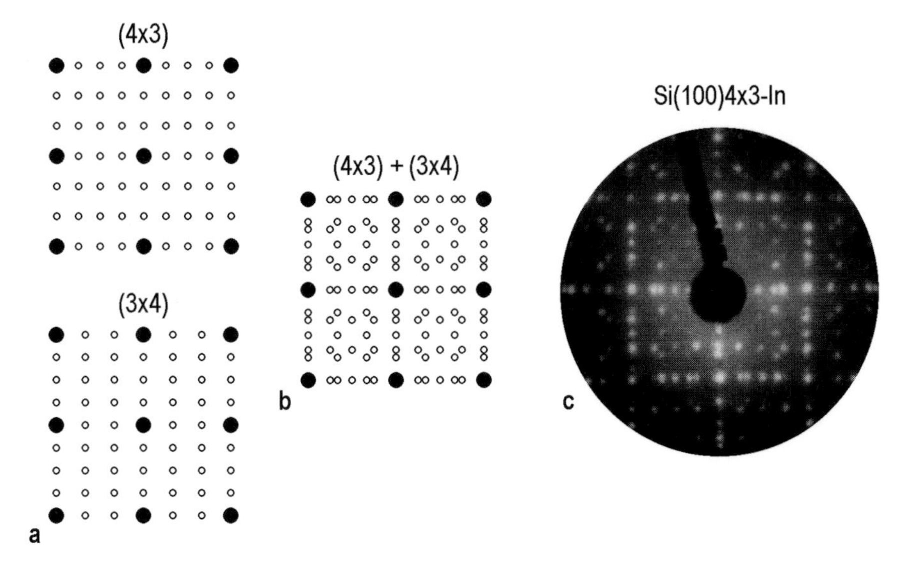

Fig. 4.7. (**a**) Sketches of LEED patterns (reciprocal lattices) for the orthogonal domains, 4×3 and 3×4, on a square lattice substrate. (**b**) Superposition of the 4×3 and 3×4 patterns. (**c**) Experimental LEED pattern ($E_{\mathrm{p}} = 54\,\mathrm{eV}$) of double-domain Si(100)4×3-In

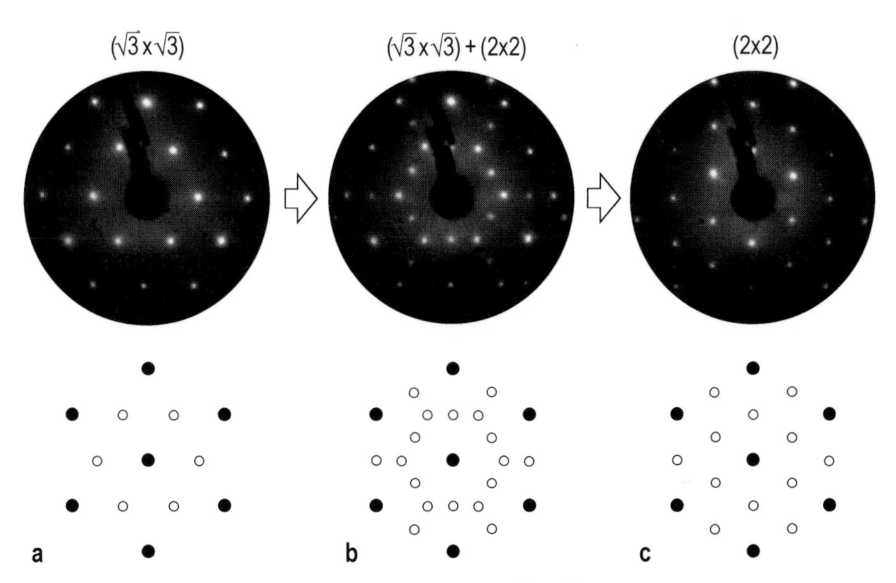

Fig. 4.8. Upon In deposition onto the Si(111)$\sqrt{3}\times\sqrt{3}$-$R30°$-In surface held at RT, the original $\sqrt{3}\times\sqrt{3}$-$R30°$ structure (**a**) transforms to the 2×2 structure (**c**). At the intermediate stage, the domains of both structures coexist (**d**). Sketches of LEED patterns and experimental LEED patterns ($E_{\mathrm{p}} = 54\,\mathrm{eV}$) are shown

the domain size results in broadening of the spots, since the spot width is inversely related to the number of regular scattering units. Note that even in the case when the whole sample surface is covered by a perfect periodic structure, the spot has a finite width due to (i) finite coherence length and (ii) instrumental limitations of the LEED apparatus (finite energy and angular spread of the primary electron beam). The instrumental contribution to spot width is characterized, in analogy to the coherence length, in terms of a *transfer width*, whose typical value for LEED is about 100 Å.

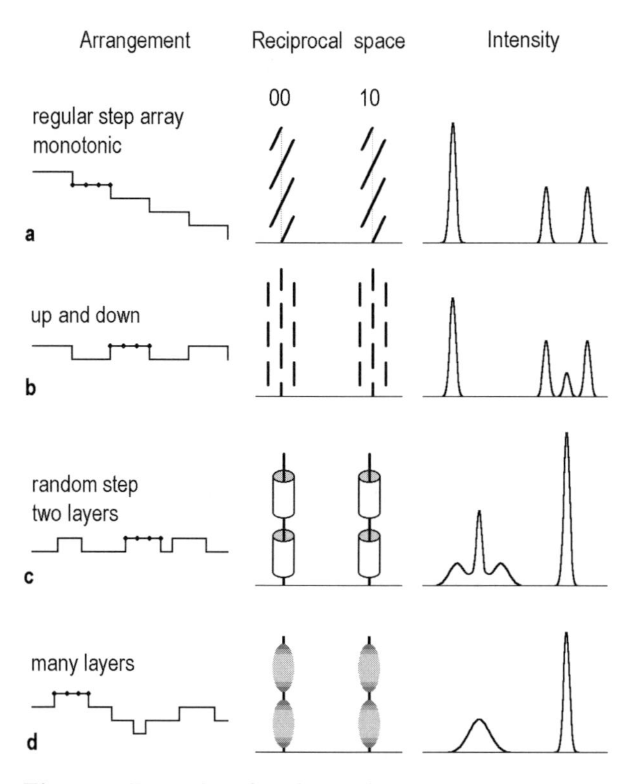

Fig. 4.9. Examples of surface defect structures, the corresponding modification of the reciprocal rods, and the resultant LEED spot profiles. (**a**) Monotonic regular step array, (**b**) regular step array in two layers, (**c**) random step array in two layers, (**d**) random step array in many layers (after Henzler [4.1])

A regular step array furnishes a specific example of surface imperfections. This surface might be thought of as that with two superposed lattices, the short-period lattice associated with the atomic lattice on the terrace and the long-period lattice associated with the periodicity of the vicinal plane (atomic-step periodicity). In the Ewald construction, the reciprocal rods of both lattices have to be taken into account, the atomic lattice rods having a finite thickness due to limited terrace width. The two set of rods are inclined to one another by the angle that the vicinal plane makes with the terrace plane. The diffraction condition has to be fulfilled by both rod systems simultaneously. The result is a spot splitting that takes place periodically with the change of primary electron energy. This is illustrated by Fig. 4.9 which shows some examples of the effect of defect structure on the resultant LEED spot profile. The expected spot pattern is obtained by the intersection of the Ewald sphere by the modified reciprocal rod.

LEED *I–V* Analysis. The analysis of the spot positions or spot profiles does not bear on the question of the atomic arrangement within a surface unit cell. In order to proceed to the evaluation of the atom positions, it is necessary to resort to analysis of the spot intensity as a function of the primary electron energy, the so-called *I–V curves*. In the majority of modern LEED systems, the experimental *I–V* curves are recorded using a TV camera accompanied by computer-controlled data handling; another way is a direct measurement of the beam current by a movable Faraday cup. Determination of the atom positions from the LEED data is not straightforward, but rather an iterative trial-and-error procedure as follows.

- As a first step, an initial trial structure is proposed which, at least, is consistent with the symmetry of the LEED pattern and is desirable to account for the data obtained by other techniques.
- Next, the *I–V* curves for a number LEED beams are calculated. It should be noted that the calculations are conducted within the framework of the *dynamic theory* which accounts for the high probability of multiple scattering of the low-energy electrons.
- Then, the calculated *I–V* curves are compared with experimental data. Depending on the result, the model may be modified and the process is continued until satisfactory agreement is reached.

The fit between the experimental and theoretical *I–V* curves might be judged by visual inspection, taking note that the peak voltages and relative heights in the experimental and theoretical curves would be generally coincident with each other. As an example, Fig. 4.10 shows the experimental *I–V* curves for a Si(100)2×2-In surface and theoretical *I–V* curves for two models of the superstructure, with Ga dimers orthogonal to the underlying Si substrate dimers ("orthodimer model") and with Ga dimers parallel to the Si dimers ("paradimer model"). The strong preference of the "paradimer model" is apparent from a comparison of the calculated curves with experiment.

However, in general, visual inspection is too subjective, especially in the cases where consistency (or inconsistency) is less evident than that seen in Fig. 4.10. For example, the different structural models might produce almost similar *I–V* curves or a change in some model parameters improves the correspondence with experimental curves for some beams but worsens it for others. For these reasons, a quantitative criterion, called the *R-factor (a reliability factor)* has been introduced. The *R*-factor takes quantitative account

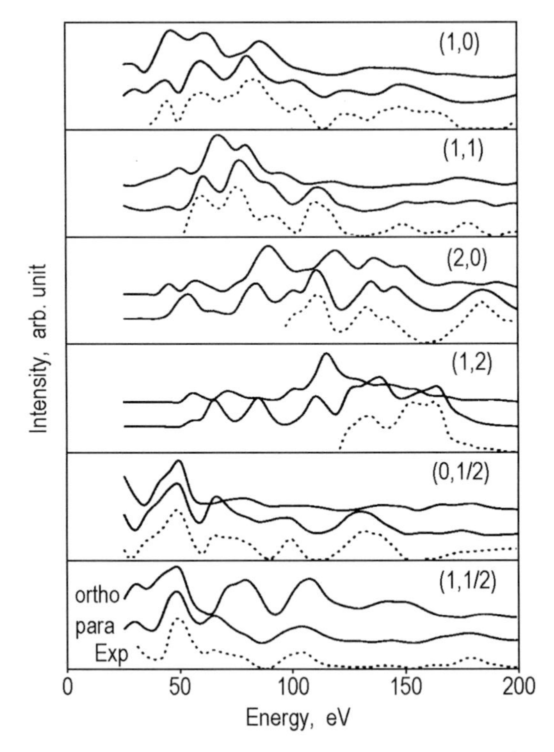

Fig. 4.10. Experimental I–V curves for a Si(100)2×2-Ga surface (dashed lines) and theoretical I–V curves optimized for the "orthodimer" model with Ga dimers orthogonal to Si substrate dimers (upper solid lines) and for the "paradimer" model with Ga dimers parallel to Si dimers (lower solid lines). The preference of the "paradimer" model is evident (after Sakama et al. [4.2])

of a lot of various features (the general shape of the curves, the presence and amount of the background, the occurrence and the positions of maxima, minima, shoulders, bumps, etc.) and cannot be expected to be very simple. In its classical form, as proposed by Zanazzi and Jona [4.3], the R-factor is defined as

$$R = \frac{A}{\delta E} \int \omega(E) |cI'_{\mathrm{th}} - I'_{\mathrm{expt}}| \mathrm{d}E \,, \tag{4.8}$$

where

$$c = \int I_{\mathrm{expt}} \Big/ \int I_{\mathrm{th}}, \qquad \omega = \frac{|cI''_{\mathrm{th}} - I''_{\mathrm{expt}}|}{|I'_{\mathrm{expt}}| + \epsilon}, \qquad \epsilon = |I'_{\mathrm{expt}}|_{\max} \,, \tag{4.9}$$

where the integrals extend over an energy range δE. A is chosen to normalize R to unity for uncorrelated curves:

$$A = \delta E \Big/ \left(0.027 \int I_{\mathrm{expt}} \right) . \tag{4.10}$$

It is generally accepted that $R \leq 0.2$ is good agreement, $R \simeq 0.35$ mediocre and $R \simeq 0.5$ is bad agreement.

An alternative R-factor which is widely used now has been proposed by Pendry [4.4]. This reliability factor is sometimes called a logarithmic R-factor, as it treats the I–V curves in the form of their logarithmic derivatives

$$L(E) = I'/I \,. \tag{4.11}$$

The logarithmic Pendry R-factor is defined as

$$R = \sum_g \int (Y_{g\text{th}} - Y_{g\text{expt}})^2 \mathrm{d}E \Big/ \sum_g \int (Y_{g\text{th}}^2 + Y_{g\text{expt}}^2) \mathrm{d}E \,, \tag{4.12}$$

where $Y(E) = L^{-1}/(L^{-2} + V_{\text{oi}}^2)$ with V_{oi} being the imaginary part of the electron self-energy (typical value is of around $-4\,\text{eV}$). The advantage of the Pendry R-factor is that it avoids the use of double derivatives. This is especially important in the case of noisy or insufficiently finely scaled experimental I–V curves.

Surface structure determination is a non-trivial procedure that involves significant computational effort and would be very time-consuming when a trial-and-error search is controlled at each step by a human being. Fortunately, modern computational programs combine the calculation of I–V curves and the search of the R-factor minimum. The recent development of the so-called *tensor LEED*, the procedure that optimizes the search for the global minimum, greatly reduces the computer time and allows the treatment of relatively complex structures.

4.2 Reflection High-Energy Electron Diffraction (RHEED)

4.2.1 Ewald Construction in RHEED Conditions

Besides LEED, reflection high-energy electron diffraction (RHEED) is the second diffraction technique that is widely used in many UHV systems for routine structural analysis. Though utilizing high-energy electrons, RHEED is a surface-sensitive technique. Surface specificity at high energies is provided by using both grazing incidence of the primary beam and grazing detection angles of the diffracted beams. As a result, on its relatively long mean free path through the sample, high-energy electrons still keep in the near-surface region of only a few atomic layers deep (for example, 50–100 keV electrons have a mean free path of about 1000 Å and at an incidence angle of about 1° penetrate to a depth of about 10 Å).

The Ewald construction in RHEED conditions is illustrated in Fig. 4.11. In contrast to the case of LEED, the Ewald sphere is now much larger than the spacing of the reciprocal rods and cuts these rods at grazing angles. The

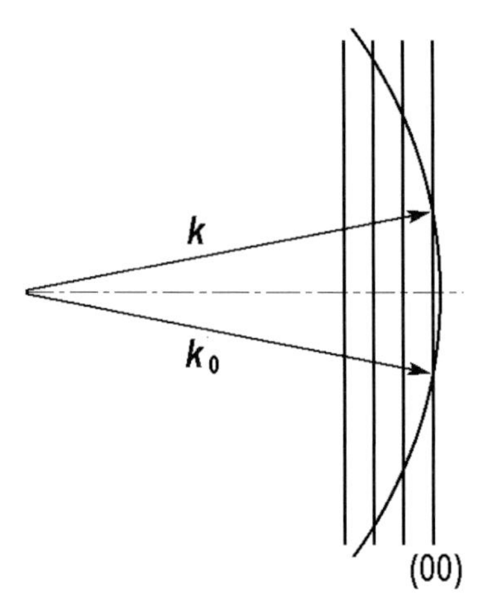

Fig. 4.11. Ewald construction in RHEED conditions

intersection of the Ewald sphere and the reciprocal rods at such grazing angles leads to a noticeable streaking of the diffraction spots, as both the Ewald sphere and the reciprocal rods have a finite thickness due to instrumental limitations and surface imperfection, respectively.

4.2.2 RHEED Set-Up

Figure 4.12 shows schematically the experimental arrangement for RHEED observations. A flux of high-energy electrons from the electron gun is incident under grazing angles of about 1–5° onto the sample surface and the diffracted electron beams produce a RHEED pattern on a fluorescent screen. A wide variety of electron guns are in use, from simple electrostatically focused guns, operating at 5–20 keV, to much more sophisticated guns that approach electron microscope quality, operating at higher voltages (up to 100 keV). Sometimes magnetic lenses are used for focusing the beam and for fine control of the beam trajectory. The sample holder with the sample is mounted in the manipulator stage, which is desirable to allow sample rotation about its surface normal to record the RHEED patterns at various azimuthal directions.

The fluorescent screen is often coated directly onto the inside of the viewport in the UHV system along with a transparent conducting film to avoid screen charging. As the primary electron energy is high enough to produce fluorescence on the screen, no extra acceleration of electrons is needed. Moreover, no energy filtering is generally required, since the intensity of the diffracted beams is much higher than that of the background. This stems from the large energy difference between the elastically scattered (i.e., diffracted) electrons and the inelastically scattered background and, thus, the energy filtering is just caused by the fact that higher energy electrons produce

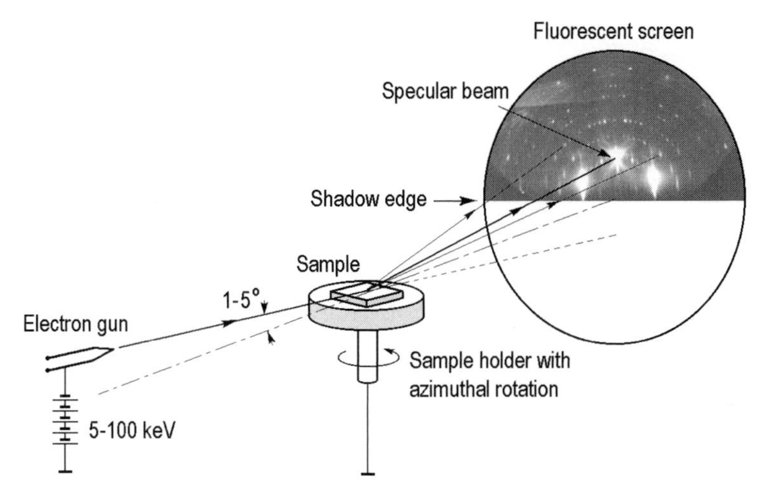

Fig. 4.12. Schematic of RHEED set-up

more light from the screen. However, some advanced systems for quantitative RHEED analysis utilize additional facilities for energy filtering.

4.2.3 RHEED Analysis

Due to the difference in the scattering geometry, the RHEED pattern clearly differs from the LEED pattern. However, for an ideal crystal this is also a projection of a 2D reciprocal lattice of a sample surface. Here the spots in the arcs correspond to the grazing intersection of reciprocal rods with the large Ewald sphere. As an example, Figs. 4.13a and 4.14a show the RHEED patterns of the Si(111)7×7 surface recorded for two azimuths of incidence. The schematic sketches in Fig. 4.13b, c and 4.14b, c serve to explain the relationship between the surface reciprocal lattice and the observed RHEED pattern.

The basic structural information gained by RHEED is very similar to that obtained by LEED, namely:

- First, surface perfection can be evaluated qualitatively by the brightness and sharpness of the diffraction spots. However, this relationship is not so apparent, as in LEED. For example, the question of whether streaks or spots are a sign of a well-ordered atomically flat surface is still widely discussed in the literature.
- Second, from the projection of the reciprocal lattice the real space lattice of the sample surface can be established. Note that, for reliable determination of the full 2D periodicity, the RHEED pattern should be recorded at several (at least, two) azimuths.
- Third, RHEED is employed for quantitative structural analysis, i.e., for checking models of the surface atomic arrangement. Here the analog of

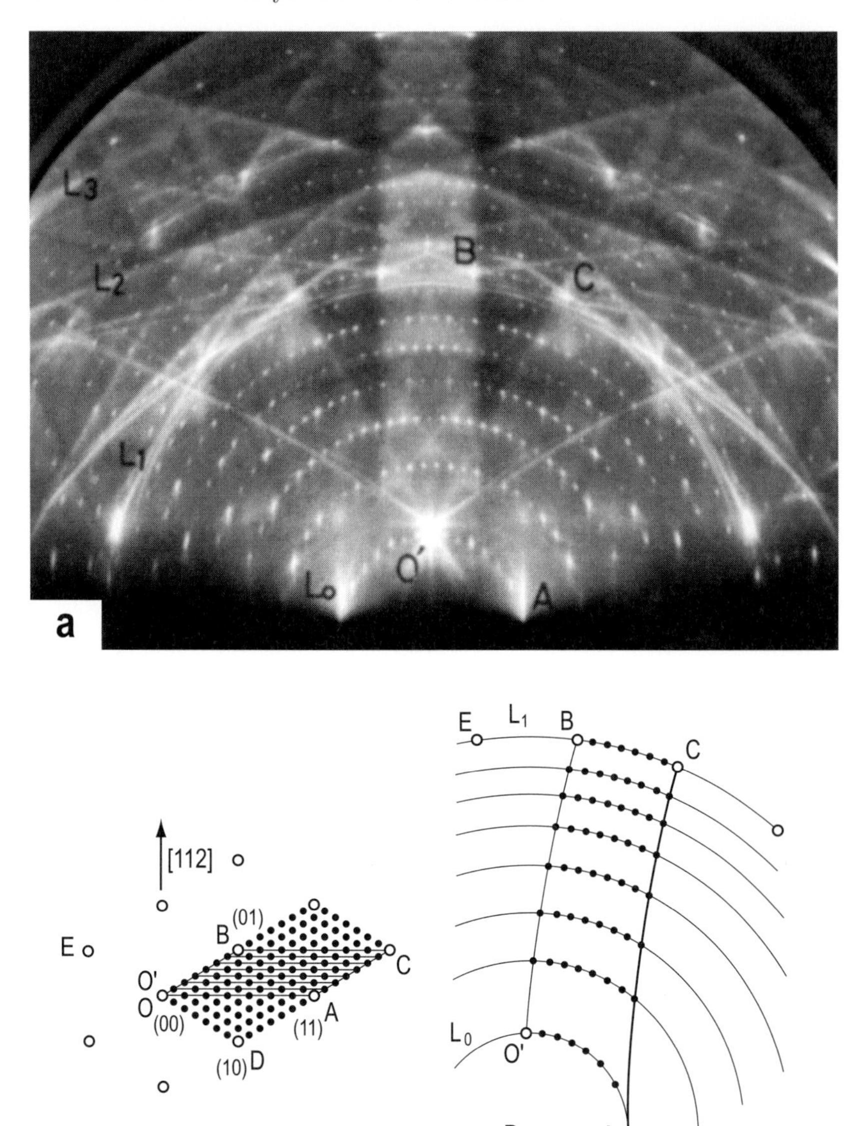

Fig. 4.13. (**a**) RHEED pattern (20 kV) from the Si(111)7×7 surface taken along the $[\bar{1}2\bar{1}]$ incidence direction. (**b**) Fragment of the 2D reciprocal lattice of the 7×7 structure. (**c**) Sketch of the RHEED pattern, where O is the direct incidence point, O′ the specular reflection, and O′ACB corresponds to that in (**b**) (after Ino [4.5])

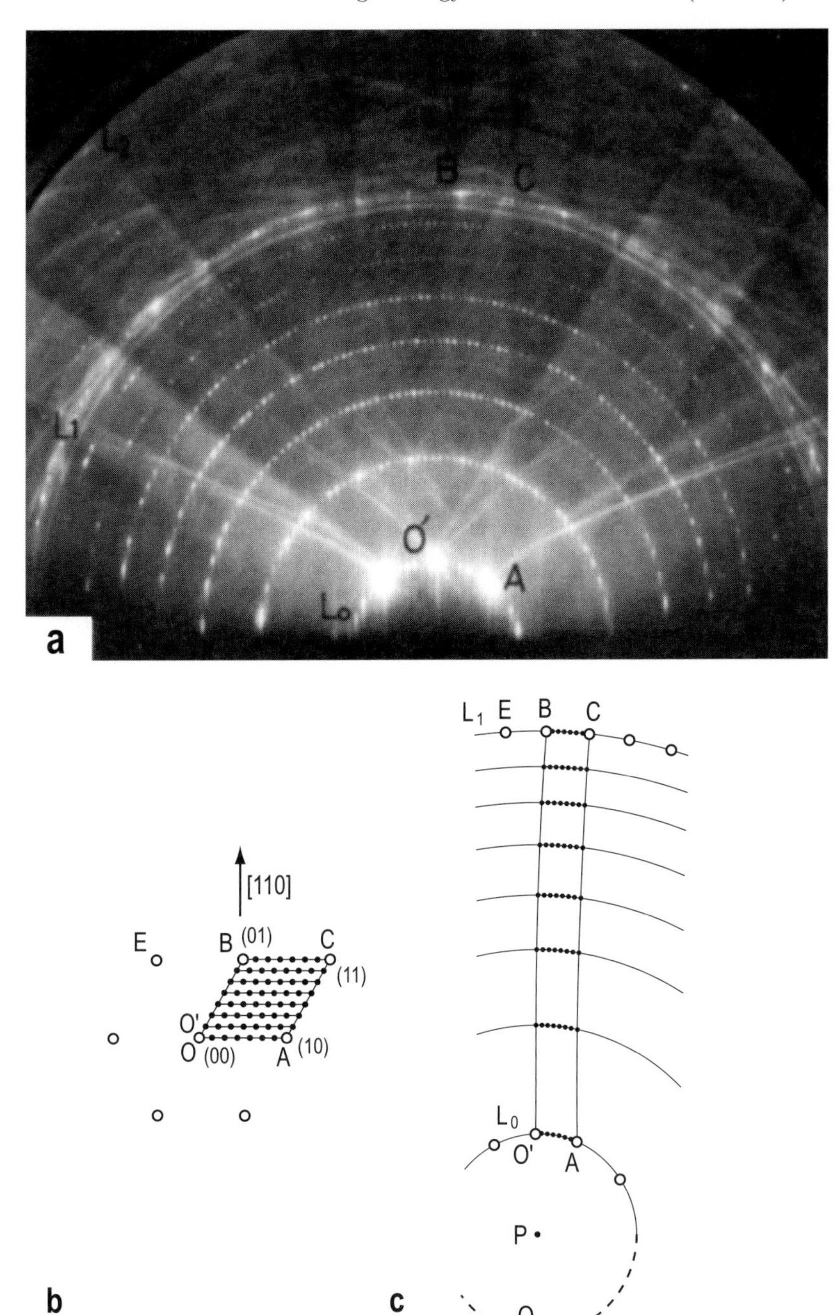

Fig. 4.14. (**a**) RHEED pattern (20 kV) from the Si(111)7×7 surface taken along the $[1\bar{1}0]$ incidence direction. (**b**) Fragment of the 2D reciprocal lattice of the 7×7 structure. (**c**) Sketch of the RHEED pattern, where O′ACB corresponds directly to the 7×7 unit mesh O′ACB in (**b**) (after Ino [4.5])

LEED I–V curves are *RHEED rocking curves*, the dependence of spot intensity on the primary beam incidence angle.

Besides the above general facilities, RHEED provides additional ones which appear to be especially useful for studying thin film growth and monitoring the formation of multilayer epitaxial structures.

- The first virtue arises from the sensitivity of RHEED to surface asperities, which expands the data set to the third dimension. If 3D crystalline islands form on the surface (for example, upon adsorbate deposition), they can be readily identified by the appearance of new spots in the RHEED pattern. These spots are produced by transmission electron diffraction in the islands (Fig. 4.15).
- The second merit resides in the fact that in the RHEED arrangement the electron gun and the screen are spaced far apart from the sample, thus leaving the front of the sample open for deposition sources. Therefore, RHEED can be used as an in situ technique to control the structure of the growing surface directly during the growth process. A beautiful example of such monitoring is the observation of the intensity oscillations of the diffracted beam in the RHEED pattern (*RHEED oscillations*) during MBE growth of GaAs(100) (Fig. 4.16). In the experiment, the growth of the GaAs film is initiated by the onset of deposition of Ga atoms on to the substrate surface which is maintained under continuous supply of As. Simultaneously, the intensity of the specular beam is recorded in the RHEED pattern from a GaAs(100)2×4 surface. The intensity demonstrates well-defined oscillations with a period that corresponds exactly to the growth of a single monolayer (as determined independently). The observed oscillations are clearly related to the periodic evolution of the surface roughness during layer-by-layer growth, as shown schematically in Fig. 4.16b: the intensity maxima correspond to the completed atomically smooth surfaces

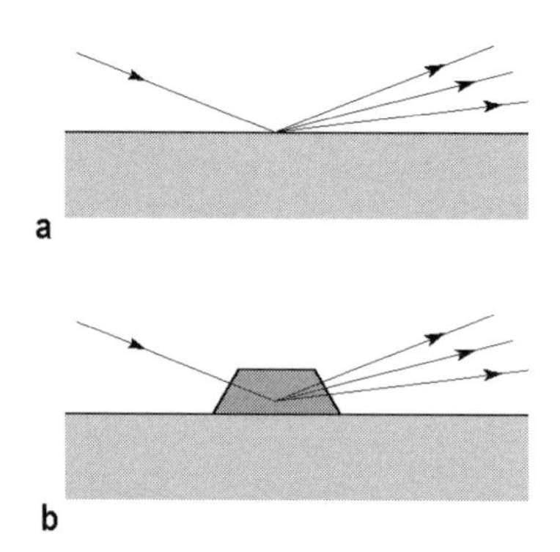

Fig. 4.15. Two possible situations for glancing incidence high-energy electrons: (**a**) surface scattering on a flat surface, and (**b**) transmission diffraction in 3D crystalline island located on the surface

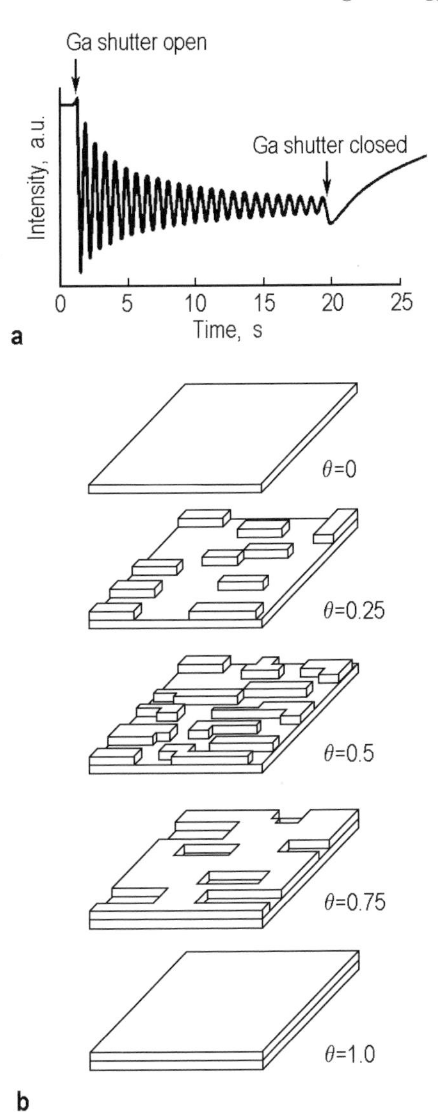

Fig. 4.16. MBE growth of GaAs(100) monitored by RHEED. (**a**) Intensity oscillations of the RHEED specular beam during the growth process. The period exactly corresponds to the growth rate of a single GaAs atomic layer. (**b**) Real space representation of the formation of a single complete layer in layer-by-layer growth mode (after Neave et al. [4.6])

near $\Theta = 0$ and $\Theta = 1$, while the minima correspond to the most disordered surface with irregularly distributed atoms or little 2D islands near $\Theta = 0.5$. The gradual decrease in oscillation amplitude indicates the decrease in surface perfection. This technique has found wide application in MBE growth of multilayer structures like quantum superlattices, since by counting the number of oscillations one can directly control the number of complete atomic layers grown.

4.3 Grazing Incidence X-Ray Diffraction (GIXRD)

Through many decades x-ray diffraction (XRD) has proved to be one of the most useful diffraction technique for bulk structure analysis. The applicability of x-rays for probing the bulk stems from the fact that the scattering cross-section of atoms to x-rays is extremely small ($\sim 10^{-6}$ Å^2 compared to that for LEED of ~ 1 Å^2) which allows x-rays to penetrate deep into the solid (of the order of μm). An additional advantageous consequence of the weak scattering of x-rays by matter is the weak multiple scattering effects, which make the single scattering (kinematical) approximation well justified.

4.3.1 Refraction of X-Rays at Grazing Incidence

In general, x-ray diffraction is not surface sensitive, as the scattered intensity from the surface is less than that from the bulk by about five orders of magnitude. However, recently the goal to exploit the merits of XRD for surface analysis has been successfully achieved. Two main ideas constitute the basis of the approach.

- The first idea relates to the fact that if the surface periodicity differs from that of the bulk, the surface superlattice reflections are placed in reciprocal space well away from the main spots. Thus, the problem of surface analysis is reduced to distinguishing such superreflections over the featureless background from the bulk crystal.
- The second idea is that an optimum signal-to-noise ratio is obtained at very grazing incidence of x-rays, namely, when the angle of incidence is equal or lower than the critical angle for *total external reflection.*

Figure 4.17 illustrates schematically the refraction of an x-ray wave incident upon the interface between the vacuum and the solid. Snell's law for refraction gives

$$\cos \alpha_{\mathrm{i}} = n \cos \alpha_{\mathrm{r}} \, , \tag{4.13}$$

where α_{i} and α_{r} are the incidence and refraction angles, respectively, and n is the refractive index of the solid, while the refractive index outside is taken to be unity. In contrast to light, for which the refractive index of most

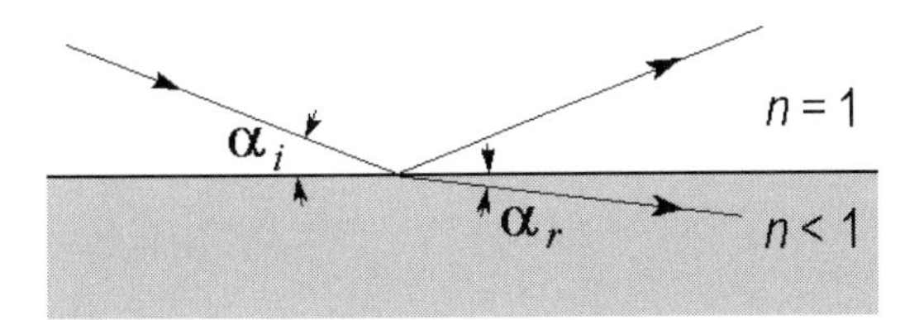

Fig. 4.17. Refraction of an x-ray wave at the interface between the vacuum and the material. For x-rays, the refractive index of most materials is slightly less than that of the vacuum, hence, total reflection occurs at glancing angles less than the critical angle

materials is greater than unity, in the case of x-rays the refractive index of most materials is *less than unity*, albeit only slightly. For a given material, it can be expressed in a simplified way as:

$$n = 1 - \delta = 1 - 2.7 \times 10^{-6}(\sum Z_j / \sum A_j)\rho\lambda^2 \,, \tag{4.14}$$

where $\sum Z_j$ and $\sum A_j$ are the sums over the atomic charges and the atomic masses, respectively, in the unit cell, ρ the density in g/cm^3, and λ the wavelength in Å. For typical values n is less than unity by only about 10^{-5}.

The critical angle $\alpha_{\mathrm{i}} = \alpha_{\mathrm{c}}$ is that when $\alpha_{\mathrm{r}} = 0$, i.e., $\cos\alpha_{\mathrm{c}} = n$, which for small angles leads to $\alpha_{\mathrm{c}} = \sqrt{2\delta}$. For x-ray wavelengths of around 1.5 Å, typical values for the critical angle lie in the range 0.2°–0.6°.

When the incidence angle becomes smaller than the critical angle, the refracted wave is exponentially damped in the depth with a characteristic length of only few dozen Å (for example, 32 Å for Si and 12 Å for Au). As a result, an *evanescent wave* is generated traveling parallel to the surface. Thus, the diffraction of evanescent waves provides information about the structure of the surface layer.

4.3.2 Ewald Construction in GIXRD Conditions and Basics of the Kinematic Approximation

The Ewald construction for GIXRD (Fig. 4.18) incorporates features of both RHEED and LEED. As in RHEED, the incidence is grazing, but the Ewald sphere radius is close to that in LEED conditions (for example, a wavelength of 1.5 Å, typical for XRD, corresponds to an electron energy of about 65 eV). As a result, besides the grazing specular beam, other diffraction beams are scattered generally at angles far away from grazing and the corresponding reciprocal lattice rods are cut by the Ewald sphere at totally different heights (i.e., different values of $\Delta k_\perp$). In the GIXRD geometry, the grazing incidence angle is kept strictly constant and the collecting of diffraction data involves an azimuthal rotation of the sample and spanning the space by a detector to collect the scattered beams.

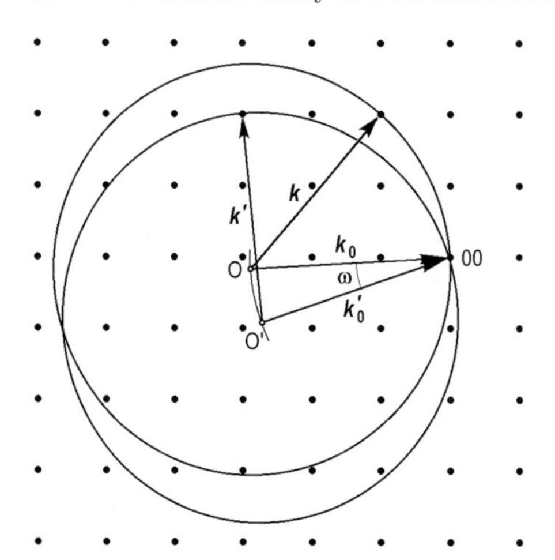

Fig. 4.18. Ewald construction for grazing-incidence x-ray diffraction (in the surface plane). The azimuthal rotation of the sample by angle ω changes the scattering geometry so that new diffraction beams are generated in the same polar angle. The scattered beams shown with wave vectors $\boldsymbol{k}$ and $\boldsymbol{k}'$ lie almost in the surface plane ($\Delta k_\perp \approx 0$). Other scattered beams (not shown) are out of the plane and the corresponding reciprocal lattice rods intersect the Ewald sphere at different heights which change also with sample rotation

If we adopt a single scattering or *kinematic* approximation [4.7, 4.8] and when the diffraction condition $\boldsymbol{k} - \boldsymbol{k}_0 = \boldsymbol{G}_{hkl}$ of (4.1) is satisfied then the scattering amplitude A_{hkl} for the particular diffracted beam for a crystal of N cells may be written as

$$A_{hkl} = NF_{hkl} \,, \tag{4.15}$$

where the quantity F_{hkl} is called the *structure factor* and is defined by a single cell as a sum over the s atoms of the basis

$$F_{hkl} = \sum_{j=1}^{s} f_j \exp(-\mathrm{i}\boldsymbol{G}_{hkl} \cdot \boldsymbol{r}_j) \tag{4.16}$$

with $\boldsymbol{r}_j = 0$ at one corner. The factor $\exp(-\mathrm{i}\boldsymbol{G}_{hkl} \cdot \boldsymbol{r}_j)$ is a phase factor between the incoming and outgoing waves and f_j is the *atomic form factor* which is a measure of the scattering power of the j-th atom of the unit cell. It takes account of interference effects within the atom, and is defined as

$$f_j = \int \rho_j(\boldsymbol{r}) \exp(-\mathrm{i}\Delta\boldsymbol{k} \cdot \boldsymbol{r}) \, dV \,, \tag{4.17}$$

where $\rho_j(\boldsymbol{r})$ is the atomic electron density, and the integral is extended over the electron density associated with a single atom. Finally, the *scattered intensity* is the square of the corresponding amplitude $I_{hkl} = |A_{hkl}|^2$.

In another formulation of (4.15), the scattering amplitude from the whole crystal can be calculated by the integral of the electron density $\rho(\boldsymbol{r})$ as

$$A(\boldsymbol{G}) = \int\limits_{\text{crystal}} \rho(\boldsymbol{r}) \exp(\mathrm{i}\boldsymbol{G} \cdot \boldsymbol{r})\, d\boldsymbol{r} \,, \tag{4.18}$$

and the electron density distribution, that is, the coordinates of all the atoms within the unit cell, can be obtained as the Fourier transform of $A(\boldsymbol{G})$:

$$\rho(\boldsymbol{r}) = \frac{1}{8\pi^3} \int A(\boldsymbol{G}) \exp(-\mathrm{i}\boldsymbol{G} \cdot \boldsymbol{r})\, d\boldsymbol{r} \,. \tag{4.19}$$

Unfortunately, even in x-ray diffraction, the problem is not so simple. From integrated intensities only the amplitudes are measured so that the phase information is lost. This is the famous problem of the missing phase information. However, it can be shown that the autocorrelation function $P(\boldsymbol{r})$, called the *Patterson function*, contains only structural factor amplitudes:

$$P(\boldsymbol{r}) = \int\limits_{\text{unit cell}} \rho(\boldsymbol{r}')\rho(\boldsymbol{r} - \boldsymbol{r}')d\boldsymbol{r}' = \frac{1}{S} \sum_{hkl} |F_{hkl}|^2 \exp(-\mathrm{i}\boldsymbol{G} \cdot \boldsymbol{r}) \,. \tag{4.20}$$

Hence, it can be directly calculated from the experimental data.

4.3.3 GIXRD Experimental Set-Up

As the experimental set-up for GIXRD measurements is very complicated and costly, their number in the world is very limited. To achieve the required surface sensitivity, extremely intense (up to 10^{12} photons/mm^2 s) and highly collimated x-ray beams from a synchrotron radiation source beamlines are employed. The sample is mounted in a sample holder inside the UHV chamber. Access for the incident x-ray beam to the sample and scattered beams to the detector is accomplished through a beryllium window. A special design is demanded to combine the UHV environment of the sample with the required $0.001°$ angular precision of the x-ray diffractometer. The UHV chamber often incorporates additional facilities, like LEED, AES, an ion-sputtering system, deposition sources, etc. LEED is extremely useful for initial assessment of the surface periodicity.

4.3.4 Structural Analysis by GIXRD

The complexity in obtaining the XRD data is offset by the relative simplicity of their processing in structural analysis, where the single-scattering (kinematic) approximation is well applicable (recall that this does not hold for LEED at all, due to the multiple-scattering effects). For a full, three-dimensional determination, the GIXRD analysis proceeds conventionally in a three-step procedure:

- First, the *in-plane* structure (i.e., the projection of the 3D surface structure onto the surface plane) is determined.

- Second, the *out-of-plane* structure (i.e., the relative heights of atoms) is found.
- Third, the *registry* of the surface superstructure to the bulk is established.

Consider each of the steps in a greater detail.

In-Plane Surface Structure. In this kind of x-ray analysis, the experimental geometry is adjusted so that the scattering of x-rays occurs at very grazing angles, hence the momentum transfer is kept almost parallel with the surface ($\Delta k_\perp \approx 0$). In this stage, only fractional-order reflections are taken into consideration. From the measured intensities of the fractional-order beams, a *Patterson function* of the in-plane structure is derived, using Fourier transform methods widely accepted in conventional bulk x-ray diffraction. Since positive peaks in a contour plot of the Patterson function correspond to vectors between atom pairs on the surface, the search for a trial structure becomes more straightforward (say, compared to the case of LEED I–V analysis). Note that the Patterson function has the same symmetry as probing structure plus an extra center of inversion. Further refinement of the model structure goes through calculation of the diffraction intensities for the trial model, comparison with the experiment, and adjusting the model parameters in a trial and error procedure until good agreement is reached.

Determination of the InSb(111)2×2 structure using GIXRD furnishes an illustrative example. InSb is a crystal with a zincblende structure, in which the InSb(111) layers comprise alternate In layers and Sb layers. Figure 4.19a shows an ideally terminated InSb(111)1×1 surface with In atoms (open circles) forming the top layer and Sb atoms (shaded circles) constituting the next layer down. This surface is defined as the InSb(111)A surface to distinguish it from the InSb($\bar{1}\bar{1}\bar{1}$)B surface in which the top layer is built of Sb. The InSb(111)A surface is known to show 2×2 periodicity. A Patterson function map obtained in the in-plane x-ray experiment (Fig. 4.20a) provides a clue to understanding the surface structure. Note that the Patterson function is mapped only in the triangle that is the smallest asymmetric unit (outlined in Fig. 4.19a). The interatomic distances and the direction of the bonds deduced from the Patterson function appear to be consistent with the structure, in which the hexagon is distorted and one In per (2×2) unit cell is missing (Fig. 4.19 and Fig. 4.20). The diffraction beam intensities calculated for this model are in good agreement with experiment.

Out-of-Plane Surface Structure. The above discussed in-plane GIXRD measurements do not provide information about displacements of surface atoms in the direction normal to the surface. Such information is gained from measuring *rodscans*, i.e., intensity distribution along the reciprocal lattice rods or, in other words, diffraction intensity versus $\Delta k_\perp$. As is evident from the Ewald construction (Fig. 4.18), the height at which a given reciprocal lattice rod intersects the Ewald sphere varies with azimuthal rotation of the sample. Thus, the rodscans are recorded using rotation of the sample

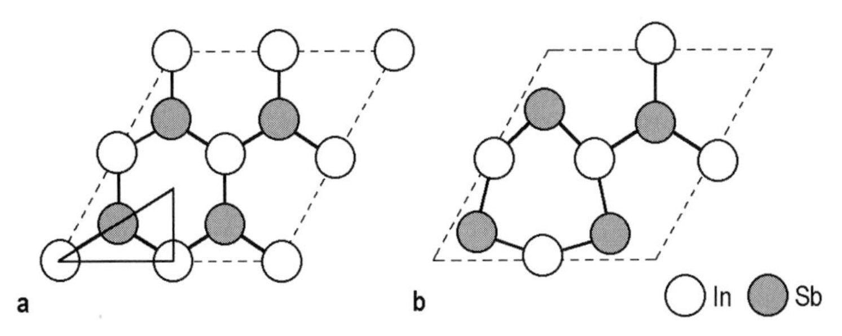

Fig. 4.19. The projection of atomic positions within a (2×2) unit cell in (**a**) the ideal bulk-like terminated InSb(111) surface and (**b**) the reconstructed InSb(111)2×2 surface structure as determined using GIXRD analysis. The triangle shown in bold lines in (**a**) is the smallest symmetrically inequivalent unit after taking into account the symmetry of the substrate and the inversion imposed in x-ray diffraction data (after Feidenhans'l [4.8])

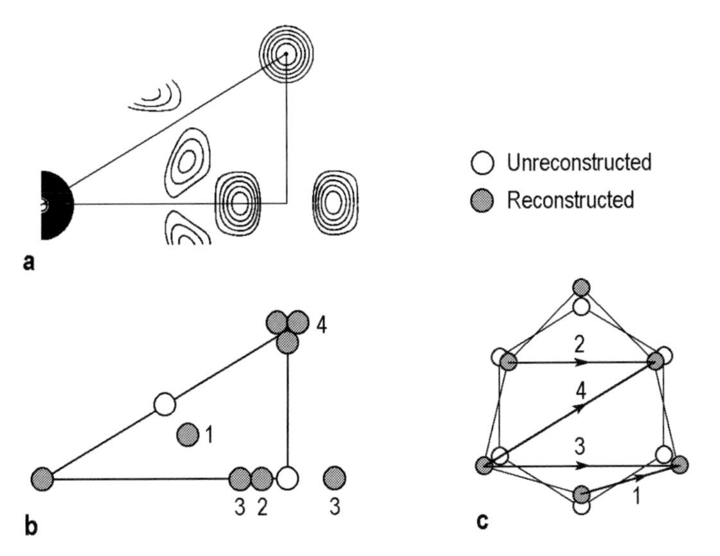

Fig. 4.20. (**a**) Contour map of the Patterson function for the InSb(111) surface within the unit triangle shown in Fig. 4.19a. (**b**) Interatomic vectors as derived from vectors 1 to 4 in (**a**). (**c**) Undistorted and distorted hexagonal arrangement of atoms producing the peaks in (**b**) (after Feidenhans'l [4.8])

with accompanied relocation of the detector to follow the diffraction beam. One can see that the x-ray rodscan is essentially similar to the LEED I–V curve and differs only in the way of acquisition, i.e., at fixed energy but at variable geometry. In structural analysis, the out-of-plane atomic coordinates in the model are adjusted so that the calculated rodscans coincide with the experimental ones.

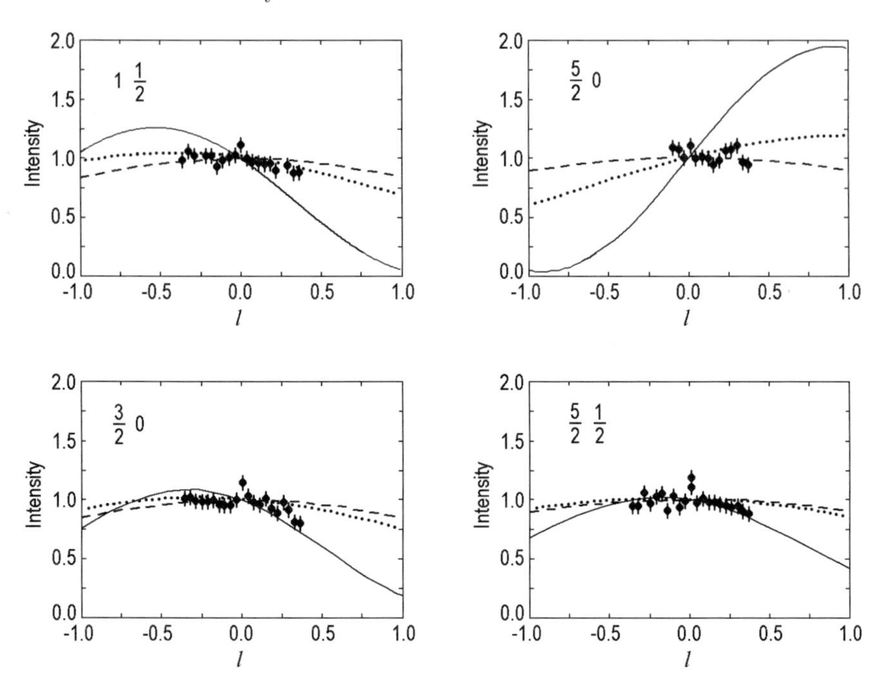

Fig. 4.21. Rodscans for (1,1/2), (3/2,0), (5/2,0), and (5/2,1/2) beams for the InSb(111)2×2 surface. Solid, dotted, and dashed curves represent calculations with the vertical distance between the In layer and the Sb layer of 0.935 Å (bulk value), 0.2 Å, and 0.0 Å, respectively (after Feidenhans'l hb:1163)

As an example, Fig. 4.21 shows rodscans for a set of half-order reflections for the InSb(111)2×2 surface, whose in-plane structure was considered above. The experimental data are shown by black circles with error bars, and the curves are the results of calculations for various vertical spacing between the In layer and the Sb layer. The solid curves represent the case of the unrelaxed surface with interlayer spacing equal to the bulk value of 0.935 Å. The dotted and the dashed curves correspond to the surface with interlayer distance reduced to 0.2 Å and 0.0 Å, respectively. It is apparent that the latter curves fit the experimental data much better, therefore, the surface structure of InSb(111)2×2 is concluded to be flat to within ∼0.2 Å.

Registry Between Surface and Bulk. In the above two steps, the intensities of only fractional-order reflections are considered and, hence, the structural information is limited to that of the surface layer with periodicity different from that of the bulk. If one requires information on the structure of a surface with 1×1 periodicity and that of the bulk, the integer-order reflection intensities should be taken into account. These data are often discussed in terms of *crystal truncation rods*, as in reciprocal space the diffuse rods normal to the surface appear for integer-order reflections as a consequence of the sharp truncation of the crystal at the surface. The analysis of of the crystal truncation rods appears to be useful for many particular problems (for example, determination of surface roughness, adsorbate residing sites, relationship between two materials).

4.4 Other Diffraction Techniques

Besides LEED, RHEED, and GIXRD, there are a number of other less common diffraction techniques used in surface science. Below we characterize some of them.

4.4.1 Transmission Electron Diffraction (TED)

In the conventional transmission electron diffraction (TED) technique, the high-energy electron flux is incident close to the sample surface normal and the diffraction pattern of the electrons passed through the sample is recorded. The TED measurements are conducted in UHV electron microscopes operating conventionally at 100–200 keV. To ensure electron transmittance, the sample should be very thin (≤ 1000 Å), which requires sample pretreatment, including, mechanical polishing, chemical etching, or ion sputtering from the backside. Usually, the pretreatment leads to formation of small holes having electron-transparent peripheries, which are used for observations. The final preparation of the sample (cleaning, adsorbate deposition, annealing, etc.) is conducted in situ in the UHV microscope.

As most of the electrons are scattered in the sample bulk, the surface contribution to the diffraction pattern is rather small, but fortunately it is distinguishable in some favourable cases. In TED scattering geometry, the surface reciprocal lattice rods are almost perpendicular to the surface of the very large Ewald sphere (which can be considered as being almost a plane). Therefore, the momentum transfer is essentially parallel to the surface ($\Delta k_{\perp} \approx 0$) for all diffraction beams and, hence, such reflections are most sensitive to in-plane atom positions. The single-scattering (kinematic) approximation is well applicable for TED and a Patterson function map for in-plane interatomic distances can be extracted from the measured intensities. Further stages of structure determination include building a trial model structure according to the Patterson function map and its refinement to reach agreement between the calculated and experimental diffraction intensities. The most pronounced result obtained by the TED technique is establishing the Si(111)7×7 DAS (dimer–adatom–stacking-fault) (see Sect. 8.5.2) structure by Takayanagi et al. [4.9, 4.10]

4.4.2 Atom Scattering

He atoms with a conventional thermal energy of about 20 meV have a de Broglie wavelength of about 1 Å and, hence, their interaction with a solid should be described in terms of diffraction. Moreover, at such low energies atoms cannot penetrate into the bulk and interact only with the outermost surface atoms. These ideas provide the basics of the He scattering technique for surface structure analysis. The experimental set-up includes a nozzle source which generates an intense and highly monoenergetic He atom beam directed towards the sample surface and the detector which is most often an ionization gauge, measuring the current of particles in the selected direction. Both the sample and detector can be rotated to obtain the full diffraction pattern. The sensitivity of the technique is rather low for smooth densely packed surfaces with small atomic corrugations. However, it appears useful in probing surfaces with adsorbed atoms or molecules. The technique is sensitive to the adsorbate structure only, which allows researchers to solve the problem of whether the origin of the superstructure is related to the adsorbate or to the substrate. Note that LEED fails to distinguish definitely between these possibilities.

4.4.3 Photoelectron Diffraction (PED) and Auger Electron Diffraction (AED)

The analysis of the anisotropy in emission of the secondary electrons is also used in structural studies. In *photoelectron diffraction (PED)*, the electrons due to the photoelectric effect (Sect. 5.4.1) are considered; in *Auger electron diffraction (AED)*, these are electrons due to the Auger recombination process (Sect. 5.2.1). The energy of the emitted electrons is a well-defined characteristic of the atom involved. If the adsorbate atoms on a substrate are being investigated, it can be directly determined at which atoms the initial "input wave" is generated. The elastic scattering of the wave from the "emitter atom" on the surrounding atoms is responsible for the anisotropy in the emission. In the experiment, angle-resolved photoelectron and Auger electron spectrometers are utilized to collect the data set. The evaluation of the surface structure proceeds through comparison of the experimental data with the results of model calculations. The structure determination is more reliable when analysing the high-energy ($\geq$400 eV) electrons emitted from the core levels, for which case the emitted wave can be approximated by a spherical wave and the kinematic approach is justified.

Problems

4.1 Consider the following real space lattices. How many symmetric domains are possible? Draw sketches of the LEED patterns for single-domain and multiple-domain (if available) surfaces?

- hexagonal $2\sqrt{3}\times 2\sqrt{3}$-$R30°$ superlattice;
- hexagonal $\sqrt{7}\times\sqrt{7}$-$R\pm 19.1°$ superlattice;
- rectangular 2×3 superlattice;
- centered rectangular $c(4\times 12)$ superlattice.

4.2 From the LEED pattern shown in the figure, identify the superstructure.

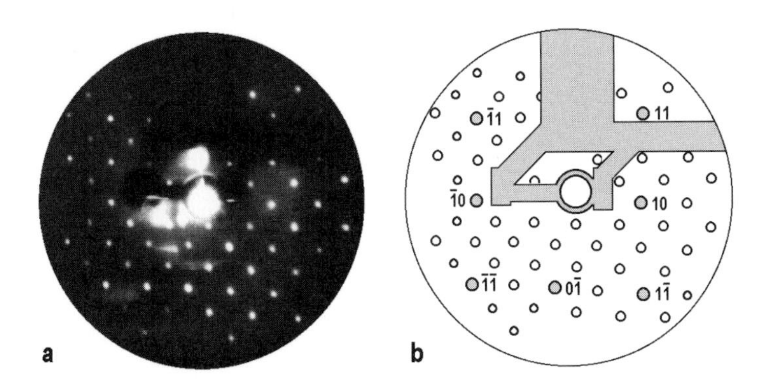

4.3 Using the Ewald construction, evaluate the scale of the recorded diffraction pattern from the surface with a square lattice of period 3 Å. How many diffraction spots are seen (analyzed) in each of the following three cases?

- In the LEED pattern at energy 50 eV in the LEED system with a 120° hemispherical screen.
- In the RHEED pattern at energy 10 keV in a system with a beam incidence angle of 5°, sample–screen distance of 30 cm, and screen diameter of 10 cm.
- In GIXD analysis with an x-ray wavelength of 1.5 Å and azimuthal rotation of the sample by 360°.

Further Reading

1. M.A. Van Hove, W.H. Weinberg, C.-M. Chan: *Low-Energy Electron Diffraction. Experiment, Theory and Surface Structure Determination* (Sringer, Berlin 1986)
2. C. Kittel: *Introduction to Solid State Physics*, 7th edn. (John Wiley, New York 1996) Chapter 2

5. Surface Analysis II. Electron Spectroscopy Methods

5.1 General Remarks

5.1.1 Surface Specificity

Electron spectroscopy probes the electronic structure of the surface through analysis of the energy spectra of the secondary electrons emitted from the sample. The secondary electrons are created generally by bombarding the surface with electrons or photons (other particles are also used (for example, ions or atoms), but seldom). The typical energies of secondary electrons analyzed in surface electron spectroscopy belong to the range 5–2000 eV. The surface sensitivity of electron spectroscopy stems from the fact that electrons with energies in that range are strongly scattered in solids. Figure 5.1 shows a plot of experimental values of the electron inelastic mean free path versus the electron kinetic energy. Though the data are energy and material dependent the magnitude of the inelastic mean free path in the whole energy range is of the order of several dozens of Å and in the favorable energy interval (~20–200 eV) is less than 10 Å.

The main electron spectroscopy techniques used for surface analysis are:

- Auger electron spectroscopy (AES);
- electron energy loss spectroscopy (EELS);
- photoelectron spectroscopy (PES).

In PES, the secondary electrons are generated by irradiation of the surface by photons. In AES and EELS, the surface is bombarded by electrons. To specify the difference between AES and EELS, consider the typical secondary electron energy distribution.

5.1.2 Spectrum of Secondary Electrons

If the solid is bombarded by monoenergetic electrons with energy E_p, a typical secondary electron spectrum $N(E)$ (Fig. 5.2) shows several features:

- A sharp elastic peak at the primary electron energy E_p;
- A broad structureless peak near $E = 0$ which extends, however, as a weak tail up to E_p (true secondary peak);

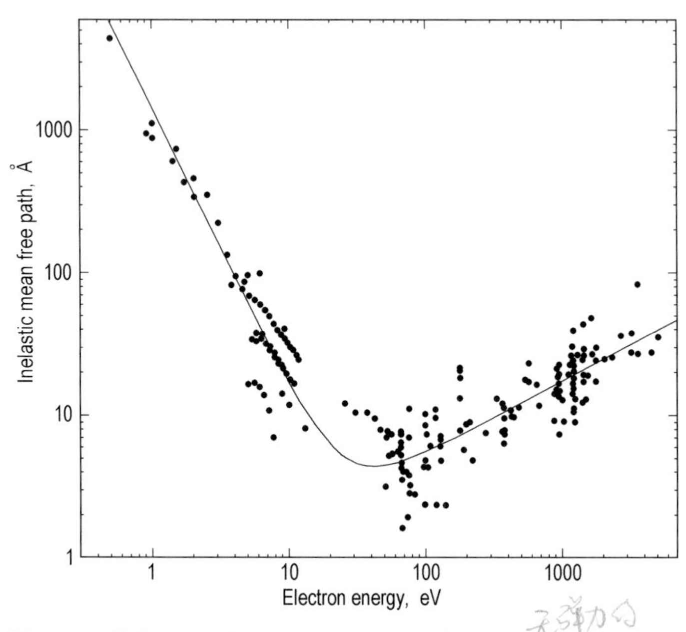

Fig. 5.1. Collection of experimental data of the inelastic mean free path of electrons as a function of their energy above the Fermi level for different materials. The full curve is the empirical least squares fit over the complete energy range [5.1]

- A long region with relatively few electrons, between E_{p} and $E = 0$, with a number of tiny peaks.

These tiny peaks fall into two categories. The first group is "tied" to the elastic peak, so that if E_{p} is raised by ΔE_{p}, all peaks in this group move up by ΔE_{p}. These are *loss peaks*, due to primary electrons that have lost discrete amounts of energy, say for discrete energy level ionization or plasmon excitation. These electrons are analyzed in EELS. The energy position of the peaks of the second group is fixed and independent of the primary energy of the excitation source. The most important of these are due to Auger electrons which are of interest for AES analysis.

5.1.3 Electron Energy Analyzers

As one can see from the above, the recording of the secondary electron energy spectra is a key point of any electron spectroscopy. Depending on the task, the spectra are recorded in the form of $N(E)$ (number of emitted electrons versus energy) and also in the form of its first and second derivatives, $\mathrm{d}N(E)/\mathrm{d}E$ and $\mathrm{d}^2N(E)/\mathrm{d}E^2$. The device used for recording the spectra is called the *electron energy analyzer*. The goal of the energy analyzer is to separate out of the whole secondary electron flux only electrons with a definite energy, called the *pass energy*. The pass energy is controlled by the voltages applied

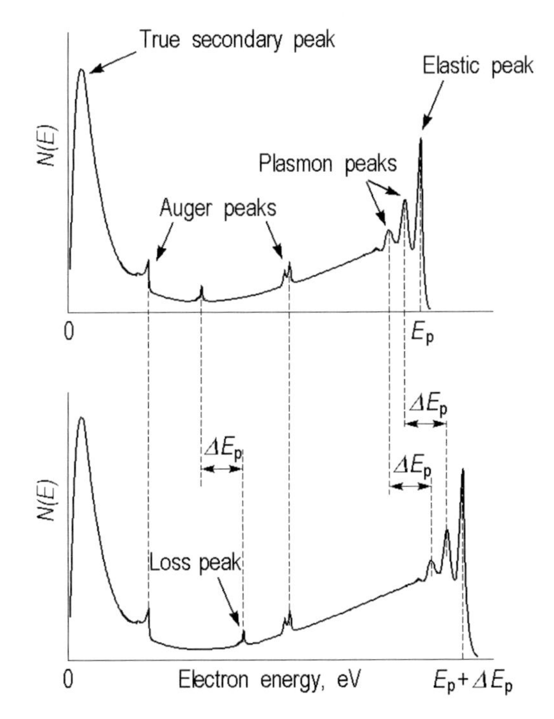

Fig. 5.2. Schematic electron energy spectrum, showing different groups of electrons backscattered from the solid

to the electrodes of the analyzer. To restore the whole spectrum, the electrode voltages are varied and the electron current is recorded as a function of the pass energy.

Most analyzers used for electron spectroscopy utilize electrostatic forces for electron separation and can be divided into two main classes:

- retarding field analyzers, and
- deflection analyzers.

The *retarding field analyzer (RFA)* functions by repelling electrons with an energy less than $E_0 = eV_0$, where V_0 is the voltage applied to the grids. Therefore, the collector receives an electron current (see Fig. 5.3a)

$$I(E) \propto \int_{E_0}^{\infty} N(E) \mathrm{d}E \, . \tag{5.1}$$

The most popular type of retarding field analyzer is a *four-grid analyzer*, which utilizes conventional four-grid LEED optics (see Fig. 5.6b). In this case, it is usually referred to as a LEED-AES device. The second and third grids of the analyzer are tiered together electrically and used as retarding grids for energy analysis. The fluorescent screen is used as a collector of

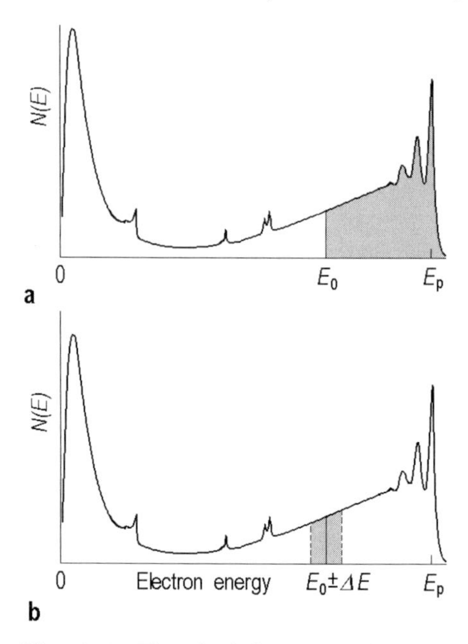

Fig. 5.3. The shaded area in the secondary electron spectrum corresponds to electrons selected by (**a**) retarding field analyzers (all electrons with energy higher than E_0) and (**b**) deflection analyzers (electrons in the energy window $E_0 \pm \Delta E$)

electrons. One can see that the four-grid analyzer collects emitted electrons over a wide solid angle.

In *deflection-type analyzers*, electrons in narrow energy window are collected (see Fig. 5.3b). The electron separation is due to the use of geometries in which only the electrons with a desired energy move along a specific trajectory leading to the collector. This is ensured by applying the electrostatic field cross-wise to the direction of the electron motion. To enhance the efficiency of the analyzers, they are designed so that the electrons having the same energy but entering the input aperture at somewhat different angles are all focused at the output aperture. The most widely used types of deflection analyzer are

- cylindrical mirror analyzer,
- concentric hemispherical analyzer,
- 127°-angle analyzer.

In the *cylindrical mirror analyzer (CMA)* (Fig. 5.4a), electrons leaving the target enter the region between two concentric cylinders through a conical annulus. For a negative voltage V_a applied to the outer cylinder with the inner cylinder grounded, electrons are deflected in the region between the cylinders and only electrons with a certain energy E_0 pass through the output aperture and are collected by an electron multiplier. The electron pass

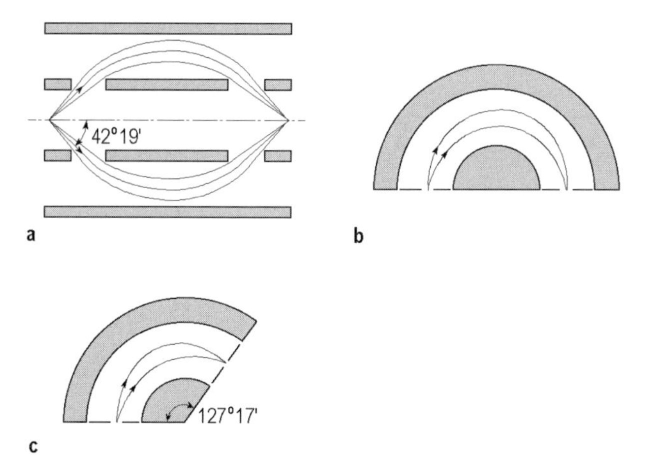

Fig. 5.4. Schematic sketches of the main deflection analyzers: (**a**) cylindrical mirror analyzer (CMA); (**b**) concentric hemispherical analyzer (CHA); (**c**) 127°-angle cylindrical sector analyzer. In all analyzers, the outer electrode is biased negatively with respect to the inner electrodes

energy E_0 is proportional to V_a and is determined by the analyzer geometry (the ratio V_a/E_0 is usually between 1 to 2). To ensure focusing of the electrons, the location of the sample and the windows in the cylindrical electrode guarantee the electron entrance angle of $42°19'$. The CMA is characterized by high sensitivity but moderate energy resolution. To enhance the resolution, a double-stage CMA is used with two successive analyzer units. Cylindrical mirror analyzers are widely used in Auger spectrometers with an electron gun often being integrated into the CMA on its central axis (Fig. 5.6a).

A *Concentric hemispherical analyzer (CHA)* is shown in Fig. 5.4b. The main elements of a CHA are two metallic concentric hemispheres. The outer hemisphere is biased negatively with respect to the inner hemisphere to produce an electrostatic field which balances the centrifugal force of the electrons on their trajectory. The entrance and exit apertures are circular holes. The efficiency of the CHA stems from the existence of a focusing condition for electrons deflected through an angle of 180°. CHA is widely applied for PES and AES analysis, especially when the measurements require angular resolution.

The *127°-angle analyzer* (Fig. 5.4c), which is often also called a *cylindrical sector analyzer*, utilizes an operational principle similar to that of CHA. As electrodes, it uses two concentric cylinder sectors with an angle of $127°17'$, which satisfies the focusing condition. The 127°-angle deflector is characterized by high energy resolution but not very good transmission. It is used mainly in high-resolution EELS measurements both as a monochromator and as an energy analyzer.

As for the energy resolution, deflection-type analyzers can be used in two modes:

- constant $\Delta E/E$ mode, and
- constant ΔE mode.

The constant $\Delta E/E$ mode is used when the electron pass energy E_0 is swept by the variation of the voltage applied to the electrodes. In this case, the energy window ΔE grows continuously with electron energy, leaving the ratio $\Delta E/E$ constant. The value of $\Delta E/E$ is determined mainly by the angular dimensions of the input and output apertures. The electron current collected in the constant $\Delta E/E$ mode is proportion to $EN(E)$, i.e.,

$$I(E) \propto EN(E)\ . \tag{5.2}$$

In the constant ΔE mode, the electron pass energy E_0 is held constant, thus preserving the constant resolution ΔE. In this case, the electron spectrum is continuously "shifted" through a fixed energy window ΔE by varying the acceleration or deceleration voltage in front of the analyzer.

5.2 Auger Electron Spectroscopy

Auger electron spectroscopy (AES) is one of the most commonly used techniques for the analysis of surface chemical composition by measuring the energies of Auger electrons. It was developed in the late 1960s, and derived its name from the effect first observed by Pierre Victor Auger (1899–1993), a French physicist, in the mid-1920s.

5.2.1 Physical Principles

Auger Process. The principle of the Auger process is illustrated schematically in Fig. 5.5. The primary electron, typically having an energy in the range 2–10 keV, creates a core hole and both electrons leave the atom. As an example, in Fig. 5.5a ionization is shown to occur by removal of a K-shell electron. This hole is filled by an electron from a higher lying level, say, L_1. The ionized atom is in a highly excited state and will rapidly relax back to a lower energy state by one of two routes:

- Auger emission (non-radiative transition),
- x-ray fluorescence (radiative transition),

as shown in Fig. 5.5a and b, respectively. x-ray fluorescence and Auger emission are alternatives. For low-energy transitions ($E < 500\,\mathrm{eV}$), especially for the lighter elements, x-ray fluorescence becomes negligible and, consequently, Auger emission is favored. Only at about 2000 eV does x-ray production become roughly comparable to the Auger efficiency.

The states involved in the Auger process, initial (one-hole state) and final (two-hole state), are both excited states. In total three electrons are involved in the transition, which means that Auger transitions can take place in all elements of the periodic table, except for H and He (which have less than three electrons per atom).

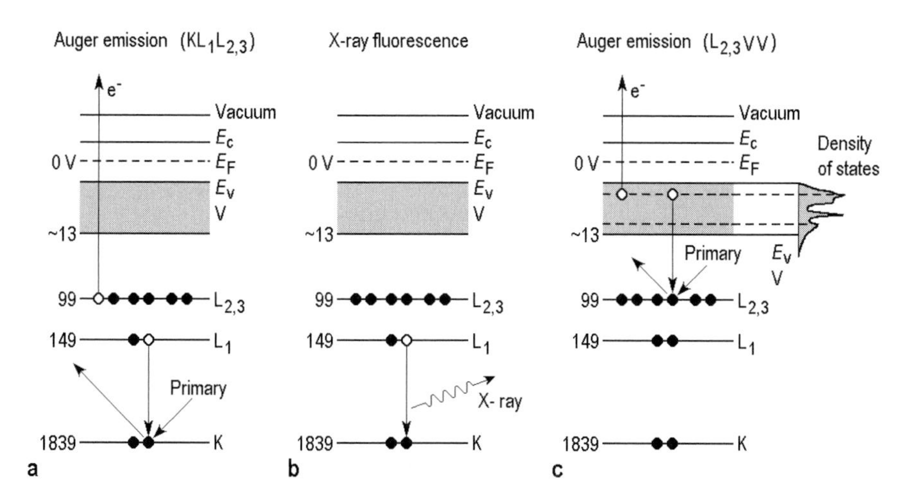

Fig. 5.5. (**a**) and (**b**) schematic diagram for the illustration of two competing paths for energy dissipation with silicon as an example. The $KL_1L_{2,3}$ Auger electron energy is about 1591 eV ($E_{KL_1L_{2,3}} = E_K - E_{L_1} - E_{L_{2,3}}$) and the x-ray photon energy is 1690 eV ($\hbar\omega = E_K - E_{L_1}$). (**c**) Schematic diagram for $L_{2,3}VV$ Auger transition (after Chang [5.2])

The nomenclature for Auger transitions uses x-ray level notation and reflects the levels involved. For example, the transition shown in Fig. 5.5a would be labeled as a $KL_1L_{2,3}$ transition. If an Auger process occurs in the solid and the valence band electrons are involved in it, the atomic level notation is often replaced by the symbol V (i.e., valence band). As an example, Fig. 5.5c shows the $L_{2,3}VV$ Auger transition.

Energy of Auger Electrons. An Auger transition is characterized primarily by the location of the initial hole and the location of the final two holes. Thus, the kinetic energy $E_{KL_1L_{2,3}}$ of the ejected electron in our particular example, shown in Fig. 5.5a, can be estimated from the binding energies of the levels involved as

$$E_{KL_1L_{2,3}} = E_K - E_{L_1} - E_{L_{2,3}} - \phi \,, \tag{5.3}$$

where $\phi = E_{\text{vacuum}} - E_{\text{Fermi}}$ is the work function of the material . Note that (5.3) is a rough estimation which does not take into account that the final emission occurs from an ion, not a neutral atom (Atom ionization results in shifting the final energy levels downwards and thus affects the energy of the Auger electron).

5.2.2 AES Experimental Set-Up

The standard equipment for AES consists of (Fig. 5.6)

- electron gun,
- energy analyzer,
- data processing electronics.

The electron gun produces the primary electron beam with a typical energy of 1 to 5 keV. The most commonly used energy analyzers for Auger electron spectroscopy are cylindrical mirror, hemispherical, and four-grid analyzers.

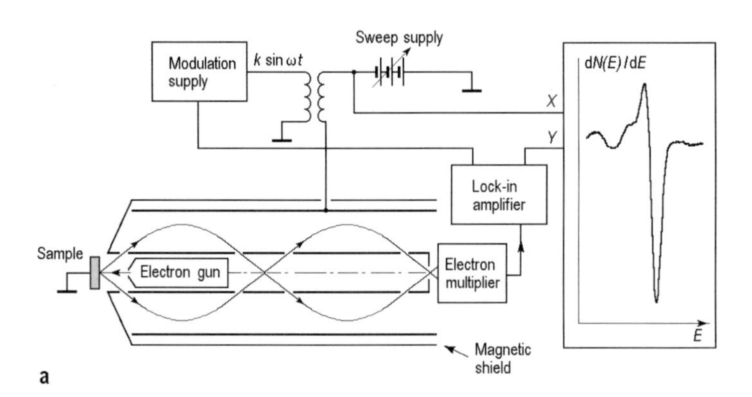

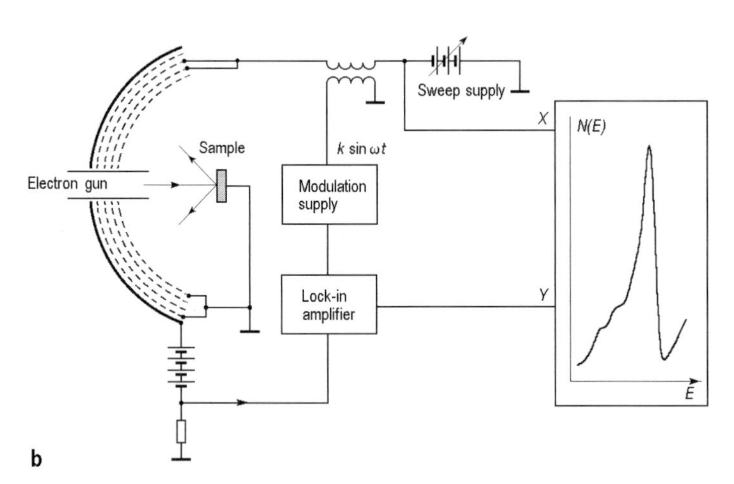

Fig. 5.6. Schematic diagram of the experimental set-up for Auger electron spectroscopy with the most widely used analyzers: (**a**) cylindrical mirror analyzer (the CMA analyzer shown is of the double-stage type with an electron gun integrated inside the inner cylinder) and (**b**) four-grid retarding field analyzer (LEED-AES device). The synchronous detection of the signal at the frequency ω of the modulation supply results in recording the spectra in the $\mathrm{d}N(E)/\mathrm{d}E$ mode in the case of the cylindrical mirror analyzer and in the $N(E)$ mode in the case of the four-grid analyzer. To obtain the $\mathrm{d}N(E)/\mathrm{d}E$ spectra in the latter case one uses detection at the frequency 2ω

Typically the relatively small Auger signals are superposed on a high background of true secondary electrons (electrons that undergo multiple losses of energy). To suppress the large background, and by doing so to improve peak visibility, Auger spectra are usually recorded in derivative mode $\mathrm{d}N(E)/\mathrm{d}E$ (Fig. 5.7). The differentiation is performed by modulating the analyzing energy by a small perturbing voltage $\Delta U \propto \sin \omega t$ applied to the two inner grids in the LEED-AES device or to the outer cylinder in the CMA and synchronously detecting the output signal. In this mode the collected current is

$$I(E_0 + k \sin \omega t) \simeq I_0 + \frac{\mathrm{d}I}{\mathrm{d}E} k \sin \omega t - \frac{d^2 I}{\mathrm{d}E^2} \frac{k^2}{4} \cos 2\omega t \,, \tag{5.4}$$

so that $\mathrm{d}N(E)/\mathrm{d}E$ spectra can be obtained with the CMA by detecting the output signal at the frequency ω, while for the LEED-AES device at the frequency 2ω.

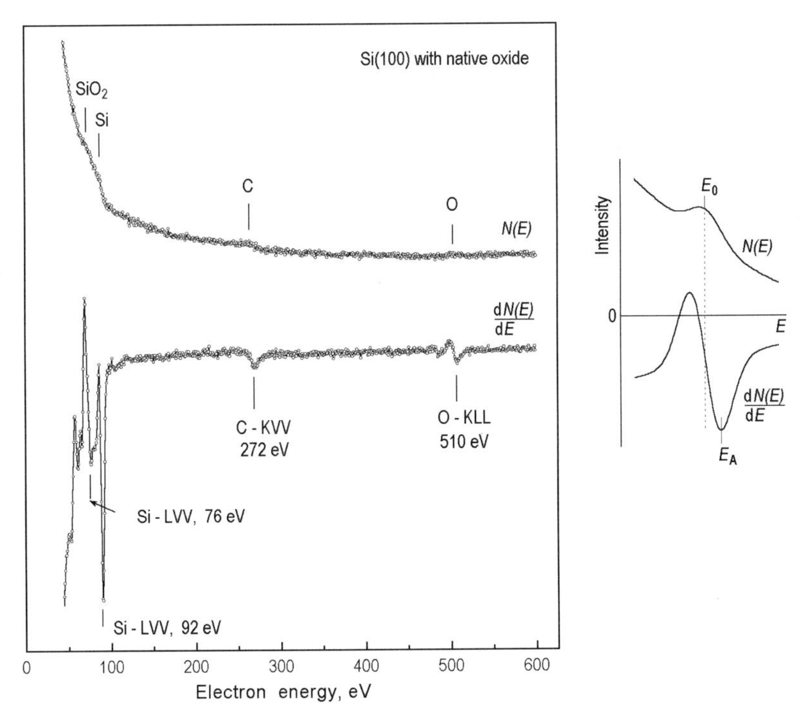

Fig. 5.7. $N(E)$ and $\mathrm{d}N(E)/\mathrm{d}E$ experimental Auger spectra of a Si(100) sample just introduced into the vacuum chamber. The right panel shows a schematic diagram illustrating the differentiating of the tiny Auger peak superposed on the intense background. Note that the minimum E_A of the derivative spectrum $\mathrm{d}N(E)/\mathrm{d}E$ corresponds to the steepest slope of $N(E)$ (not to the maximum E_0 of the Auger peak). Nevertheless, E_A is usually accepted in reference works as the Auger line energy

5.2.3 AES Analysis

General Information. Since the emitted Auger electron has a well-defined kinetic energy directly related to the nature of the atom-emitter, elemental identification is possible from the energy positions of the Auger peaks. The Auger electron energies are widely tabulated for all elements of the periodic table (except of cause for H and He) and are plotted in Fig. 5.8. In addition, the Auger spectrum contains certain information on the chemical bonding states of the atoms. The change in the chemical environment of a certain atom changes the electron binding energies (the so-called *chemical shifts*, to be discussed in Sect. 5.4.3) and leads to redistribution of the electron density of states in the valence band. These changes are reflected in the change of the Auger peak position and/or the peak shape. However, the quantitative

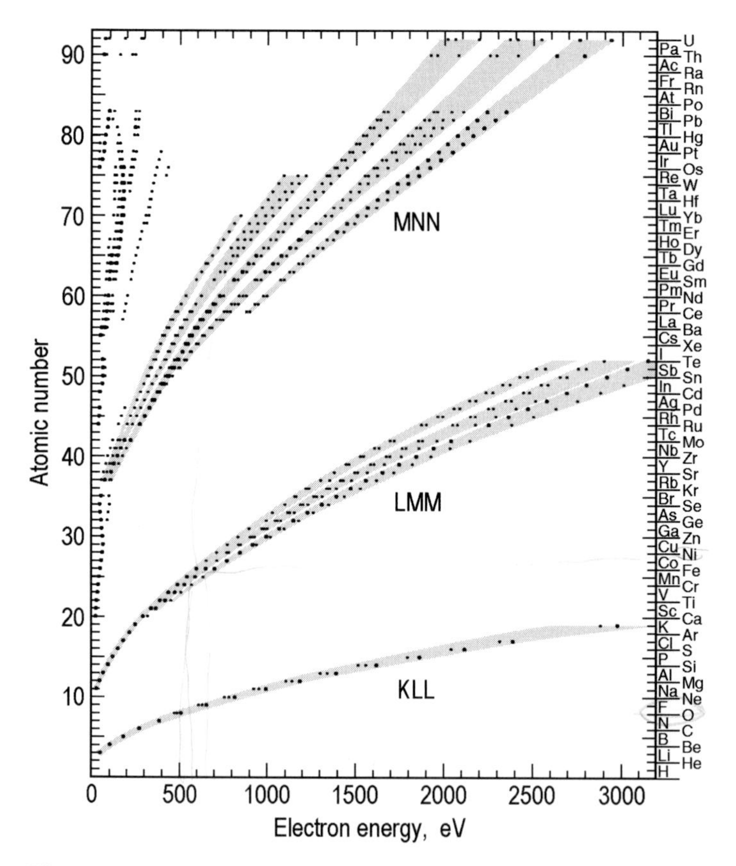

Fig. 5.8. The principal Auger electron energies for the elements used for qualitative analysis. Three main series KLL, LMM and MNN are indicated. The dots indicate the strongest and most characteristic peaks and the gray bands indicate the rough structure of less intense peaks [5.3]

interpretation of these data is hampered by the fact that three electrons are involved in the Auger process.

Other applications of AES include Auger mapping and Auger compositional depth profiling analysis. In Auger mapping, the spatial distribution of an element across the sample surface can be determined. To acquire an Auger map, the intensity of an Auger peak is monitored as a function of electron beam position as the electron beam is scanned across the sample. The usage of electrons as a primary beam has the advantage that the incident electron beam might be focused, thus giving spatial resolution down to 10 nm. The variation of the sample composition with depth can be obtained by combining AES with ion beam sputtering. The intensities of particular Auger peaks are monitored as a function of sputter time. The intensity scale can be converted to composition and the time scale to depth.

Although AES is usually considered as a non-destructive method, in some cases, electron-stimulated processes such as electron-stimulated desorption or heating of the target may result in undesirable side effects.

Quantitative Analysis. Auger electron spectroscopy provides the possibility of quantitative analysis, i.e., the number of atoms of a certain species present on the surface can be calculated based on the measured Auger intensities. The general equation for the Auger current I_i from an atom species i can be written as (considering as an example the KLM Auger transition)

$$I_i = I_{\mathrm{p}} \sigma_i \gamma_i (1 + r_i) T \frac{1}{4\pi} \iiint n_i(z) \exp \frac{-z}{\lambda_i \cos\theta} \sin\theta \, \mathrm{d}\theta \, \mathrm{d}\varphi \, \mathrm{d}z \,, \qquad (5.5)$$

where

I_{p} is the intensity of the beam of the primary electrons of energy E_{p};

$\sigma_i(E_{\mathrm{K}}, E_{\mathrm{p}})$ the ionization cross-section of the core level K by electrons of energy E_{p};

$\gamma_i(\mathrm{KLM})$ the probability of relaxation by the KLM Auger transition;

$(1 + r_i)$ a term, called the backscattering factor, which accounts for the effect of backscattering of the primary electrons (ionization of the core levels is produced not only by the primary electron beam but also by backscattered or secondary electrons which, in turn, results in an increase of the total number of core holes by a factor of $[1+r_i(E_{\mathrm{K}}, E_{\mathrm{p}}, \theta_0)]$, where θ_0 is the incident angle);

$n_i(z)$ the number of atoms i as a function of depth z;

$\exp(-z/\lambda_i(E_{\mathrm{KLM}}) \cos\theta)$ describes the probability of no-loss escape of Auger electrons from depth z. (Here $\lambda_i(E_{\mathrm{KLM}})$ is the electron *attenuation length* and θ is the escape angle of the Auger electrons with respect to the surface normal. The attenuation length depends on the electron energy and to some extent on the material. Its use is now generally accepted, instead of the inelastic mean free path, to characterize the sampling depth of electron spectroscopy. The attenuation length is shorter than the inelastic mean free path and accounts for the role of elastic scattering, which

increases the average path of the electron in the solid and in the presence of inelastic scattering eventually increases the probability of loss.);
T characterizes the transmission of the energy analyzer;
the integration is over the azimuthal φ and polar θ angles and the depth z.

Expression (5.5) in its complete form is a general illustration of what factors contribute to the intensity of the AES signal, rather than a formula suitable for practical needs. Its disadvantages are as follows.

- The exact values of many factors (for example, σ_i, γ_i) for a given particular system are often not available.
- The formula concerns absolute current measurements, which are not convenient in practice.
- In general, the distribution of atom species in the sample $n_i(z)$ is not known and integration over z cannot be done.

The first two items can be overcome to some extent if one measures an Auger spectrum in arbitrary intensity units, and then uses a normalization procedure (to be considered below in the illustrative examples). The usual way to overcome the third item is to make some reasonable assumptions about the character of the $n_i(z)$ distribution and then to conduct an evaluation within the chosen model. As an example, let us consider two commonly occurring cases, a homogeneous binary material and a uniform layer on the substrate.

Homogeneous Binary Material AB. In this case, $n_i(z)$ $(i = \mathrm{A, B})$ is constant and we can do the integration over z in (5.5). Then, by using the so-called elemental sensitivity factors I_A^∞ and I_B^∞ for the elements A and B (i.e., the intensity of AES signals from the "semi-infinite" bulk samples A and B, which can be found in reference handbooks [5.3] or obtained in calibration experiments), the ratio of Auger intensities can be written as [5.4]:

$$\frac{I_\mathrm{A}/I_\mathrm{A}^\infty}{I_\mathrm{B}/I_\mathrm{B}^\infty} = \frac{[1 + r_\mathrm{AB}(E_\mathrm{A})]}{[1 + r_\mathrm{AB}(E_\mathrm{B})]} \frac{\lambda_\mathrm{AB}(E_\mathrm{A})}{\lambda_\mathrm{AB}(E_\mathrm{B})} \times \frac{[1 + r_\mathrm{B}(E_\mathrm{B})]}{[1 + r_\mathrm{A}(E_\mathrm{A})]} \frac{\lambda_\mathrm{B}(E_\mathrm{B})}{\lambda_\mathrm{A}(E_\mathrm{A})} \frac{X_\mathrm{A}}{X_\mathrm{B}} \left(\frac{a_\mathrm{A}}{a_\mathrm{B}}\right)^3, \tag{5.6}$$

where a_A^3 and a_B^3 are the atomic volumes for A and B and X_A and X_B are the atomic fractions of A and B, respectively. The greatest source of uncertainty in the use of (5.6) is the electron attenuation length. It is energy and material dependent and can be estimated by using a general formula known as TPP-2 and derived by Tanuma, Powell and Penn [5.5]. Backscattering factors are usually calculated using generic equations provided by Shimizu [5.6].

For a very rough estimation of the surface composition, sometimes one ignores backscattering and attenuation-length terms [5.3], which simplifies (5.6) to:

$$X_{\mathrm{A}} = \frac{1}{1 + (I_{\mathrm{A}}^{\infty}/I_{\mathrm{B}}^{\infty})/(I_{\mathrm{A}}/I_{\mathrm{B}})} . \tag{5.7}$$

Here it is also taken into account that $X_{\mathrm{A}} + X_{\mathrm{B}} = 1$. It should be noted, however, that this simplification is rather hazardous and may result in significant errors.

Layer of Material A on a Substrate of B. If a uniform layer has a thickness d_{A} (Fig. 5.9a), it means that:

$$X_{\mathrm{A}}(z) = \begin{cases} 1 & \text{if } 0 < z \leq d \\ 0 & \text{if } z > d \end{cases} \qquad X_{\mathrm{B}}(z) = \begin{cases} 0 & \text{if } 0 < z < d \\ 1 & \text{if } z \geq d \end{cases}$$

After integration of (5.5) over z under these conditions and after some calculations, one obtains that

$$I_{\mathrm{A}} = I_{\mathrm{A}}^{\infty} \frac{1 + r_{\mathrm{B}}(E_{\mathrm{A}})}{1 + r_{\mathrm{A}}(E_{\mathrm{A}})} \{1 - \exp[-d_{\mathrm{A}}/\lambda_{\mathrm{A}}(E_{\mathrm{A}}) \cos\theta]\} , \tag{5.8}$$

$$I_{\mathrm{B}} = I_{\mathrm{B}}^{\infty} \exp[-d_{\mathrm{A}}/\lambda_{\mathrm{A}}(E_{\mathrm{B}}) \cos\theta] . \tag{5.9}$$

If the layer of material A is not continuous but covers a surface fraction ϕ_{A}, as shown in Fig. 5.9b, the exponential terms in (5.8) and (5.9) change to $\phi_{\mathrm{A}}\{1 - \exp[-d_{\mathrm{A}}/\lambda_{\mathrm{A}}(E_{\mathrm{A}}) \cos\theta]\}$ and $\{1 - \phi_{\mathrm{A}} \exp[-d_{\mathrm{A}}/\lambda_{\mathrm{A}}(E_{\mathrm{B}}) \cos\theta]\}$, respectively. A typical application of the latter equations is to distinguish layer-by-layer from other types of growth.

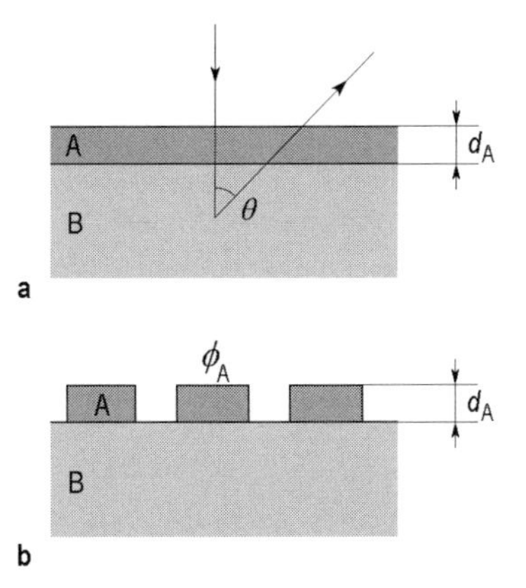

Fig. 5.9. Schematic illustration of (**a**) a uniform layer of material A with thickness d_{A} on a substrate of B; (**b**) layer of material A covers the fraction ϕ_{A} of the substrate surface

5.3 Electron Energy Loss Spectroscopy

Inelastically scattered electrons, which have lost well-defined energies in the course of interaction with the solid surface, are considered in *electron energy loss spectroscopy (EELS)*. These losses cover a wide energy range from 10^{-3} to 10^4 eV and can originate from different scattering processes:

- Core level excitation — 100–10^4 eV
- Excitation of plasmons and electronic interband transitions — 1–100 eV
- Excitation of vibrations of surface atoms and adsorbates — 10^{-3}–1 eV

Study of the first group of these losses is usually referred as *core level electron energy loss spectroscopy (CLEELS)*. Depending on the core levels studied it requires excitation sources with a relatively high primary energy of several keV or higher. As a result, the contribution of the bulk to the CLEELS signal is very significant.

Conventional EELS deals with the second group of losses which are related to excitation of plasmons and interband transitions. These losses are studied using electrons with a mediocre primary energy from 100 eV to a few keV. EELS spectra usually contain both bulk and surface components.

When loss spectroscopy is performed at low primary energies ($E_\mathrm{p} \lesssim 20$ eV) and with high energy resolution, it is called *high-resolution electron energy loss spectroscopy (HREELS)*. This provides the possibility of studying surface phonons and vibrational modes of the adsorbed atoms and molecules.

Typically, conventional electron energy analyzers like CMA or CHA are used in CLEELS and EELS, while high-resolution operation for HREELS requires a more sophisticated set-up with cylindrical sector deflectors used both as a monochromator for the primary electron beam and as an energy analyzer for the secondary electrons.

5.3.1 Core Level Electron Energy Loss Spectroscopy

An electron passing through the material can lose some of its kinetic energy to induce an electron transition from the core level to the empty state. For metals, these empty states are those above the Fermi level, while in semiconductors and insulators, they are located above the band gap in the conduction band. If the primary electron excites a transition, say, of the K electron to the final state ε_c in the conduction band (Fig. 5.10), the energy loss ΔE can be written as

$$\Delta E = E_\mathrm{K} + \varepsilon_\mathrm{c} \,, \tag{5.10}$$

and an inelastically scattered primary electron will have an energy of

$$E_\mathrm{s} = E_\mathrm{p} - \Delta E = E_\mathrm{p} - E_\mathrm{K} - \varepsilon_\mathrm{c} \,. \tag{5.11}$$

Note that this energy does not depend on the material work function as the primary electron passes through the surface barrier twice. Figure 5.11 shows an illustrative example of the CLEELS spectrum of oxidized silicon. As the intensity of the CLEELS peaks is typically 10 times lower than that in AES,

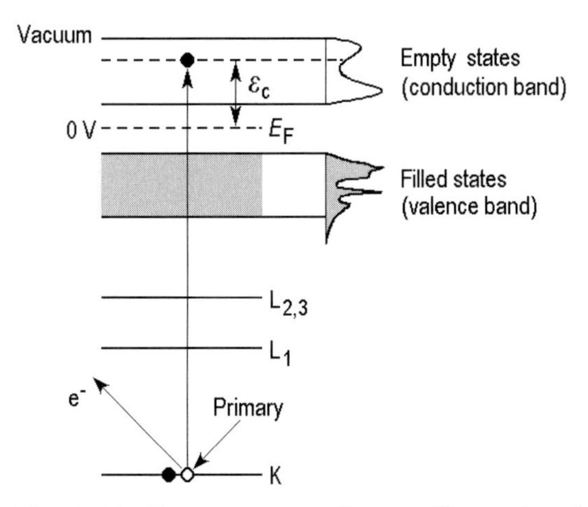

Fig. 5.10. Electron energy diagram illustrating the excitation of the K level in a semiconductor by a primary electron

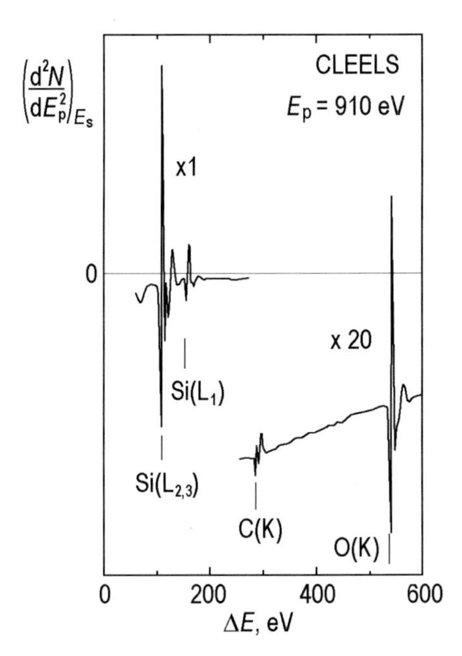

Fig. 5.11. Core level electron energy loss spectrum of oxidized silicon. In the measurements, the pass energy of the analyzer E_s was fixed, while the primary energy E_p was scanned. To improve the visibility of the CLEELS peaks, the spectrum is recorded in the second derivative mode. The loss peaks due to excitation of K levels of oxygen (O) and carbon (C) and L_1 and $L_{2,3}$ levels of silicon (Si) are seen (after Gerlach [5.7])

the second derivative mode ($\mathrm{d}^2N(E)/\mathrm{d}E^2$) is used to improve the visibility of the peaks.

One can see that the loss energy identifies the transition and so the energy loss spectrum can be used for elemental analysis. Quantification of the data is also possible as the intensity is proportional to the elemental concentration. As the transition probability of the electron depends on the density of final (empty) states, the fine structure of the CLEELS spectrum provides information on the energy distribution of the density of empty states.

5.3.2 Electron Energy Loss Spectroscopy

The term *electron energy loss spectroscopy (EELS)* has a double meaning. On one hand, it is used as a general term to characterize all the techniques dealing with electron energy losses (i.e., including CLEELS and HREELS). On the other hand, it is used as a particular term to characterize the electron energy loss technique that studies only the losses in the range from a few eV to several dozen eV. The losses in this range originate primarily from plasmon excitation and electronic interband transitions (Fig. 5.12). It should

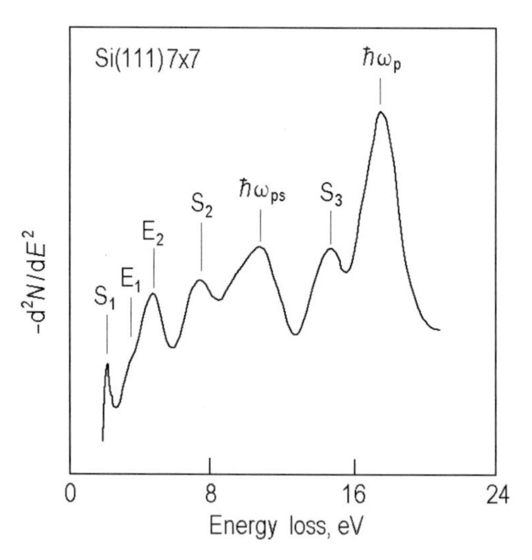

Fig. 5.12. Electron energy loss spectrum of a clean silicon Si(111)7×7 surface. The loss peaks in the spectrum are due to excitation of the bulk plasmon ($\hbar\omega_\mathrm{p}$), surface plasmon ($\hbar\omega_\mathrm{ps}$), bulk interband transitions (E_1, E_2) and transitions from occupied surface states (S_1, S_2, S_3). Primary energy is 100 eV (after Ibach and Rowe [5.8])

be noted that it is not a simple task to interpret conclusively the origin of a given loss peak, which is often a debated subject.

Excitation of Electronic Interband Transitions. An electronic interband transition in a semiconductor or insulator involves the excitation of an electron from an occupied state in the valence band or occupied surface state to the normally empty state in the conduction band. Thus, the process is basically similar to that used in CLEELS, but the loss energies are essentially lower.

Plasmon Excitation. A primary electron can lose some energy to induce oscillations of the electron density (or *plasma oscillation*) in a solid. The quantum of plasma oscillations is called *a plasmon.* A classical calculation [5.9] shows that the eigenfrequency of the bulk oscillations of a homogeneous electron gas with respect to a positively charged ionic skeleton is given by (in CGS units)

$$\omega_\mathrm{p} = \left(\frac{4\pi n e^2}{m}\right)^{1/2} \tag{5.12}$$

(the so-called Langmuir formula), where e and m are the electron charge and mass, respectively, and n is the electron concentration. Accordingly,

$$\Delta E = \hbar\omega_\mathrm{p} \tag{5.13}$$

is the energy of a *bulk plasmon.*

The Langmuir formula (5.12) is derived for a homogeneous electron gas, the model assumption seemingly suitable only for simple metals. However, the formula appears to be a reasonable approximation for many metals and even semiconductors (for example, Si, Ge, InSb) and insulators (for example, SiC, SiO_2). In the latter cases, valence electrons participate in the plasma oscillations.

The presence of a surface manifests itself by *a surface plasmon.* This type of oscillations is localized at the surface and its amplitude decays rapidly with depth. In the classical case of an abrupt interface between a homogeneous electron gas and a vacuum, the surface plasmon frequency is related to the bulk plasmon frequency by

$$\omega_\mathrm{sp} = \frac{\omega_\mathrm{p}}{\sqrt{2}}\,. \tag{5.14}$$

As an example, Fig. 5.13 shows the EELS spectrum of aluminum. The two right-most loss peaks at 10.3 and 15.3 eV correspond to surface and bulk plasmons, respectively. Other peaks are associated with losses due to multiple plasmon excitations.

In EELS analysis, the energy of the loss peak is of prime interest. Therefore, EELS spectra are recorded in $N(E)$ or $\mathrm{d}^2N(E)/\mathrm{d}E^2$ mode. In the latter case, though the peak shape is disturbed the peak visibility is enhanced and

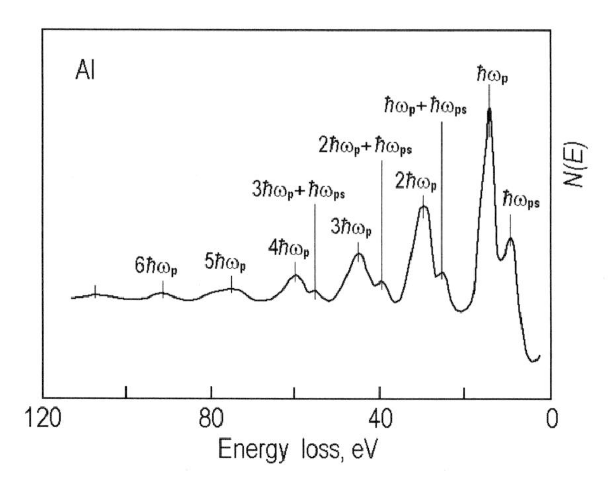

Fig. 5.13. Electron energy loss spectrum of aluminum. Losses due to bulk (15.3 eV) and surface (10.3 eV) plasmons are indicated. Losses due to multiple plasmon excitations are clearly seen (after Powell and Swan [5.10])

its energetic position is preserved. The widest application of EELS is concerned with the study of surface properties as follows.

- *Electron Density Determination.* Using Langmuir formula (5.12), one can evaluate the density of electrons participating in plasma oscillations (for example, the density of valence electrons in a semiconductor). However, one should bear in mind that the usage of (5.12) implies that the effective mass of an electron in a solid is close to that of a free electron and that the electron density is constant throughout the whole probing depth.
- *Chemical Analysis.* As the electron density is an individual characteristic of a given solid, the measured energy of the plasmon can be used to identify the surface species. The example shown in Fig. 5.14 demonstrates that one can distinguish between Si with a bulk plasmon at 17 eV and SiO_2 with a bulk plasmon at 22 eV.
- *Analysis of Depth Distribution of Species.* The probing depth in EELS is tied to the energy of the primary electrons. Thus, by varying the primary electron energy one can get information about the depth distribution of the species. As an illustrative example, Fig. 5.14 shows a set of EELS spectra from a 18 Å-thick SiO_2 film on a Si(111) substrate recorded at various primary electron energies from 100 to 1500 eV. One can clearly see that by increasing the primary electron energy the contribution of the Si substrate component increases and that of the SiO_2 film consequently diminishes. In principle, the quantification of such data is possible in order to evaluate the thickness of the film.

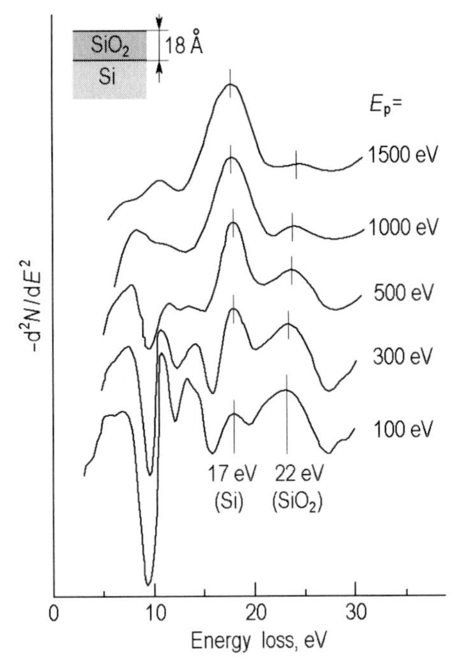

Fig. 5.14. Electron energy loss spectra from a SiO_2 film (18 Å) on a Si(111) substrate recorded at various primary electron energies (100, 300, 500, 1000 and 1500 eV). The loss peaks due to bulk plasmons in Si (17 eV) and bulk plasmons in SiO_2 (22 eV) are indicated

5.3.3 High-Resolution Electron Energy Loss Spectroscopy

High-resolution electron energy loss spectroscopy deals with the losses due to excitation of vibrational modes of the surface atoms and of the adsorbates on the surface. The latter are of most interest. As the losses are generally of a fraction of an eV, the energy separation between the elastic peak and the loss features is very small. As a consequence, to observe these features a detection technique with extremely high energy resolution is required. This goal is reached with the experimental set-up shown in Fig. 5.15.

It consists of a cathode system, a first energy dispersive element (a monochromator), a lens system between the monochromator and the sample, a second lens system between the sample and the analyzer, a further dispersive element (an analyzer), and finally the electron detector. One or two sequential 127°-angle cylindrical sector deflectors fixed at a constant pass energy are used as a monochromator. From the broad Maxwell distribution of hot electrons emitted by the cathode (with energetic half-width of about 0.5 eV), it selects electrons in a narrow energy window (typically 1–10 meV and down to a fraction of a meV in the most advanced systems). Two lens systems are used to focus the primary electron beam onto the sample and

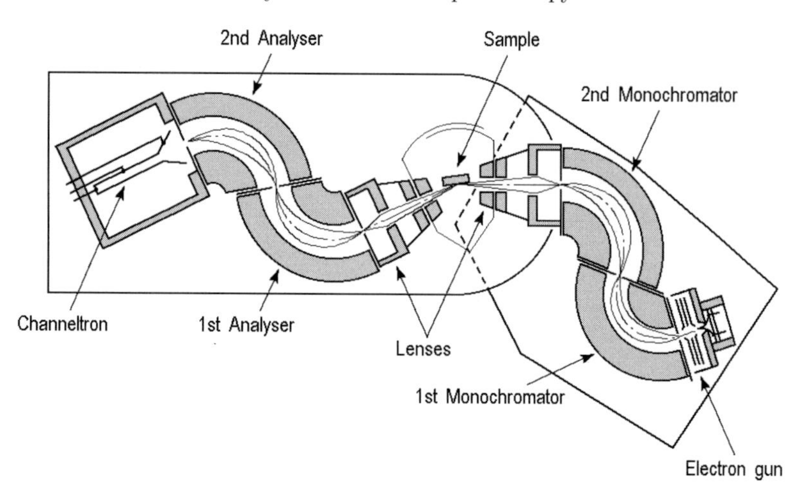

Fig. 5.15. A typical experimental set-up used for high-resolution electron energy loss spectroscopy. It comprises a cathode emission system, a monochromator, two lens systems, an analyzer, and an electron detector. Both analyzer and monochromator utilize 127°-angle cylindrical sector deflectors as the energy dispersive elements (after Ibach [5.11])

the backscattered electrons onto the entrance slit of the analyzer. Like the monochromator, the analyzer utilizes cylindrical sector deflectors. It operates usually in a constant ΔE mode (i.e., the pass energy is held constant and the acceleration voltage between the sample and analyzer is varied). A channeltron electron multiplier is a conventional electron detector.

HREELS has been proved to be a powerful tool to study the adsorption of atoms and, especially, molecules on solid surfaces. It allows the identification of the adsorbed species and provides information on their bonding geometries. The consideration is usually based on comparison of the vibrational modes detected in the HREELS experiments with the known vibrational spectra of molecules in the gas phase determined in IR absorption or Raman measurements. In analogy to the common convention in optical spectroscopy, the vibration energies in HREELS are often expressed in wave numbers, cm^{-1}, which is related to eV as

$$100\,\mathrm{cm}^{-1} = 12.41\,\mathrm{meV}\ . \tag{5.15}$$

Applications of HREELS address the following questions:

- *Identification of the adsorbed species.* As each molecule is characterized by a set of certain stretch modes, the identification of the adsorbed species is possible. The example shown in Fig. 5.16 demonstrates that the cases of atomic O and molecular O_2 adsorption can be clearly distinguished. In particular, one can see whether a given molecule dissociates upon adsorption or not.

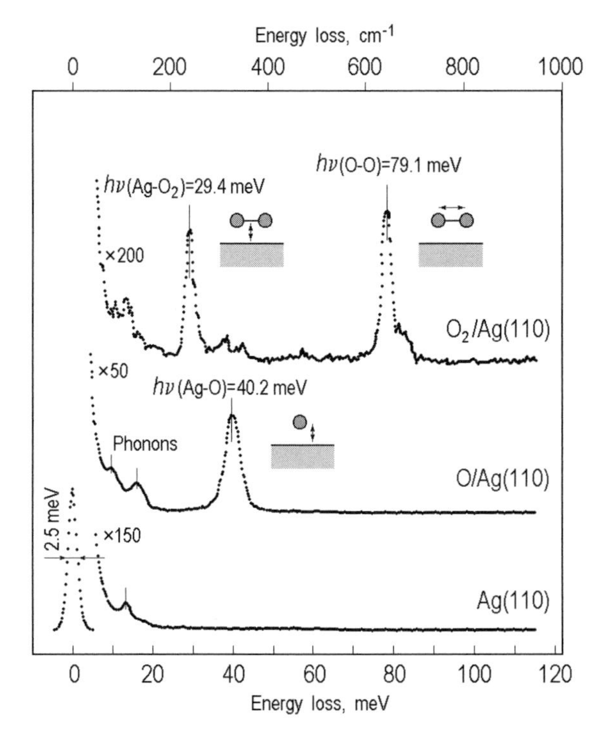

Fig. 5.16. HREELS study of oxygen adsorption on a silver surface. In the case of RT adsorption (middle spectrum), a characteristic peak at 40.2 eV which corresponds to the O-Ag mode is observed indicating dissociation of the O_2 molecules. In the case of O_2 exposure at 100 K (upper spectrum), molecular O_2 adsorption takes place: characteristic peaks at 29.4 meV and 79.1 meV correspond to the Ag-O_2 and O-O stretch modes, respectively. To detect the stretch mode (which is parallel to the surface) the 5° off-specular geometry was used. The HREELS spectrum (lower curve) of the clean Ag(110) surface is given for comparison (after Stietz et al. [5.12] and Vattuone et al. [5.13])

- *Identification of the adsorption sites.* Sometimes this can be judged from the observation of a certain adsorbate–substrate atom vibrational mode. For example, in the case of the adsorption of an organic molecule on the GaAs surface, occurrence of the As-H stretch mode indicates that the surface As atoms are the bonding sites.
- *Identification of the spatial orientation of the adsorbed molecule.* The consideration is based on the dipole selection rule in HREELS: in specular geometry (Fig. 5.17a), only dipoles oriented normal to the surface give rise to significant losses. Dipoles oriented parallel to the surface can be detected only in the off-specular geometry (Fig. 5.17b). Therefore, one can discriminate between the chemical bonds oriented normal and parallel to the surface.

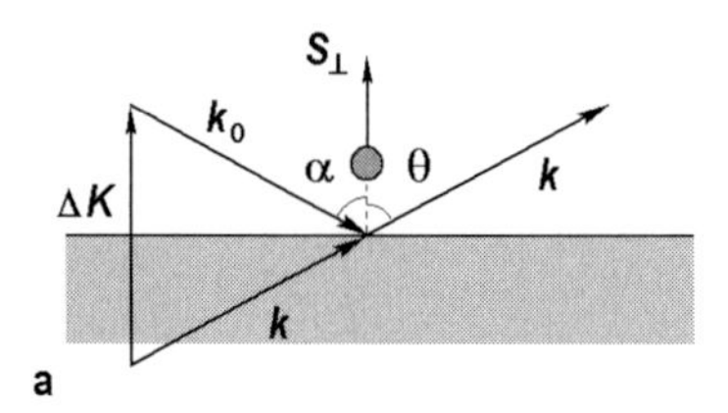

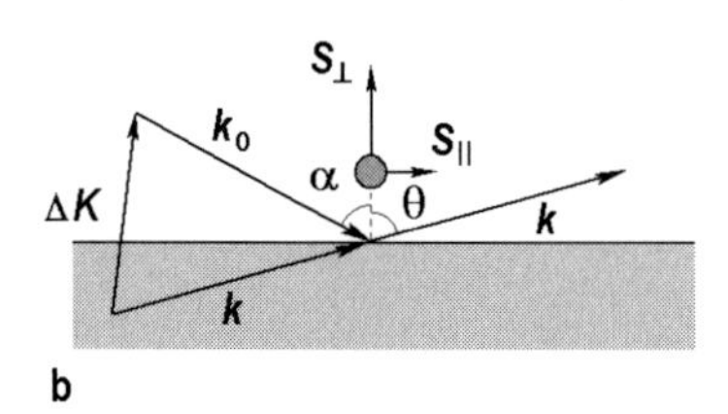

Fig. 5.17. (**a**) Specular ($\alpha = \theta$) and (**b**) off-specular ($\alpha \neq \theta$) geometries in HREELS measurements. The energy loss $\hbar\omega$ is assumed to be small in comparison with the primary energy E, i.e., $|\boldsymbol{k}| = |\boldsymbol{k_0}|$. In (**a**) the scattering vector $\Delta\boldsymbol{k}$ is normal to the surface, therefore atom vibration $\boldsymbol{S}$ parallel to the surface is not detectable, only the vibration $\boldsymbol{S}$ normal to the surface. In (**b**) $\Delta\boldsymbol{k}$ has components parallel and normal to the surface, thus both vibrations $\boldsymbol{S}$ and $\boldsymbol{S}$ can be studied (after Lüth [5.14])

5.4 Photoelectron Spectroscopy

5.4.1 Photoelectric Effect

Photoelectron spectroscopy (PES) is the most commonly used analytical technique to probe the electronic structure of occupied states at the surface and near-surface region. It is based on the *photoelectric effect*, in which the electron, initially in a state with binding energy E_i, absorbs a photon of energy $\hbar\omega$ and leaves the solid with kinetic energy

$$E_\mathrm{kin} = \hbar\omega - E_\mathrm{i} - \phi \,, \tag{5.16}$$

where $\phi = E_\mathrm{vacuum} - E_\mathrm{Fermi}$ is the work function of the material (Fig. 5.18). The necessary conditions for detecting the escaping elastic electron are as follows.

- The energy of the photon is sufficient to allow the electron to escape from the solid, i.e., $\hbar\omega \geq E_\mathrm{i} + \phi$.
- The electron velocity is directed towards the outer surface.
- The electron does not lose energy in collisions with other electrons on its way to the surface.

Depending on the energy (wavelength) of the photons used for electron excitation, photoelectron spectroscopy is considered conventionally as

- *XPS (x-ray photoelectron spectroscopy)* or *ESCA (electron spectroscopy for chemical analysis)*, when x-ray radiation is used with the photon energy in

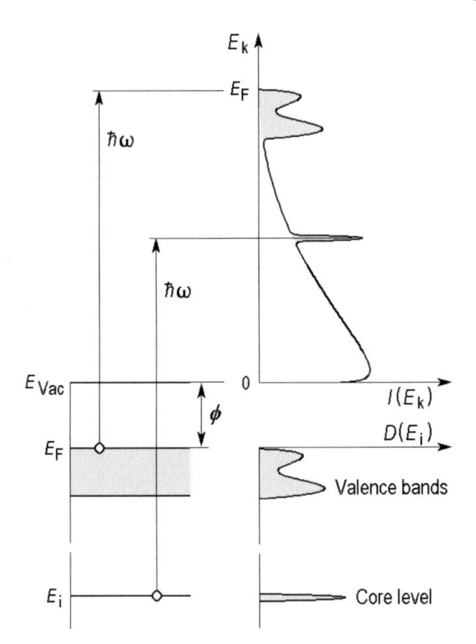

Fig. 5.18. Schematic illustration of the photoemission process on a metal surface. The correspondence between the density of occupied states $D(E_{\mathrm{i}})$ in the solid and the angle-integrated photoemission spectrum $I(E_{\mathrm{k}})$ is shown. The peaks of the elastic photoelectrons are superposed on the continuous background of inelastic secondary electrons

the range 100 eV–10 keV (corresponding to wavelengths from 100 to 1 Å). As a consequence, XPS probes the deep core levels.

- *UPS (ultraviolet photoelectron spectroscopy)*, when the photons are in the ultraviolet spectral range 10–50 eV (corresponding wavelengths from 1000 to 250 Å). As a result, UPS is used for studying valence and conduction bands.

It should be noted that this subdivision is rather a matter of convention: both from the viewpoint of the object of investigation (the division of energy levels into core levels and valence levels is itself loose) and from the viewpoint of radiation sources used (with synchrotron radiation one has a continuum for photoemission from the softest ultraviolet to hard x-rays). Moreover, both techniques are based on the same physical process.

5.4.2 PES Experimental Set-Up

The experimental set-up for photoemission experiments (Fig. 5.19) includes a monochromatic source of photons, the sample which is maintained in ultra-high vacuum and an electron energy analyzer to record the spectra of the photoemission electrons. The photon sources can be subdivided into *labora-*

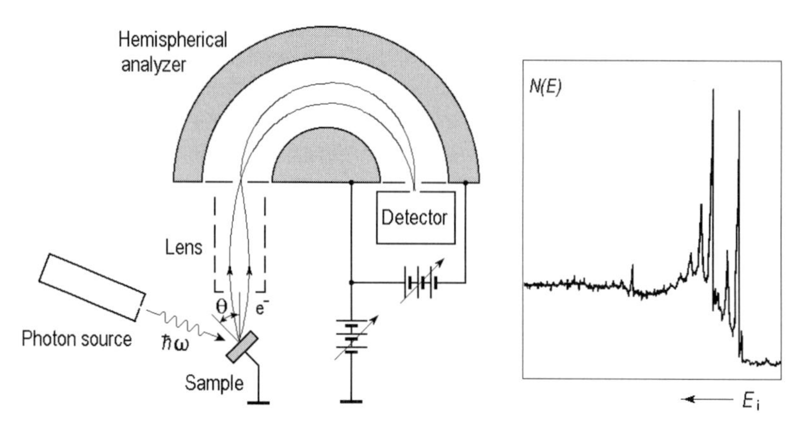

Fig. 5.19. Schematic diagram of the experimental set-up for photoemission experiments including photon source, sample, and concentric hemispherical analyzer. Only electrons with the appropriate energy and an acceptable entry angle reach the collector. Biasing the whole CHA with respect to ground is used to decelerate the electrons as they enter. As an example, the recorded XPS spectrum of aluminum is shown in the right panel

tory sources, which are specific for UPS and XPS, and *synchrotron radiation sources*, which cover the whole photoemission energy range.

The laboratory sources for XPS are x-ray tubes, in which the x-ray flux is created by bombarding a target with high-energy electrons. Common target materials are Mg and Al, for which the emission spectrum is dominated by the unresolved doublet $K_{\alpha 1,2}$ ($2p_{1/2}\rightarrow 1s$ and $2p_{3/2}\rightarrow 1s$ transitions) at 1253.6 eV (Mg $K_{\alpha 1,2}$) and 1486.6 eV (Al $K_{\alpha 1,2}$). The linewidth of the $K_{\alpha 1,2}$ doublet amounts to ∼700–800 meV for both materials. This fact and the existence of satellites with noticeable intensity (for example, intensity of the Mg $K_{\alpha 3,4}$ is ∼10% that of $K_{\alpha 1,2}$) demand the use of x-ray sources in combination with monochromators.

As laboratory sources for UPS, gas discharge lamps are used. The filling gas is conventionally an inert gas, a He discharge source being the most important. In this source, depending on the gas pressure and discharge current, one of two intense lines with photon energy of 21.2 eV (He I) and 40.8 eV (He II) can be generated. The advantageous of characteristics, narrow linewidth (3 meV for He I and 17 meV for He II) and low intensities of satellites, allow He discharge source to be used without a monochromator.

The modern alternative to laboratory sources is the use of synchrotron facilities. The source of photons is the radiation emitted from the accelerated beam of charged particles. The spectrum of the synchrotron radiation is essentially a broad continuum from a few eV to several keV. By using appropriate monochromators, photons with any required energy can be selected. The further advantages of high intensity and stability, 100% polarization in the plane of the accelerator, and high degree of beam collimation, make synchrotron radiation the most powerful tool for modern PES experiments.

Various types of analyzers are employed depending on the task. These are hemispherical (shown in Fig. 5.19) and 127° deflectors used for angle-resolved measurements, cylindrical mirror analyzers, which are used when angular resolution is not required, and retarding field analyzers used to collect photoelectrons over a large acceptance angle for studying the density of occupied states.

5.4.3 PES Analysis

X-ray Photoelectron Spectroscopy. Referring back to the schematic diagram of the photoemission process shown in Fig. 5.18 (which represents the XPS case), one can see that the photoemission spectrum $I(E_\mathrm{k})$ is a fingerprint of the density of occupied states $D(E_\mathrm{i})$ in the probed material. In reality, the correspondence, of course is not so unambiguous as in an ideal scheme. Besides the peaks due to elastic photoelectrons, there is a number of additional features in the XPS spectrum, like the continuous background of inelastic secondary electrons, Auger peaks, and peaks due to plasmon losses (on the low-energy side of each photoemission peak), as one can see in Fig. 5.20. Moreover, one should bear in mind that the cross-section for excitation is different for different electron levels, which greatly affects the shape of the spectra. For example, x-ray photons also eject electrons from the valence band, not merely from the deep core levels. However the valence band features in the XPS spectrum are very weak (see Fig. 5.20) due to very low photoelectric cross-section for shallow valence band levels at typical XPS photon energies. In general, photoemission has a maximum probability at photon energies close to threshold and drops off rapidly when the photon energy exceeds sufficiently the electron binding energy. Therefore, XPS is a tool to probe mainly deep core levels.

The core levels show up in XPS spectra as sharp peaks whose locations are defined by the electron binding energies, which, in turn, is essentially characteristic of atomic species. In other words, the presence of peaks at certain energies can be treated as a sign of the presence of a particular elemental species at the surface region and, thus, the XPS spectrum contains information on the *surface composition.*

On the qualitative level, one can distinguish what atomic species are present by comparing the peak energies on the experimental spectrum with the tabulated binding energies of electrons in elements (Fig. 5.21).

On the quantitative level, one can evaluate the concentration of the atomic species constituting the surface layer from the XPS peak heights. By analogy with quantitative AES (see (5.5)), we can write a general equation for the intensity of the photoelectrons from a core level of an atom species i as

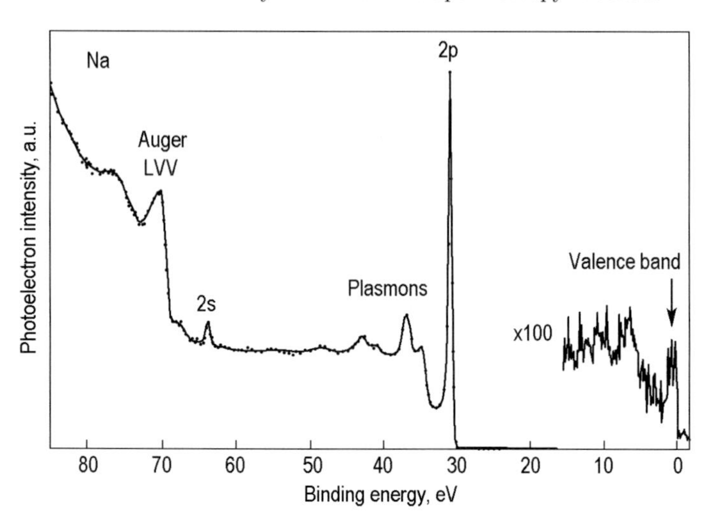

Fig. 5.20. XPS spectrum of Na excited by synchrotron radiation at 100 eV photon energy. Auger electron emission and plasmon losses are also indicated. Note the very weak intensity of the valence band features at binding energies from 0 to about 3 eV (after Kammerer [5.15])

$$I_i = J_0 \sigma_i(h\nu) T \int_{\gamma=0}^{\pi} \int_{\varphi=0}^{2\pi} L_i(\gamma) \int_{z=0}^{\infty} n_i(z) \exp \frac{-z}{\lambda_i \cos\theta} \mathrm{d}z \, \mathrm{d}\gamma \, \mathrm{d}\varphi \,, \qquad (5.17)$$

where

J_0 is the intensity of the primary x-ray beam;
γ is the angle between the incident x-ray beam and the direction of the ejected photoelectron;
$L_i(\gamma)$ is the angular dependence of photoemission;
$\sigma_i(h\nu)$ is the photoionization cross-section of the core level K by a photon with an energy $h\nu$.

All other notations are the same as in (5.5). In contrast to AES analysis, there is no backscattering factor in (5.17), but the angular anisotropy of photoemission is added. For the two simple cases, a homogeneous binary material and a uniform layer on a substrate, one can express the photoelectron intensities as follows.

Homogeneous binary material AB:

$$\frac{I_{\mathrm{A}}/I_{\mathrm{A}}^{\infty}}{I_{\mathrm{B}}/I_{\mathrm{B}}^{\infty}} = \frac{\lambda_{\mathrm{AB}}(E_{\mathrm{A}})}{\lambda_{\mathrm{AB}}(E_{\mathrm{B}})} \frac{\lambda_{\mathrm{B}}(E_{\mathrm{B}})}{\lambda_{\mathrm{A}}(E_{\mathrm{A}})} \frac{X_{\mathrm{A}}}{X_{\mathrm{B}}} \left(\frac{a_{\mathrm{A}}}{a_{\mathrm{B}}}\right)^3 . \qquad (5.18)$$

Layer A of thickness d_{A} on a substrate B:

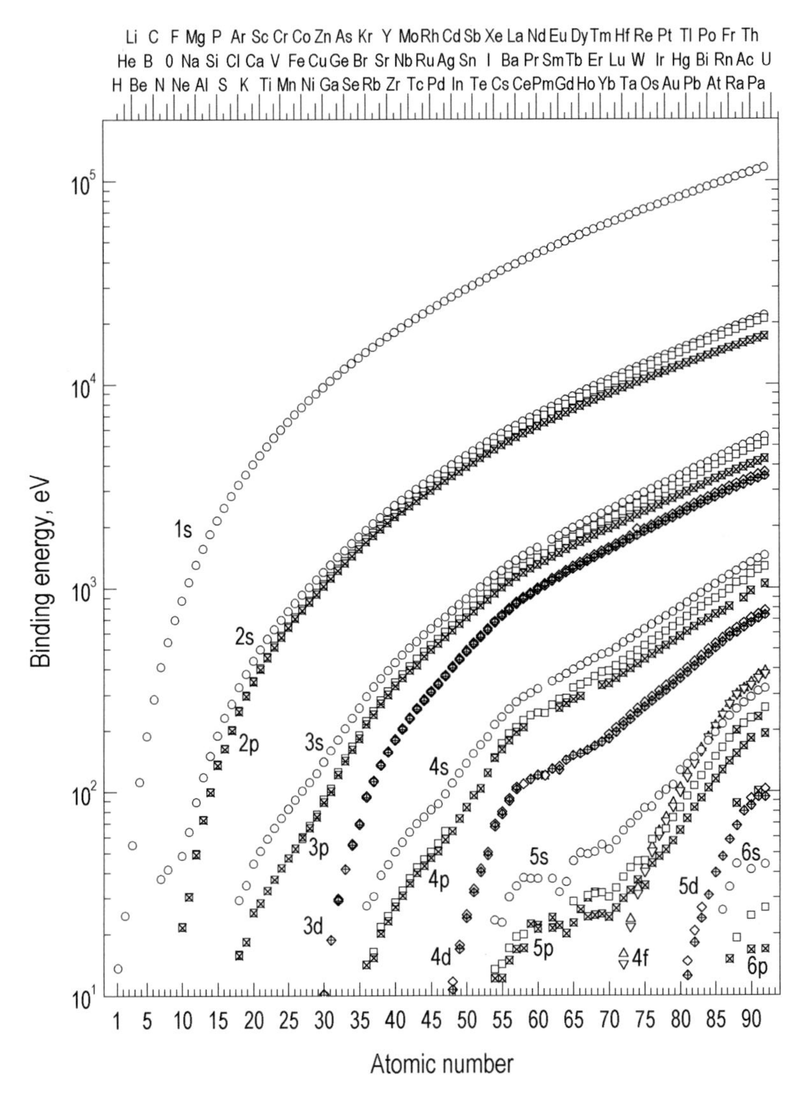

Fig. 5.21. Binding energies of filled core levels of the elements (after Wertheim [5.16])

$$I_{\mathrm{A}} = I_{\mathrm{A}}^{\infty} \left\{1 - \exp[-d_{\mathrm{A}}/\lambda_{\mathrm{A}}(E_{\mathrm{A}}) \cos\theta]\right\} , \tag{5.19}$$

$$I_{\mathrm{B}} = I_{\mathrm{B}}^{\infty} \exp[-d_{\mathrm{A}}/\lambda_{\mathrm{A}}(E_{\mathrm{B}}) \cos\theta] . \tag{5.20}$$

One can see that, compared to the appropriate equations used in AES analysis (5.6), (5.8), and (5.9), the equations for PES do not include backscattering terms which makes quantitative PES analysis somewhat more accurate than AES analysis.

Using XPS with high-energy resolution one can visualize the fine structure of core levels. In particular, the spin–orbit splitting is well resolved (Fig. 5.22). Accurate measurements of the core level peak energies for a given element reveal that they vary, depending on its chemical environment (Fig. 5.23). These are, so-called, *chemical shifts.* Their typical values are in the range from 1 to 10 eV. The origin of the chemical shifts can be understood as follows.

- The binding energy of the electron at a given level is defined by the interplay between the Coulomb attraction to the nucleus and the screening of this attraction by other electrons in the atom.
- Formation of the chemical bond involves electron transfer, thus the charge density on the atom is changed, which results, in turn, in changing the electron binding energy.
- Electron charge transfer to a given atom enhances the electron screening and, hence, weakens the electron binding energy (the corresponding peak shifts to shallower binding energies (relative to the Fermi level)). Conversely, electron charge transfer from a given atom weakens the electron screening and enhances the electron binding to the nucleus (the peak shifts to the deeper binding energies).

The ability of XPS to provide information on chemical composition and chemical bonding states warrants the alternative name *electron spectroscopy for chemical analysis (ESCA).* The name was introduced by the founder of the technique Kai M. Siegbahn, who received the Nobel Prize in Physics in 1981 "for his contribution to the development of high-resolution electron spec-

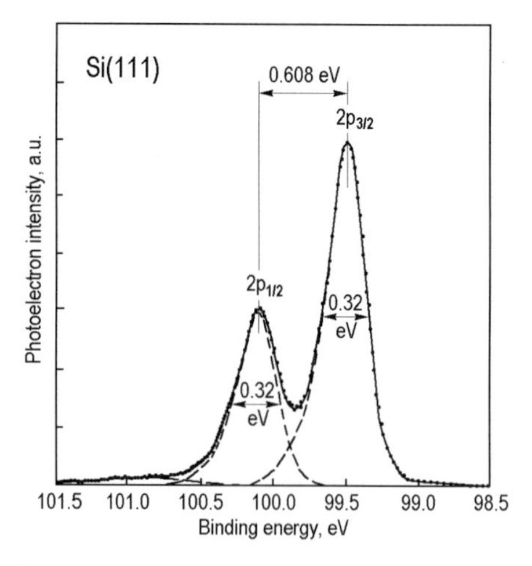

Fig. 5.22. The spin–orbit splitting of the Si 2p core level is indicated by observation of $2p_{1/2}$ and $2p_{3/2}$ partner lines. The splitting of 0.608 eV and the $2p_{1/2}$ to $2p_{3/2}$ intensity ratio of 1:2 are atomic properties and practically independent of the chemical environment (after Siegbahn [5.17])

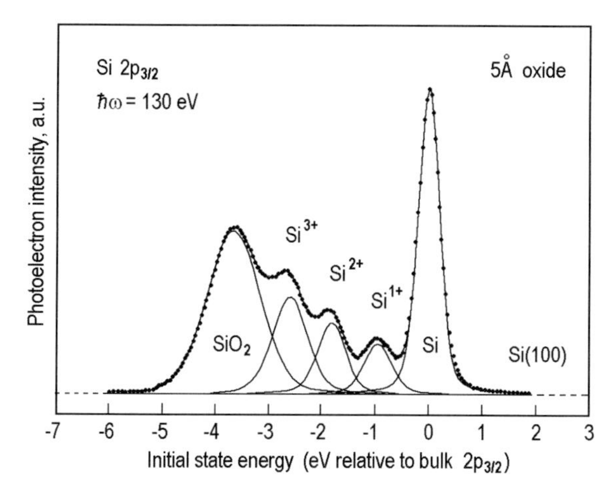

Fig. 5.23. Si $2p_{3/2}$ core-level spectrum from an ultrathin SiO_2 overlayer on a Si(100) surface (note that the Si $2p_{1/2}$ partner line is already subtracted). Besides the peak due to clean Si, the peaks due Si in different oxides are seen, showing a shift to lower energies with increasing oxidation states (after Himpsel et al. [5.18])

troscopy." However, nowadays the term ESCA is not in common use due to the abundance of other types of electron spectroscopy with the same object.

The atomic environment of the atoms at the surface differs from that in the bulk even in the case of adsorbate-free atomically clean surfaces. This difference is reflected particularly in the shifts of the core levels. In XPS spectra, the *surface components* are superposed with bulk components due to the finite penetration depth of x-rays and the electron mean free path. To enhance the surface sensitivity of XPS, grazing incidence radiation (refraction limits the penetration depth) and collection of photoemission electrons also at grazing angles (which reduces the effective mean free path normal to the surface) are usually used. In addition, by changing the polar angle of the detector, the fraction of the signal coming from various depths can be altered for analyzing the depth distribution of a given atom species.

Ultraviolet Photoelectron Spectroscopy. Since UPS utilizes relatively low photon energies (typically less than ~50 eV), only the valence levels become excited in the photoelectric process. Note that besides levels which correspond to the occupied band states of a solid surface these are also the filled bonding orbitals of the adsorbed molecules. If one takes into account the large photoemission cross-section of the valence states under UPS excitation energies, it becomes clear why UPS has proved to be a powerful tool for studying the valence band structure of surfaces and its modification during various surface processes (adsorption, thin film growth, surface chemical reactions, etc.).

Depending on the task, UPS is employed conventionally in one of two regimes:

- angle-integrated UPS,
- angle-resolved UPS.

In *angle-integrated UPS*, the electrons are collected in the ideal case over the whole half-space above the sample surface. Usage of the retarding field (LEED) analyzer is a good approximation. The obtained data are used to evaluate the density of states within the surface valence band.

In *angle-resolved UPS (ARUPS)*, the photoemission electrons are collected only in a chosen direction. The hemispherical and 127° deflectors are suited for this task. In this kind of measurement, one deals not only with electron energy, but also with the corresponding wave vector, which provides access to the dispersion of the surface states.

The kinetic energy of the photoemission electron (see (5.16)) can be written also as

$$E_{\mathrm{kin}} = \frac{\hbar^2 ({k_\perp^{\mathrm{ex}}}^2 + {k_\parallel^{\mathrm{ex}}}^2)}{2m} , \tag{5.21}$$

where $k_\parallel^{\mathrm{ex}}$ and $k_\perp^{\mathrm{ex}}$ are the components of the wave vector $\boldsymbol{k}^{\mathrm{ex}}$ of the photoelectron in vacuum parallel and perpendicular to the surface, respectively. The superscript *ex* (external) indicates that the wave vector $\boldsymbol{k}^{\mathrm{ex}}$ is referred to a free electron which has already escaped from solid to vacuum. If $\boldsymbol{k}^{\mathrm{ex}}$ makes an angle θ with the surface normal, one can write

$$k_\parallel^{\mathrm{ex}} = k^{\mathrm{ex}} \sin\theta = \sqrt{\frac{2mE_{\mathrm{kin}}}{\hbar^2}} \sin\theta . \tag{5.22}$$

To consider the wave vector of an electron inside the solid $\boldsymbol{k}^{\mathrm{in}}$, one should recall that, when passing through the solid–vacuum interface, only the parallel component of the electron momentum is preserved as

$$\boldsymbol{k}_\parallel^{\mathrm{ex}} = \boldsymbol{k}_\parallel^{\mathrm{in}} + \boldsymbol{G}_{hk} , \tag{5.23}$$

where $\boldsymbol{G}_{hk}$ is a vector of the 2D surface reciprocal lattice. In contrast, the perpendicular component $k_\perp$ is not preserved and, thus, does not bear any particular relationship to $k_\perp^{\mathrm{in}}$.

In experiments, to restore the dispersion curve $E_{\mathrm{i}}(k_\parallel^{\mathrm{in}})$ of the surface states along a certain surface direction, the photoemission spectra are recorded as a function of the polar angle, the azimuth being fixed. For each polar angle θ, the binding energy E_{i} and the parallel wave vector component $k_\parallel^{\mathrm{in}}$ are extracted from the UPS spectra using (5.16) and (5.22)–(5.23), respectively. As an example, Fig. 5.24 demonstrates the experimental determination of the energy dispersion relation for the surface states of Cu(111).

As a final remark, one should bear in mind that UPS is a surface-sensitive technique, but not a surface-specific one. This means that some efforts are required to distinguish between surface and bulk contributions to the photoemission spectrum. There are some tests which help to verify whether particular peaks are actually due to the surface states.

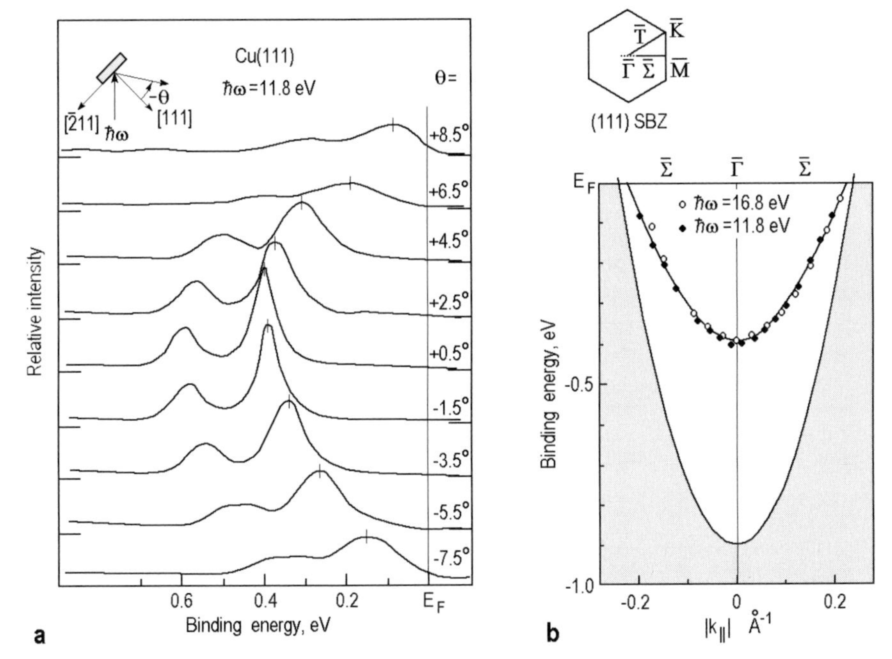

Fig. 5.24. ARUPS determination of the dispersion for the Cu(111) sp surface states. (**a**) Experimental photoemission energy distribution curves from Cu(111) for several angles near normal emission (the scattering geometry is shown in the inset). The location of the main maximum is of interest, the second peak being due to the Ar I doublet. (**b**) Evaluated dispersion of Cu(111) surface states plotted with a projection of bulk continuum of states (shaded region). Note that the dispersion curve is invariant with the change of the photon energy (open circles correspond to 16.8 eV, closed circles to 11.8 eV) (after Kevan [5.19])

- The surface state dispersion curve is one and the same at different photon energies used for excitation (for example, see Fig. 5.24b). This is a consequence of the fact that for surface states only the parallel component of the wave vector is essential without regard to the wave vector value itself. Clearly, this does not hold for bulk bands.
- Surface states reside in the bulk band gap.
- Surface states are more sensitive to surface treatment. For example, if the peaks in a spectrum from a clean surface vanish with gas adsorption, they are likely to refer to surface states.

Problems

5.1 How can one distinguish between Auger and loss peaks in the secondary electron spectrum?

5.2 Aluminum is deposited onto the Si(100) substrate at low temperatures to produce a continuous uniform Al film without Al–Si intermixing. Estimate the decay of the Si LVV Auger signal (92 eV) after deposition of the Al film with a thickness of 0.25, 1, 5 and 10 ML. Take the attenuation length of 92 eV electrons in aluminum to be $\lambda_{\mathrm{Al}}(92\ \mathrm{eV}) = 4.03\ \text{Å}$ and the thickness of 1 ML of Al to be $d_{\mathrm{Al}}(1\ \mathrm{ML}) = 1.13\ \text{Å}$.

5.3 Consider the shape of the spectrum in the second derivative mode ($-\mathrm{d}^2N/\mathrm{d}E^2$) for the case of a small peak on the intense background in the $N(E)$ spectrum (see Fig. 5.7). Prove that the energy position of the main peak in the $-\mathrm{d}^2N/\mathrm{d}E^2$ spectrum coincides with that in the $N(E)$ spectrum.

5.4 Prove the validity of Langmuir formula (5.12) for silicon which has four valence electrons per atom and displays a bulk plasmon loss peak at about 17 eV.

Further Reading

1. D.P. Woodruff, T.A. Delchar: *Modern Techniques of Surface Science*, 2nd ed. (Cambridge University Press, Cambridge 1999) chapter 3 (About applications of AES and PES)
2. D. Briggs, M.P. Seach (Ed.): *Practical Surface Analysis by Auger and X-Ray Photoelectron Spectroscopy* (John Wiley, Chichester 1983) (about practical aspects of AES and XPS, including quantitative chemical analysis in detail)
3. C. Kittel: *Introduction to Solid State Physics*, 7th ed. (John Wiley, New York 1996) Chapter 10 (about plasma oscillations in solids, derivation of expressions for bulk and surface plasmon energies)
4. H. Ibach: *Electron Energy Loss Spectrometers* (Springer, Berlin, Heidelberg 1991) (about the HREELS set-up in great detail)
5. H. Lüth: *Surfaces and Interfaces of Solid Materials*, 3rd ed. (Springer, Berlin, Heidelberg 1995) Chapter 4 (on the dielectric theory of inelastic electron scattering)
6. S. Hüfner: *Photoelectron Spectroscopy: Principles and Applications*, 2nd ed. (Springer, Berlin, Heidelberg 1996)

6. Surface Analysis III. Probing Surfaces with Ions

There is a set of various analytical techniques which employ an ion beam to probe a surface. The most widely used techniques are as follows.

- *Ion scattering spectroscopy* (ions elastically scattered from a surface are analyzed). This, in turn, is subdivided depending on the ion energies into
  - *Low-energy ion scattering spectroscopy* (~1–20 keV);
  - *Medium-energy ion scattering spectroscopy* (~20–200 keV);
  - *High-energy ion scattering spectroscopy* or *Rutherford backscattering spectroscopy* (~200 keV–2 MeV).
- *Elastic recoil detection analysis* (target atoms or ions elastically recoiled by primary ions are analyzed).
- *Secondary ion mass spectroscopy* (ions sputtered from a surface by a primary beam are mass analyzed).

The major applications of the above techniques are associated with the evaluation of the elemental composition and atomic structure of surfaces. Structural analysis is based on real space considerations and concerns mainly short-range arrangement of neighboring surface atoms.

6.1 General Principles

6.1.1 Classical Binary Collisions

In the first approximation, ion surface scattering can be described by elastic binary hard-sphere collision between a projectile particle (incident ion) and a stationary target (surface atom). This model is reasonable as the interaction time for the collision is short and the kinetic energies of the particles far exceed the interatomic binding energies. During the collision, the projectile loses a certain portion of its kinetic energy and is scattered away from its original direction of flight through a scattered angle ϑ_1, while the target gains kinetic energy and recoils away from its initial position through a recoil angle ϑ_2 as shown in Fig. 6.1. If a projectile particle has mass m_1 and energy $E_0 = \frac{1}{2} m_1 v_0^2$ and a surface atom at rest has mass m_2, the final energy $E_1 = \frac{1}{2} m_1 v_1^2$ of the projectile and the final energy $E_2 = \frac{1}{2} m_2 v_2^2$ of the target are obtained from the conservation laws of energy and momentum:

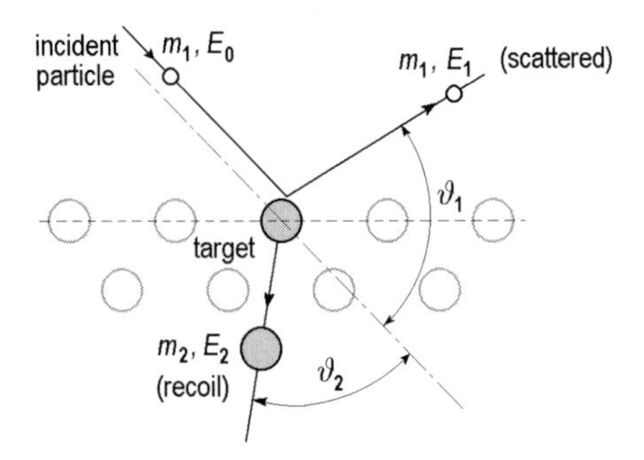

Fig. 6.1. Schematic diagram of a binary elastic collision of an incident ion (projectile) with a surface atom (target). The projectile of mass m_1 has initial energy E_0. The projectile's final scattering angle is ϑ_1 and its final energy is E_1. The target particle of the mass m_2, initially at rest, recoils at an angle ϑ_2 with energy E_2

$$\frac{m_1 v_0^2}{2} = \frac{m_1 v_1^2}{2} + \frac{m_2 v_2^2}{2} , \tag{6.1}$$

$$m_1 v_0 = m_1 v_1 \cos\vartheta_1 + m_2 v_2 \cos\vartheta_2 , \tag{6.2}$$

$$0 = m_1 v_1 \sin\vartheta_1 - m_2 v_2 \sin\vartheta_2 . \tag{6.3}$$

Thus, the scattered projectile has an energy E_1, given by

$$E_1 = E_0 \left(\frac{m_1 \cos\vartheta_1 \pm \sqrt{m_2^2 - m_1^2 \sin^2\vartheta_1}}{m_1 + m_2} \right)^2 . \tag{6.4}$$

When the projectile mass is less than the target mass ($m_1 < m_2$), the plus sign is taken in (6.4). In this case, the scattered angle ϑ_1 can vary from 0° to 180°, i.e., from no scattering to complete backscattering. However, if the projectile is heavier than the target ($m_1 > m_2$), both signs have to be considered in (6.4) and only forward scattering for angles smaller than $\vartheta_{1,\max} = \arcsin(m_2/m_1)$ is possible. The relation (6.4) is represented graphically in Fig. 6.2, where the scattered energy ratio E_1/E_0 is plotted versus the scattered angle ϑ_1 for various values of the mass ratio $A = m_2/m_1$.

In a similar way, the energy of the recoil atom E_2 is given by

$$E_2 = E_0 \frac{4 m_1 m_2 \cos^2\vartheta_2}{(m_1 + m_2)^2} \tag{6.5}$$

and Fig. 6.3 shows the recoil energy ratio E_2/E_0 as a function of the recoil angle ϑ_2 (which is $\leq 90°$) for various values of the mass ratio. Note that the E_2/E_0 curves for m_1/m_2 values are identical to those for the inverse values, i.e., m_2/m_1.

The relations (6.4) and (6.5) constitute the basis for surface elemental analysis. Indeed, the energy spectrum of the scattered or recoil particles,

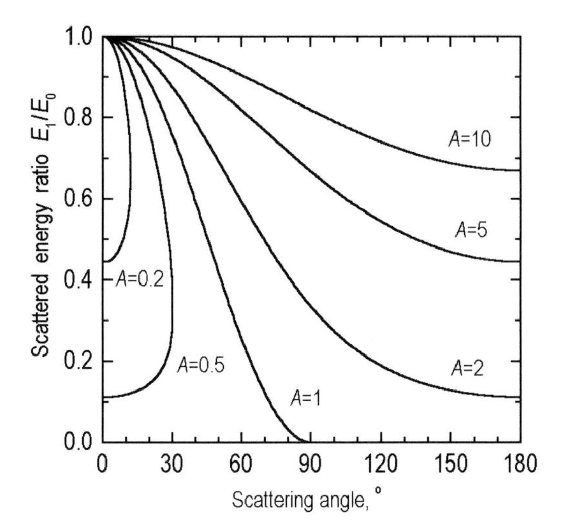

Fig. 6.2. Graphical representation of relation (6.4): Energy ratio E_1/E_0 for the scattered particles as a function of the scattering angle ϑ_1 for various values of the mass ratio $A = m_2/m_1$

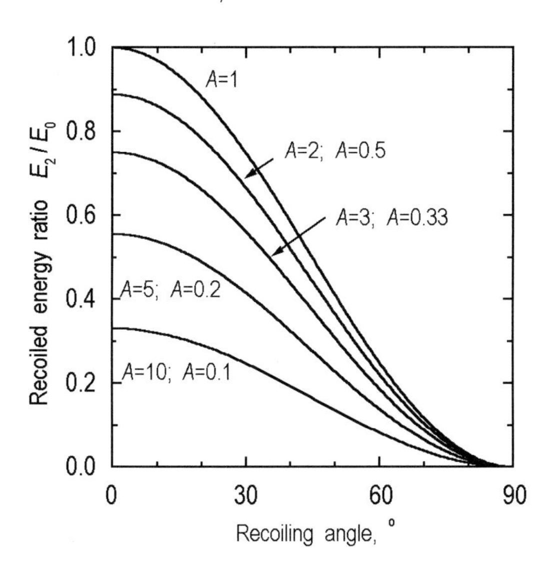

Fig. 6.3. Graphical representation of relation (6.5): Energy ratio E_2/E_0 for the recoil particles as a function of the recoiling angle ϑ_2 for various values of the mass ratio $A = m_2/m_1$. The $E_2/E_0(\vartheta_2)$ curves for A are identical to those for A^{-1}

combined with the knowledge of the primary ion mass and energy and the scattering geometry, provides direct information on the masses of atomic species present at the probed surface. Using (6.4) and (6.5), one can convert the energy of the spectral peak to the corresponding atomic mass. This simple approach proves its validity for ion energies in a wide range from hundreds of eV to several MeV (Fig. 6.4).

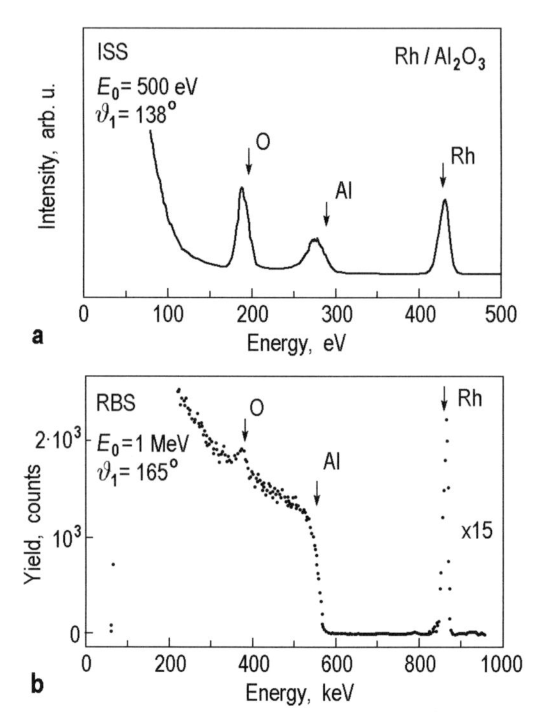

Fig. 6.4. Surface elemental analysis using ion scattering techniques. (**a**) ISS (500 eV He^+) and (**b**) RBS (1 MeV He^+) spectra of a Rh/Al_2O_3 sample (an alumina film, formed by oxidation in air, with a deposited ~1 ML of rhodium). The arrows indicate the peak positions calculated with the binary collision model (after Linsmeier et al. [6.1])

The hard-sphere model describes quite well the energetics of the scattering, but, in fact, it ignores the nature of particle interaction and is unable to describe accurately the true trajectories and probability of scattering (Fig. 6.5). A more adequate description is based on Coulomb repulsion between ion nucleus with charge Z_1e and the target atom nucleus with charge Z_2e. However, only for high energies (MeV),can the collision be described by a pure Coulomb force. At lower energies (keV), the screening of the nucleus by surrounding electrons has to be taken into account and the interaction

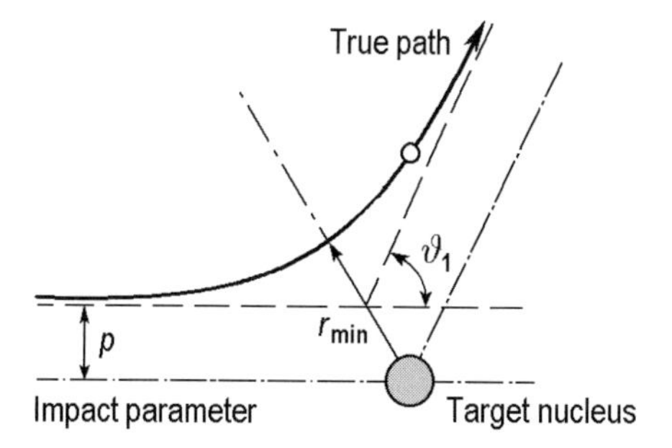

Fig. 6.5. A schematic view of the trajectory of a scattered particle repulsed from the target nucleus by the repulsive Coulomb force. The impact parameter p and the distance of the closest approach $r_{\min}$ are denoted

is usually described by the Coulomb potential multiplied by the screening function $\Phi(r)$ (CGS units) [6.2]

$$V(r) = \frac{Z_1 Z_2 e^2}{r} \Phi(r) . \tag{6.6}$$

A variety of screening functions have been derived [6.2]. The choice of an adequate one depends on the ion energy range and the nature of the ion–target combination. The potentials are routinely used for numerical calculations of the trajectories and the scattering cross-sections.

6.1.2 Scattering Cross-Section

The probability density that an ion is scattered over a certain scattering angle is given by the *differential cross-section* $\mathrm{d}\sigma/\mathrm{d}\Omega$. In general, the expression for the cross-section depends on the interaction potential and for the Coulomb potential this is the famous cross-section formula of Rutherford (CGS units) [6.3]

$$\frac{\mathrm{d}\sigma}{\mathrm{d}\Omega} = \left(\frac{Z_1 Z_2 e^2}{4E_0}\right)^2 \frac{1}{\sin^4(\vartheta_1/2)} g(\vartheta_1, m_1, m_2) , \tag{6.7}$$

where $g(\vartheta_1, m_1, m_2)$ is a transformation factor from the center of mass to the laboratory frame of reference:

$$g(\vartheta_1, m_1, m_2) = 1 - 2\left(\frac{m_1}{m_2}\right)^2 \sin^4(\vartheta_1/2), \qquad \text{for } m_1 \ll m_2 . \tag{6.8}$$

Equation (6.7) highlights that the cross-section is proportional to the square of Z_1 and Z_2. Thus, the scattering intensity strongly increases for heavier ions and targets. Table 6.1 illustrates the cross-section of $^4\mathrm{He}^+$ ion scattering by various target atoms at a fixed scattering angle. The dependence on $\sin^4(\vartheta_1/2)$ means that the scattering cross-section is a strong function of the scattering angle. Thus, the scattering intensity is much higher for forward scattered particles (small angles) than for backscattered particles (large angles). However, mass separation in the energy spectrum is larger at larger scattering angles (see Fig. 6.4). So, one should select what requirement is more essential for a particular experiment: resolution or sensitivity.

Table 6.1. Energies and cross-sections for the scattering of 2 MeV $^4He^+$ ions through 170° by various target nuclei [6.4]

Element	Mass	E_1 MeV	$d\sigma/d\Omega$ mbarn*/sr
C	12	0.50	38
O	16	0.73	74
Si	28	1.13	248
Fe	56	1.51	881
Au	197	1.85	8206

* 1 barn = 10^{-24} cm

6.1.3 Shadowing and Blocking

When a parallel beam of ions impinges on a target atom, the ion trajectories are bent by the repulsive potential of the target such that there is a region behind the target, called the *shadow cone*, for which the trajectories are totally excluded (Fig. 6.6). Most trajectories excluded from inside the shadow cone are concentrated at its edges (the so-called *focusing effect*). For the case of a repulsive Coulomb potential, the shadow cone has the shape of paraboloid whose radius R_s at a distance L is given by

$$R_s = 2\sqrt{\frac{Z_1\, Z_2\, e^2\, L}{E_0}} \,. \qquad (6.9)$$

Expression (6.9) is a good approximation for high-energy (MeV) light ions. However, it does not hold for low-energy (keV) ions, where more sophisticated screened potentials are used. From (6.9), one can see that the shadow cone size decreases dramatically with increasing incident ion energy (Fig. 6.7).

Atoms located inside the cone are shielded from impinging ions and do not contribute to the backscattering signal. If we consider ion scattering on the linear atomic chain which makes an angle α with the direction of the primary ion flux, one can see that a *critical angle* α_c exists starting from which the scattered projectiles can intercept a second atom with a small impact parameter (Fig. 6.7). It is apparent that the critical angle α_c is a

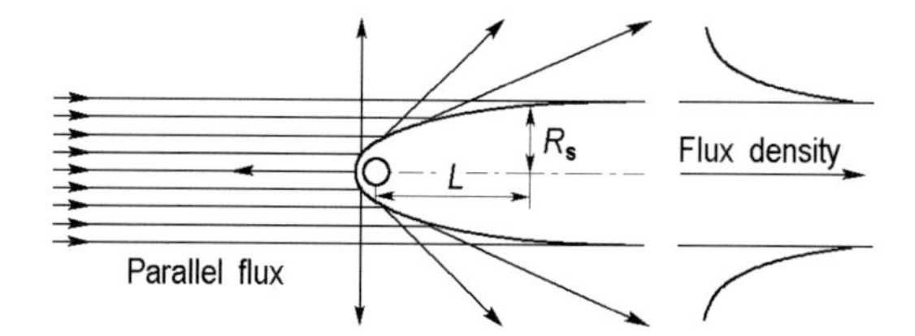

Fig. 6.6. Shadow cone formed from trajectories of projectile ions scattered from a target atom

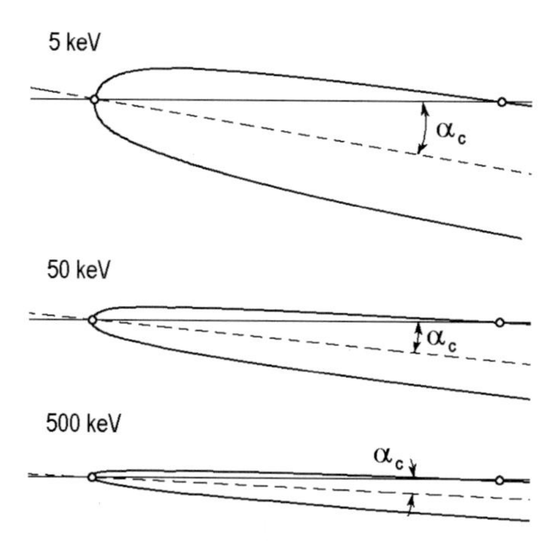

Fig. 6.7. Calculated shadow cones for Li^+ ions with energy of 5 keV, 50 keV, and 500 keV scattering from Ag atoms. The critical angles of shadowing α_c are indicated. The shadow cone width and critical angle decrease substantially with increasing ion kinetic energy (after Williams [6.3])

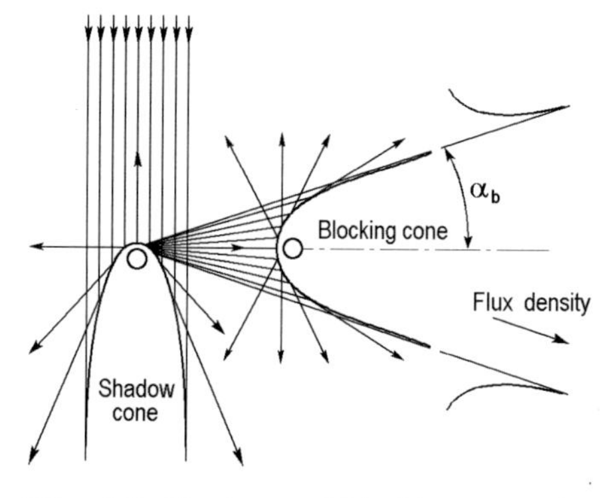

Fig. 6.8. Shadow and blocking cones for scattering from a pair of atoms

function not only of the atomic spacing but also of the ion energy: the greater the ion energy, the lower the critical angle.

The *blocking effect* is a phenomenon related to shadowing. It occurs when the trajectory of a scattered ion or a recoil atom is directed towards a neighboring (blocking) atom (Fig. 6.8). The blocking atom deflects these projectiles. For a large number of scattering or recoiling events, a *blocking cone* is formed behind a blocking atom where there are no allowed trajectories. In contrast to primary ions, the scattered ions or recoil atoms move initially along trajectories that are not parallel, but diverging. The critical angle α_b for blocking is the minimal possible angle for which the scattered projectile can exit to reach the detector. A fascinating example of the blocking process is the blocking photograph shown in Fig. 6.9.

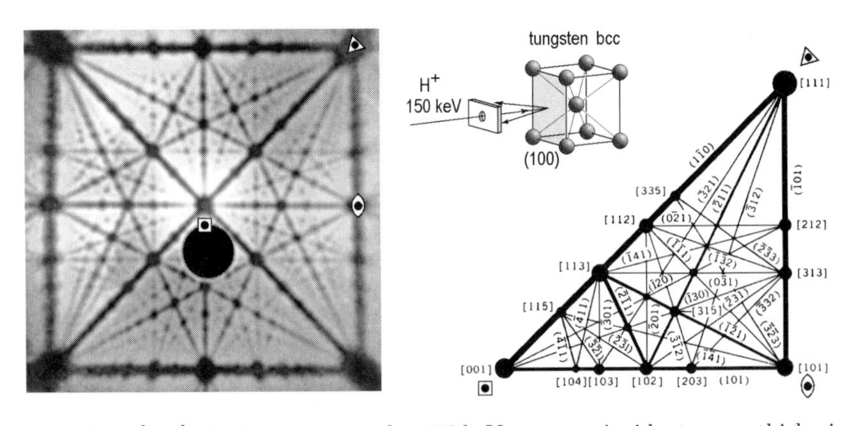

Fig. 6.9. [100] blocking pattern for 150 keV protons incident on a thick single crystal of tungsten in the non-channeling ($\sim 5°$) direction and backscattered along various high symmetry blocking directions. The angular distribution of scattered protons was monitored with radiation sensitive film. In this print from the film the black lines and spots correspond to proton-deficient regions in the film (After Barrett, Müller and White [6.5])

Note that when the mass of the projectile particle is less than that of the target ($m_1 < m_2$), shadowing (or blocking) is complete. In the opposite case of $m_1 > m_2$, the shadowing (or blocking) is incomplete, i.e., the density of allowed trajectories behind the shadowing (or blocking) atom is decreased but not eliminated completely. However, there is still a considerable increase in the effective flux density at the edge of the partial shadow (or blocking) cone [6.3].

The effects of shadowing and blocking are the main factors responsible for the anisotropy in the yield of scattered and recoiled particles, and this relationship is used in ion spectroscopies to gain information on the positions of the atoms near the surface with respect to each other.

6.1.4 Channeling

When the ion beam is aligned with a major symmetry direction of a single crystal, those ions that escape the close encounter with surface atoms can penetrate deep into the crystal (up to thousands of Ångstroms) moving through the channels formed by the rows of atoms (Fig. 6.10). Hence, this effect is referred to as *channeling*. Channeling takes place when the size of shadowing cones of the surface atoms are small in comparison with the interatomic distances, i.e., for high-energy light ions (for example, MeV-He^+ ions). The tolerance limits for the deviation of the entrance angle from an ideal symmetry direction is characterized by the critical angle for channeling ψ_c. The critical angle depends on the atomic numbers of the ion and target atoms, Z_1 and Z_2, the ion energy and interatomic spacing along the row, d:

$$\psi_c \propto \sqrt{\frac{Z_1 Z_2}{E_0 d}} \, . \tag{6.10}$$

Typical values for ψ_c lie in the range from fractions of a degree to a few degrees (for example, for 1 MeV He ions incident at room temperature along the $\langle 100 \rangle$ direction the critical angle is 0.6° for Si and 1.9° for W). It is clear that the critical angle for channeling ψ_c and the critical angle for shadowing α_c have much in common and are of the same order.

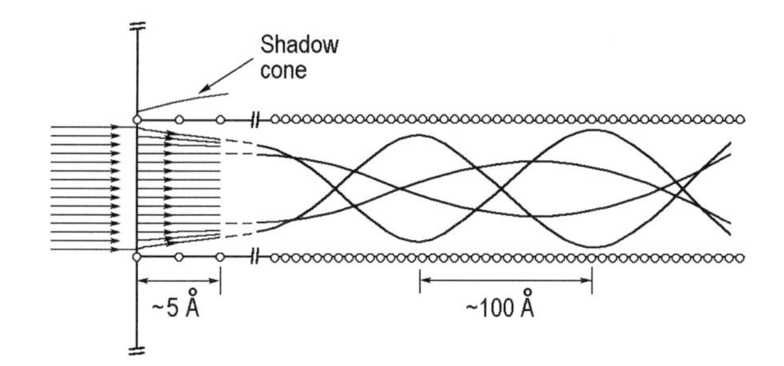

Fig. 6.10. Schematic illustration of the particle trajectories undergoing scattering at the surface and channeling within the crystal. The depth scale is compressed relative to the width of the channel in order to display the shape of the trajectories (after Feldman, Mayer and Picraux [6.6])

An important feature of ion motion along the channel is that the ion cannot get close enough to the atomic nuclei to undergo large-angle scattering. Small-angle scattering leads to oscillatory trajectories with a wavelength of hundreds of Ångstroms. On its motion through solid, the ion continuously loses energy due to inelastic interaction with the electrons of the solid. The rate of energy loss is often referred to as the *stopping power*, $\mathrm{d}E/\mathrm{d}x$. For MeV-He ions, the stopping power is roughly proportional to Z_2/E. Often stopping powers are expressed as $(1/N)\mathrm{d}E/\mathrm{d}x$, i.e., in units of $\mathrm{eV}/(10^{15}\ \mathrm{atoms/cm^2})$.

For example, for 1 MeV He^+ ions in Si, which has a density $N = 5 \times 10^{22}$ atoms/cm^3, the stopping power equals 60 eV/(10^{15} atoms/cm^2). This means that an ion will lose 60 eV of its energy when it passes through a layer containing 10^{15} atoms/cm^2 along the ion path, in the case of the $\langle 100 \rangle$ direction in Si ~2 Å.

6.1.5 Sputtering

In the preceding sections, we have considered elastic collisions in which the motion of the scattered ion or recoiled atom is directed out of the sample. However, another destiny of an energetic ion impinging on the surface is possible, namely, it penetrates into the bulk, dissipates its kinetic energy in a series of collisions with atoms in the solid and comes to rest. As a result of energy dissipation, a number of atoms within the solid gain energy sufficient to displace them from the original sites and to generate recoil cascades. Certain of these cascades terminate at the surface, causing the surface atoms to leave the surface as neutrals or as positive or negative ions (see Fig. 6.11). The process of removal of material from the surface by ion bombardment is called *sputtering*.

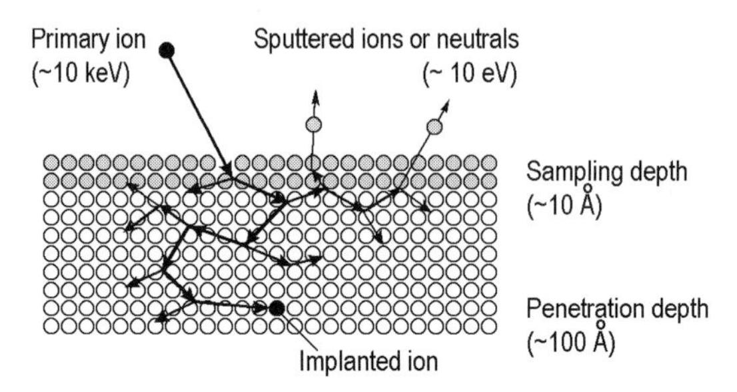

Fig. 6.11. Schematic representation of the processes taking place upon penetration of the impinging ion into the solid. The cascade of collisions results in ion implantation and sputtering of surface species. The shown numerical values provide a feeling for the orders of magnitude for the ion penetration depth, escape depth, and energy of sputtered species when bombarding the surface with 10 keV ions

The *sputtering yield* is defined as the number of sputtered atoms per incident ion. In general, this is a complicated function of the mass, kinetic energy, and incidence angle of ions, as well as of the chemical composition and crystallographic orientation of the bombarded sample. Figure 6.12 shows the sputtering yield from a polycrystalline Ni target as a function of the primary energies for various ions. The common trend is that, after exceeding the threshold, determined by the binding energy, the sputtering yield for a given ion increases rapidly, than continues to grow, albeit not so fast and finally slightly decreases. Another obvious regularity resides in the fact that ions

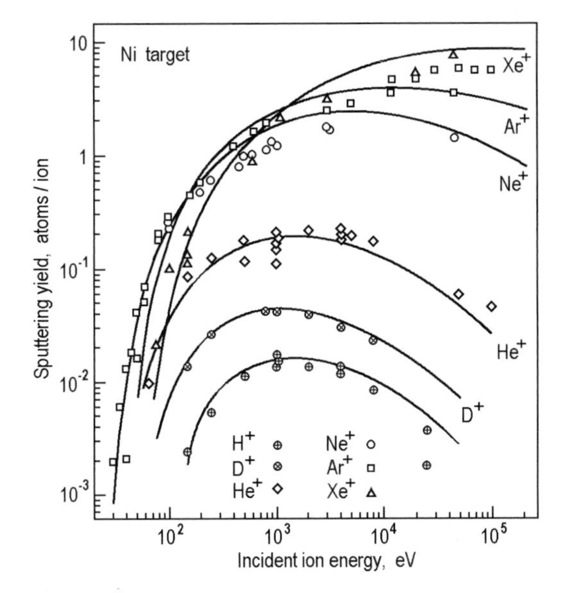

Fig. 6.12. Sputtering yield as a function of the primary ion energy for bombardment of a polycrystalline Ni sample by various ions at normal incidence (after Ziegler et al. [6.7])

with a greater mass demonstrate greater sputtering efficiency. The angular dependence of the sputtering yield is shown in Fig. 6.13. The shape of the dependence can be understood from the following consideration. When the incident direction deviates from the normal by an angle θ, the ion penetration depth is decreased by $\cos\theta$ and the scattering cascade becomes concentrated in the thinner near-surface region. As a consequence, the sputtering yield increases approximately as $1/\cos\theta$. The increase in sputtering yield takes place up to $\theta \sim$70–80° and then the sputtering yield falls down steeply. The fall is a consequence of the fact that most of the ions impinging at grazing angles are scattered off the surface; they do not penetrate into the solid, nor transfer sufficient energy to recoiled surface atoms. It should be noted that the example in Fig. 6.13 and the above discussion concerns isotropic polycrystalline samples. In the case of single-crystal samples, the effect of ion channelling becomes essential and the angular dependence of the sputtering yield is more complicated.

The application of ion sputtering for surface analysis concerns two major areas.

- First, particles sputtered from a surface carry information about the sample elemental composition. The detection and identification of the secondary ions and neutrals is the basis for *secondary ion mass spectroscopy (SIMS)*.

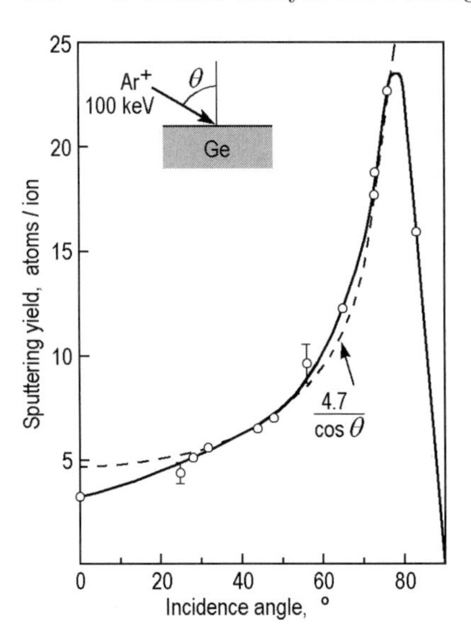

Fig. 6.13. Variation of the sputtering yield with ion incidence angle θ for 100 keV argon bombardment of germanium which amorphizes readily under ion impact (after Wilson et al. [6.8])

- Second, sputtering leads to controlled etching of the surface which can be used to peel off successive layers of the sample to strip the underlying layers. When coupled with facilities for chemical analysis (for example, AES or XPS), this provides a way to study the subsurface composition as a function of depth. This technique is conventionally referred to as *depth ion profiling.*

6.1.6 Ion-Induced Electronic Processes

When impinging on a solid surface, an ion interacts with the electronic system of the surface in different ways. The result of these interactions is a number of electronic processes of which the major are neutralization/deionization of the projectile, ionization of the target atom, and emission of an electron. Ion-induced electron emission can be ascribed roughly to one of two possible mechanisms:

- kinetic emission, and
- potential emission.

Kinetic Emission. This mechanism is associated with the transfer of a certain portion of the ion kinetic energy to the electron in the solid. As the ion projectile is much heavier than the electron, only a small portion of its energy is transferred to the electron and the projectile energy should be much greater than the electron binding energy to give appreciable ionization of a given electron level (for example, for ionization of K and L shells that are characterized by binding energies of ∼100–1000 eV, MeV protons are used). The kinetic energy and mass of the projectile are of foremost importance for kinetic emission, while "internal" projectile properties (for example, chemical configuration, charge state, excitation state) are not of essential concern. The target atom excited in the kinetic emission process can relax by emission of an x-ray photon or Auger electron. These processes are used for surface characterization in *particle-induced x-ray emission spectroscopy* and *particle-induced Auger electron spectroscopy*, respectively. Towards low projectile velocities, the kinetic emission clearly disappears.

Potential Emission. Potential emission processes do not require any kinetic energy of the projectile and dominate at relatively low projectile velocities. The processes are associated with electron exchange between an ion and a surface, when their electronic wave functions overlap. The main charge-exchange processes are as follows:

- One-electron processes:
  - resonance neutralization,
  - resonance ionization,
  - quasi-resonant neutralization;
- Two-electron processes:
  - Auger neutralization,
  - Auger de-excitation.

Resonance neutralization is illustrated in Fig. 6.14a and labeled RN. It resides in tunneling of the electron from the surface valence band into the ion. As resonance neutralization involves long-range delocalized valence electron states, it takes place at fairly large distances of ∼5 Å.

Resonance ionization (labeled RI in Fig. 6.14a) is the process inverse to resonance neutralization, i.e., the neutralized particle is reionized, when the electron occupying the state in the projectile, above the Fermi level of the surface, tunnels back to the surface.

Quasi-resonant neutralization (labeled QRN in Fig. 6.14a) takes place when an occupied core level of the surface lies energetically close to the hole state of the ion (for example, Pb 5d-levels (−24.8, −22.2 eV) and He 1s-level (−24.6 eV)). At very short ion–surface distances ($\lesssim$0.5 Å), quasi-resonant electron exchange between these levels occurs. Charge transfer is possible in both directions. The competition between quasi-resonant neutralization and ionization shows up in the spectacular effect of the oscillatory ion yield observed in ion scattering experiments (see Fig. 6.15).

Auger neutralization (labeled AN in Fig. 6.14b) takes place at distances of ∼1–2 Å and involves two electrons. The first electron tunnels into the ion well and falls down to fill the core level hole. The released energy is transferred to the second electron of the surface which is emitted as an Auger electron, provided that the ionization energy of the ion is more than twice the

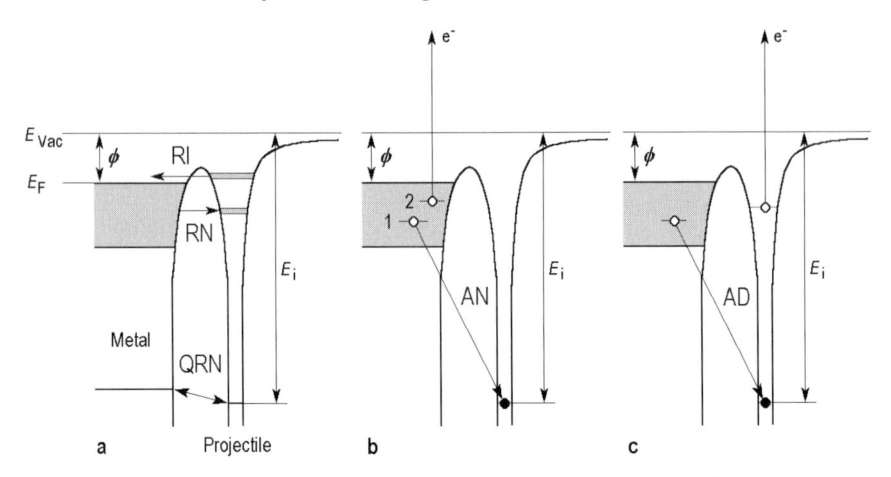

Fig. 6.14. Schematic energy level diagrams showing the charge exchange processes between a solid and an ion. (**a**) One-electron processes are represented by resonance neutralization (RN), resonance ionization (RI), and quasi-resonant neutralization (QRN). Two-electron processes are represented by (**b**) Auger neutralization (AN) and (**c**) Auger de-excitation (AD). E_F is the Fermi energy of the solid, E_Vac is the vacuum energy, $\phi = E_\mathrm{Vac} - E_\mathrm{F}$ is the work function of the solid, and E_i is the ionization energy of the ion (after Hagstrum [6.9])

work function of the solid ($E_\mathrm{i} > 2\phi$, see Fig. 6.14b). Analysis of the energy spectra of the electrons emitted in the neutralization process is used in a technique which is called *ion neutralization spectroscopy (INS)* or *metastable de-excitation spectroscopy (MDS)* and yields information on the density of states of the solid surfaces.

Auger de-excitation (labeled AD in Fig. 6.14c) is a rather special process, in which the projectile is not an ion but an excited neutral. The process involves the filling of the projectile core hole by an electron from the solid and the ejection of the electron from the excited state of the projectile. So far, Auger de-excitation has been found to be an important process only in the case of incident excited He atoms and then usually only with adsorbate covered surfaces [6.11].

The resultant efficiency of the charge-exchange processes occurring in ion scattering is characterized conventionally by the *ion survival probability* P^+. The value of P^+ may vary in a great range with variation in scattering conditions. For example, for scattering 1 keV $^4\mathrm{He}^+$ ions from metals $P^+_\mathrm{He} \simeq 10^{-3}$–$10^{-1}$ (depending on scattering geometry), while in the case of alkali ions, for example, $^7\mathrm{Li}^+$, the ion survival probability is almost unity, $P^+_\mathrm{Li} \simeq 1$. The difference can be qualitatively understood if one bears in mind that the condition for occurrence of the Auger neutralization, $E_\mathrm{i} > 2\phi$, is held for He^+ ion scattering, as $E_\mathrm{i}(\mathrm{He}) = 24.6\,\mathrm{eV}$ and typical values of the work function ϕ for metals are in the 4–5 eV range, but this is not held for Li^+ ions, as $E_\mathrm{i}(\mathrm{Li})$ is only 5.4 eV. However, although most charge-exchange mechanisms

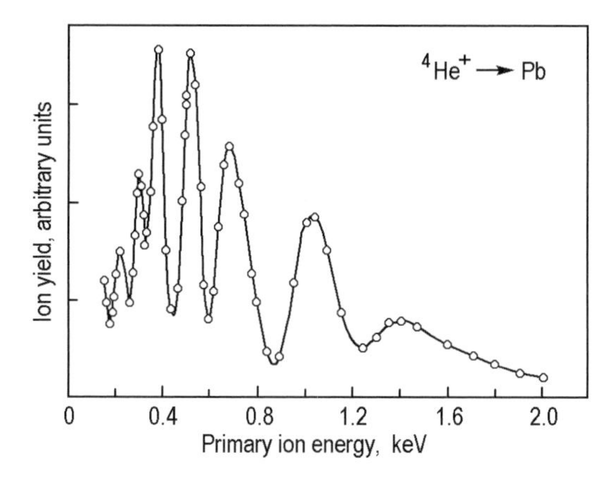

Fig. 6.15. Ion yield as a function of the primary ion energy for $^4He^+$ scattering from Pb targets. The oscillatory behavior of the ion yield is explained by *quasi-resonant neutralization processes* as follows. Electron exchange processes are dependent upon the interaction time, which is a function of the primary ion velocity (energy). For a sufficiently high ion velocity, the collision time is so short that the electron has just enough time to transfer to the incident ion; the ion then scatters away, neutralized. For a somewhat lower ion velocity, there will be sufficient time for the electron to transfer to the incident ion and back again to its parent atom, resulting in no neutralization. For progressively lower ion velocities, a succession of electron-exchange events takes place. The minima in the ion yield correspond to those ion velocities (collision times) where charge exchange results in scattering of a neutral. The period of oscillation is constant when the data are plotted against inverse velocity of the incident ion (after Erickson and Smith [6.10])

have been identified, it is not yet possible to have a universal evaluation of P^+ for the purposes of quantitative analysis.

6.2 Low-Energy Ion Scattering Spectroscopy

6.2.1 General Remarks: Merits and Problems

Low-energy ion scattering (LEIS) spectroscopy or, simply, *ion scattering spectroscopy (ISS)* employs ion energies from the range 1–20 keV and is well suited for laboratory surface studies. The extreme surface sensitivity of the technique arises from the large cross-section and shadow cone radius (both ~ 1 Å), which guarantees that most ions never get deeper than a few surface atomic layers. The major applications of LEIS concern evaluation of the surface composition and structure. Based on a binary collision model, the qualitative identification of the surface species is rather simple and straightforward (see (6.4) and Fig. 6.4), although the quantitative chemical analysis is highly conjectural. The shadow and blocking concepts constitute the basis for LEIS structural analysis which actually becomes quantitative in the impact-collision geometry.

Major problems of LEIS are associated with undesired effects as follows:

- Charge transfer between the ion and the surface (ion neutralization) is highly probable. The problem concerns the case when only scattered ions are analyzed and is especially tough for noble gas ions.
- The exact form of the ion–target interatomic potential, in general, is unknown, which results in uncertainty in the scattering cross-section and the size of the shadowing cone.
- Multiple scattering effects are essential.

Quantitative elemental analysis suffers from the first two drawbacks, as uncertainties in ion survival probabilities and scattering cross-section hamper the establishing of a universal relationship between surface concentration of species and the scattered ion beam intensity. In the case of detecting ions with a low ion survival probability, the primary ion dose should be large enough to ensure a sufficient scattered ion signal, which might cause surface damaging. The angular dependence of ion neutralization might plague the analysis of shadowing and blocking. The uncertainty in the size of the shadowing cone hinders accurate evaluation of the surface structure. Multiple scattering complicates considerably the data analysis.

6.2.2 Alkali Ion Scattering and Time-of-Flight Techniques

Consider now the ways to overcome the above listed difficulties. To solve the problem associated with the perturbing charge-exchange effects two approaches are in use. The first one is to suppress ion neutralization by replacing the noble gas ions (which demonstrate a very low ion survival probability) by alkali ions (for which the ion survival probability is almost unity). The technique is sometimes referred to as *Alkali ISS*. The advantage of using alkali ions is illustrated in Fig. 6.16. The second approach resides in detecting *all* scattered projectiles, i.e., both ions and neutrals (see Fig. 6.17). This goal is reached by using, instead of electrostatic analyzers (that detect only ions) the *time-of-flight (TOF)* technique (which provides an energy analysis of all particles irrespective of their charge state).

The elements of a typical TOF arrangement are shown in Fig. 6.18. The measurement cycle starts with the generation of the primary ion beam pulse by applying a voltage to the chopping plates and is repeated with a typical frequency of 20–100 kHz. The bunches of primary ions strike the target and after being scattered pass through a grounded drift tube about 1 m long and finally hit the detector. The flight time t is simply determined by the drift length L, projectile mass m_1, and energy E_1 as

$$t = L\sqrt{\frac{m_1}{2E_1}} , \tag{6.11}$$

thus, it is different for particles with different scattered energies E_1. The typical flight time is on the order of several µs. For instance, for $L = 1$ m and 1 keV $^4He^+$ ions the flight time is about 6.5 µs.

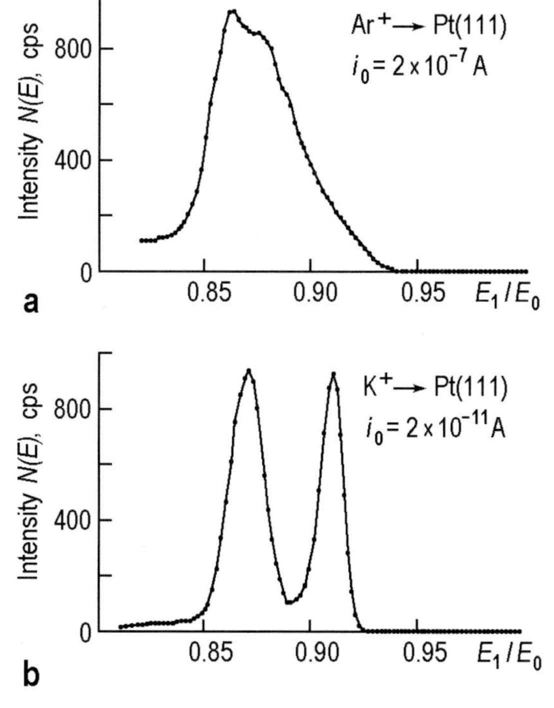

Fig. 6.16. Comparison of the energy spectra for (**a**) Ar^+ and (**b**) K^+ ions scattered from a Pt(111) surface. The two peaks are due to single and double scattering. Experimental parameters ($E_0 = 2\,\mathrm{keV}$, $\psi = 20°$, $\vartheta_1 = 47°$, $\varphi_{\mathrm{out}} = 30°$) are the same in both cases, except for the intensity of the primary beam which is four orders of magnitude greater for Ar^+ ions (after Niehus [6.12])

The most common detectors used in TOF experiments are microchannel plates and secondary electron multipliers, as they are sensitive to both ions and fast neutrals. Neutrals with energies above 1 keV are detected with the same efficiency as ions. The signal from the detector is amplified and the TOF spectrum is generated by the fast digitizer and histogramming unit and is acquired with the multichannel analyzer. The TOF spectrum $F(t)$ can be converted into the energy spectrum $N(E)$ according to

$$N(E) = \frac{F(t)t^3}{m_1 L^2} \,. \tag{6.12}$$

The energy resolution $\Delta E/E$ is defined primarily by the duration of the ion pulse and is ~1% for a typical value of about 50 ns.

Two additional merits of the TOF technique are worth mentioning. First, if desired, the ions may be separated from the neutrals by applying an appropriate potential to the acceleration tube located at the beginning of the flight path. This ability extends the range of possible applications of the tech-

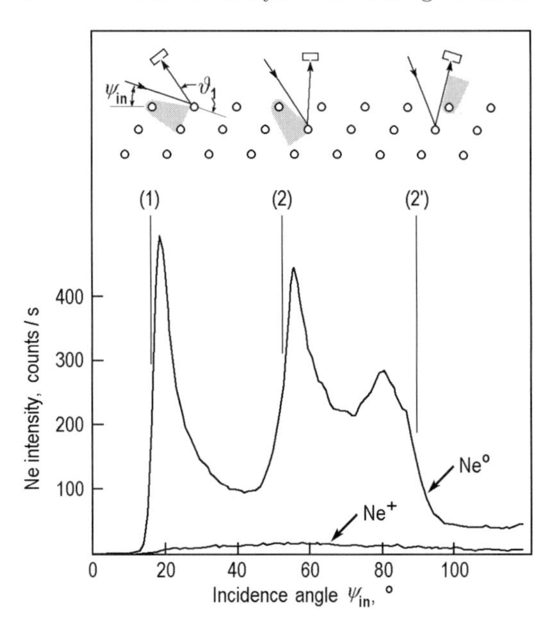

Fig. 6.17. Comparison of the ISS spectra (scattered particle yield versus incidence angle ψ_{in}) for Ne^0 and Ne^+ particles backscattered from a Pt(111) surface. In the experiment the primary energy was 2 keV and the scattering angle ϑ was fixed at 165°. One can see that data obtained with Ne^+ ions suffer from high neutralization during the scattering process. The inset in the upper part of the figure illustrates the origin of the features seen in the Ne^0 spectrum (after Niehus [6.13])

nique (for example, the charge-exchange processes can be studied). Second, the TOF technique has a very high efficiency in spectra recording, as all ions and neutrals are registered in parallel with the multichannel technique. Consequently, the primary dose required to measure the spectrum can be greatly reduced (down to 10^{11} ions/cm^2), making the ISS method virtually non-destructive [6.2].

6.2.3 Quantitative Structural Analysis in Impact-Collision Geometry

In principle, the shadow and blocking concept is applicable for the determination of surface structures in any scattering geometry. However, in general, most results remain in a qualitative state due to multiple scattering effects and uncertainty in the interatomic potential. Fortunately, multiple scattering effects are eliminated and quantitative structural analysis becomes relatively simple and straightforward when the 180°-scattering geometry is used. This special type of ISS is called *impact-collision ion scattering spectroscopy (ICISS)* [6.14]. In the most accurate, almost ideal, way, this mode is realized in the *coaxial impact-collision ion scattering spectroscopy (CAICISS)* tech-

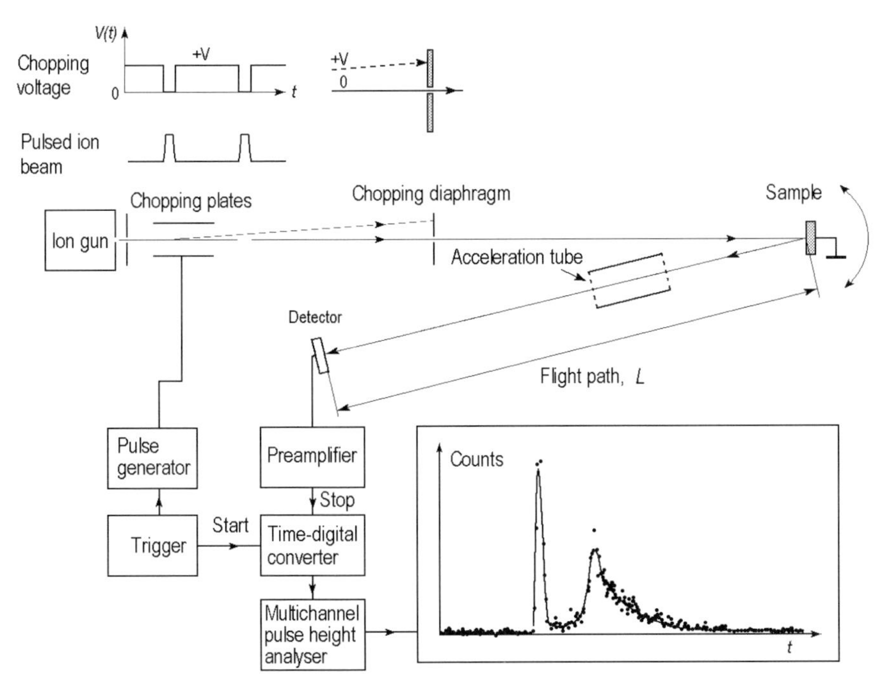

Fig. 6.18. Schematic of the time-of-flight (TOF) set-up. The primary ion pulse is produced by a square-wave voltage at the chopping plates. The scattered ions and neutrals after passing over the flight path are detected by a microchannel plate detector. The detector signal is amplified and after the time-to-pulse height conversion registered with a multichannel analyzer. The acceleration tube is used to separate the ion fraction, if desired

nique [6.15] in which the pulsed primary ion beam passes through a small hole in the center of the microchannel detector. With a microchannel plate diameter of 32 mm, a hole diameter of 3.5 mm, and microchannel plate–sample distance of 60 cm, the half-angle of the detecting cone is 1.5° and the deviation from the exact 180° scattering direction is only 0.2°.

The basic idea of the ICISS concept is to detect only 180°-scattered ions, i.e., those that have undergone impact (head-on) collision with target atoms. As an example, consider the scattering on the chain of atoms at a crystal surface as shown in Fig. 6.19. When the incidence angle α_{in} is below a critical value, head-on collisions are impossible due to the shadowing effect and primary ions are forward-scattered by a series of small-angle deflections, hence, the ICISS signal is zero. Figure 6.19 illustrates the situation for the impact collision just at critical angle α_{c}, namely, when the edge of the shadow cone of each atom passes the center of its neighboring atom. (Recall that impact collision implies practically zero impact parameter.) One can see in Fig. 6.19 that the shadow cone radius R_{s} at a distance L is simply related to the interatomic distance d as

$$R_s = d \sin \alpha_c \quad \text{and} \quad L = d \cos \alpha_c \; . \tag{6.13}$$

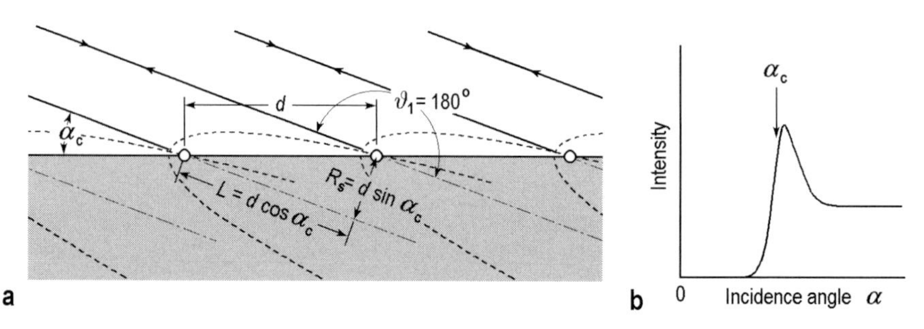

Fig. 6.19. Schematic of the shadow cone concept in the 180°-scattering geometry used in ICISS. (**a**) Situation for scattering at the critical incidence angle α_c. (**b**) Intensity of scattered particles as a function of incidence angle variation (after Aono et al. [6.14])

Thus, by measuring the critical angles at surfaces with well-established lateral interatomic distances the shape of the shadow cone can be determined experimentally. As an example, Fig. 6.20 shows the experimental shadow cone for 1 keV He^+ scattering at Ti. In the experiment, the (111) and (100) surfaces of TiC were taken, since both surfaces have the 1×1 structure and the Ti-Ti distances along surface atomic rows are readily calculated from the bulk lattice constant.

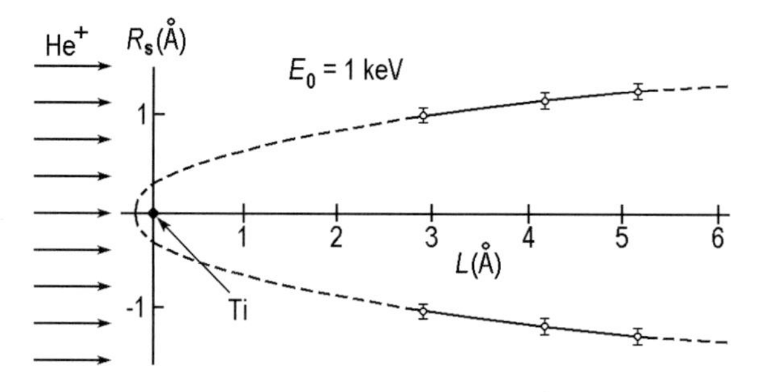

Fig. 6.20. Shadow cone for 1 keV He^+ scattering at Ti determined using ICISS measurements of critical angles along selected azimuths on TiC(100)1×1 and TiC(111)1×1 surfaces (after Aono et al. [6.14])

When the shape of the shadow cone is known, quantitative analysis of the surface atomic geometry becomes possible. As an example, consider two neighboring atoms (labeled A and B in Fig. 6.21a), which may be bond to each other, and suppose that we measure the intensity of particles scattered from atom B as a function of incidence angle α in the ICISS geometry (see

Fig. 6.21b). The intensity is zero around $\alpha = \alpha_0$ due to shadowing of atom B by atom A, it grows at critical angles α_{c1} and α_{c2}, goes through maxima due to the focusing effect, and becomes almost constant at large deviations from α_0. One can see that the bond direction of atoms A and B is given by α_0, while the difference between α_{c1} and α_{c2}, $\Delta\alpha_c$, corresponds to their interatomic distance d (the greater d, the smaller $\Delta\alpha_c$). Thus, by measuring the intensity curves like in Fig. 6.21b experimentally, one can determine both the bond directions and interatomic distances, i.e., conduct a quantitative structural analysis of the surfaces.

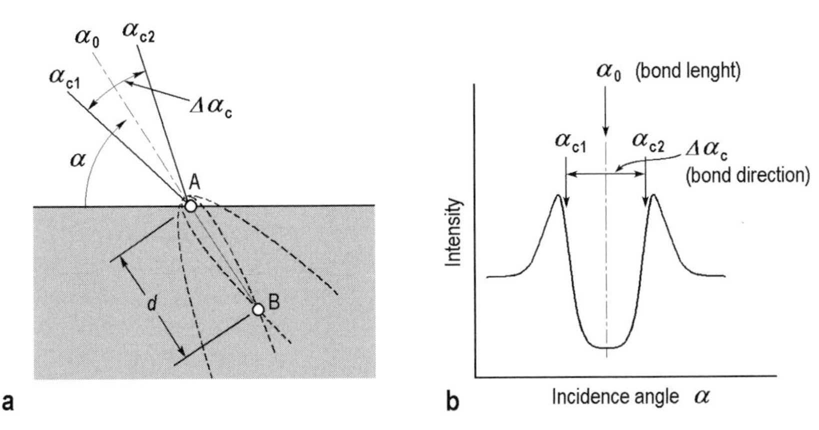

Fig. 6.21. Schematic diagram showing how the bond direction and interatomic spacing of two neighboring atoms A and B can be determined at the same time by CAICISS. (**a**) Scattering geometry. (**b**) Intensity of scattered particles versus incidence angle (after Aono et al. [6.16])

An experimental example, the determination of the Si-Si dimer length, is shown in Fig. 6.22. The principle of evaluation is as discussed above, though the azimuthal (not the polar) angle is varied and the situation is complicated by the presence of two orthogonal domains in the Si(100)2×1/1×2 surface. From the experimental value of the critical angle $\delta = 13 \pm 1°$, the intradimer atomic distance parallel to the surface was found to be 2.4±0.1 Å.

6.3 Rutherford Backscattering and Medium-Energy Ion Scattering Spectroscopy

6.3.1 General Remarks

High-energy ion scattering (HEIS) spectroscopy, more often referred to as *Rutherford backscattering spectroscopy (RBS)*, conventionally employs He^+

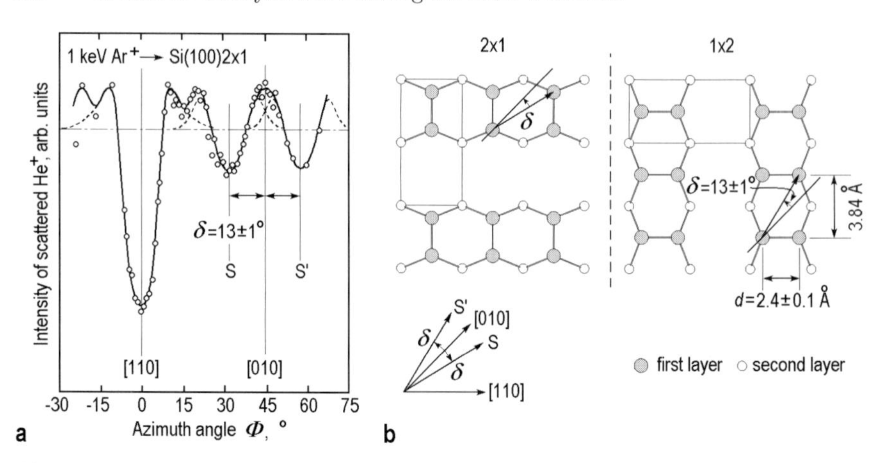

Fig. 6.22. Determination of the Si-Si intradimer distance using ICISS. (**a**) The intensity of 1 keV He^+ ions scattered from Si(100) in the ICISS geometry as a function of azimuth angle Φ. The polar angle $\alpha = 4°$. The intensity drops at $\Phi = 0°$, 32°, and 58° are due to shadowing effects. (**b**) Schematic diagram showing the directions for critical shadow angles on 1×2 and 2×1 Si(001) domains (after Aono et al. [6.17])

ions or protons with typical energies of ~1–3 MeV. The typical range of energies used in *medium-energy ion scattering (MEIS)* experiments is between 50 and 500 keV. Most physical principles underlying the two techniques are the same (the reason for discussing RBS and MEIS together), though their experimental set-ups, possibilities, and limitations often differ (to be outlined in the appropriate places in the text).

The basic features of ion spectroscopy in the hundreds of keV to MeV range are as follows.

- The scattering cross-section is small and, hence, the shadow cone is narrow compared to interatomic distances (for example, for He^+ ions incident on Si the shadow cone radius $R_s \cong 0.2$ Å at 100 keV and $R_s \cong 0.09$ Å at 1 MeV).
- The neutralization probability is low.
- The role of electron screening on the scattering potential is weak and a simple Coulomb interaction may be used. The approximation is better for higher ion energies, i.e., more appropriate for RBS, than for MEIS.
- Multiple-scattering effects are negligible (again, rather for RBS, than for MEIS).

As the scattering cross-section for high-energy ions is small and the ions penetrate deep into the bulk, it might be thought that they are a likely tool to probe the bulk rather than the surface. Indeed, probing the structure and composition as a function of depth is an important application of RBS (with a greater probing depth range but poorer depth resolution) and MEIS (conversely, with smaller depths but better resolution). However, the techniques become essentially surface-sensitive, when the primary ion beam is aligned with a major crystal direction and, in this way, the contribution of the bulk is minimized due to the channeling effect.

6.3.2 Surface Peak

Consider the situation when a parallel flux of primary ions is impinging on the crystal surface along a direction for which channeling occurs (Fig. 6.23a). In the case of an ideal static lattice, ions will "see" only the top-layer atoms, while deeper atoms along the aligned atomic rows will be shielded and, hence, cannot contribute to backscattering. The surface backscattered yield is expressed conventionally as the number of visible atoms per row and in the ideal case would be equal to one. However, for non-zero temperature the thermally displaced atoms expose the deeper atoms to the ion beam. Note that thermal vibration of atoms in a solid is not correlated and is much slower than ion velocities. The backscattering yield I thus includes contribution from the deeper layers:

$$I = \sum_i P_i \,. \tag{6.14}$$

Here P_i is the individual close-encounter probability of the ith atom in the atomic row. By definition $P_1 = 1$. The greater the vibrational amplitude compared to the shadow cone radius, the greater the contribution of the deeper atoms and, hence, the backscattering yield (see Fig. 6.23b).

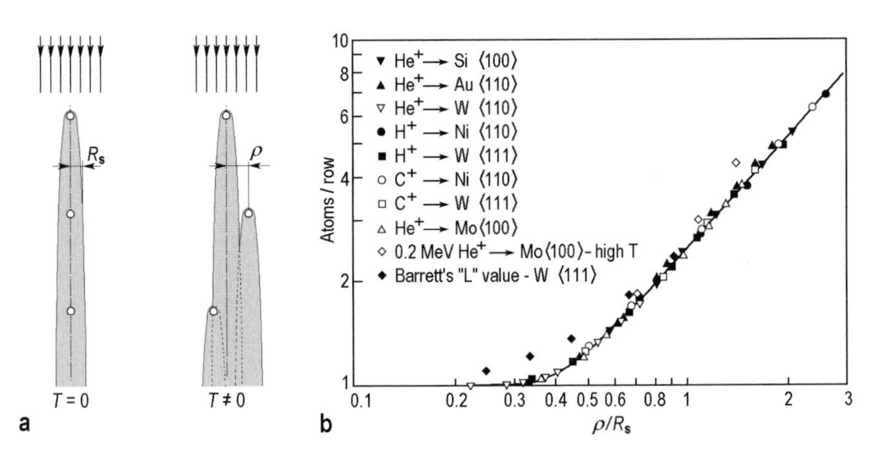

Fig. 6.23. (**a**) Schematic illustration of the shadow effect along an atomic row at zero and non-zero temperature. R_s indicates the radius of the shadow cone at interatomic distance, ρ is the amplitude of the thermal atomic vibrations. (**b**) Number of atoms per row visible to the incident ion beam as a function of ρ/R_s calculated for a large set of ion–target combinations (after Stensgaard et al. [6.18])

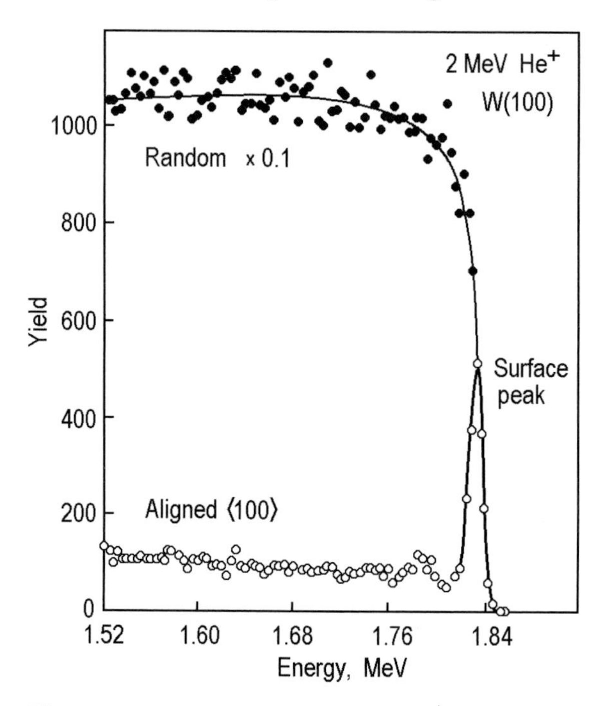

Fig. 6.24. RBS spectra for 2 MeV He^+ ions incident on W along the ⟨100⟩ axial direction ("aligned" spectrum, open circles) and in the non-channeling direction ("random" spectrum, closed circles) (after Feldman et al. [6.19])

The classical example of the experimental "aligned" backscattering spectrum for 2 MeV He^+ ions incident on W(100) along the ⟨100⟩ axial direction is shown in Fig. 6.24. The backscattering surface yield manifests itself in the spectrum as a *surface peak* at the energy determined by (6.4). Note that the experimental surface peak combines, in fact, the contributions from several outermost layers owing to the finite depth resolution of the experimental system. To the lower energies from the surface peak, one can see a plateau due to the ions that have been scattered deeper in the bulk and on their way forward and back have lost an amount of energy according to the stopping power of the solid. Because of channeling, the yield of these ions is low, ~0.01 of that when the incidence angle does not coincide with the channeling direction (compare the "aligned" and "random" spectra in Fig. 6.24).

Analysis of the surface peak intensity provides information on the surface structure. A qualitative overview of some simple examples is represented in Fig. 6.25. The dashed spectra correspond to the backscattering yield from a crystal with an ideal surface (Fig. 6.25a). In the absence of thermal vibrations, the surface peak intensity in this ideal case corresponds to one atom per row. Figure 6.25b illustrates the case of a reconstructed surface, when the top atoms are laterally displaced and, hence, the shadowing of the second-layer atoms is no longer complete. This leads to an increase of surface peak intensity (at maximum, twice that of the ideal crystal). In the case of a relaxed surface, i.e., when the top atomic layer is displaced normal to the surface, (Fig. 6.25c) the surface peak intensity at normal incidence is equivalent to one monolayer. The relaxation is revealed by using non-normal incidence, so that shadow cones established by the surface atoms are not aligned with atomic rows in the bulk. An adsorbate layer (Fig. 6.25c) manifests itself by the appearance of a new peak. The atomic mass sensitivity of ion scattering allows the discrimination between substrate and adsorbate. If adsorbate atoms reside atop substrate atoms (as shown in Fig. 6.25c), the intensity of the substrate surface peak is greatly reduced.

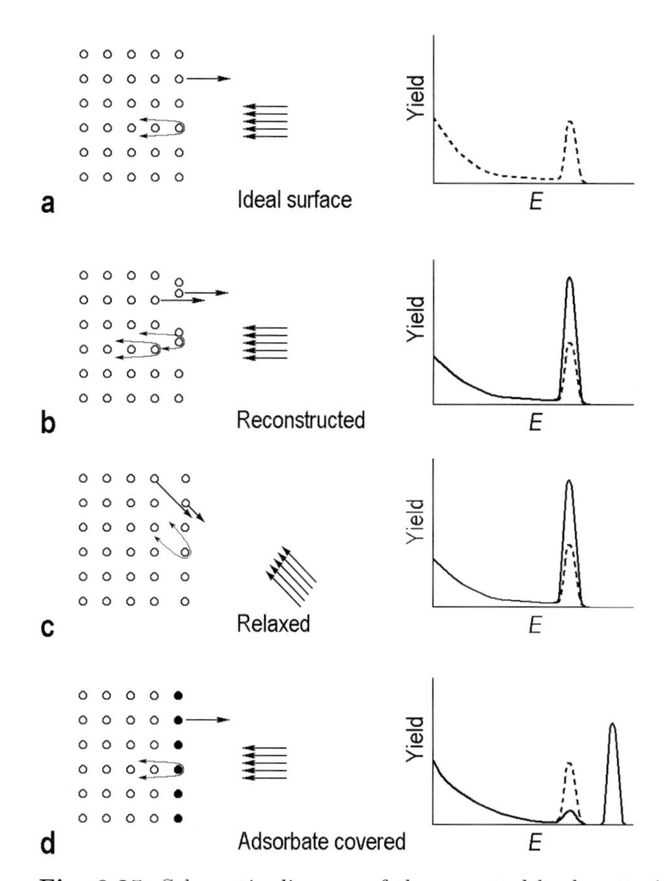

Fig. 6.25. Schematic diagram of the expected backscattering signal from various surface structures: (**a**) ideal crystal surface; (**b**) surface with lateral reconstruction; (**c**) surface relaxed in the normal direction; (**d**) surface with an adsorbate layer (after Feldman et al. [6.6])

Quantitative determination of the surface structure parameters is based conventionally on measurements of the surface peak angular dependence. As an example, consider the evaluation of surface relaxation, i.e., modification of the interlayer spacing between the top atomic layers (see Sects. 8.1 and 8.2). The method is illustrated schematically in Fig. 6.26. The primary beam is aligned so that it can only hit the atoms in the first two atomic layers, while deeper lying atoms are shadowed. Hence, only the two top layers contribute to the intensity of the surface peak. The ions scattered from the top atoms can leave the surface in any arbitrary direction. In contrast, for ions scattered from the second-layer atoms some directions are blocked by atoms of the first layer. In these directions, minima in the backscattering yield (so-called, *surface blocking minima*) are observed. The direction at a which surface blocking

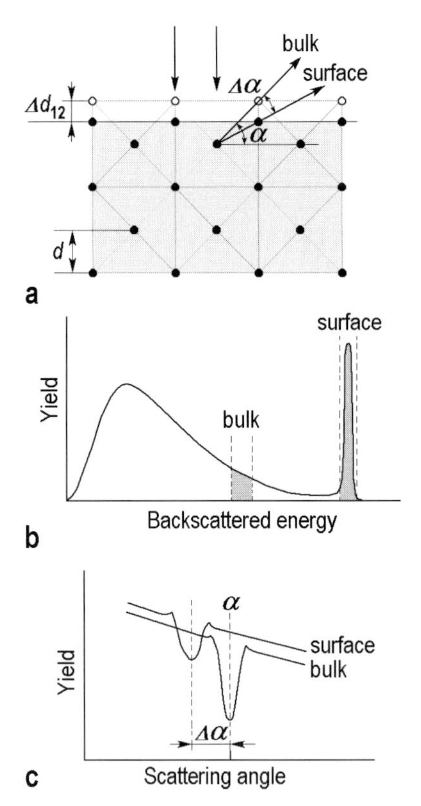

Fig. 6.26. Schematic illustration of the method to study surface relaxation. In scheme (**a**) actual positions of atoms are shown by closed circles, while open circles show the positions of the atoms in the unrelaxed ideal first atomic layer. Due to relaxation by Δd_{12}, the surface blocking minimum direction is tilted by $\Delta\alpha$ with respect to the bulk axis. In the experiment, $\Delta\alpha$ is measured by plotting the surface and bulk scattering intensities (**b**) as a function of scattering angle (**c**). Δd_{12} is calculated from $\Delta\alpha$ according to (6.15) (after Turkenberg et al. [6.20])

minimum occurs coincides with the axes passing through the atoms of the first and second layers (Fig. 6.26a). The change in the distance between the two top layers (surface relaxation) manifests itself in the tilt of the surface blocking minimum from the position for an ideal unrelaxed surface. The latter can be determined experimentally by measuring the yield of the ions with a lower energy, corresponding to bulk scattering (Fig. 6.26b and c). The blocking minimum in this case arises along the bulk crystallographic direction. If the angle of the bulk axis is α and the tilt angle is $\Delta\alpha$, the relaxation Δd_{12} can be readily found from simple trigonometry (see Fig. 6.26a) as

$$\frac{\Delta d_{12}}{d} = \frac{\tan(\alpha + \Delta\alpha)}{\tan\alpha} - 1 , \tag{6.15}$$

where d is the bulk interlayer spacing. $\Delta\alpha$ and Δd_{12} are negative for contraction and positive for expansion relaxations.

A spectacular example of surface relaxation determination is shown in Fig. 6.27 for a clean Pb(110)1×1 surface. In the experiment, the beam of 97.6 keV protons was aligned along the $[1\bar{1}1]$ direction of the Pb crystal. The sample was cooled to 29 K. As a result, the shadowing of the atoms below the first two layers was essentially complete (the yield equals one atom/row in the blocking minima and otherwise two atoms/row). The observed tilt of the surface blocking minimum to a smaller scattering angle by 4.9° corresponds to the contraction of the outer interlayer spacing by 17.2±0.5%. Detailed Monte Carlo simulations (the best fit is shown by the solid line through the

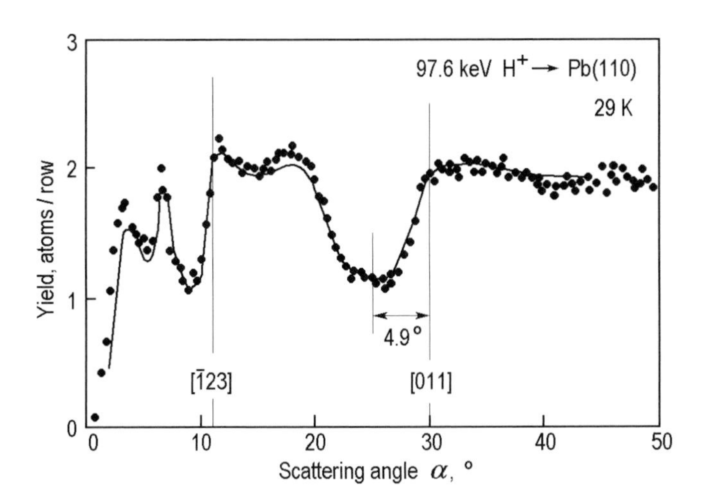

Fig. 6.27. Angular dependence of the surface backscattering peak from a Pb(110)1×1 surface at 29 K. The blocking minimum at 25.1° is shifted from the bulk [011] direction due to the inward relaxation of the outermost atomic layer of the crystal. The best fit of Monte Carlo simulations (solid curve) is obtained for $\Delta d_{12}/d = -(17.2 \pm 0.5)\%$ and $\Delta d_{23}/d = +(8.0 \pm 2.0)\%$ (after Frenken et al. [6.21])

experimental points) reveal that, in addition to the contraction of the first layer, the second layer is expanded by 8.0±2.0%.

6.3.3 Thin Film Analysis

A light energetic ion can penetrate into a solid and backscatter from a deep lying atom. The amount of energy lost by an ion in this process is the sum of two contributions:

- First, continuous loss of energy on the inward and exit paths of the ion in the solid. The rate of energy loss, called the stopping power, $\mathrm{d}E/\mathrm{d}x$, is tabulated for most materials. Hence, the energy scale can be transferred into the depth scale.
- Second, the loss of energy at the scattering event. The amount of this loss is determined by the mass of the scattering atom according to (6.4).

Hence, the backscattering spectrum of a sample provides information on its chemical composition as a function of depth. As an example, consider the case of a thin film on a substrate (Fig. 6.28). The film of thickness d shows up as a plateau of energy width ΔE in the spectrum. The right edge of the plateau corresponds to ions elastically scattered from the surface and having

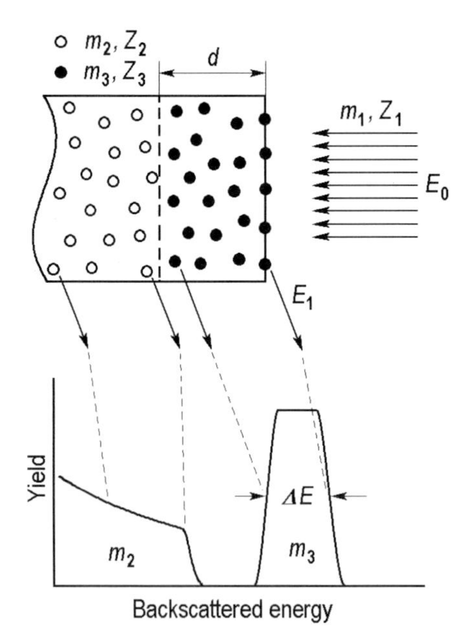

Fig. 6.28. Schematic diagram of the energy spectrum of ions (m_1, Z_1, E_0) scattered from a sample composed of a substrate (m_2, Z_2) and a film (m_3, Z_3) of thickness d. For simplicity, both film and substrate are assumed to be amorphous to neglect the structural effects. (after Feldman et al. [6.6])

an energy E_1 determined by (6.4). The left edge is due to ions scattered from the film atoms located at the film–substrate interface. Scattering from the substrate atoms at the interface corresponds to the onset of the substrate signal.

The depth resolution in the experimental spectra is determined basically by the resolution of the energy analyzer. In RBS experiments, surface-barrier detectors are used with a typical resolution of 15 keV, which results in a depth resolution of 100 Å or more. Electrostatic analyzers used in MEIS experiments provide an energy resolution of 500 eV or better at 100 keV, consequently, a depth resolution of 3–10 Å can be achieved.

To obtain information on the crystal structure, backscattering experiments are conducted in a channeling mode. In the case of a perfect crystal or high-quality epitaxial film, the bulk backscattering yield in the "aligned" spectrum (the so-called minimum yield, χ_{min}) is low, typically ~1–4% of the random yield. In the other extreme case of an amorphous or finely polycrystalline material, $\chi_{min} = 100\%$. In other cases, i.e., when the crystal or the film contain some defects, the value of χ_{min} is somewhere in between and serves as a measure of the crystallinity.

6.4 Elastic Recoil Detection Analysis

In *elastic recoil detection analysis (ERDA)*, which is also referred to as *direct recoil spectroscopy (DRS)*, the target particles, which have been recoiled by primary ions, are energy analyzed. The technique uses the same physics of binary elastic collisions as ion scattering spectroscopy. Moreover, both techniques basically utilize the same experimental equipment and are even sometimes tied together under the term *scattering and recoiling spectroscopy (SARS)*.

The peculiar features of ERDA are as follows.

- ERDA uses a forward-scattering geometry with glancing incidence of the primary ion beam, since the recoil angle $\vartheta_2 < 90°$ (see Fig. 6.1 and Fig. 6.3)
- For useful measurements the projectile ion should have a higher mass than the target atom, since the cross-section for recoil increases with the target-to-projectile mass ratio as $\mathrm{d}\sigma/\mathrm{d}\Omega \propto (1 + m_1/m_2)^2$.

ERDA nicely complements ISS techniques, since it is sensitive to exactly those atoms for which ISS gives only poor results, namely, *light atoms.* In particular, the ability of recoil analysis to detect hydrogen is especially valuable. Hydrogen is an important element in science and technology, but the analysis for hydrogen is difficult or impossible by most conventional analytical techniques: H has no Auger transition and is characterized by small cross-sections for electron scattering, x-ray scattering and photoionization.

Like ISS techniques, ERDA mainly concerns the study of the surface composition and structure. Relation (6.5) and Fig. 6.3 provide the basis for the elemental analysis. When high-energy (MeV) primary ions are used (for which the recoil cross-sections are well established), the absolute determination of the surface coverages is possible. In the case of low-energy (keV) ions, relative coverage values are obtained, which can be converted into absolute values by using appropriate calibration on the samples with a known composition.

An example of the application of ERDA for quantitative structural analysis is presented in Fig. 6.29. The object of the investigation is the Si(100)2×1 surface after exposure to atomic hydrogen at 350°. Under these conditions, H atoms are known to terminate the dangling bonds of Si-Si dimers. The goal of the study is to determine the H–Si–Si bond angle (labeled β in Fig. 6.29a). In the experiment, the Si(100)2×1-H surface was irradiated by 593 eV Ne^+ at an incidence angle of 20° and the recoiled hydrogen was detected. In the recoil energy spectrum (Fig. 6.29b), two peaks are seen. The peak at the higher energy corresponds to *direct recoil.* The peak at the lower energy is due to so-called *surface recoil,* i.e., due to the recoiled hydrogen atoms, which have been scattered on the Si atoms. The dependence of the energy of the direct recoil H atoms versus recoil angle is described by (6.5) and shown in Fig. 6.29c by the solid line. The curves for the surface recoil, calculated within elastic binary collision model, are shown as dashed lines. The corresponding H–Si–Si bond angle β, used as a parameter in the calculations, is indicated for each curve. One can see that the energies of the surface recoil H atoms remain nearly constant, in contrast to the energies of the direct recoil H atoms, which decrease with recoil angle. Thus, two recoiling process can be clearly separated. The experimental peak energies for direct recoil are denoted by circles and those for surface recoil by triangles. Comparison of the experimental data with calculations reveal that H–Si–Si bond angle lies in the range from 108° to 111°.

In the above example, the quantitative structural evaluation uses purely trigonometric considerations. However, a more general approach in studying surface structures by ERDA relies on the use of the shadow and blocking concept, similar to that described above for LEIS. Recall that in this case one should know the exact form of the interaction potential for conducting the model calculations of the recoil yield angular dependence.

6.5 Secondary Ion Mass Spectroscopy

Bombardment of a sample surface with a primary ion beam followed by mass spectrometry of the emitted secondary species constitutes *secondary ion mass spectroscopy (SIMS).* The Cs^+, O_2^+, and Ar^+ ions with energy between 1 and 30 keV are typically used for bombardment. The sputtered particles are single atoms and clusters of the sample material. They carry negative, positive, and neutral charges and have kinetic energies that range from zero to hundreds of eV. Composition analysis by SIMS relies on the fact that only those types of atoms will be emitted which are present in the near-surface region. Moreover,

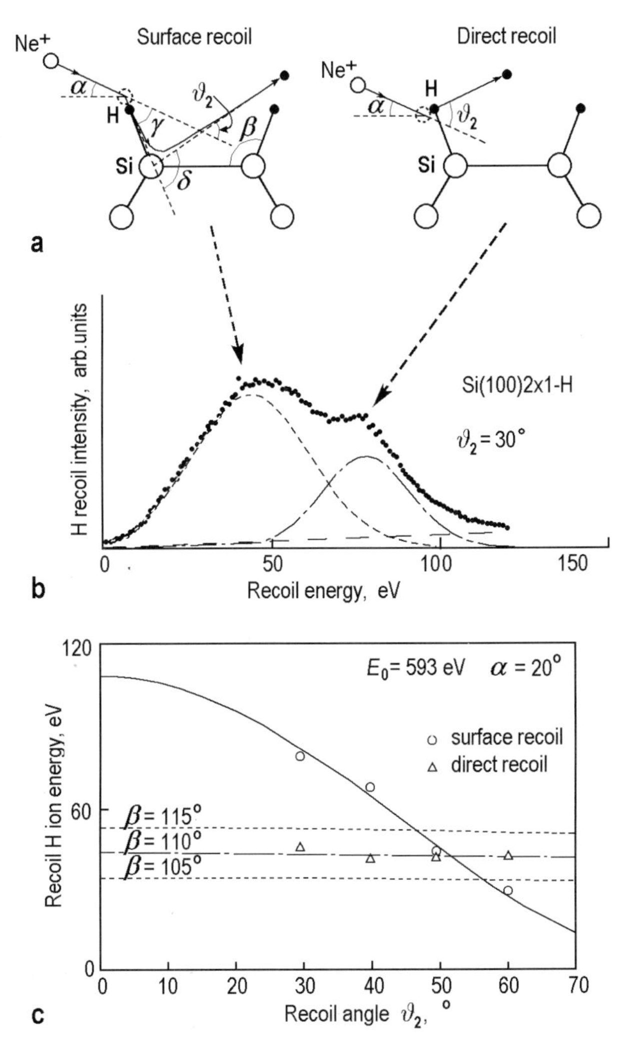

Fig. 6.29. Evaluation of the H–Si–Si bond angle β in the monohydride Si(100)2×1-H phase by ERDA. (**a**) Schematic diagrams illustrating the processes of surface recoil and direct recoil, when a primary Ne^+ ion hits the H atom terminating the dangling bond in the Si–Si dimer. (**b**) Experimental H recoil spectrum acquired with 593 eV Ne^+ incident at $\alpha = 20°$ and recoiled H detected at $\vartheta_2 = 30°$. The correspondence of the two peaks to the two recoil processes is indicated. (**c**) Comparison between the experimental data and the curves calculated within the elastic binary collision model. The open circles denote the peak energies for direct recoil and the triangles denote the peak energies for surface recoil (after Shoji et al. [6.22])

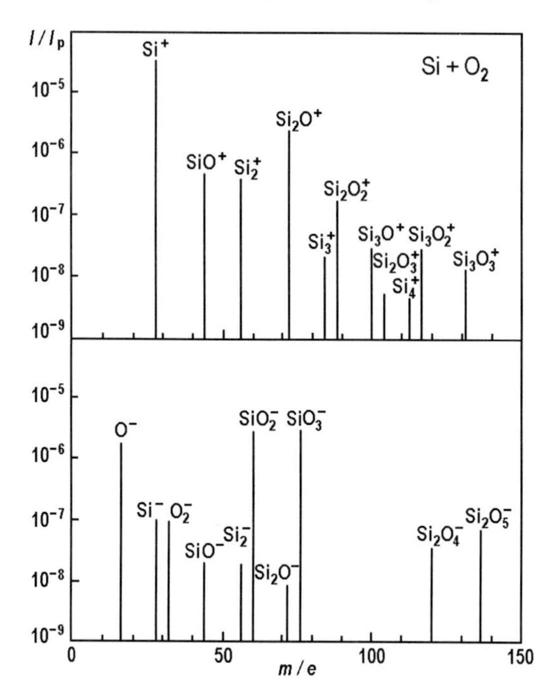

Fig. 6.30. SIMS spectra for positive (upper panel) and negative (lower panel) ions of a Si(111) surface after oxygen exposure of 100 L at room temperature. The primary ion flux is 4×10^{-9} A/cm^2; the total recording time is 10^3 s (after Benninghoven et al. [6.23])

the atoms which are emitted as a cluster are only those which have occupied adjacent sites on the surface. An example of an experimental SIMS spectrum is shown in Fig. 6.30.

Secondary ion mass spectroscopy provides a principal possibility of quantitative analysis, i.e., the number of atoms of a certain species present in the near-surface region can be calculated from the measured ion current. The general equation for the positive ion current I_{i}^{+} from an atomic species i in a sample j can be written as

$$I_i^+ = I_{\mathrm{p}} C_i S_{i,j} \alpha_{i,j}^+ T \,, \tag{6.16}$$

where

I_{p} is the primary ion current,
C_i is the relative volume concentration of species i,
$S_{i,j}$ is the sputtering yield (the number of sputtered atoms of species i per incident ion),
$\alpha_{i,j}^+$ is the ion yield (the fraction of sputtered atoms of species i that become ionized),
T characterizes the overall transmission of the mass spectrometer.

The combined subscript i, j for a given parameter emphasizes the fact that this parameter depends on the nature of both the sputtered species and the sample. Naturally, the similar relation holds for negative ions.

Being the most critical parameter in SIMS quantitative analysis, the *ion yield* is controlled primarily by the species, sample matrix, and type of primary ions. The general tendency, which is held for many species, is that the positive ion yield is greater for species with a lower ionization potential and the negative ion yield is greater for species with a higher electron affinity. Hence, alkali metals exhibit high positive ion yield (for example, K^+, Rb^+, Cs^+), while halogens exhibit high negative ion yield (for example, F^-, Cl^-, Br^-). The most spectacular example of the matrix effect is the great increase in the metal ion yield on replacing the pure metal by its oxide (for example, the positive ion yield for oxidized Cr is a factor of 1000 higher than that for clean Cr, though for oxidized Si and clean Si this factor is 70 [6.24]). The type of incident ions is also essential. For example, with oxygen primary ions the yield of the positive ions is enhanced, while cesium ions enhance the yield of the negative ions. The variations can range up to four orders of magnitude. Therefore, practical quantitative SIMS analysis is usually carried out after experimental determination of the ion yield from reference samples or standards and often by using relative sensitivity factors.

The are two principal modes of operation in SIMS, namely,

- static SIMS, and
- dynamic SIMS.

In *static SIMS*, very low primary ion current densities of 10^{-10}–10^{-9} A/cm^2 are used and, hence, the sputtering rate is also extremely low, typically of a fraction of a monolayer per hour. In this instance, destruction of the surface is minor and the method is used to study the surface composition as well as adsorption and surface chemical reactions. Though the secondary ion current density in this mode is low (due to the low sputtering yield), nevertheless the detection limit of static SIMS can be as low as 10^{-8} ML in the most favorable cases.

In *dynamic SIMS*, the primary ion current densities are 10^{-5}–10^{-4} A/cm^2 and the sputtering rate is high (typically, several monolayers per second). Thus, continuous analysis allows depth profiling, i.e., the determination of the concentration versus depth for one or more of the species present in the sample (Fig. 6.31). The detection limit is typically between 10^{12} and 10^{16} atoms/cm^2. The achievable SIMS depth resolution is affected by basic sputtering processes (for example, atomic mixing, selective sputtering, microroughening of the crater bottom) as well as by instrumental factors (for example, sputtering rate, uniformity of the primary ion beam, crater edge effects).

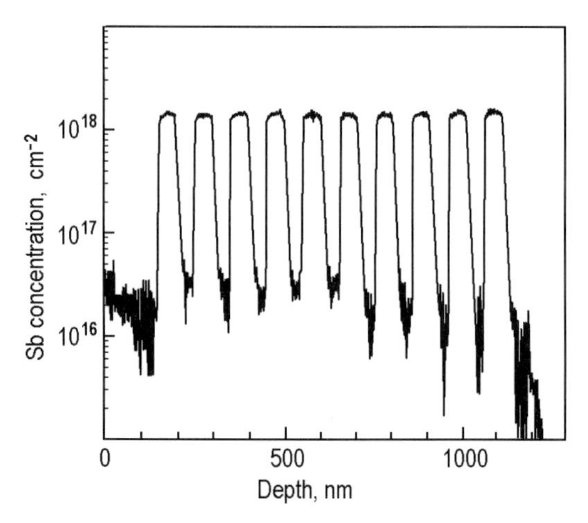

Fig. 6.31. SIMS depth profile of Sb in the modulation-doped silicon multilayer structure grown by molecular beam epitaxy (after Casel et al. [6.25])

Problems

6.1 Are diffraction effects essential for ion interaction with a solid surface when the ion has an energy in the range ~1 keV to ~1 MeV (typical ion energies employed in ion spectroscopies)? Estimate at what energy the de Broglie wavelength of the ion is comparable to the interatomic distance in a solid (~3 Å). Does this energy fall into the above range? As an example, consider $^4He^+$ ions.

6.2 $^4He^+$ ions with an initial energy of $E_0 = 2\,\mathrm{MeV}$ hit a target made of unknown material. After elastic impact collision ($\vartheta_1 = 180°$) ions return with a final energy $E_1 = 1.1\,\mathrm{MeV}$. What is the atomic weight of the unknown material? What element is it?

6.3 A 2 MeV $^4He^+$ ion impinges at normal incidence on a Si(100) target and backscatters ($\vartheta_1 = 180°$) at a depth of 2000 Å. Estimate the energy of the ion when it leaves the sample. For estimation, consider the stopping power for the He ion in silicon to be $60\,\mathrm{eV}/(10^{15}\,\mathrm{atoms/cm^2})$ independent of ion energy. The atomic weight of Si is 28.

6.4 Hydrogen (^1H) and oxygen (^{16}O) are recoiled by 1 keV $^4He^+$ ions at a recoil angle of 45°. Compare the energies and velocities of the recoiled particles. Specify which type of analyzer, time-of-flight or electrostatic, is more suitable for element identification in each particular case.

Further Reading

1. J.W. Rabalais (Ed.): *Low Energy Ion–Surface Interactions* (John Wiley, Chichester 1994) (about LEIS)
2. J.F. Van der Veen: *Ion Beam Crystallography of Surfaces and Interfaces.* Surf. Sci. Rep. **5**, 199–288 (1985) (about MEIS and RBS crystallography)
3. L.C. Feldman, J.W. Mayer, S.T. Picraux: *Materials Analysis by Ion Channeling* (Academic Press, New York 1982) (on the physics underlying RBS)
4. D. Briggs, M.P. Seah (Ed.): *Practical Surface Analysis. Vol. 2. Ion and Neutral Spectroscopy* (John Wiley, Chichester 1992) (about LEIS, MEIS and SIMS)
5. P. Varga, H. Winter: 'Slow Particle-Induced Electron Emission from Solid Surfaces'. In: *Particle Induced Electron Emission II* (Springer, Berlin 1991) (about charge exchange between ions and surfaces)
6. D.P. Woodruff, T.A. Delchar: *Modern Techniques of Surface Science*, 2nd edn. (Cambridge University Press, Cambridge 1994) (about charge exchange between ions and surfaces).
7. A. Benninghoven, F.G. Rudenauer, H.W. Werner: *Secondary Ion Mass Spectrometry: Basic Concepts, Instrumental Aspects, Applications, and Trends* (John Wiley, New York 1987) (the best SIMS reference)

7. Surface Analysis IV. Microscopy

Microscopy techniques are used to produce real space magnified images of a surface showing what it looks like. In general, microscopy information concerns surface crystallography (i.e., how the atoms are arranged at the surface), surface morphology (i.e., the shape and the size of topographic features making the surface), and surface composition (the elements and compounds that the surface is composed of). The operational principles vary greatly from one type of microscopy to another and include electron-beam transmission (transmission electron microscopy) and reflection (reflection electron microscopy, low-energy electron microscopy, scanning electron microscopy), field emission of electrons (field emission microscopy, scanning tunneling microscopy) and ions (field ion microscopy), and scanning the surface by a probing electron beam (scanning electron microscopy) or a probing tip (scanning tunneling microscopy, atomic force microscopy). Most microscopy techniques used in surface science ensure resolution on the nm scale, while field ion microscopy, scanning tunneling microscopy and atomic force microscopy allow acquisition of images with atomic resolution.

7.1 Field Emission Microscopy

The *field emission microscope (FEM)* was invented by Erwin Müller in 1936. The FEM design is very simple. It contains a metallic sample in the form of a sharp tip and conducting fluorescent screen inside an evacuated volume (Fig. 7.1a). A large negative potential is applied to the tip with respect to the screen. Typically, the potential is 1–10 keV, the radius of curvature of the tip is $\sim$1000 Å, hence, the electric field near the tip apex is on the order of 1 V/Å. At such high fields, the field emission of electrons takes place (for details see Sect. 11.6.3, page 287).

Electrons emerging from the tip diverge radially along the lines of the force and the magnification of the microscope is simply

$$M = L/r \,, \qquad (7.1)$$

where r is the tip apex radius and L is the tip–screen distance (Fig. 7.1b). Linear magnifications of about 10^5–10^6 are possible. The resolution limit of

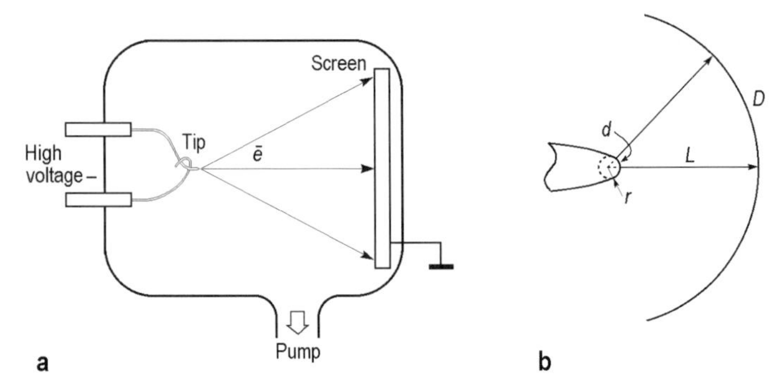

Fig. 7.1. (**a**) Experimental set-up for field emission microscopy. (**b**) Schematic diagram of FEM optics. An object with linear dimension d on the tip surface will be magnified by a factor of L/r and appears on the screen with dimension D

the instrument is about 20 Å and is determined by the electron tangential velocity, which is on the order of the Fermi velocity of the electron in metal.

Figure 7.2 shows the FEM image of the clean W tip with the W wire axis perpendicular to the (110) plane. The emission current varies strongly with the local work function in accordance with the Fowler–Nordheim relation (11.20), hence, the FEM image displays the projected work function map of the emitter surface. The closely packed faces, {110}, {211} and {100} , have higher work functions than atomically rough regions, and thus they show up in the image as dark spots on the brighter background.

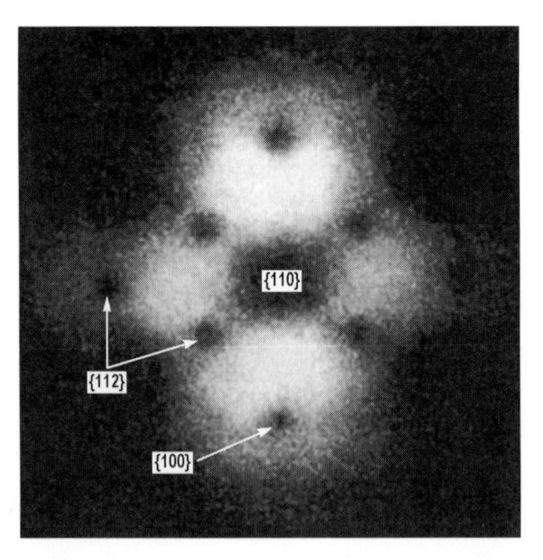

Fig. 7.2. Field emission microscopy pattern of a clean (110) oriented tungsten surface, showing the location of various crystal planes (after Müller [7.1])

Application of FEM is limited by the materials which can be fabricated in the shape of a sharp tip, can be cleaned in a UHV environment, and can withstand the high electrostatic fields. For these reasons, refractory metals with high melting temperature (for example, W, Mo, Pt, Ir) are conventional objects for FEM experiments. FEM allows the measurements of the work function for various crystallographic planes on a single sample. In cases when the adsorbate affects the work function, the technique is applicable for studying adsorption–desorption kinetics and surface diffusion.

7.2 Field Ion Microscopy

The *field ion microscope (FIM)* is also an invention of Ervin Müller (1951). It is an outgrowth of FEM in an effort to improve resolution by using the field desorption technique. The FIM apparatus is very similar to that of FEM with a sharp sample tip and a fluorescent screen (replaced currently by a multichannel plate) being the main elements (Fig. 7.3a). Similarly, the field strength at the tip apex is a few V/Å. However, there are some essential differences as follows.

- The tip potential is positive.
- The chamber is filled with "imaging" gas (typically, He or Ne at 10^{-5} to 10^{-3} Torr).
- The tip is cooled to low temperatures (∼20–80 K).

The principle of FIM image formation is illustrated in Fig. 7.3b. The imaging gas atoms in the vicinity of the tip become polarized by the field and, as the field is non-uniform, the polarized atoms are attracted towards the tip surface. Reaching the surface, they lose their kinetic energy through a series of hops and accomodate to the tip temperature. Eventually, the atoms are ionized by tunneling electrons into the surface and the resulting ions are accelerated to the screen to form a field ion image of the emitter surface. Resolution of the FIM is determined by the thermal velocity of the imaging ion. Effective cooling of the tip to low temperature can achieve a resolution of ∼1 Å, i.e., atomic resolution (see Fig. 7.4).

The limitations of the tip material are similar to those in FEM, hence most FIM experiments have been done with refractory metals. Metal tips for FEM and FIM are prepared by electrochemical polishing of thin wires. However, these tips usually contain many asperities. The final preparation procedure resides in the in situ removal of these asperities by field evaporation just by raising the tip voltage. For most materials, *field evaporation* (i.e., field-induced desorption of surface atoms in the form of ions) occurs in the range 2–5 V/Å. This process is self-regulating as the most protruding atoms desorb first making the surface smooth. The tips used in FIM is sharper (tip radius is ∼100–300 Å) compared to those used in FEM experiments with a radius of ∼1000 Å.

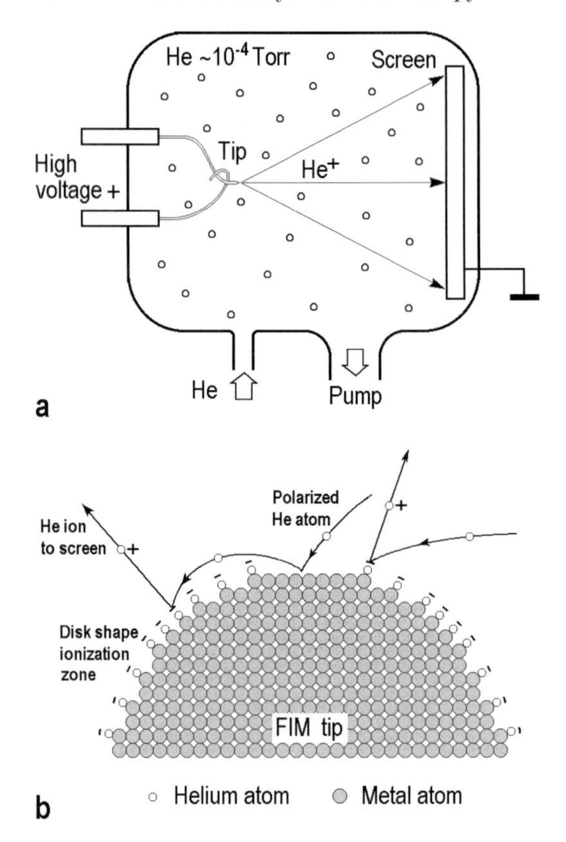

Fig. 7.3. (**a**) Experimental set-up for field ion microscopy. (**b**) Schematic diagram of the image formation process in FIM. In the field of a few V/Å, polarized He atoms are attracted to the tip and a monolayer of He atoms is field adsorbed on the surface. Other hopping He atoms may be field ionized above the adsorbed He atoms and accelerated as positive ions to a screen to produce a FIM image (after Tsong and Chen [7.2])

The most spectacular results obtained by FIM are associated with the investigations of the dynamical behavior of surfaces and the behavior of adatoms on surfaces. The problems under investigation include adsorption–desorption phenomena, surface diffusion of adatoms and clusters, adatom–adatom interactions, step motion, equilibrium crystal shape, etc. However, one should bear in mind that the results might be affected by the limited surface area (i.e., by edge effects) and by the presence of the large electric field.

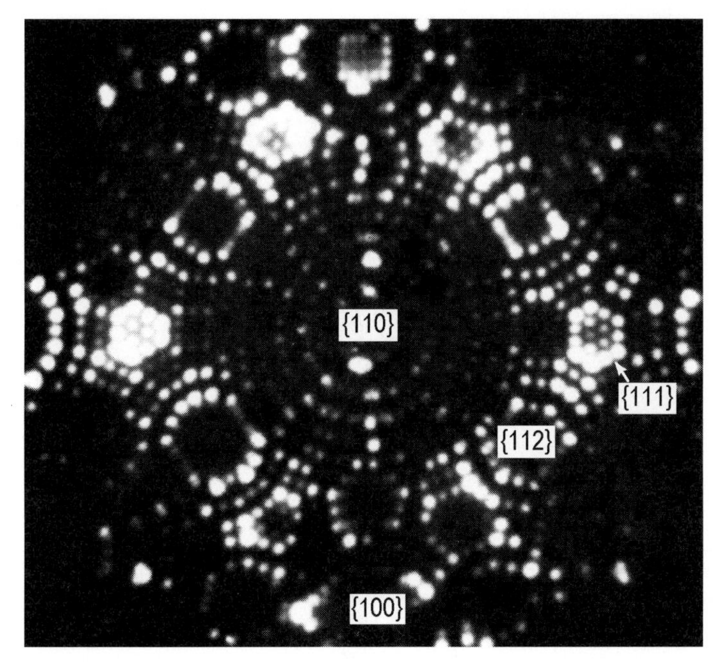

Fig. 7.4. FIM image of a W tip of radius ∼120 Å acquired at 21 K using a He-H_2 mixture as the imaging gas (after Tsong and Sweeney [7.3])

7.3 Transmission Electron Microscopy

In *transmission electron microscopy (TEM)*, the image is formed by electrons passing through the sample. The principle of operation of the transmission electron microscope is almost the same as that of an optical microscope, using magnetic lenses instead of glass lenses and electrons instead of photons. A beam of electrons emitted by an electron gun is focused by a condenser lens into a small spot (∼2–3 μm) on the sample and after passing through the sample is focused by the objective lens to project the magnified image onto the screen (Fig. 7.5). A very essential element is the aperture located at the back focal plane of the objective lens. As will be shown below, this determines the image contrast and resolution limit of the microscope. Note that this simple scheme illustrates only the principle of image formation in TEM, not the actual TEM set-up, which is much more sophisticated.

Due to the limited penetration depth of electrons in solids, the samples should be very thin: the acceptable thickness is 100–1000 Å for conventional microscopes with accelerating voltages of 50–200 keV and a few thousand Å for high-voltage microscopes with acceleration voltages up to 3 MeV. Of cause, the required sample thickness depends on the sample material: the larger the atomic number, the greater the electron scattering, hence a thinner sample should be.

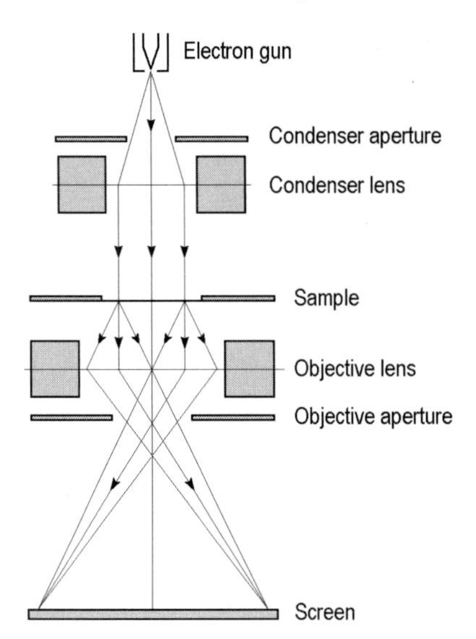

Fig. 7.5. Schematic diagram illustrating image formation in the transmission electron microscope

The diffraction limit for TEM resolution can be estimated from $\Delta = 0.5\lambda/\sin\alpha$, where λ is the electron wavelength and α equals one-half the angular aperture, which can be approximated by the ratio of the objective diaphragm radius to the objective focal length. For a voltage of 100 keV ($\lambda = 0.037$ Å), diaphragm radius of 20 μm, and focal length of 2 mm, the estimation yields $\Delta \approx 2$ Å. In practice, the resolution is usually worse due to non-ideality of the electronic optic system.

The formation of the TEM image contrast can be understood as follows. When passing through a sample, the electron flux loses part of its intensity due to scattering. This part is greater for thicker regions or regions with species of higher atomic number. If the objective aperture effectively cuts off the scattered electrons, the thicker regions and the regions of higher atomic number appear dark. The smaller aperture enhances the contrast (but leads to the loss of resolution, as shown above). In crystals, the elastic scattering of electrons results in the appearance of *diffraction contrast.*

In classical TEM experiments on studying surface phenomena, metals were deposited onto alkali halide surfaces cleaved in vacuum. At the thickness of metal deposit used (~10 Å), a continuous film is not formed but, instead, a large number of small nuclei form on the surface. A thin film of carbon was then deposited onto the sample surface to fix the metal nuclei and the sample was taken off the chamber. A carbon film with the embedded metal nuclei was stripped by gentle immersion of the alkali halide crystal in water and subsequently used for TEM observations. Two main subjects were treated in those experiments. The first one concerns the mechanisms of the nucleation, growth, and coalescence of metal islands. The second one is related to the step structure of the alkali halide surfaces, making use of the *step decoration*, i.e., preferential nucleation of islands along the step edges (Fig. 7.6). It should be noted that presently the technique is not so popular and the use of SEM, STM, or AFM for similar tasks is more common.

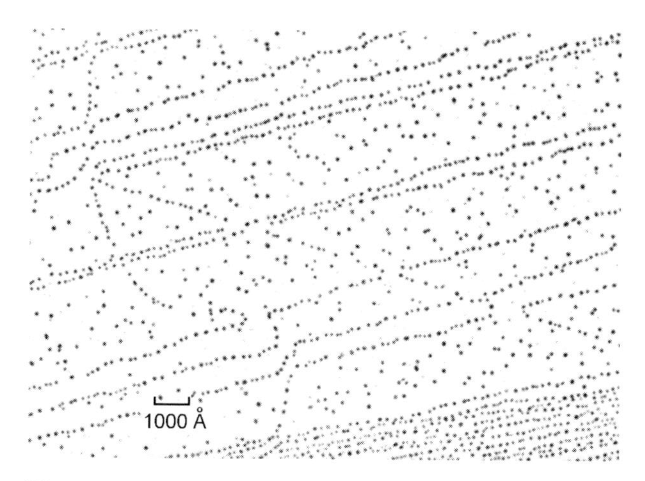

Fig. 7.6. The NaCl cleavage surface annealed at 250°C. Decoration of the surface by Au reveals that interaction between migrating steps leads to formation of rounded corners where initially there were linear intersections (after Bassett et al. [7.4])

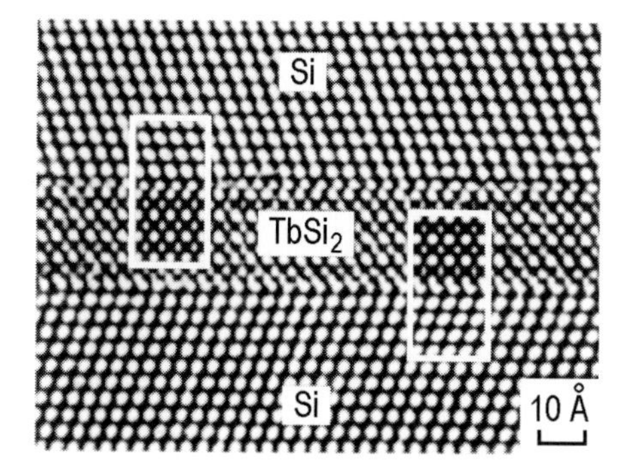

Fig. 7.7. High-resolution TEM image of Si/$TbSi_2$/Si double heterostructure interfaces with simulated images pasted for direct comparison (after Luo et al. [7.5])

For studying buried interfaces, cross-section TEM observations are widely used. For this purpose, the sample is cut normal to the surface and thin slices are prepared by chemical etching and ion milling. With high-resolution TEM, images showing the atomic structure can be obtained. However, the contrast seen in the image does not correspond necessarily to single atoms and, hence, numerical image simulations are required for the reliable interpretation of the image features in terms of atomic structure (Fig. 7.7).

7.4 Reflection Electron Microscopy

In *reflection electron microscopy (REM)*, high-energy electrons are incident at glancing angles to the sample and reflected electrons are used to form a REM image of the sample surface (see Fig. 7.8). The elastically scattered electrons produce a RHEED pattern at the back focal plane of the objective lens, where one or several of the RHEED reflections are selected by the objective aperture. A magnified image of the surface is projected on the observation screen of the microscope.

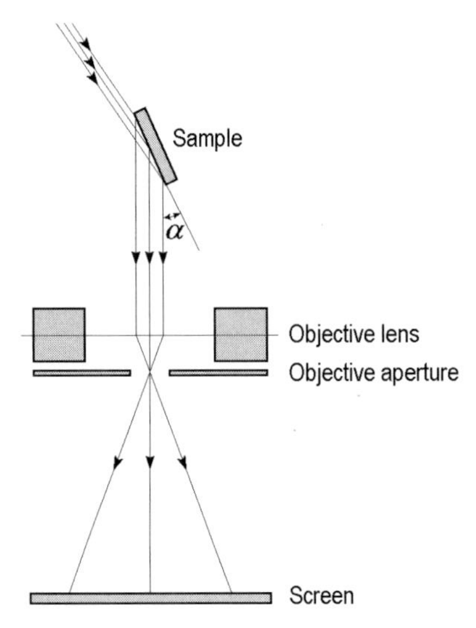

Fig. 7.8. Schematic diagram illustrating image formation in the reflection electron microscope

The REM image is foreshortened by a factor of $\sin\alpha$, where α is the angle of the imaging electrons from the surface (typically, $\sim$1/40–1/70). The image is generally reproduced to show the view of the surface along the imaging beam direction, hence the foreshortening is in the vertical direction of the image. As a consequence of the foreshortening, only the central region of the image is in focus, while the upper and lower parts are over-focused and under-focused, respectively. Another consequence of the foreshortening is the poor resolution along the beam direction.

Provided that the UHV conditions around the sample are ensured, REM is suitable for studying well-defined crystalline surfaces. Its advantageous feature resides in the ability to resolve atomic steps and to distinguish regions of

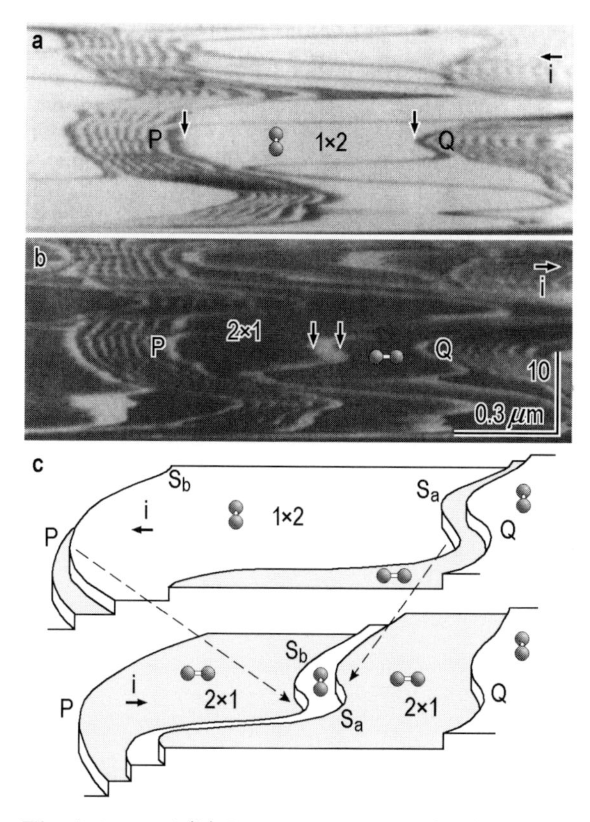

Fig. 7.9. **a**and (**b**) REM images of a Si(100)2×1 surface which display the conversion of the major domain from 1×2 in (**a**) to 2×1 in (**b**) by a reversal of the current direction. The vertical arrows indicate the initial and final locations of the steps. (**c**) shows schematically the conversion processes. The orientation of the surface Si-Si dimers on a terrace, which defines the type of reconstruction, 2×1 or 1×2, is indicated. The diffraction contrast, which distinguishes the areas of different reconstruction is obtained by selecting an appropriate RHEED spot by an objective apperture (after Yagi et al. [7.6])

different surface reconstructions by the use of diffraction contrast (Fig. 7.9). The most prominent applications of REM concern the study of surface dynamic processes as follows:

- step motion
- surface phase transformations
- adsorption
- nucleation and growth of thin films
- surface reactions.

Complementary REM techniques, which employ the same apparatus, are RHEED and reflection high-energy loss spectroscopy (REELS).

7.5 Low-Energy Electron Microscopy

Low-energy electron microscopy (LEEM) was invented by Ernst Bauer in the early 1960s, but it became a viable surface imaging technique only 20 years later. In LEEM, primary low-energy electrons hit the sample and the reflected electrons are used to form a focused, magnified image of the surface. The spatial resolution limit of the technique is on the order of several tens of Å.

A typical LEEM set-up is shown in Fig. 7.10. Electrons leave the electron gun with a relatively high energy (10–20 keV) and pass a set of lenses. Then the electrons are deflected by 90° by the magnetic prism array, pass the objective lens and are decelerated to a final low energy of 0–100 eV. The low-energy electrons are scattered at the sample surface as in LEED and are reflected back to be re-accelerated to the gun energy on their return to the objective lens. To accomplish this, the sample is maintained at a potential close to that of the emitter in the gun, while the objective lens is grounded. Finally, the prism deflects the electrons into the imaging column, where the projector lenses and the screen are located.

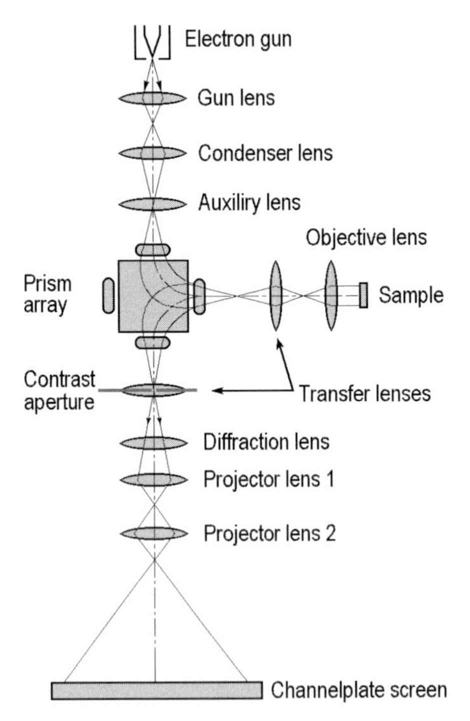

Fig. 7.10. Experimental set-up for low-energy electron microscopy (after Tromp [7.7])

When a parallel beam of low-energy electrons hits the crystal surface, the electrons are diffracted at the surface and the result is a LEED pattern. With a contrast aperture in the imaging column, one of the diffracted beams can be chosen for imaging. If the specular (0,0) beam is selected, the resulting image is called a *bright-field image*. Images taken with any other beam are called *dark-field images*.

In a bright-field image, the contrast is caused by local differences in the low-energy electron reflectivity due to the differences in crystal orientation, surface reconstruction, adsorbate coverage, etc. The LEEM image of the Si(111) surface at the $7\times7\leftrightarrow1\times1$ phase transition occurring at $\sim$860°C furnishes a classical example for reconstruction contrast (Fig. 7.11). One can clearly distinguish the regions of the 7×7 structure (bright) and those of the 1×1 structure (dark). In addition, the atomic steps to which the 7×7 regions adhere are also visible.

Fig. 7.11. Bright-field LEEM image of the Si(111) $7\times7\leftrightarrow1\times1$ phase transition. The 7×7 phase (bright) decorates the atomic steps during the phase transition, while the terraces are mostly covered by the 1×1 structure (dark). Contrast is due to the difference in the (0,0) structure factor for the two phases. The field of view is 5 μm (after Tromp [7.7])

An example of dark-field imaging is illustrated in Fig. 7.12 which shows the images of the same region of a clean Si(100) surface taken at different diffraction beams. The Si(100) surface is reconstructed in Si dimer rows and has a 2×1 periodicity. Due to the properties of the Si crystal lattice, the bond geometry is rotated by 90° on each atomic step and so the reconstruction rotates from 2×1 to 1×2 upon crossing an atomic step. Thus, 2×1 terraces, diffracted into the (1/2,0) or (−1/2,0) beams, appear bright in Fig. 7.12b, while negative contrast is obtained in Fig. 7.12c where the (0,1/2) or (0,−1/2) spot is selected by the contrast aperture.

As LEEM images can be acquired very quickly, the technique is suitable for studying dynamic processes at surfaces, such as thin film growth, strain relief, etching, adsorption, and phase transitions, in real time.

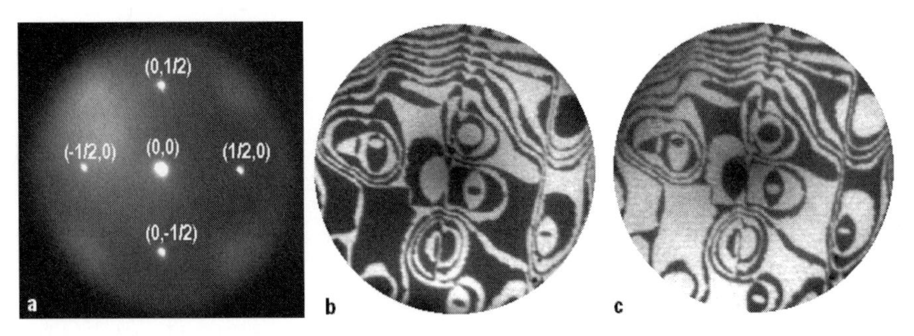

Fig. 7.12. (**a**) LEED image of the Si(100)2×1 clean surface. (**b**) Dark-field LEEM image of the surface in (**a**) taken using one of the (1/2,0) or (−1/2,0) spots for imaging. (**c**) Same as (**b**) using one of the (0,1/2) or (0,−1/2) spots. The field of view is 5 μm (after Tromp [7.8])

7.6 Scanning Electron Microscopy

In *scanning electron microscopy* (Fig. 7.13a), an electron beam with primary electron energy typically of ∼1–10 keV is focused by a lens system into a spot of 1–10 nm in diameter on the sample surface. The focused beam is scanned in a raster across the sample by a deflection coil system in synchronism with an electron beam of a video tube, which is used as an optical display. Both beams are controlled by the same scan generator and the magnification is just the size ratio of the display and scanned area on the sample surface. A variety of signals can be detected, including secondary electrons, backscattered electrons, x-rays, cathodoluminescence, and sample current (Fig. 7.13b). The two-dimensional map of the signal yields a SEM image.

The main applications of SEM concerns visualization of the surface topography and elemental mapping (Fig. 7.14). To consider the nature of the contrast in various SEM modes, recall first the energy spectrum of the electrons emitted from the surface which is irradiated by an electron beam of energy E_0 (see Fig. 5.2). Besides the peak of elastically scattered electrons at E_0, the spectrum contains a broad peak of secondary electrons (SE) from 0 to about 50 eV and the region of the inelastic backscattered electrons (BSE) (from 50 eV to E_0). The Auger peaks and loss peaks due to plasmon excitation and interband transitions also fall into the BSE range. By appropriate choice of the detector, the signal of the electrons from a desired energy range can be monitored.

Secondary Electrons (SE). In SE mode, electrons with an energy of a few eV are collected by a directional detector. Due to the angular dependence of the SE yield and the shadowing effect, the SE image shows a surface topography. The contrast due to the chemical composition is not so essential. Most of the secondary electrons originate from a shallow (a few nanometers in depth) layer within primary beam, which allows a resolution of the order

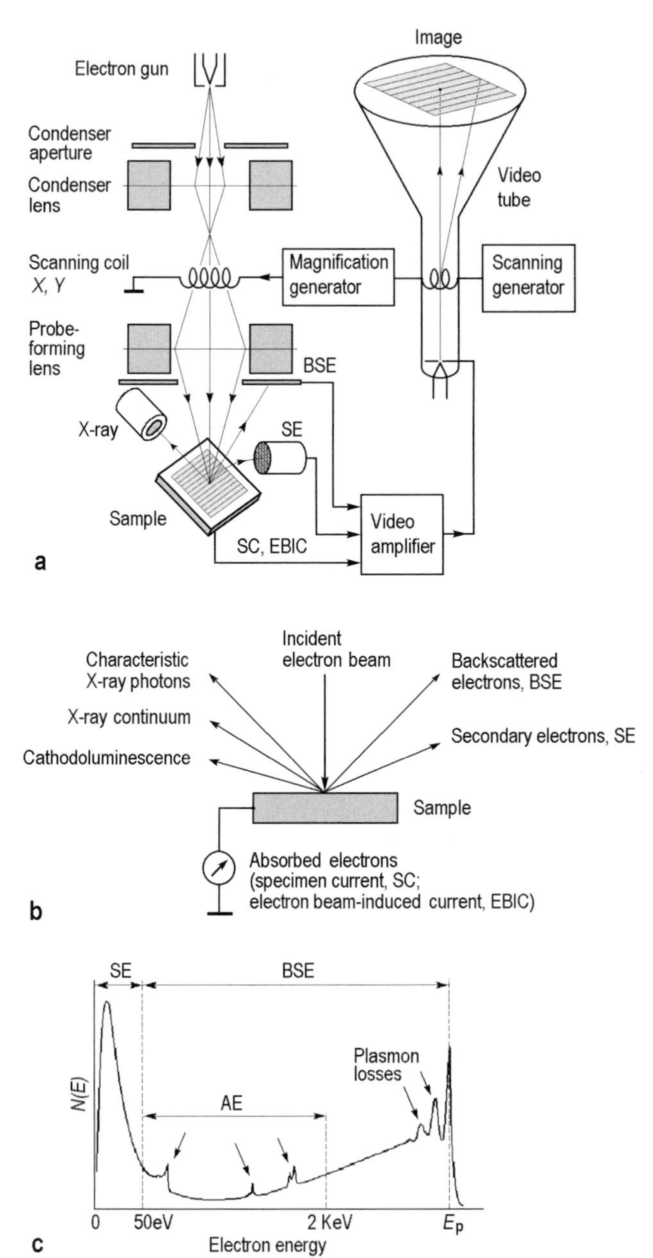

Fig. 7.13. (**a**) Schematic of a secondary electron microscope. (**b**) The types of signals generated by irradiation of the specimen by a primary electron beam. (**c**) Energy spectrum of electrons emitted from the sample which is irradiated by primary electrons with energy E_p. The energy regions corresponding to secondary electrons (SE), backscattering electrons (BSE), and Auger electrons (AE) are indicated

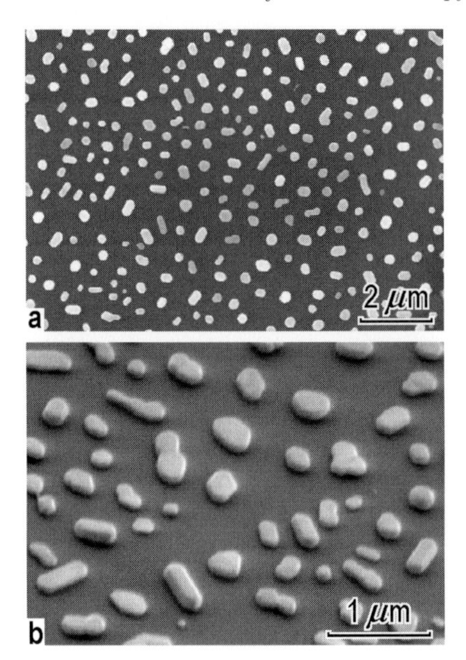

Fig. 7.14. SEM image of 120 Å Au film deposited at RT onto $TiO_2(110)$ and annealed at 500°C, showing Au islands (**a**) at normal incidence and (**b**) with the surface normal tilted by 45° (after Cosadney et al. [7.9])

of 5–20 nm to be reached. However, some secondary electrons are excited by the backscattered electrons (for which the information region exceeds the primary beam spot), hence typical BSE contrast is superimposed on every SE micrograph.

Backscattered Electrons (BSE). For collecting backscattered electrons of relatively high energies, detectors with a large acceptance angle are used. The most essential contrast in BSE mode is due to the dependence of the backscattering coefficient on the atomic number of the species. The higher the atomic number, the greater the backscattering, and consequently the brighter the corresponding area in the image. Furthermore, if using energy filtering one selects an Auger peak or a characteristic loss peak for a given atomic species, a map of the spatial distribution of this species on the surface can be obtained.

Sample Current (SE). The sample current is essentially the primary-beam current minus the emission currents of SE and BSE. If the SE emission is suppressed by a positive bias applied to the sample, the map of the sample current will correspond to the map of the backscattered coefficient in the inverted contrast. Note that SE mode does not require an additional detector.

X-Rays. Besides secondary and backscattered electrons, the primary electron beam induces the emission of x-rays, whose signal can also be employed for surface imaging. With the detector tuned to the characteristic x-ray energy, a map of the spatial distribution of a given element is obtained. This is quantitatively similar to elemental mapping using an AES signal. The difference is the greater information depth of the x-ray probe (0.1–10 μm) compared to that of AES (a few nm). As a consequence, x-ray probe has worse spatial resolution, but offers better depth analysis. In contrast to AES, x-ray probing does not require UHV conditions.

Electron-Beam-Induced Current (EBIC). The electron irradiation of a semiconductor generates a lot of electron–hole pairs (a few thousand per incident electron). The generated charge carriers are separated in the depletion layers of the *p-n* junction. This process results in an electron-beam-induced current (EBIC), which can be amplified and used for inspection of semiconductor devices (for example, to image *p-n* junctions, to localize avalanche breakdowns, to visualize the electrically active defects).

Cathodoluminescence (CL). Cathodoluminescence (CL), i.e., emission of ultraviolet or visible light induced by electron bombardment, is another source of analytical information. The CL signal has a very low intensity, hence a sensitive detector with a large collection solid angle is required. CL is often used in combination with EBIC for characterization of semiconductor devices, in particular, for imaging of lattice defects which affect the recombination rate of charge carriers.

7.7 Scanning Tunneling Microscopy

The scanning tunneling microscope (STM) was developed in the early 1980s by Gerd Binnig and Heinrich Rohrer, who were awarded the Noble Prize in Physics in 1986 for this invention.

As main components, STM contains (Fig. 7.15):

- An *atomically sharp tip*. The STM tips are typically fabricated from metal (for example, W, Pt-Ir, Au) wires. The preparation procedure of the atomically sharp tip includes preliminary ex situ treatments, like mechanical grinding, cleavage, or electrochemical etching and subsequent in situ treatments, like annealing, field emission/evaporation, and even a "soft crash" of the tip by touching a sample surface.
- A *scanner* to raster the tip over the studied area of the sample surface. Piezoelectric ceramics are used in scanners as electromechanical transducers, as they can convert electric signals of 1 mV to 1 kV into mechanical motion in the range from fractions of Å to a few μm.
- *Feedback electronics* to control the tip–sample gap.
- A *computer system* to control the tip position, to acquire data, and to convert the data into an image.

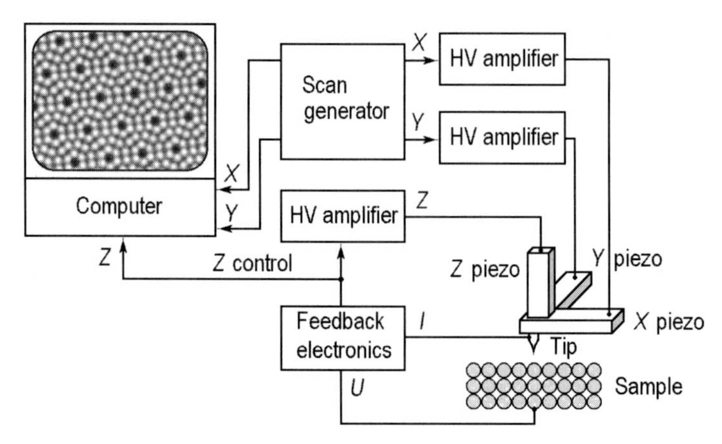

Fig. 7.15. Schematic illustration of the STM set-up

Auxiliary but also indispensable components of STM are:

- A *Coarse positioning system* to bring the tip to within the tunneling distance of the sample and, when required (say, for changing the sample), to retract it back to a sufficient distance (a few mm).
- *Vibration isolation.* For stable operation of the STM, changes of the tip–sample separation caused by vibrations should be kept to less than ∼0.01 Å. The required vibration damping is achieved by suspending the inner STM stage with the tip and sample on very soft springs and by using the interaction between the eddy current induced in cooper plates attached to the inner stage and the magnetic field of the permanent magnets mounted on the outer stage.

The concept of the operation of STM is as follows. The very sharp tip of the microscope is placed so close to the probed surface that the wave functions of the closest tip atom and the surface atoms overlap. This takes place at tip–sample gaps of ∼5–10 Å. If one applies a bias voltage V between the tip and the sample, a tunneling current will flow through the gap. In simplified form, the tunneling current density j is given by:

$$j = \frac{D(V)V}{d} \cdot \exp(-A\,\phi_{\mathrm{B}}^{1/2}\,d)\;, \tag{7.2}$$

where d is the effective tunneling gap, $D(V)$ reflects the electron state densities, A is a constant, and ϕ_{B} is the effective barrier height of the junction (Fig. 7.16).

The sharp dependence of the tunneling current on the gap width determines the extremely high vertical resolution of STM. Typically, a change of the gap by $\Delta d = 1$ Å results in a change in the tunneling current by an order of magnitude, or, if the current is kept constant to within 2%, the gap width remains constant to within 0.01 Å. As for the lateral resolution of STM, this is determined by the fact that up to 90% of the tunneling current flows through

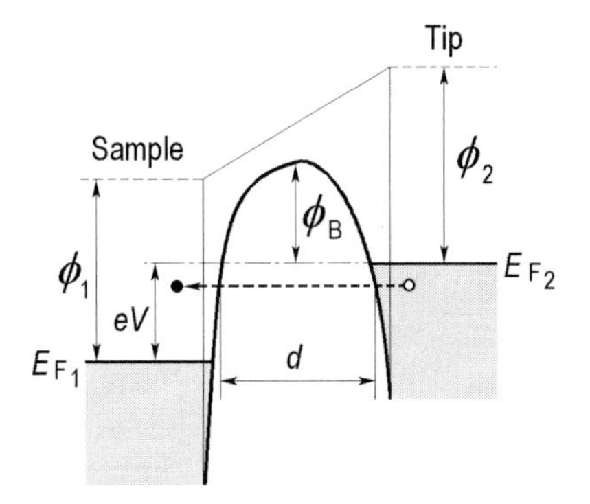

Fig. 7.16. Energy diagram of the tunneling contact of the STM tip and the metallic sample. E_{F1} and E_{F2} are the Fermi levels of the surface and tip, ϕ_1 and ϕ_2 are the work functions of the surface and tip, ϕ_B is the effective barrier height, d is the effective tunneling gap, and V is the bias voltage. The diagram illustrates the situation when STM probes the empty states of the surface (after Wilson [7.10])

the gap between the "last" atom of the tip and the atom of the surface, which is the closest to it. Surface atoms with an atomic separation down to ~2 Å can be resolved.

By scanning the tip along the surface, one obtains the pattern of the surface topography. However, one should bear in mind that STM is not primarily sensitive to atomic positions, but rather to the local density of electronic states. When the tip bias voltage is positive with respect to the sample, the STM image corresponds to the surface map of the filled electronic states. With a negative tip bias voltage, the empty-state STM image is obtained. Hence, the maxima in the STM image might correspond to both the topographical protrusions on the surface and the increased local density of states.

There are five main variable parameters in STM. These are the lateral coordinates, x and y, the height z, the bias voltage V, and the tunneling current I. Three main modes of STM operation are defined, depending on the manner in which these parameters are varied:

- *Constant-current mode.* In this mode, I and V are kept constant, x and y are varied by rastering the tip and z is measured.
- *Constant-height mode,* which is also called *current imaging.* In this mode, z and V are kept constant, x and y are varied by rastering the tip, and I is measured.
- *Scanning tunneling spectroscopy (STS).* This is rather a series of various modes, in which V is varied.

Consider each of the modes in greater detail.

Constant-Current Mode. This is the most widely used technique for acquiring STM images. In this mode, the tip is scanned across the surface at constant voltage and current. To maintain the tunneling current at a preset value, a servo system continuously adjusts the vertical position of the tip by

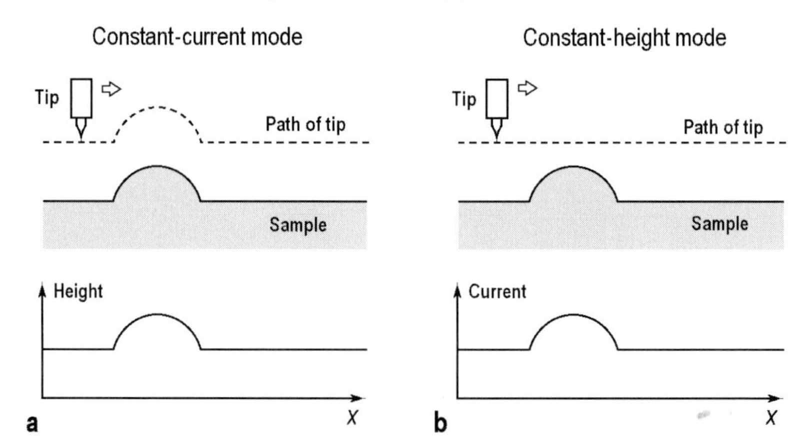

Fig. 7.17. Schematic illustration of STM imaging in (**a**) constant-current mode and (**b**) constant-height mode

variation of the feedback voltage V_z on the Z-piezoelectric driver. In ideal circumstances of an electronically homogeneous surface, constant current essentially means constant gap, i.e., the scanning tip tracks all the features of the surface topography (Fig. 7.17a). The height of the surface features is derived from V_z. Thus, the surface relief height $z(x, y)$ is acquired as a function of the tip position at the surface. The advantages of constant-current mode reside in the possibility of probing surfaces which are not necessarily atomically flat and in the ability to determine the surface height quantitatively from V_z and the sensitivity of the piezoelectric driver. The disadvantage is the limited scan speed due to the finite response time of the servo system

Constant-Height Mode. In this mode, the tip is scanned across the surface with V_z kept constant and the variations of tunneling current are recorded as function of a tip position (Fig. 7.17b). The bias voltage is fixed and the feedback circuitry is slowed or turned off completely. The rastering of the tip can be done at much greater speed than that in constant-current mode, as the servo system does not have to respond to the surface features passing under the tip. This ability is especially valuable for studying real-time dynamic processes, in particular, for recording the STM video. The disadvantages are as follows: the technique is applicable only for relatively flat surfaces; and quantitative determination of the topographic heights from variation of the tunneling current is not easy because a separate determination of $\phi_B^{1/2}$ is needed to calibrate z.

Scanning Tunneling Spectroscopy. Since the tunneling current is determined by summing over electron states in the energy interval determined by the bias voltage V, by varying V one can obtain information on the local density of states as a function of energy. One way is to acquire a set of constant-current STM images of the same surface area at various values of

V and both polarities. Another way is to conduct, at each pixel of the scan, measurements of the tunneling current I versus bias voltage V at constant tip–sample separation. From the measured $I-V$ curves, one can calculate $(\mathrm{d}I/\mathrm{d}V)/(I/V)$, which corresponds closely to the sample density of states. Thus, the spatial distribution of particular states can be mapped. This technique is referred to as *current-imaging tunneling spectroscopy (CITS)* [7.11].

STS allows the probing of electronic properties of a very local preselected area, even of an individual adatom at the surface (Fig. 7.18). This provides the significant possibility of distinguishing surface atoms of different chemical natures. In general, spectroscopic information is valuable for consideration of the energy gap, band bending, and chemical bonding.

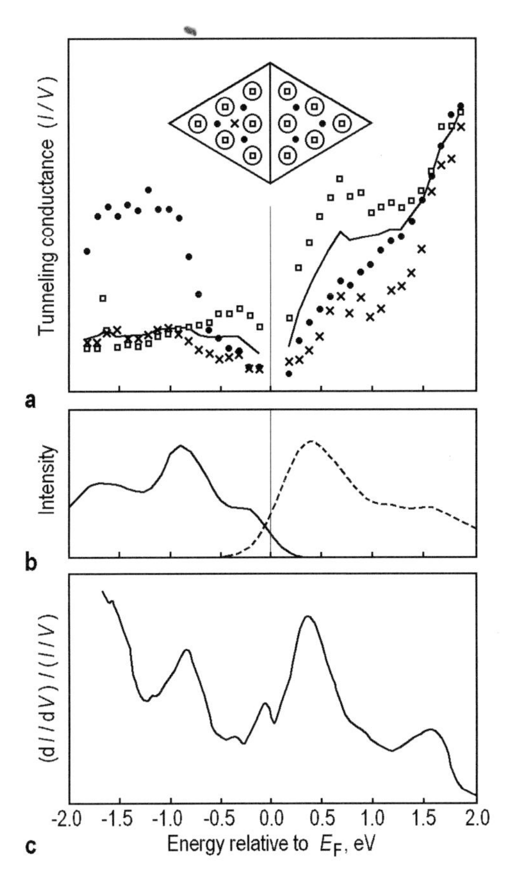

Fig. 7.18. A schematic illustration of STM imaging in (**a**) tunneling conductance (I/V) spectra measured for the Si(111)7×7 surface averaged over one unit cell (solid line) and at selected locations in the unit cell: (□) adatom; (•) rest atom; (×) center position. (**b**) Surface states on the Si(111)7×7 surface observed using ultraviolet photoemission spectroscopy (solid line) and inverse photoemission spectroscopy (dashed line). (**c**) Area-averaged $(\mathrm{d}I/\mathrm{d}V)/(I/V)$ tunneling spectra (after Tromp [7.12])

However, in general there is a lack of a sure and simple way for the interpretation of STS data. For meaningful consideration, thorough theoretical calculations are required and, even in this case, one should be cautious about the conclusions. For example, it is usually assumed that the tip is featureless and its electron density of states resembles that of the free electron gas in an ideal metal. In fact, the validity of the assumption for a real tip requires special testing. Another problem is associated with the fact that the tunneling current depends also on the tunneling transmission probability density. As a result, the large density of states of the sample is not accessible in the measurements, if these states do not overlap with the tip.

7.8 Atomic Force Microscopy

The successful achievements of scanning tunneling microscopy have inspired the development of a set of novel scanning probe microscopy methods (for example., atomic force microscopy, lateral force microscopy, magnetic force microscopy, ballistic-electron-emission microscopy, scanning ion-conductance microscopy, near-field scanning optical microscopy, etc.) Like STM, all these techniques are based on the usage of piezoelectric transducers that provide the ability to control the spatial position of the probing tip relative to the sample surface with great accuracy and, thus, to map the measured surface property on an atomic or nanometer scale. Among such methods, atomic force microscopy has found the widest application.

The atomic force microscope (AFM) was invented by Binnig, Quate, and Gerber in 1986 [7.14]. AFM senses the forces between the tip and the sample. The principle of AFM operation is illustrated in Fig. 7.19. A sharp tip a few microns long is located at the free end of the cantilever (usually, 100–200 microns long). The interatomic forces between the tip and the sample surface atoms cause the cantilever to deflect. The cantilever displacement is measured by a deflection sensor. Several techniques are in use for detection of the small displacements of the cantilever. Originally Binnig et al. used an STM as the deflection sensor and measured the tunneling current between the STM tip and the conductive rear side of the lever (Fig. 7.20a). Other sensors utilize optical interferometry (Fig. 7.20b), reflection of the laser beam from the rear side of the cantilever (Fig. 7.20c), or measurement of the capacitance between the lever and the electrode located close to the rear side of the cantilever (Fig. 7.20d). Typically, sensors can detect deflections as small as 10^{-2} Å. Measuring the deflection of the cantilever while the tip is scanned over the sample (or the sample is scanned under the tip) allows the surface topography to be mapped. The evident advantage of AFM is that it is applicable for studying all types of surfaces, conducting, semiconducting and, insulating.

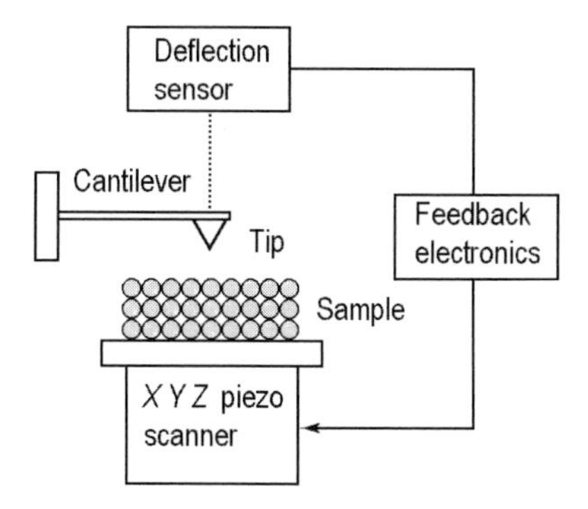

Fig. 7.19. Schematic illustration of AFM operation (after Meyer and Heinzelmann [7.13])

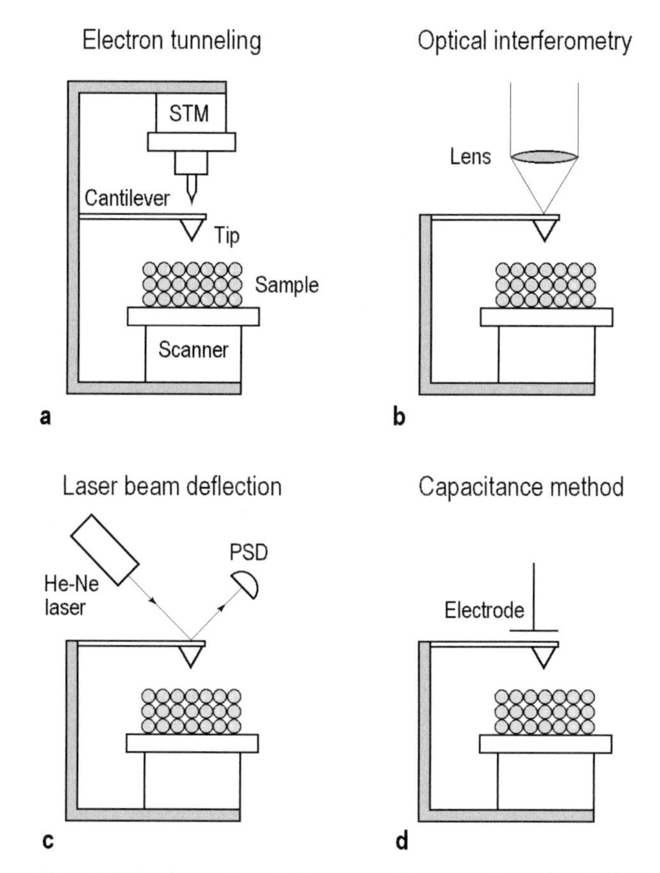

Fig. 7.20. Several techniques used to monitor the deflection of the cantilever in AFM: (**a**) electron tunneling; (**b**) optical interferometry; (**c**) laser-beam deflection, and (**d**) capacitance method (after Meyer et al. [7.15])

To consider the tip–sample interaction, refer to the interatomic force versus distance curve in Fig. 7.21. When the tip-to-sample separation is relatively large (right side of the curve), the cantilever is weakly attracted to the sample. With decreasing distance, this attraction increases until the separation becomes so small that the electron clouds of the tip and sample atoms begin to repel each other electrostatically. The net force goes to zero at a distance on the order of the length of a chemical bond (a few Å) and at closer distances the repulsion dominates.

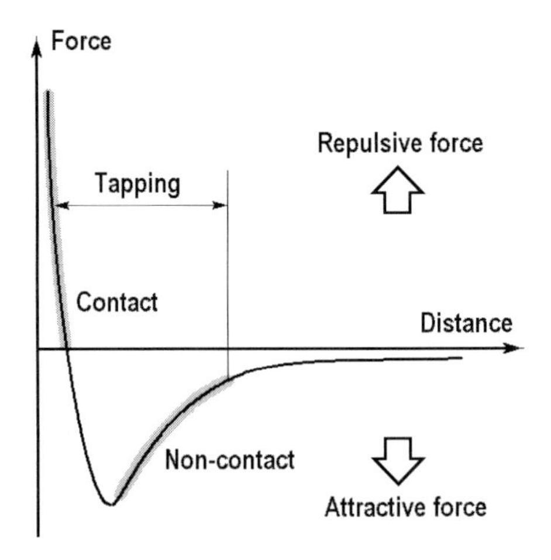

Fig. 7.21. The qualitative dependence of the van der Waals force on the interatomic distance. The regions of the tip-to-sample separation used in AFM contact and non-contact modes are indicated

The range of tip-to-sample separation used for AFM imaging defines the mode of operation, namely,

- contact mode;
- non-contact mode;
- tapping mode.

AFM Contact Mode. In contact mode, the tip–sample separation is on the order of a few Å. Thus, an AFM tip is in soft physical contact with the sample and is subjected to repulsive forces. To avoid damaging of the probed surface, the cantilever should not be stiff, i.e., the cantilever spring constant should be lower than the effective spring constant of the sample atom bonding. Therefore, the tip–sample interaction causes the cantilever to bend following the changes in surface topography. Topographic AFM images are acquired typically in one of two modes:

- constant-height mode or
- constant-force mode.

In *constant-height mode*, the scanner height is fixed and the cantilever deflection is monitored to generate the topographic image. Constant-height mode is preferred for the acquisition of atomic-scale images of atomically flat surfaces (the case where the cantilever deflections are small) and for fast recording of real-time images of changing surfaces (the case where rapid scanning is required).

In *constant-force mode*, the cantilever deflection is fixed (which means that the net force applied to the cantilever is constant) through the continuous adjustment of the scanner height by a servo system. The image is generated from the scanner motion. Constant-force mode is the most widely used mode as the net force is well controlled and the data set is easy to interpret. The disadvantage of this mode is the limited scan speed due to the finite response time of the feedback circuit.

AFM Non-Contact Mode. The tip–sample separation in non-contact mode is conventionally on the order of tens to hundreds of Å, i.e., the cantilever is affected by weak attractive forces. In this technique, the stiff cantilever is kept vibrating near its resonance frequency. Typical frequencies are from 100 to 400 kHz, the typical amplitude is of a few tens of Å. Due to interaction with the sample, the cantilever resonance frequency f_1 changes according to

$$f_1 \propto \sqrt{c - F'} \,, \tag{7.3}$$

where c is the cantilever spring constant and F' is the force gradient. If the resonance frequency (or vibrational amplitude) is kept constant by a feedback system which controls the scanner height, the probing tip traces lines of constant gradient. The motion of the scanner is used to generate the data set. This mode is also referred to as *constant-gradient mode*. The efficiency of this detection scheme has been proved by reaching the atomic resolution in AFM images acquired in non-contact mode (Fig. 7.22).

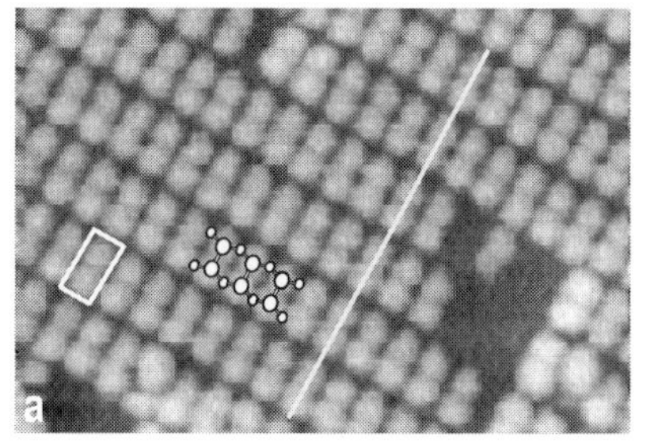

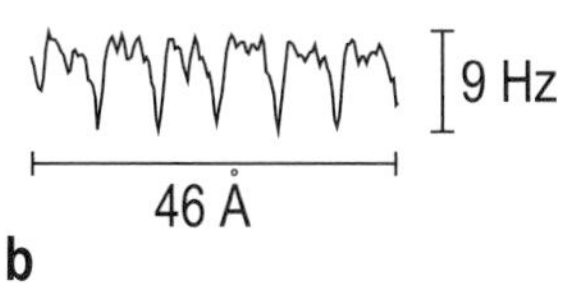

Fig. 7.22. (**a**) Non-contact AFM image (69×46 Å^2) of a Si(100) clean 2×1 reconstructed surface. The image was taken by the beam deflection technique in the variable frequency shift mode. The oscillation amplitude of the cantilever is about 65 Å. A 2×1 unit cell is outlined. (**b**) The cross-sectional profile, as indicated by the white line. (after Yokoyama et al. [7.16])

AFM Tapping Mode. This mode is similar to non-contact mode, the only difference being that the cantilever tip at the bottom point of its oscillation barely touches (taps) the sample surface. Tapping mode does not provide atomic resolution but appears to be advantageous for imaging rough surfaces with high topographical corrugations.

Problems

7.1 In FEM, the apex of a tungsten tip (work function 4.5 eV) has a radius of curvature of 500 Å and is maintained at a potential of -5 kV relative to its surroundings.
(a) Estimate the electric field strength just outside the tip.
(b) Over what distance do electrons have to tunnel in order to leave the tip?
If you need help, see Fig. 11.22.

7.2 A TEM objective lens has a focal length of 2 mm and forms an image at a distance of 10 cm from the center of the lens. Calculate the image magnification assuming thin-lens behavior.

7.3 Explain why an I/V $((\mathrm{d}I/\mathrm{d}V)/(I/V))$ versus V dependence is used in STS to characterize the density of states, but an I versus V dependence is insufficient.

7.4 A silicon cantilever used in a non-contact AFM has a spring constant $k \sim 50\,\mathrm{N/m}$ and a mechanical resonant frequency $\nu_0 = 175\,\mathrm{kHz}$. Estimate the mass of the cantilever.

Further Reading

1. E.W. Müller, T.T. Tsong: *Field Ion Microscopy. Principles and Applications* (Elsevier, New York 1969)
2. R.D. Heidenreich: *Fundamentals of Transmission Electron Microscopy* (Interscience, New York 1964)
3. K. Yagi: *Reflection Electron Microscopy: Studies of Surface Structures and Surface Dynamic Processes*, Surf. Sci. Rep. **17**, 305–362 (1993)
4. E. Bauer: *Low Energy Electron Microscopy.* Rep. Prog. Phys. **57**, 895–938 (1994)
5. L. Reimer: *Scanning Electron Microscopy. Physics of Image Formation and Microanalysis. Springer Series in Optical Sciences. Vol. 36* (Springer, Berlin 1985)

6. C. Bai: *Scanning Tunneling Microscopy and its Application Springer Series in Surface Science. Vol. 32* (Springer, Berlin 1995) (various aspects of scanning tunneling microscopy and related scanning probe techniques)
7. R.M. Tromp: *Spectroscopy with the Scanning Tunneling Microscope: A Critical Review.* J. Phys.: Condensed Matter **1**, 10211–10228 (1989) (about scanning tunneling spectroscopy)
8. R. Wiesendanger, H.-J. Güntherodt (Ed.): *Scanning Tunneling Microscopy II. Springer Series in Surface Science. Vol. 28.* 2nd edn. (Springer, Berlin 1995) (a set of papers on novel applications of STM and on various scanning probe microscopy techniques)

8. Atomic Structure of Clean Surfaces

Most studies in surface science start with atomically clean substrate surface. Hence, knowledge of the atomic structure of clean surfaces is of great importance. It appears that the majority of surfaces, especially those of semiconductors, are essentially modified with respect to the corresponding atomic planes in the bulk. In this chapter, the general types of such modifications, called relaxation and reconstruction, are first introduced and then illustrated by several examples. Of course, these examples cannot cover the diversity of atomic structures occurring at clean surfaces of crystals, but they should give an impression of what the surface might look like and what kind of physics might stand behind a particular surface structure.

8.1 Relaxation and Reconstruction

Imagine that an infinite crystal is broken along a certain crystallographic plane and consider the surface atomic structure of the semi-infinite crystal formed in this process. Due to the absence of neighboring atoms on one side, the forces acting on the surface atoms are modified. Therefore, one expects that the equilibrium structure of the top atomic layer differs from that of the corresponding atomic plane in the bulk. Indeed, as it happens, the ideal bulk-like arrangement of the surface atoms appears to be the exception rather than the rule. Two general types of atomic rearrangements are identified as follows:

- relaxation and
- reconstruction.

Relaxation. The case when the atomic structure of the topmost layer is the same as in the bulk but the first interlayer spacings are modified, is defined as *normal relaxation* (Fig. 8.1a). In the uncombined form, normal relaxation is observed in metals. In most cases, contraction of the first interlayer spacing occurs. For deeper layers, the deviations from the bulk interlayer spacings are damped with depth and often in an oscillatory way. In addition to normal relaxation, sometimes uniform displacement of the topmost layers parallel to the surface takes place. This case is defined as *parallel* or *tangential relaxation* (Fig. 8.1b) and occurs mainly at high-index surfaces with low atomic density.

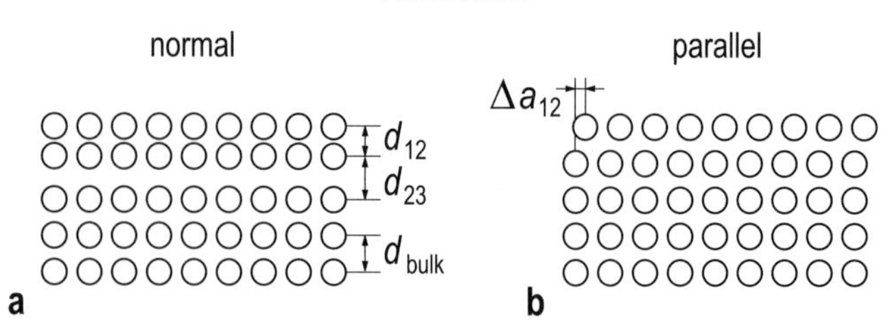

Fig. 8.1. Schematic illustration of (**a**) normal and (**b**) lateral relaxation in the top atomic layers of a semi-infinite crystal

Reconstruction. The abundance of cases when the atomic structure of the top layer is modified is referred to as *reconstruction*. In most instances, the reconstructed surface is characterized by symmetry and periodicity different from those of the bulk. However, even if the 1×1 lattice is preserved but the atomic arrangement within a 2D unit mesh (i.e., the basis) is modified, the case is also assigned to reconstruction.

Depending on whether the number of atoms in the top layer (or layers) is preserved or not, each reconstruction can be attributed to one of two types, namely,

- conservative reconstructions or
- non-conservative reconstruction.

In *conservative reconstructions*, the number of atoms is conserved and reconstruction involves only displacement of surface atoms from the ideal sites. Therefore, conservative reconstruction is sometimes called *displacive reconstruction* in the literature. A simple example of conservation reconstruction is the pairing of the top atoms, which results in a surface lattice with double periodicity (Fig. 8.2a). A more sophisticated example of conservative multilayer reconstruction is shown in Fig. 8.2c: though the number of atoms in each of the top three layers differs from that in the bulk atomic layer, their total corresponds to an integer number (here, two) of bulk-like layers.

In *non-conservative reconstructions*, the number of atoms in the reconstructed layer is changed in comparison with the bulk. An illustrative example is the missing-row reconstruction (Fig. 8.2b), in which every other atomic row is removed leaving half of the topmost atoms at the surface. Figure 8.2d demonstrates the case of non-conservative reconstruction, which involves reordering of the top three layers. Here, the total number of atoms in the reconstructed layers is not equal to the number of atoms in an integer number of bulk layers.

While reconstructions occur only for a limited number of metal surfaces, they are a common feature for most semiconductor surfaces. In semiconductors, ideal bulk-like termination of the crystal is unstable due to the presence of the high surface density of dangling (unsaturated) bonds. To minimize the

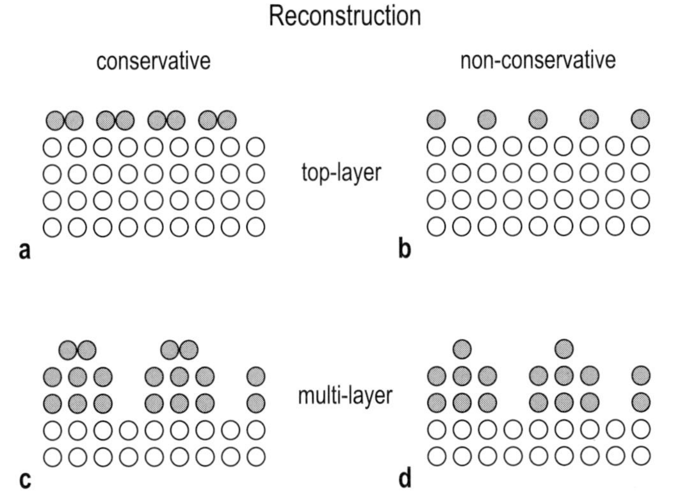

Fig. 8.2. Schematic illustration of possible types of reconstruction. (**a**) and (**c**) represent the conservative reconstruction, i.e., the case when the number density of surface atoms is preserved (**b**) and (**d**) represent the non-conservative reconstruction, i.e., the case when the number density of surface atoms is modified. In (**a**) and (**b**) only topmost atoms are involved in reconstruction; in (**c**) and (**d**) atoms of several top atomic layers are involved in reconstruction

surface free energy, the surface atoms are displaced from their original positions in order to saturate the dangling bonds by forming the new bonds with each other. A further gain in surface energy is achieved due to electric charge transfer between residual dangling bonds, making some of them empty and others filled. This mechanism is referred to as *autocompensation* [8.1]. On the other hand, atom displacements induce lattice stress which increases the surface free energy. The interplay of these trends leads to the formation of a particular reconstruction. Reconstruction in the top surface layers is generally accompanied by relaxation of the deeper layers.

8.2 Relaxed Surfaces of Metals

With a few exceptions, the surfaces of most metals are non-reconstructed and subjected to relaxation. For low-Miller-index surfaces, a purely normal relaxation typically occurs. Some high-Miller-index surfaces display concomitant normal and lateral relaxations.

8.2.1 Al(110)

The Al(110) surface (Fig. 8.3) furnishes a typical example of relaxation at a low-index metal surface. It exhibits a purely normal relaxation. Like in most

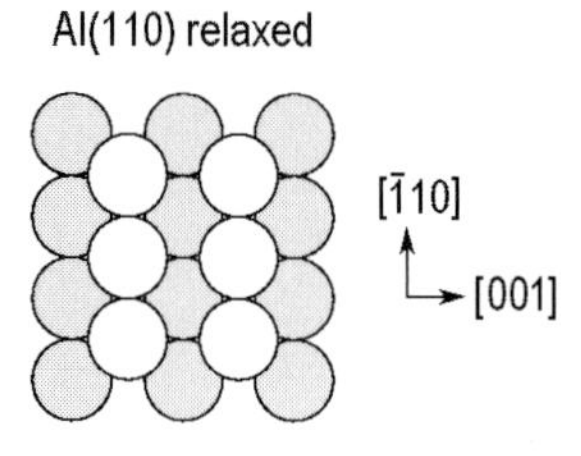

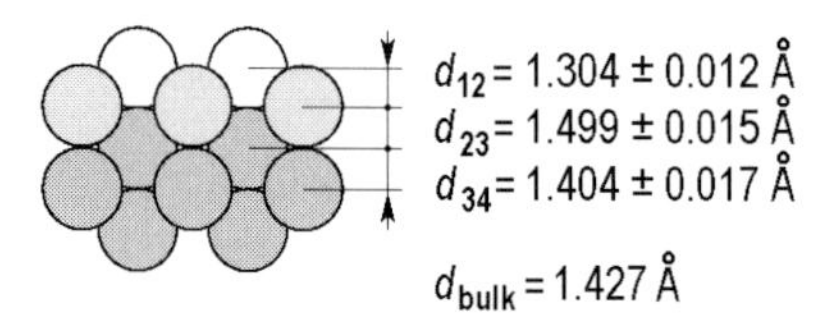

Fig. 8.3. Schematic diagram showing top and side views of the relaxed Al(110) surface. The relaxation of the first three interlayer spacings, expressed as deviations from the bulk value, were determined by I–V LEED to be $\Delta_{12} = -8.6\%$, $\Delta_{23} = +5.0\%$ and $\Delta_{34} = -1.6\%$ (after Andersen et al. [8.2])

metals, the first interlayer spacing, d_{12}, is contracted. The determined variation of the first interlayer spacing normalized to the bulk interlayer spacing, $\Delta_{12} = (d_{12} - d_{\text{bulk}})/d_{\text{bulk}}$, equals -8.6%. In general, the values for fcc and bcc metal surfaces fall into the range from zero to a few dozen percent, being greater for surfaces with lower packing density of atoms (Fig. 8.4). The variations in the interlayer spacings are damped with depth and often in an oscillatory way (Table 8.1). In the particular case of Al(110), the second interlayer spacing is expanded by $+5.0\%$ and the third interlayer spacing is again constricted, albeit slightly, by -1.6% [8.2].

8.2.2 Fe(211)

The ideal and relaxed (211) surfaces of iron, a bcc metal, are shown in Fig. 8.5. An essential feature of the Fe(211) surface is that the top atoms in the bulk-like positions are not symmetrically located with respect to the second-layer atoms. Shifting the first-layer atoms to symmetric locations would cause each atom to have four nearest neighbors in the second layer compared with only two for the bulk-like structure. However, such a shift would tend to shorten the interatomic distance between the atoms in the first and third layers. Thus, the lateral shift of the top layer is affected by the relaxation of the interlayer spacings. LEED I–V analysis yields the following values for the relaxed Fe(211) surface: first-to-second layer lateral shift is $-14.5 \pm 1.8\%$ and second-to-third is $+2.2 \pm 1.8\%$, the first interlayer spacing d_{12} is contracted by $10.4 \pm 2.6\%$, the second interlayer spacing d_{23} is expanded by $5.4 \pm 2.6\%$ and the third interlayer spacing d_{34} is contracted by $1.3 \pm 3.4\%$ [8.12]

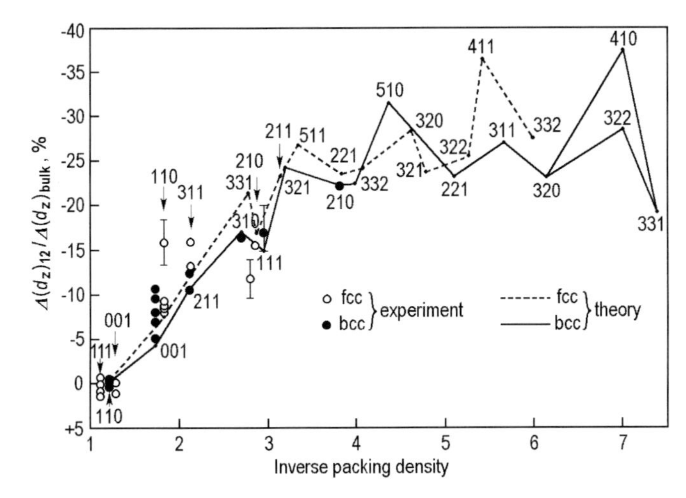

Fig. 8.4. Normal top-layer relaxation (compression) versus inverse packing density of surface atoms for various surfaces of fcc and bcc metals. Packing density is defined as the fraction of the area in one plane occupied by atoms of radii equal to one half the bulk nearest-neighbor distance. The lower the packing density, the greater the compressive relaxation (after Jona and Marcus [8.3])

Table 8.1. Multilayer relaxation of bcc Fe and fcc Ni metal surfaces. The relaxation values are given as a percentage of the spacing between layers i and j, where $i, j =$ layer numbers. The bulk value of d indicates whether the relaxation values written on the same line belong to normal (d_z) or tangential (d_x) relaxation

Surface	Bulk value, Å	Relaxation, %				Ref.
		Δ_{12}	Δ_{23}	Δ_{34}	Δ_{45}	
Fe						
(100)	$d_z = 1.433$	-5 ± 2	$+5 \pm 2$			[8.4]
(110)	$d_z = 2.027$	unrelaxed				[8.5]
(111)	$d_z = 0.827$	-16.9 ± 3	-9.8 ± 3	$+4.2 \pm 4$	-2.2 ± 4	[8.6]
(210)	$d_z = 0.641$	-22.0 ± 5	-11.1 ± 5	$+17.0 \pm 5$	-4.8 ± 5	[8.7]
	$d_x = 1.923$	$+7.1 \pm 1.6$	$+1.4 \pm 2.6$	$+0.0 \pm 2.6$	$+4.0 \pm 0.6$	
Ni						
(100)	$d_z = 1.762$	unrelaxed				[8.8]
(110)	$d_z = 1.246$	-8.5 ± 1.5	$+3.5 \pm 1.5$	$+1.0 \pm 1.5$		[8.9]
(111)	$d_z = 2.033$	$+3.0 \pm 1.5$				[8.10]
(311)	$d_z = 1.062$	-15.9 ± 1.0	$+4.1 \pm 1.5$	-1.6 ± 1.6		[8.11]
	$d_x = 1.878$	-0.8 ± 1.9	-1.4 ± 1.9	-0.5 ± 3.2		

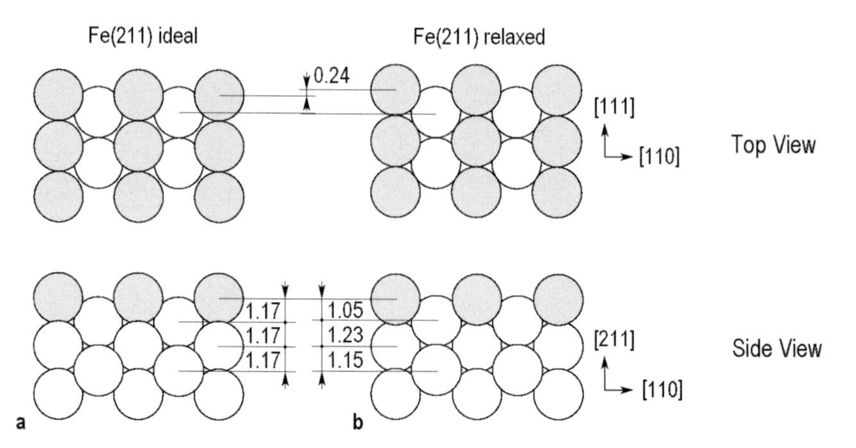

Fig. 8.5. Schematic diagram showing top and side views of the (**a**) unrelaxed and (**b**) relaxed Fe(211) surface. The relaxation includes a lateral shift of the top layer by 14.5% and oscillatory normal relaxation of the interlayer spacings, $\Delta_{12} = -10.4\%$, $\Delta_{23} = +5.4\%$ and $\Delta_{34} = -1.3\%$ (after Sokolov et al. [8.12])

8.3 Reconstructed Surfaces of Metals

In contrast to most metal surfaces, which are not reconstructed, the surfaces of some noble and near-noble fcc metals, Au, Ir, and Pt, and bcc transition metals, W and Mo, display reconstructions.

8.3.1 Pt(100)

Platinum is a fcc metal and its ideal non-reconstructed (100) surface comprises an array of atoms forming a square lattice. An ideal Pt(100) surface is loosely packed and is subjected to a large tensile stress and, hence, unstable. It prefers a higher in-plane atomic density and, indeed, becomes reconstructed into a close-packed quasi-hexagonal layer (Fig. 8.6 and Fig. 8.7). The reconstruction results in an excess of ∼20% atomic density. Another consequence of top layer reordering is the modification of its bonding to the underlying atomic layer. Thus, the reconstruction is controlled by the balance between the energy gain associated with the increase in atomic density and the energy loss due to misfit stress between mismatched top and subsurface layers. Because of the different symmetries of overlayer and substrate, the structure has either very large periodicity (for example, 20×5 or 29×5) or is incommensurate, though in early LEED studies it was considered as 1×5. Qualitatively similar (though not identical) quasi-hexagonal reconstructions have been reported also for Au(100) and Ir(100) surfaces.

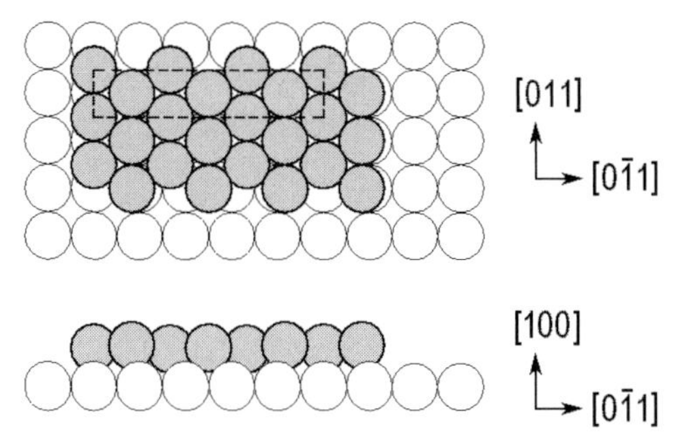

Fig. 8.6. Schematic diagram showing hexagonal packing of top-layer Pt atoms (shown by hatched circles) on the square Pt(100) atomic plane (shown by open circles). In the ideal scheme shown, the surface is characterized by 1×5 periodicity. In practice, the superstructure unit cell is much larger or might even be incommensurate (see Fig. 8.7)

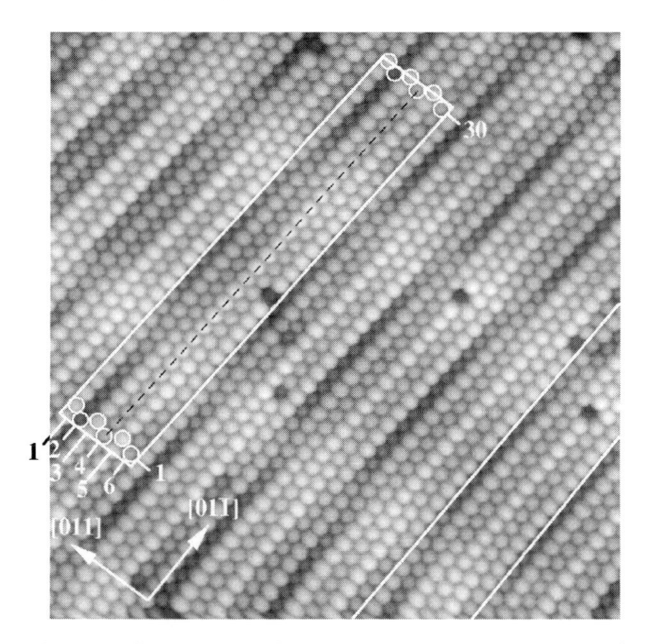

Fig. 8.7. STM image of a hexagonally reconstructed Pt(100) surface. The superstructure unit cell extends over 30 atoms in [011̄] and six atoms in the [011] direction. However, there is a long-range modulation visible in the [011] direction, which indicates that the reconstructed surface does not have an exact fivefold periodicity (after Ritz et al. [8.13])

8.3.2 Pt(110)

An ideal (110)-terminated fcc structure consists of atoms arranged in rows along the $[\bar{1}10]$ direction (Fig. 8.8a). For a real atomically clean Pt(110) surface (for example, prepared by sputter cleaning followed by annealing), a doubling of the period along the [001] direction was found. Various experimental and theoretical investigations have shown that the doubled periodicity at Pt(110) is due to the missing-row structure (Fig. 8.8b). In particular, the missing-row arrangement of the top atoms is clearly seen in STM images (Fig. 8.9). Detailed structural analysis has revealed noticeable distortions in the subsurface layers, including contraction in the first interlayer spacing, slight row pairing in the second layer, and buckling in the third layer.

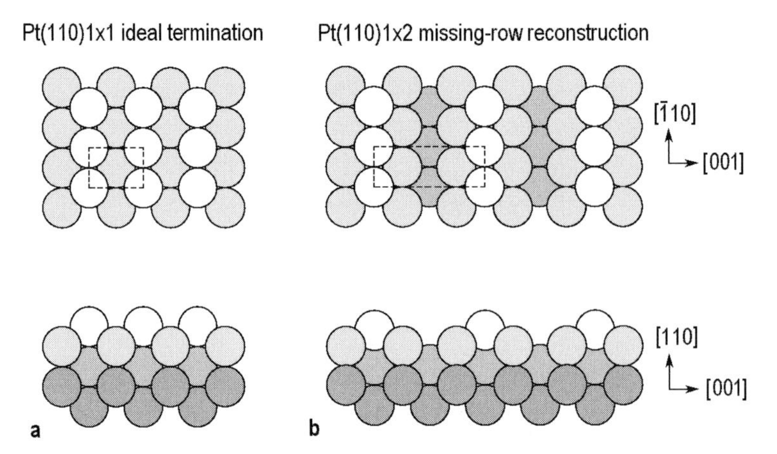

Fig. 8.8. Schematic diagram showing atomic arrangement of (**a**) an ideal non-reconstructed Pt(110)1×1 surface and (**b**) reconstructed Pt(110)2×1 with a missing-row structure

The formation of the missing-row structure can be qualitatively understood, if one takes into account that, among other low-Miller-index surfaces, the (111) surface has the lowest surface energy. The missing-row structure is essentially built of (111) microfacets. As a result, the surface energy is lowered. A missing-row 2×1 structure also occurs at a clean Au(110) surface, while at the Ir(110) and Pd(110) surfaces it seems to be stabilized by some impurity.

8.3.3 W(100)

Tungsten is a bcc metal, hence, the top atoms at the non-reconstructed W(100)1×1 surface make a square structure with a period of the bulk lattice constant, 3.16 Å (Fig. 8.10a). At low temperatures (below ~300 K), the

Fig. 8.9. STM image of a Pt(110)2×1 surface with a deposited submonolayer amount of Pt. The deposited Pt atoms occupy the sites in the troughs of the missing-row reconstruction (after Linderoth [8.14])

$c(2\times2)$ ($\sqrt{2}\times\sqrt{2}$-$R45°$) reconstruction is detected at the W(100) surface. The top atoms in the low-temperature structure are shifted alternately from their ideal positions by a small amount (~0.2 Å) along the [011] and $[0\bar{1}\bar{1}]$ directions, thus forming zigzag chains along the $[01\bar{1}]$ direction (Fig. 8.10b). As a result, the surface gains $c(2\times2)$ periodicity. There are two domains, since atoms may also be shifted along the $[0\bar{1}1]$ and $[01\bar{1}]$ directions. The transition from 1×1 to $c(2\times2)$ is reversible.

8.4 Graphite Surface

Graphite furnishes a vivid example of a layered material (Fig. 8.11). Within an atomic layer, each C atom bonds three neighboring C atoms, thus forming a plane hexagonal structure with two C atoms in the unit mesh. The lateral bonds are short (1.415 Å) and extremely strong. In contrast, the interlayer spacing is large (3.354 Å) and interlayer bonding is weak. As a result, the outer (0001) surface of the graphite preserves the non-reconstructed bulk-like structure and relaxation (contraction) of the distance between the first and second layers is small (0.05 Å, i.e., 1.5% of the bulk value) [8.15]. The graphite surface is chemically inert and is a suitable substrate for studying physisorption phenomena. The clean graphite surface can be easily prepared using "Scotch" tape: just press it onto the flat surface and then pull it off. The tape removes a thin graphite layer, leaving a freshly cleaved surface.

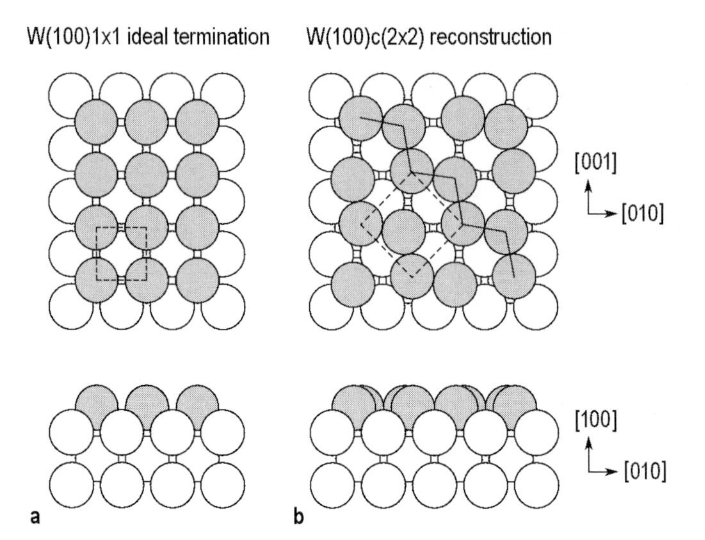

Fig. 8.10. Schematic diagram showing atomic arrangement of the (**a**) ideal non-reconstructed W(100)1×1 surface and (**b**) reconstructed W(100)$\sqrt{2}\times\sqrt{2}$-$R45°$ surface with zigzag chains (indicated by a solid line) formed by alternative shifting of top atoms from their ideal positions. The unit meshes are outlined in dashed lines

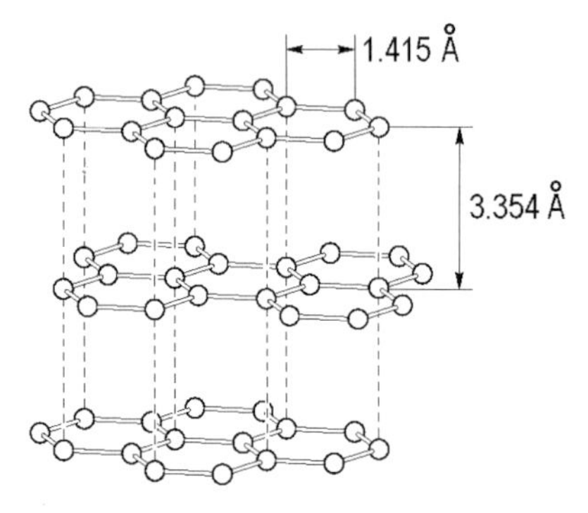

Fig. 8.11. Structure of graphite is built of plane atomic layers weakly bonded to each other. The outer (0001) surface is non-reconstructed and the first interlayer contraction is only 0.05 Å (1.5% of the bulk value)

8.5 Surfaces of Elemental Semiconductors

Silicon and germanium have a similar *diamond structure*. Recall that the space lattice of the diamond structure is fcc and the basis contains two atoms at 000 and $\frac{1}{4}\frac{1}{4}\frac{1}{4}$ of the non-primitive fcc unit cube. Atoms in the diamond structure are tetrahedrally coordinated: each atom has four nearest neighbors. The lattice constant is 5.43 Å for Si and 5.65 Å for Ge.

8.5.1 Si(100)

The ideal bulk-like Si(100) surface comprises a square lattice of the top Si atoms, where each atom is bonded to two subsurface atoms and has two dangling bonds (Fig. 8.12a). In the reconstructed Si(100) surface, the surface atoms are paired to form dimers, thus halving the number of dangling bonds (Fig. 8.12b). The dimers are arranged in rows and the surface has a 2×1 periodicity. The dimer structure was proposed in 1959 in the pioneering work of Schlier and Farnsworth [8.16], where the 2×1 periodicity of the Si(100) surface was first revealed by LEED.

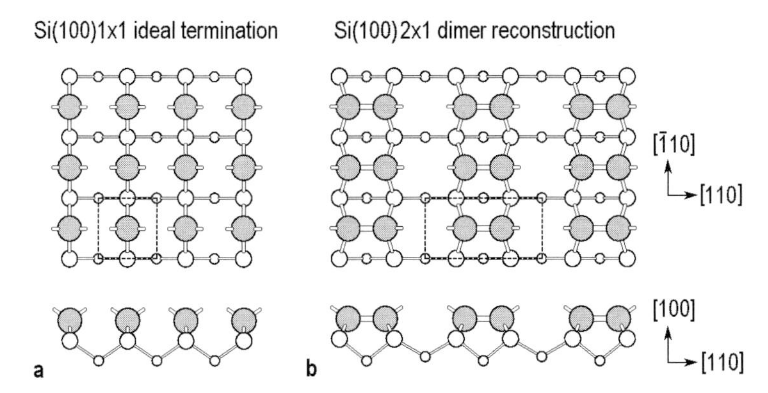

Fig. 8.12. Schematic diagram of (**a**) an ideal non-reconstructed Si(100)1×1 surface and (**b**) a reconstructed Si(100)2×1 surface. The topmost Si atoms are hatched and the surface unit meshes are outlined

The rows of dimers are clearly resolved in the STM images displayed in Fig. 8.13. Dimers show up as oblong protrusions in the filled-state image and as pairs of round protrusions in the empty-state image. Though most dimers are seen in STM as being symmetric, it is well established that, in fact, they are buckled by about 18°. Moreover, the geometric buckling is coupled to electronic charge transfer from the lower atom to the upper atom. At room temperature, the dimers flip dynamically between their two possible orientations. The flip-flop motion of the buckled dimers is random and occurs on a time scale shorter than the STM measurement time. Thus, STM images reflect the time-average position of dimers and show the asymmetric dimers as symmetric. However, one can see that dimer buckling is fixed in the rows adjacent to the step edge and in the vicinity of some surface defects. These rows show up in the filled-state STM image as zigzag chains.

If the Si(100) surface is cooled below ~200 K, the dimer flipping is frozen and the number of buckled dimers seen in the STM images considerably increases. The interaction between rows results in ordering of the buckled dimers into an antisymmetric structure with dimers in the neighboring rows

Fig. 8.13. Filled-state STM image of a Si(100)2×1 surface. The inset shows a high-resolution STM image of the dimer row structure acquired at dual polarity, in filled states (upper part) and empty states (lower part). The 2×1 unit mesh is outlined

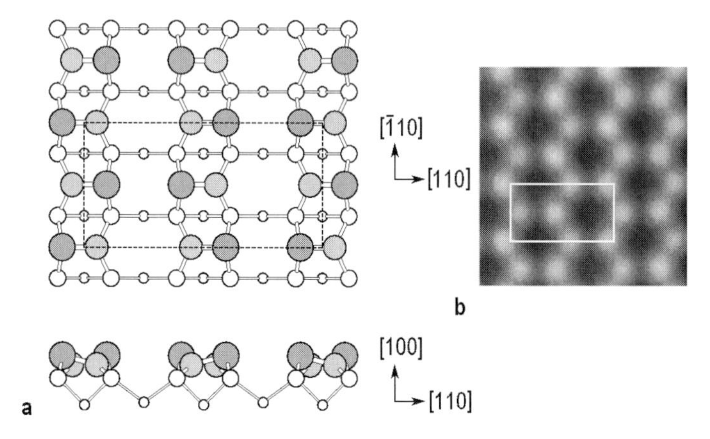

Fig. 8.14. (**a**) Schematic diagram of the Si(100)$c(4\times2)$ structure formed by correlated buckling of adjacent dimers. The upper and lower atoms of the buckled dimer are shown by dark-gray and light-gray circles, respectively. (**b**) Field-state STM image of the Si(100)$c(4\times2)$ surface obtained by cooling the sample to 63 K (after Yokoyama and Takayanagi [8.17]). The round protrusions in the STM image correspond to the upper atoms in the buckled dimers. The $c(4\times2)$ unit mesh is outlined

being buckled in the opposite directions. This structure has a $c(4\times2)$ periodicity (Fig. 8.14) and is regarded as a ground-state structure of the atomically clean Si(100) surface. It should be noted that the atomically clean Ge(100) surface is also built of buckled dimers (Table 8.2), whose ordering at low temperatures also results in the $c(4\times2)$ reconstruction.

Table 8.2. Bulk and (100) dimer reconstruction surface structural parameters for elemental semiconductors. a_0 is the lattice constant, a the a bond length (see Fig. 8.12)

Material	a_0, Å	a, Å	Dimer length, Å	Buckling angle, °	Ref.
Si	5.431	2.35	2.28 ± 0.05	18 ± 1	[8.18]
Ge	5.658	2.45	2.34 ± 0.01	19 ± 1	[8.19]

8.5.2 Si(111)

Atomically clean Si(111) surface exhibits two main reconstructions, 2×1 and 7×7, although some other 7×7 related reconstructions can occur under appropriate preparation conditions. The 2×1 surface structure is produced by cleavage of the Si crystal along the (111) plane. The 2×1 structure is metastable and converts irreversibly to the 7×7 reconstruction upon heating to about 400°C. The 7×7 reconstruction is stable up to about 850°C, where it undergoes order–disorder transition to the "1×1" structure. This transition is reversible and, upon slow cooling through the transition temperature region, the 7×7 reconstruction is restored.

Si(111)2×1. The *π-bonded chain model* is accepted as a valid description of the cleaved Si(111)2×1 surface structure. The π-bonded chain model was proposed by Pandey [8.20], who suggested a drastic restructuring of the surface in which the bonds between half of the second and third layer atoms are rearranged so that the six-member rings of the bulk-like terminated surface (see side view of Fig. 8.15a) convert to a succession of seven- and five-member rings (Fig. 8.15b). As a result, Si atoms of the first and second top layers become arranged into alternating zigzag chains. The Si atoms in the upper chains are bonded to each other by π-bonds and are shifted upward and downward making the π-bonded chains slightly buckled [8.21]. Note that Pandey's π-bonded chain model is also valid for a description of the cleaved Ge(111)2×1 surface.

Si(111)7×7. Si(111)7×7 is a remarkable structure as, on the one hand, it shows how sophisticated and fascinating surface reconstruction might be and, on the other hand, yields a vivid example of how the combined efforts of many research groups can result in conclusive solution of a very complicated structure.

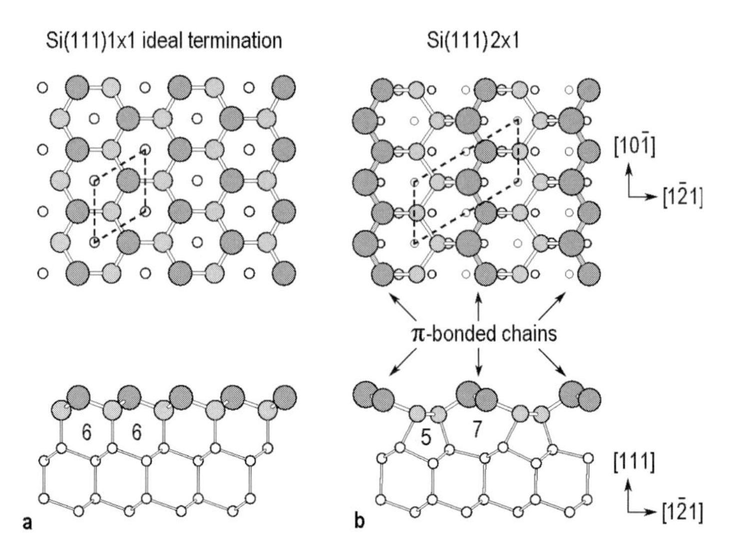

Fig. 8.15. (**a**) Schematic diagram of the ideal non-reconstructed Si(111)1×1 surface. The Si atoms of the first, second, and third top layers are shown by dark gray, light gray and white circles, respectively. Note that atoms of the first and second layers constitute the Si(111) double layer. The six-member rings are indicated in the side view. (**b**) Pandey's π-bonded chain model of the reconstructed Si(111)2×1 surface as occurs upon cleavage in vacuum. The Si atoms forming the π-bonded chains are shown by dark gray circles. The five- and seven-member rings are indicated in the side view. The surface unit meshes are outlined

The first LEED observation of Si(111)7×7 dates back to 1959 [8.16], but it was only in 1985 when Takayanagi et al. [8.22] proposed their famous *dimer–adatom–stacking fault (DAS) model* of the 7×7 reconstruction, summing up research activity of the surface science community over period of 25 years. The concept of adatoms in the outermost layer was proposed by Harrison [8.23]. The LEED [8.24] and RBS [8.25, 8.26] data revealed the presence of a stacking fault in the surface layers. Binnig et al. [8.27] obtained the first STM image of the 7×7 surface showing a deep hole in the corner and 12 protrusions inside the unit cell (Fig. 8.16). Based on the above observations, Himpsel [8.28] and McRae [8.29] predicted that the boundary between the faulted and unfaulted triangular subunits incorporates an array of dimers and deep holes. Finally, Takayanagi et al. using TED (Fig. 8.17) [8.22] correctly pieced together a model with dimers, adatoms and stacking faults.

Figure 8.18 shows the DAS model of the Si(111)7×7 structure. The basic features of the model, constituting each unit cell, are as follows.

- 12 adatoms,
- two triangular subunits of which one has a stacking fault,
- nine dimers per unit cell which border the triangular subunits,
- one deep corner hole.

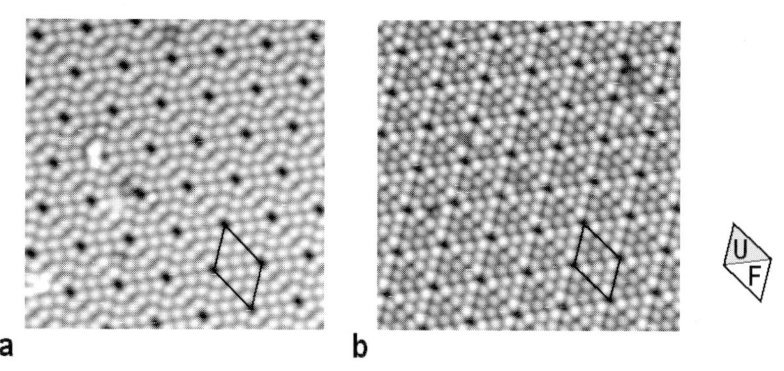

Fig. 8.16. 200×200 Å^2 constant-current STM images of the Si(111)7×7 surface acquired (**a**) at +1.6 V tip bias voltage (empty states) and (**b**) at −1.6 V tip bias voltage (filled states). Round bright protrusions correspond to Si adatoms, round dark depressions to corner holes. One can see that in the filled state image the faulted unit halves look slightly higher (brighter)

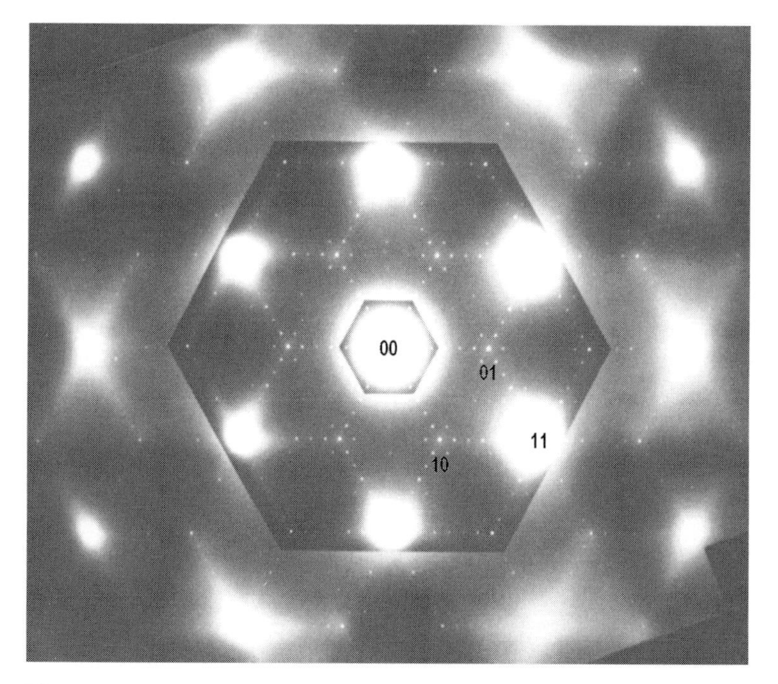

Fig. 8.17. Transmission electron diffraction pattern (TED pattern) of the reconstructed Si(111)7×7 surface taken by an incident electron beam almost exactly parallel to the surface normal. Hexagonal areas are reproduced from films of different exposures (after Takayanagi et al. [8.22])

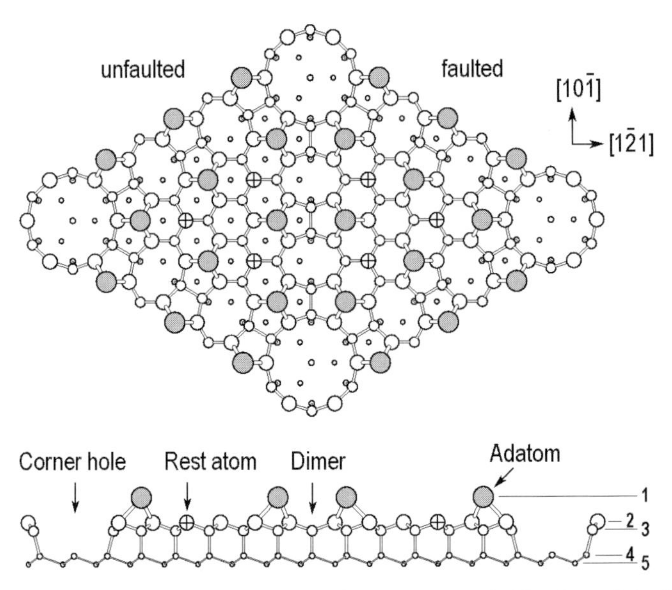

Fig. 8.18. Schematic diagram showing top and side view of the 7×7 dimer–adatom–stacking fault (DAS) model. Adatoms are shown by gray circles, the atoms of the deeper layers are shown by white circles of size decreasing with depth. The dangling-bond rest atoms are shown by circles enclosing a cross. The main structural elements (corner hole, rest atom, dimer, adatom) and numbering of the atomic layers are indicated (after Takayanagi et al. [8.22])

Adatoms occupy the T_4 sites above the second layer and form a local 2×2 structure. The Si(111) double layer which incorporates the second and third layers is made up of *triangular subunits.* Subunits are alternately faulted and unfaulted with respect to the underlying bulk and bounded by *dimer rows.* At the corner of the cell, a 12-member ring of atoms surrounds a *corner hole.* 36 of the 42 first-layer atoms are bonded to adatoms and their dangling bonds are saturated. The remaining six atoms, which preserve their dangling bonds, are called *rest atoms.* One can see that the 7×7 DAS structure contains in total 19 dangling bonds per unit cell, including those of 12 adatoms, six rest atoms and one in the corner hole. Note that in the case of an ideal bulk-like termination the 7×7 mesh contains 49 dangling bonds.

Besides the 7×7 structure, there is a family of other $(2n+1)\times(2n+1)$ DAS reconstructions, 3×3, 5×5, 9×9, 11×11, etc. (Fig. 8.19). The 7×7 has a lower energy than other DAS reconstructions. The latter occur under non-equilibrium conditions, for example, at an intermediate stage of the 2×1-to-7×7 transition, under epitaxial Si growth at temperatures below the one required for perfect epitaxy, or upon rapid quenching of a high-temperature disordered "1×1" structure.

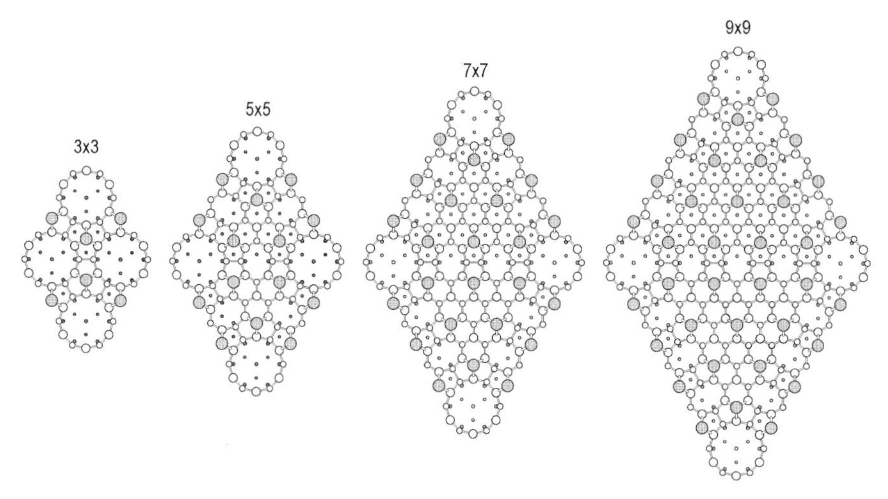

Fig. 8.19. Members of the family of $(2n+1)\times(2n+1)$ DAS reconstructions, 3×3, 5×5, 7×7, 9×9

8.5.3 Ge(111)

The cleaved Ge(111) surface displays a 2×1 π-bonded chain structure similar to that of the cleaved Si(111)2×1 surface. Like Si(111)2×1, Ge(111)2×1 reconstruction is also metastable but, in contrast to the former, converts irreversibly to quite a different structure, Ge(111)c(2×8), upon heating to about 200°C. A well-ordered Ge(111)c(2×8) surface can also be prepared by several cycles of Ar^+ ion sputtering and annealing. At about 300°C, the Ge(111)c(2×8) surface undergoes a reversible disordering transition to a "1×1" structure.

Ge(111)*c*(2×8). The atomic structure of Ge(111)c(2×8) is described by a simple *adatom* model in which the bulk-like terminated Ge(111)1×1 surface is decorated by Ge adatoms occupying T_4 sites (Fig. 8.20). The c(2×8) periodicity is constructed of alternating stacking of the 2×2 (hexagonal) and c(2×4) (rectangular) subunits. Ge adatoms saturate 3/4 of the ideal surface dangling bonds, leaving 1/4 of the dangling bonds unsaturated. The surface atoms which preserve their dangling bonds are referred to as *rest atoms*. Nominally, there are in total four dangling bonds per c(2×8) primitive unit mesh of eight 1×1 sites (two of adatoms and two of rest atoms). The charge transfer between adatoms and rest atoms leads to filled states mostly localized on the rest atoms and empty states on the adatoms, which eliminates the partly filled dangling-bond states and stabilizes the surface. The difference in the environment of the rest atom in the 2×2 and c(2×4) subunits leads to their slight asymmetry: the 2×2-subunit rest atoms is raised more (by $\sim$ 0.03 Å) and acquires slightly more electron charge than the other.

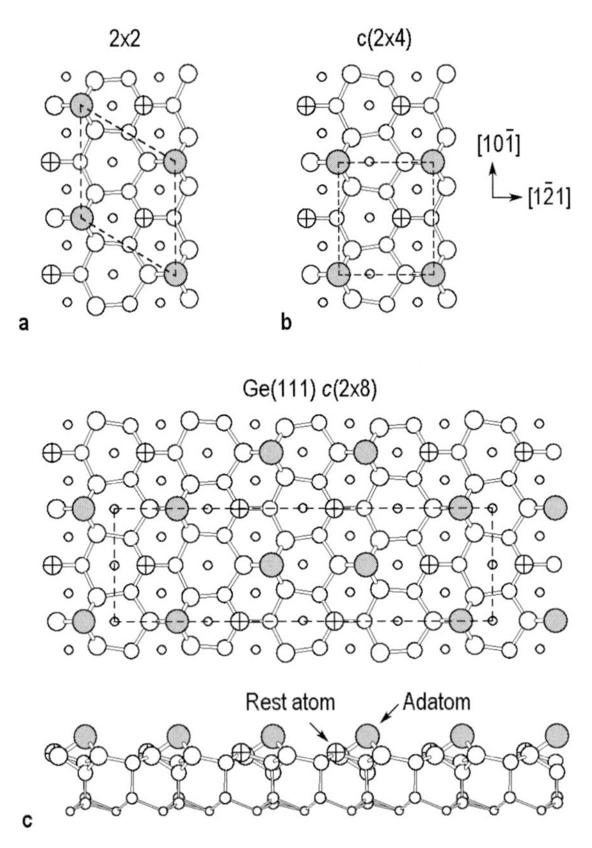

Fig. 8.20. Schematic diagram of the Ge $c(2\times8)$ atomic structure. The structure is built of Ge adatoms (shown by gray circles) residing on the bulk-like terminated Ge(111)1×1 surface (shown by white circles, including the dangling-bond rest atoms marked by crosses). Alternative stacking of the (**a**) hexagonal 2×2 and (**b**) rectangular $c(2\times4)$ subunits produces (**c**) the $c(2\times8)$ structure

8.6 Surfaces of III-V Compound Semiconductors

Like other III-V compound semiconductors, GaAs has a *zinc blende structure* with an fcc space lattice and two atoms in the basis: a Ga atom at 000 and an As atom at $\frac{1}{4}\frac{1}{4}\frac{1}{4}$ of the non-primitive fcc unit cube.

8.6.1 GaAs(110)

The GaAs(110) is the cleavage surface of the crystal. An ideal GaAs(110) surface (Fig. 8.21a) consists of zigzag chains of alternating gallium and arsenic atoms. Each surface Ga (As) atom has two bonds to the neighboring surface As (Ga) atoms within the chain and one backbond to the As (Ga) atom in the layer beneath, leaving one dangling bond. In the reconstructed

GaAs(110) surface, the electronic charge is transferred from Ga to As and the occupied state density is concentrated around the surface As atoms and the unoccupied density around the Ga atoms (Fig. 8.22). The charge transfer is accompanied by an approximately bond-length-conserving rotation with As atoms moving upward and Ga atoms moving downward (Fig. 8.21b). Note that the reconstructed surface still preserves an ideal 1×1 periodicity. This (110) surface reconstruction is typical for other zinc blende structure III-V compound semiconductors with buckling angle $\alpha = 29° \pm 3°$ being independent of the material (Table 8.3).

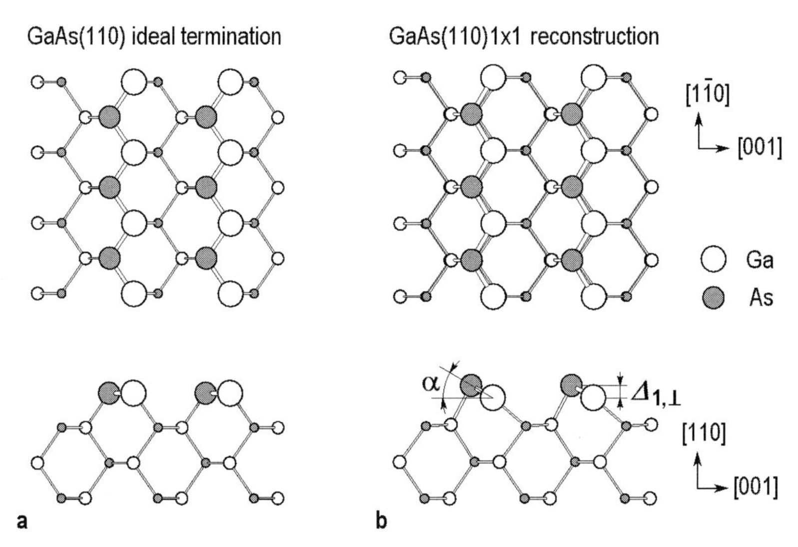

Fig. 8.21. Schematic diagram of the atomic arrangement of (**a**) an ideal non-reconstructed GaAs(110)1×1 surface and (**b**) reconstructed GaAs(110)1×1 surface as it appears after cleavage in vacuum. As atoms are shown by hatched circles and Ga atoms by open circles

8.6.2 GaAs(111) and GaAs($\bar{1}\bar{1}\bar{1}$)

GaAs(111) is a *polar surface*, which means that it can be terminated ideally either by pure Ga or As atoms. By convention, the Ga-terminated (111) surface is referred to as a GaAs(111) surface or sometimes as a GaAs(111)A surface. Accordingly, the As-terminated surface is called a GaAs($\bar{1}\bar{1}\bar{1}$) or GaAs(111)B surface. Thus, the $\langle 111 \rangle$-oriented wafer has a GaAs(111)A plane on the one side and a GaAs(111)B surface on the other side. This is a natural consequence of the fact that, in order to generate an alternate face below one layer, one needs to break three bonds instead of one (Fig. 8.23). It appears that both GaAs(111) and GaAs($\bar{1}\bar{1}\bar{1}$) surfaces exhibit 2×2 reconstructions, but these reconstructions are essentially different.

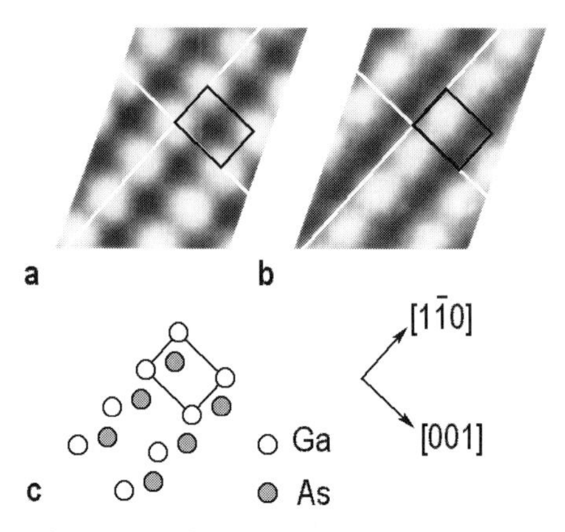

Fig. 8.22. Constant-current STM images of the GaAs(110) surface acquired at sample voltages of (**a**) +1.9 V (empty states) and (**b**) −1.9 V (filled states). (**c**) Top view of the surface atoms. Arsenic atoms are represented by hatched circles and gallium atoms by open circles. The rectangule indicates a unit cell, whose position is the same in all three figures. Thus, the empty state density is localized around Ga atoms and the filled state density around As atoms (after Feenstra et al. [8.30])

Table 8.3. Bulk and (110) surface structural parameters for zinc blende III-V compound semiconductors. a_0 is the bulk lattice constant, $\Delta_{1,\perp}$ is a height difference between anion and cation atoms, α is the buckling angle (see Fig. 8.21). The buckling parameters were determined using LEED I–V analysis [8.31]

Material	a_0, Å	$\Delta_{1,\perp}$, Å	α, °
GaP	5.451	0.54	28.4
AlP	5.467	0.63	27.5
GaAs	5.653	0.65	27.4
AlAs	5.660	0.65	27.3
InP	5.869	0.69	30.4
InAs	6.058	0.78	31.0
GaSb	6.096	0.77	30.0
InSb	6.479	0.78	28.8

GaAs(111)2×2. An atomically clean GaAs(111) surface is prepared conventionally by repeating the Ar ion sputtering/annealing cycles until a sharp 2×2 LEED pattern is observed. The 2×2 reconstruction of the Ga-terminated GaAs(111) surface is described by a model in which one out of every four Ga surface atoms is missing and the spacing between the surface Ga layer and subsurface As layer is strongly contracted to make an almost flat atomic configuration (Fig. 8.24).

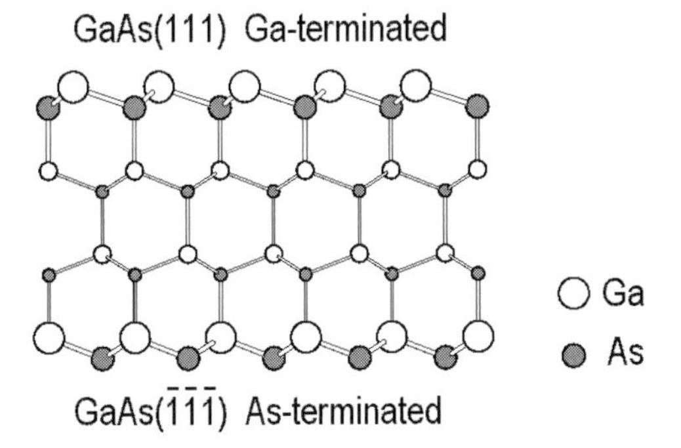

Fig. 8.23. Schematic diagram showing that the ⟨111⟩-oriented GaAs wafer has a Ga-terminated surface (referred to as GaAs(111) or GaAs(111)A) on the one side and an As-terminated surface (referred to as GaAs($\bar{1}\bar{1}\bar{1}$) or GaAs(111)B) on the other side. Ga atoms are shown by open circles and As atoms by hatched circles

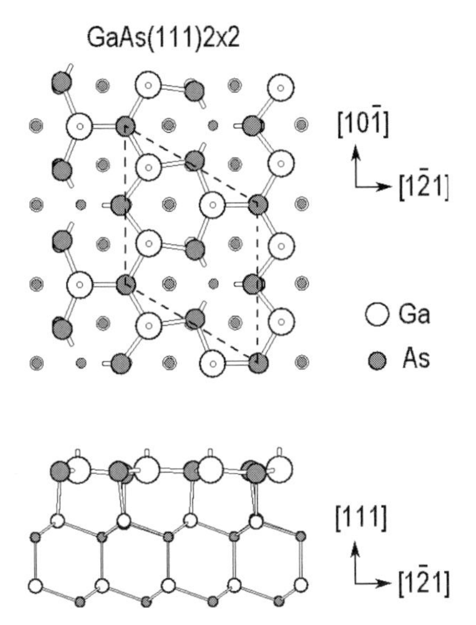

Fig. 8.24. Schematic diagram of the Ga-vacancy model of the GaAs(111)2×2 surface structure. Ga atoms are shown by open circles and As atoms by hatched circles. Surface atom dangling bonds are indicated

This reconstruction can be explained by the reduction in energy that is achieved by the complete charge transfer of electrons from Ga dangling bonds into As dangling bonds [8.32]. In the case of an ideal GaAs(111)A bulk-like surface, each Ga atom is bonded to three As atoms in the layer below, leaving one dangling bond normal to the surface partially filled, which is energetically unfavorable. The removal of the surface Ga atom produces three Ga associated dangling bonds and three As associated dangling bonds. The electronic charge transfer from Ga dangling bonds to As dangling bonds gives a stable surface with empty Ga dangling bonds and filled As dangling bonds. This 2×2 reconstruction has been found to be common for the (111)A (Group III atom-terminated) surface of zinc blende structure compound semiconductors (for example, GaP, GaSb, InSb).

GaAs($\bar{1}\bar{1}\bar{1}$)2×2. The GaAs($\bar{1}\bar{1}\bar{1}$) surface with a 2×2 reconstruction develops during and after molecular beam epitaxy growth under As-rich conditions. In contrast, the As deficiency (say, induced by heating of the sample) results in a surface with $\sqrt{19}\times\sqrt{19}$-$R23.4°$ structure. The 2×2 surface is accepted to consist of As trimers located in T_4 sites directly over the Ga atoms in the lower layer (Fig. 8.25). Each As atom in the trimer is bonded to the other two As trimer atoms and to a first-layer As atom. The trimers form a hexagonal array with a 2×2 periodicity. Besides the As trimer, one As rest atom is present in the 2×2 unit cell. A similar trimer structure was found at InAs($\bar{1}\bar{1}\bar{1}$) and InSb($\bar{1}\bar{1}\bar{1}$) surfaces.

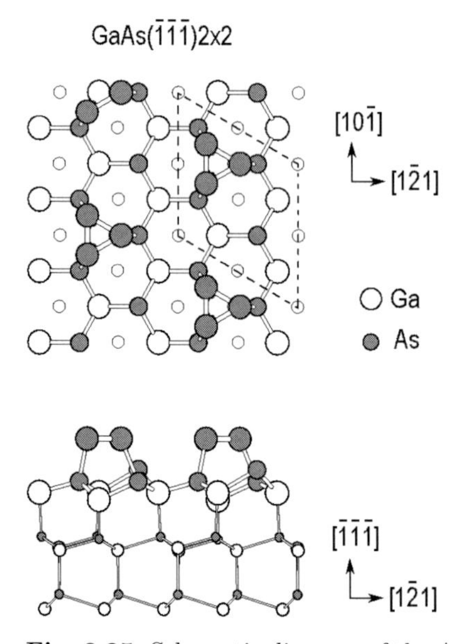

Fig. 8.25. Schematic diagram of the As-trimer model of the GaAs($\bar{1}\bar{1}\bar{1}$)2×2 surface structure. Ga atoms are shown by open circles and As atoms by hatched circles

Problems

8.1 Consider a family of DAS reconstructions (Fig. 8.18 and Fig. 8.19) and fill in Table 8.4.

Table 8.4. DAS model parameters. The number of dangling bonds, adatoms, and rest atoms is referred to a 7×7 unit cell. Total number of atoms includes all atoms incorporated in the first three top layers (see Fig. 8.18)

Unit cell	Number of				Total	Si atom
	dangling bonds	adatoms	rest atoms	dimers	number of atoms	density, ML
3×3						
5×5						
7×7	19	12	6	9	102	$\frac{102}{49} = 2.08$
9×9						
$(2n+1)\times(2n+1)$						

8.2 If a cleaved Ge(111)2×1 surface is converted by annealing to a Ge(111) $c(2\times8)$ surface, (111)-double-layer holes are created on the large terraces far from the steps, as shown in Fig. 8.26. Taking into account the atomic structure of the 2×1 and $c(2\times8)$ phases, explain the origin of the holes. What is the area fraction occupied by holes in the ideal case?

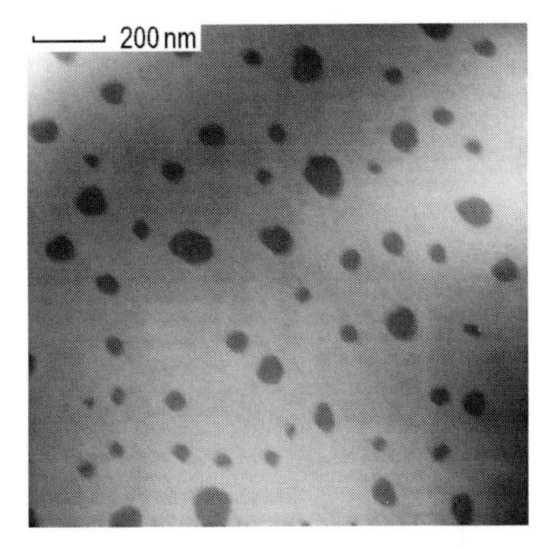

Fig. 8.26. Large-scale topographic STM image of the Ge(111)$c(2\times8)$ surface formed by annealing of the cleaved Ge(111)2×1 surface. Black areas are bilayer-deep holes (after Feenstra and Slavin [8.33])

8.3 Upon prolonged annealing, the 3×1 superstructure is formed at a Pt(110) surface. Like the Pt(110)2×1 reconstruction (Fig. 8.8), the 3×1 reconstruction is also of the missing-row type, consisting of (111) microfacets. Construct the possible structural models of the Pt(110)3×1 reconstruction.

Further Reading

1. C.B. Duke: *Surface Structures of Tetrahedrally Coordinated Semiconductors: Principles, Practice, and Universality.* Appl. Surf. Sci. **65/66**, 543–552 (1993) (general principles of semiconductor surface reconstructions)
2. G.A. Somorjai: *Introduction to Surface Chemistry and Catalysis* (John Wiley, New York 1994) Chapter 2 (vast list of structures occurring at clean crystal surfaces in tabular form with brief comments)
3. J.A. Stroscio, W.J. Kaiser (Ed.): *Scanning Tunneling Microscopy. Methods of Experimental Physics. Vol. 27* (Academic Press, San Diego 1993) Chapters 5 and 6 (clean semiconductor and metal surfaces as seen in STM)
4. A. Fasolini, A. Selloni, A. Shkrebtii: 'Surface Reconstruction and Relaxation'. In: *Physics of Solid Surfaces. Landolt–Börnstein. Vol. III/24a.* ed. by G. Chiarotti (Springer, Berlin 1993) Chapter 2.2 (detailed data on relaxation and reconstruction of clean surfaces)

9. Atomic Structure of Surfaces with Adsorbates

In studies of clean surfaces, the presence of any foreign species is absolutely undesirable. However, a large number of investigations concern surfaces on which a controlled amount of certain foreign atoms or molecules are intentionally added. The foreign species can be added to the surface in different ways, including condensation from vapor phase (*adsorption*), segregation from the sample bulk, or diffusion along the surface. Taking into account that adsorption is the most widely used technique, the added species is conventionally called the *adsorbate*. The material of the host surface is called the *substrate*. In the present chapter, the atomic structure of clean surfaces with adsorbates is discussed. The consideration is limited to adsorbate layers with an effective coverage of up to one atomic layer. Thus, multilayer thin films are beyond the scope of the chapter. Already formed (in most cases, equilibrium) structures are treated, while the dynamic processes involved in their formation will be discussed elsewhere.

9.1 Surface Phases in Submonolayer Adsorbate/Substrate Systems

Depending on the strength of the interaction between an adsorbate and a substrate, adsorption is subdivided by convention into

- physisorption (weak interaction) and
- chemisorption (strong interaction).

Note that the distinction between physisorption and chemisorption is not sharp and is conventionally accepted to be at an adsorbate–substrate binding energy of around 0.5 eV per molecule (or atom) (1 eV/molecule = 23.060 kcal mol^{-1} = 96.485 kJ mol^{-1}).

Physisorption. The term *physisorption* refers to the case of the weakest adsorbate–substrate interaction due to van der Waals forces. The typical binding energies are on the order of 10–100 meV. As the interaction is weak, the physisorbed atom or molecule does not disturb the structural environment near the adsorption site to any significant extent. Physisorption can only be detected on condition that the stronger chemisorption is absent and the substrate temperature is low (Recall that at room temperature the thermal energy $k_{\mathrm{B}}T \simeq 25\,\mathrm{meV}$, as $1\,\mathrm{eV} = 11\,604\,\mathrm{K}$). A typical example of physisorption is the low-temperature adsorption of noble-gas atoms on metal surfaces.

Chemisorption. *Chemisorption* corresponds to the case when an adsorbate forms strong chemical bonds with substrate atoms. These bonds can be either covalent (with sharing of electrons) or ionic (involving electronic charge transfer). The typical binding energies are on the order of 1–10 eV. The strong interaction changes the adsorbate chemical state and, in the case of chemisorption of molecules, can cause their dissociation and formation of new adsorbate species. The substrate structure is changed as well: these changes range from relaxation of the interlayer spacings in the top substrate layers to substrate reconstruction involving complete rearrangement of the top-layer atomic structure. In the latter case, thermal activation is inevitably required. A typical example of chemisorption is adsorption of metal atoms on metal and semiconductor surfaces at elevated temperatures.

On single crystal surfaces, mutual interactions between adsorbates often lead to the appearance of long-range order at the adsorbate/substrate interface. The ordered two-dimensional layer so formed may only consist of the adsorbate (as in the case of physisorption) or may include substrate atoms, which is equivalent to reconstruction of the original surface (as in the case of chemisorption). There is a no commonly accepted term to denote this layer, and various terms, like "impurity-induced reconstruction," "two-dimensional structure," "ordered adsorbate layer," or "ordered surface phase" are in use. We believe that the term *surface phase* is the most adequate, as it indicates that the layer is essentially a kind of two-dimensional phase and, like the bulk phase, it is characterized by its own atomic and electronic structure, composition, temperature range of stability, properties, etc.

Study of the atomic structure of a given surface phase includes, in general, several steps. LEED or RHEED observations allow us to establish the 2D lattice of the superstructure associated with a surface phase. Note that the surface phase is usually labeled in accordance with its periodicity with respect to the underlying substrate plane (for example, Si(111)$\sqrt{3}\times\sqrt{3}$-$R30°$-Ag). If in addition to the periodicity, the adsorbate coverage is known, say, from AES or XPS data, one can construct a tentative structural model. STM provides a real space image of the surface, in favorable cases, with atomic resolution. Quantitative analysis using LEED I–V, XRD, or ISS, as well as theoretical calculations, is applied to make a proper choice among the proposed models and to refine the coordinates of atoms constituting the unit cell of the surface phase.

9.2 Surface Phase Composition

The chemical composition of a surface phase is described by a combination of two values:

- coverage of adsorbate;
- coverage of substrate atoms.

9.2.1 Coverage of Adsorbate

The *adsorbate coverage* characterizes the surface concentration of adsorbed species expressed in *monolayer (ML)* units. *One monolayer corresponds to one adsorbate atom or molecule for each 1×1 unit cell of the ideal non-reconstructed substrate surface.* Note that a monolayer is a relative value associated with a given substrate. It can be converted to an absolute surface number density of atoms, if one divides the coverage in monolayer units by the area occupied by a 1×1 unit mesh.

Figure 9.1a illustrates schematically the hypothetical surface phases with an adsorbate coverage of 1.0, 0.5, and 0.25 ML. Some examples of actually occurring surface phases are presented in Table 9.1. Note that the phases listed have the same $\sqrt{3}\times\sqrt{3}$ periodicity, but differ by the adsorbate coverage and, hence, by the atomic structure. Though most of the ordered surface phases are characterized by a definite value of adsorbate coverage, some exceptions exist. These are *surface phases with a variable composition.* The so-called "mosaic" Ge(111)$\sqrt{3}\times\sqrt{3}$-Pb phase furnishes such an example (Fig. 9.2). The phase is a truly two-dimensional solution, Ge_xPb_{1-x}/Ge(111), where a fraction x of the Pb adatoms in the $\sqrt{3}\times\sqrt{3}$-Pb structure is replaced by Ge adatoms ($0 < x < 0.5$).

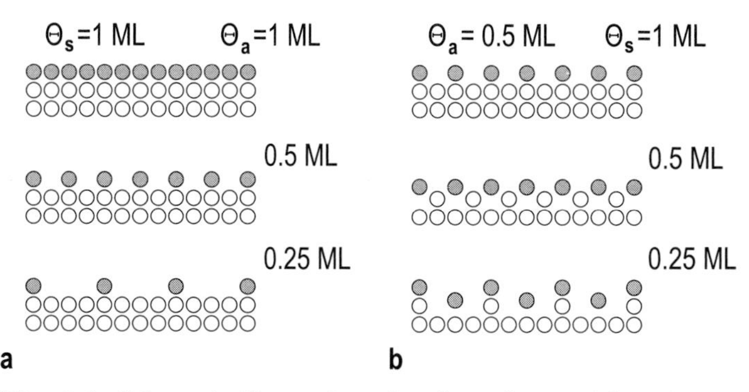

Fig. 9.1. Schematic illustration of surface phases with various compositions. (**a**) Surface phases with the same coverage of substrate atoms (1.0 ML) and different coverages of adsorbate atoms (1.0, 0.5, and 0.25 ML). (**b**) Surface phases with the same coverage of adsorbate atoms (0.5 ML) and different coverages of substrate atoms (1.0, 0.5, and 0.25 ML). Adsorbate atoms are shown by gray circles, substrate atoms by white circles

Table 9.1. Adsorbate coverage Θ_a for selected $\sqrt{3}\times\sqrt{3}$ surface phases

Phase	Θ_a, ML
Pt(111)$\sqrt{3}\times\sqrt{3}$-Sn	1/3
Pt(111)$\sqrt{3}\times\sqrt{3}$-Xe	1/3
Ni(111)$\sqrt{3}\times\sqrt{3}$-S	1/3
Si(111)$\sqrt{3}\times\sqrt{3}$-B,Al,Ga,In	1/3
Si(111)$\sqrt{3}\times\sqrt{3}$-Bi	1/3
Si(111)$\sqrt{3}\times\sqrt{3}$-Pb	1/3
Ag(111)$\sqrt{3}\times\sqrt{3}$-Cl	1/3
Ag(111)$\sqrt{3}\times\sqrt{3}$-Cl	2/3
Si(111)$\sqrt{3}\times\sqrt{3}$-Bi,Sb	1
Ge(111)$\sqrt{3}\times\sqrt{3}$-Ag	1
Si(111)$\sqrt{3}\times\sqrt{3}$-Ag	1
Si(111)$\sqrt{3}\times\sqrt{3}$-Pb	4/3

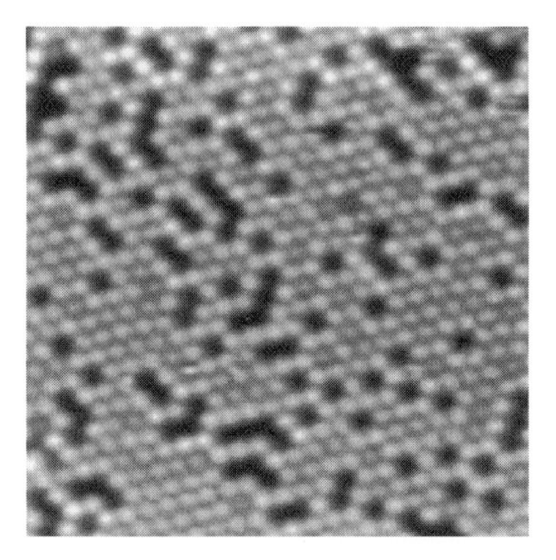

Fig. 9.2. An example of a surface phase with variable compostion. Filled-state STM image of the "mosaic" Ge(111)$\sqrt{3}\times\sqrt{3}$-Pb phase. Pb adatoms are seen bright and Ge adatoms are seen dark. The replacement of Pb adatoms by Ge adatoms results in a variation of composition over a wide range, ideally, from 100% Pb (0% Ge) down to 50% (50% Ge), and, actually, from ~10% to ~40%. The phase is adequately labeled as a Ge_xPb_{1-x}/Ge(111) solid solution. The surface shown in the STM image is characterized by $x = 0.19$. A similar "mosaic" structure is observed also in Pb/Si(111), Sn/Si(111) and Sn/Ge(111) systems (after Carpinelli et al. [9.1])

9.2.2 Coverage of Substrate Atoms

Coverage of substrate atoms is also referred to in the literature as *density of substrate top atoms*. To specify the meaning of the term *coverage of substrate atoms*, let us in our mind remove the adsorbate species from the surface phase and consider the remaining substrate surface. For convenience, let us assume here that an ideal substrate surface contains one atom per 1×1 unit mesh. After mentally removing the adsorbate atoms in this gedanken experiment, the substrate surface can be attributed to one of three possible surface types:

- non-reconstructed surfaces (including relaxed surfaces);
- conservatively reconstructed surfaces;
- non-conservatively reconstructed surfaces.

In the case of a purely *non-reconstructed (relaxed) surface*, the number of top atoms is the same as in an ideal bulk plane, i.e., the coverage of substrate atoms is 1 ML. In the case of a surface which has undergone *conservative reconstruction*, the number of atoms in the top layer (layers) is preserved, hence, the substrate atom coverage is also 1 ML. If the substrate surface displays *non-conservative reconstruction*, the substrate atom coverage is not an integer. Figure 9.1b shows schematically some examples of surface phases with the same coverage of adsorbate (0.5 ML), but different coverages of substrate atoms, namely, 1.0, 0.5, and 0.25 ML. The two latter substrate surfaces display essentially missing-row (i.e., non-conservative) reconstructions: when every other top substrate atom is removed, the substrate atom coverage is 0.5 ML; when three of every four atoms are removed, the substrate atom coverage is 0.25 ML. Some actually occurring surface phases with non-integer substrate atom coverages are listed in Table 9.2.

Table 9.2. Adsorbate Θ_a and substrate Θ_s atom coverages for selected surface phases

Phase	Θ_a, ML	Θ_s, ML
Al(100)c(2×2)-Li	1/4	3/4
Pt(100)2×2-Sn	1/4	3/4
Cu(100)3×3-Li	5/9	4/9
Ni(100)c(6×2)-Na	1/6	2/3
Au(110)c(2×2)-K	1/2	1/2
Cu(110)4×1-Bi	3/4	3/4
Al(111)$\sqrt{3}\times\sqrt{3}$-Li,K	1/3	2/3

Phase	Θ_a, ML	Θ_s, ML
Cu(111)2×2-Li	3/4	3/4
Pt(111)2×2-Sn	1/4	3/4
Si(100)(2×3)-Na	1/3	1/3
Ge(111)$\sqrt{3}\times\sqrt{3}$-Ag	1	1
Si(111)$\sqrt{3}\times\sqrt{3}$-Ag	1	1
Si(111)(3×1)-Li,Na	1/3	4/3
Si(111)(6×1)-Ag	1/3	4/3

9.2.3 Experimental Determination of Composition

Adsorbate Coverage. Several techniques are in use to characterize the adsorbate coverage. A *quartz crystal monitor* (see Sect. 3.5.3) allows indirect measurements of the adsorbate coverage received by a substrate surface. However, one should be aware that the sticking coefficient of the adsorbate at the studied surface is unity, otherwise the coverage values will be overestimated. *AES and XPS quantitative analysis* is a routine technique widely used to evaluate the actual adsorbate coverage (see Sect. 5.2.3 and 5.4.3). Usually the accuracy of the measurements is limited to ~10–20%. Less often, *ion scattering and recoiling* is used for precise absolute measurements. The additional merit of the latter technique is the ability to detect hydrogen, whose detection is difficult or impossible for most conventional analytical techniques (see Sect. 6.4).

Coverage of Substrate Atoms. If the composition of the surface phase is such that it adopts only a fraction of the top substrate atoms, the surplus substrate atoms become liberated in the process of surface phase formation. Quantitative consideration of the mass transport associated with the redistribution of the liberated atoms allows the determination of the substrate atom coverage of the forming surface phase.

As an example, consider the formation of the Si(111)$\sqrt{3}\times\sqrt{3}$-$R30°$-Ag surface phase upon Ag deposition onto a Si(111)7×7 surface held at about 500°C. Note that the ideal non-reconstructed Si(111)1×1 surface is essentially a bilayer: it contains two atoms in a unit mesh, hence, the ideal coverage is 2.0 ML. Recall also that the Si(111)7×7 surface has 102 atoms per 7×7 unit cell, i.e., the coverage is $102/49 \simeq 2.08$ ML. The structure of the Si(111)$\sqrt{3}\times\sqrt{3}$-Ag surface phase is known to correspond to the honeycomb-chained-trimer model, which will be discussed in detail later. Here, it is only important that it has a Si atom coverage of 1.0 ML.

The STM image in Fig. 9.3 shows the early stage of Si(111)$\sqrt{3}\times\sqrt{3}$-Ag surface phase formation. The most interesting feature is that the phase grows in two associated domains called a *"hole–island" pair*. Both the "hole" bottom (seen as a dark area in the STM image) and the "island" top (seen as a bright area) display the same $\sqrt{3}\times\sqrt{3}$-Ag structure, but these domains are one atomic step (i.e., one Si(111) bilayer, 3.14 Å) apart, as shown schematically in Fig. 9.3b. The "hole–island" pair is a natural consequence of the difference in Si coverage of 7×7 (2.08 ML) and $\sqrt{3}\times\sqrt{3}$-Ag (1.0 ML), and its formation process can be understood as follows. Deposited Ag atoms migrate across the 7×7 surface and react at a suitable site with Si atoms to form the $\sqrt{3}\times\sqrt{3}$-Ag structure in the region of the hole. For the reaction, only 1.0 ML out of 2.08 ML of Si is required, thus, Si atoms of the surplus 1.08 ML are ejected onto the surrounding region of 7×7. These Si atoms react with Ag and become consumed by the $\sqrt{3}\times\sqrt{3}$-Ag "island" domain.

Consider Si mass transport in "hole–island" pair formation (Fig. 9.3c). If $\Theta_{\mathrm{Si}}^{\sqrt{3}}$ and $\Theta_{\mathrm{Si}}^{7\times7}$ are the Si atom coverages in Si(111)$\sqrt{3}\times\sqrt{3}$-Ag and in

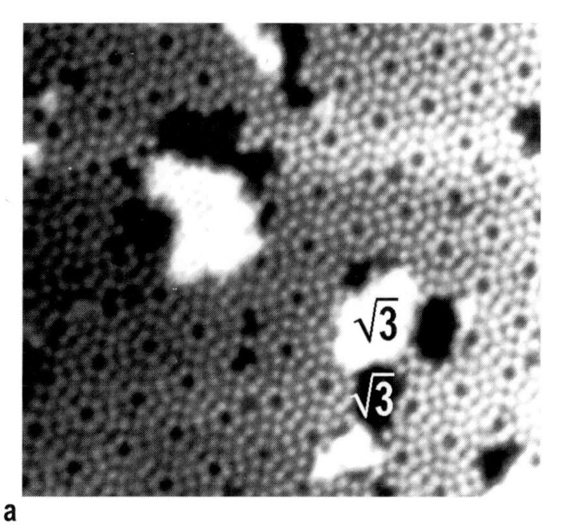

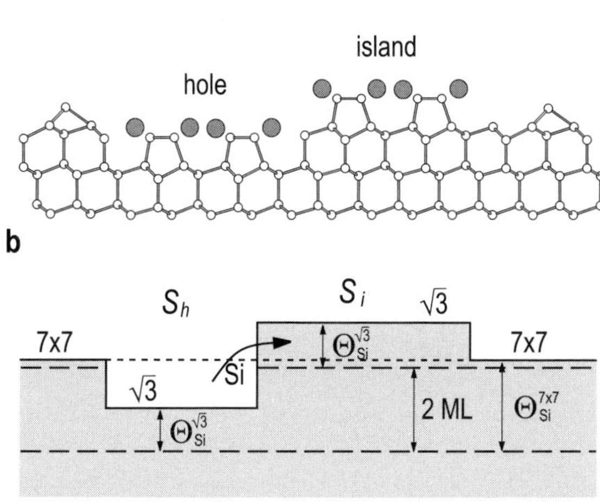

Fig. 9.3. (**a**) STM image of "hole–island" pairs of Si(111)$\sqrt{3}\times\sqrt{3}$-Ag surface phase domains formed at the Si(111)7×7 surface. The $\sqrt{3}\times\sqrt{3}$-Ag "holes" are seen as dark areas and the $\sqrt{3}\times\sqrt{3}$-Ag "island" as bright areas (after Shibata et al. [9.2]. (**b**) Schematic diagram of structure of a "hole–island" pair. Ag atoms are shown by gray circles, Si atoms by white circles (after McComb et al. [9.3]). (**c**) Schematic diagram illustrating Si mass transport involved in the formation of a "hole–island" pair

Si(111)7×7, respectively, and S_h and S_i are the areas occupied by the hole and island, respectively, one can write that

$(\Theta_\mathrm{Si}^{7\times7} - \Theta_\mathrm{Si}^{\sqrt{3}})S_\mathrm{h}$ is the number of Si atoms, which have been ejected from a "hole";

$(\Theta_\mathrm{Si}^{\sqrt{3}} + 2 - \Theta_\mathrm{Si}^{7\times7})S_\mathrm{i}$ is the number of Si atoms, which become consumed by an "island".

Assuming the conservation of Si atoms during the restructuring process, one can equate these Si amounts and obtain:

$$\Theta_\mathrm{Si}^{\sqrt{3}} = \Theta_\mathrm{Si}^{7\times7} - \frac{2}{1 + S_\mathrm{h}/S_\mathrm{i}} \; . \tag{9.1}$$

Thus, from the measured areas of "hole" and "island" one can determine the coverage of substrate atoms in the forming surface phase using (9.1).

With Ag deposition, $\sqrt{3}\times\sqrt{3}$-Ag "holes" and "islands" increase in number and size at the cost of a decrease in 7×7 area and, eventually, the whole surface becomes occupied by the Si(111)$\sqrt{3}\times\sqrt{3}$-Ag surface phase in two levels (Fig. 9.4). Taking this surface as an example, there is another way to construct the balance equation, as compared to the above. Consider the surface before and after the transition from Si(111)7×7 to Si(111)$\sqrt{3}\times\sqrt{3}$-Ag. The original Si(111)7×7 surface contains $\Theta_\mathrm{Si}^{7\times7}S$ silicon atoms in the top layers, where S is the total surface area (Fig. 9.5a). Upon formation of the Si(111)$\sqrt{3}\times\sqrt{3}$-Ag surface phase in two levels, these Si atoms are redistributed between the lower and upper $\sqrt{3}\times\sqrt{3}$-Ag levels (Fig. 9.5b). The lower level (originated from "holes"), occupying the area S_l, contains $\Theta_\mathrm{Si}^{\sqrt{3}}S_\mathrm{l}$ silicon atoms. The upper level (originated from "islands") of area S_u accumulates $(\Theta_\mathrm{Si}^{\sqrt{3}} + 2)S_\mathrm{u}$ silicon atoms. Under the condition of Si atom conservation, the total number of Si atoms in the upper and lower $\sqrt{3}\times\sqrt{3}$-Ag levels should be same as at the original Si(111)7×7 surface and the balance equation is written as

$$\Theta_\mathrm{Si}^{\sqrt{3}}S_\mathrm{l} + (\Theta_\mathrm{Si}^{\sqrt{3}} + 2)S_\mathrm{u} = \Theta_\mathrm{Si}^{7\times7}S \; . \tag{9.2}$$

Taking into account that $S_\mathrm{l} + S_\mathrm{u} = S$, one obtains

$$\Theta_\mathrm{Si}^{\sqrt{3}} = \Theta_\mathrm{Si}^{7\times7} - 2S_\mathrm{u}/S \; . \tag{9.3}$$

Note that (9.1) and (9.3) are essentially identical.

As the atomic steps present at the original surface might serve as traps or sources for mobile atoms, the above considerations are only suitable for terrace regions located far apart from the step edges. It should be noted, however, that the study of the surface phase growth at the steps allows us by itself to evaluate the substrate atom density of the forming phase. Figure 9.6 illustrates the formation of the Si(111)$\sqrt{3}\times\sqrt{3}$-Ag domain at the step edge. One can see that the surplus Si atoms released during $\sqrt{3}\times\sqrt{3}$-Ag formation

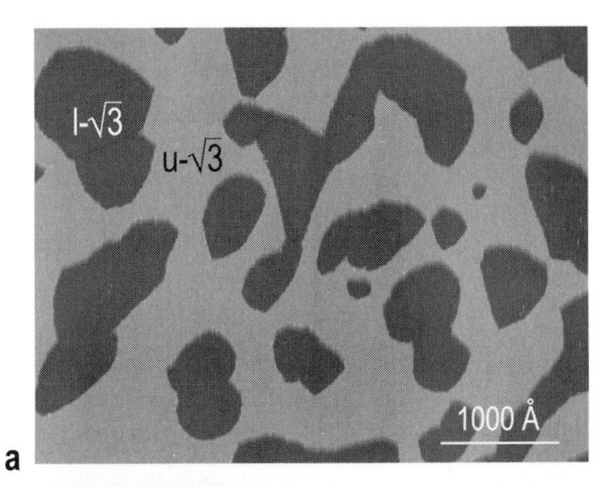

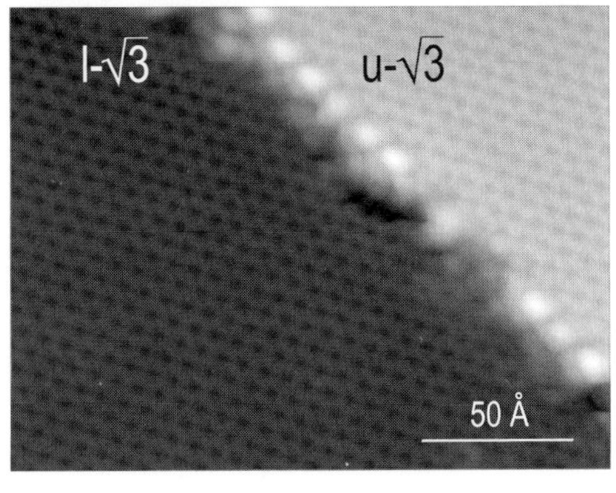

Fig. 9.4. (**a**) Large-scale STM images of the Si(111)$\sqrt{3}\times\sqrt{3}$-Ag surface occurring in two levels one (111) bilayer apart. The bright area corresponds to the upper $\sqrt{3}\times\sqrt{3}$-Ag level (u-$\sqrt{3}$) and dark regions to the lower $\sqrt{3}\times\sqrt{3}$-Ag level (l-$\sqrt{3}$). (**b**) High-resolution STM image showing that the $\sqrt{3}\times\sqrt{3}$-Ag structure in the lower and upper levels is the same (after Saranin et al. [9.4])

on the upper terrace become incorporated in the $\sqrt{3}\times\sqrt{3}$-Ag forming at the lower terrace. By analogy with (9.3), one can write

$$\Theta_{\mathrm{Si}}^{\sqrt{3}} = \Theta_{\mathrm{Si}}^{7\times7} - 2s/S \,, \tag{9.4}$$

where S stands for the total area of the $\sqrt{3}\times\sqrt{3}$-Ag domain formed and s is the area that extends across the original step location. If one monitors the step displacement (for example, as in REM experiments [9.5]), s/S can be replaced by the length ratio l/L.

For the evaluation of the substrate atom coverage, the so-called *atomic-layer titration* method [9.6] employs the pre-deposition of a controlled amount of the substrate material prior to the deposition of the adsorbate species. When the pre-deposited amount exactly equals the amount required for surface phase formation (i.e., corresponds to the substrate atom coverage of this

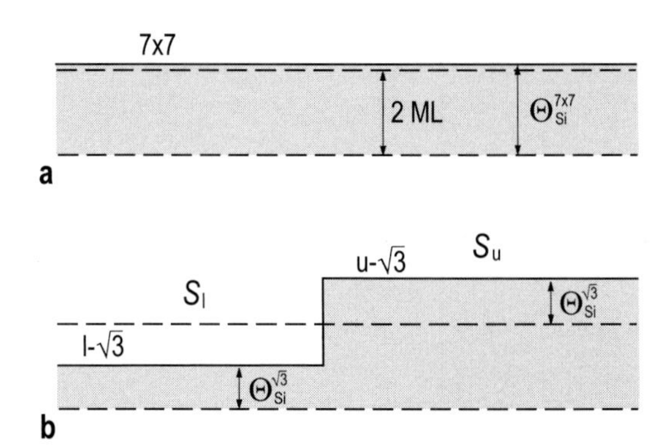

Fig. 9.5. Schematic diagram showing redistribution of Si atoms upon transition from (**a**) a flat Si(111)7×7 surface to (**b**) a Si(111)$\sqrt{3}\times\sqrt{3}$-Ag surface phase occurring in two levels: (upper, u-$\sqrt{3}$) and (lower, l-$\sqrt{3}$). The dashed lines show the levels of the complete Si(111) bilayers (2 ML) (after Saranin et al. [9.4])

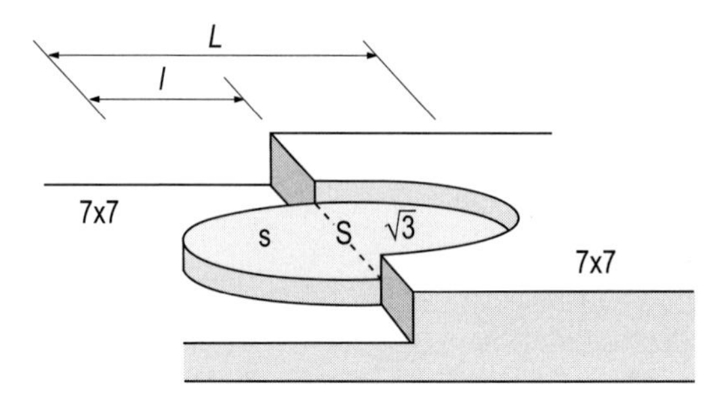

Fig. 9.6. Schematic illustration of the growth of a Si(111)$\sqrt{3}\times\sqrt{3}$-Ag domain at the step edge (after Tanishiro et al. [9.5])

phase), a flat island-free surface is obtained. Otherwise, the surface phase always occurs in two levels.

Figure 9.7 illustrates schematically the results of atomic-layer titration experiments on Si(100)2×3-Ag [9.6]. The Si(100)2×3-Ag surface phase is formed by deposition of Ag onto the Si(100)2×1 surface held at 550°C. When Ag is deposited directly onto the flat Si(100)2×1 surface, the final Si(100)2×3-Ag surface is rough (Fig. 9.7a). However, when Ag is deposited onto the Si(100)2×1 surface with pre-deposited 0.5 ML of Si, the final Si(100)2×3-Ag surface is atomically smooth (Fig. 9.7). This means that the Si(100)2×3-Ag surface phase has a Si coverage of 0.5 ML.

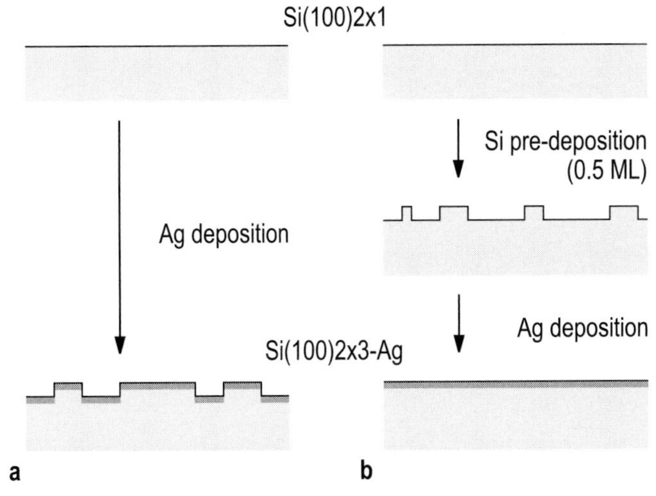

Fig. 9.7. Schematic illustration of the results of atomic-layer titration experiments on Si(100)2×3-Ag [9.6]. (**a**) Direct deposition of Ag onto a smooth Si(100) surface produces a rough Si(100)2×3-Ag surface. (**b**) If 0.5 ML of Si is pre-deposited, the final Si(100)2×3-Ag surface is atomically smooth

9.3 Formation Phase Diagram

To specify in compact form the formation conditions and range of stability for the surface phases of a given adsorbate/substrate system, *formation phase diagrams* are used. The phase diagram shows the phase occurrence regions in *coverage–temperature* coordinates. Figure 9.8 displays a schematic formation phase diagram.

The formation phase diagram can be constructed using various experimental procedures. First, the adsorbate can be deposited step by step onto the substrate held at a fixed temperature. After each deposition step, LEED or RHEED observations are conducted to elucidate the surface structure. The corresponding coverage value can be determined using a quartz crystal thickness monitor or AES (XPS). With a quartz crystal, this is rather the nominal coverage, which coincides with the actual value only when the sticking probability of adsorbate equals one. AES and XPS measurements yield the actual coverage. The set of experimental dots in such a measurement occupies pass **A** in Fig. 9.8. This procedure if repeated at various temperatures allows one to localize the coverage–temperature regions of the phase occurrence. The formation phase diagram is the graphical presentation of these data.

Second, the adsorbate layer of a fixed coverage can be deposited at low temperature and then annealed at progressively higher temperatures. In this case, the LEED or RHEED patterns are recorded and indexed after each annealing step. This procedure is illustrated by pass **B** in Fig. 9.8. High-temperature annealing might result in the partial desorption of adsorbate

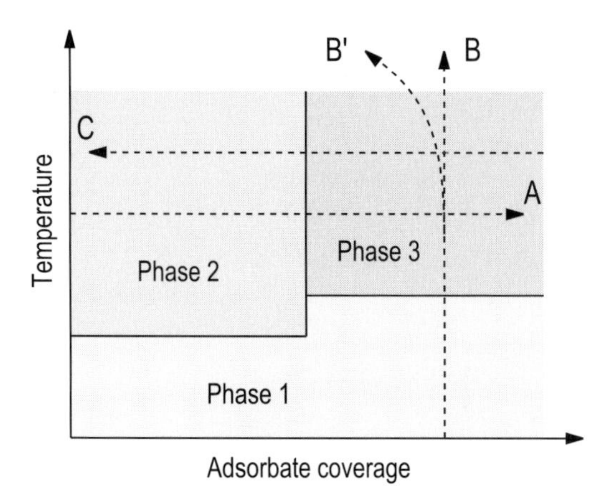

Fig. 9.8. Schematic illustration of formation phase diagrams. Possible pathways are indicated: **A** – adsorbate deposition at constant temperature; **B′** and **B** – isochronal annealing of low-temperature deposit with and without desorption of adsorbate; **C** – isothermal desorption of adsorbate

from the surface (sometimes, in adsorbate diffusion into the bulk) and the actual pass (marked **B′**) would deviate from the vertical pass **B**. Thus, it is essential to distinguish between the case when the coverage axis shows nominal adsorbate coverage of a low-temperature deposit before annealing and the case when the actual adsorbate coverage left at the surface after annealing is shown.

Third, a relatively thick layer of adsorbate can be deposited at low temperature and then annealed to cause isothermal desorption of the adsorbate. The corresponding pass is indicated by **C**. One can see that pass **C** is essentially a reversal of pass **A**.

Crossing the boundaries in the phase diagram corresponds to transitions from one structure to another. The transitions can be caused by the change in adsorbate coverage (path **A**) or by variation in temperature (path **B**). Consider, first, the structural transitions occurring with the change of adsorbate coverage and consider the path **A** in greater detail (Fig. 9.9). Note that with rare exceptions, each surface phase is characterized by a definite stoichiometric coverage of adsorbate (indicated in Fig. 9.9 by Θ_2 for Phase 2 and Θ_3 for Phase 3). Hence, for any coverage in between Θ_2 and Θ_3, a surface is composed of coexisting domains of two phases (Fig. 9.10). With increasing coverage, the area fraction of Phase 2 decreases, while that of Phase 3 increases accordingly. Thus, the coverage range attributed to a given phase is rather a range where this phase prevails (i.e., occupies the greater area fraction and produces the major LEED (RHEED) pattern) and boundaries correspond to the adsorbate coverages where the neighboring phases occupy about the same area fraction.

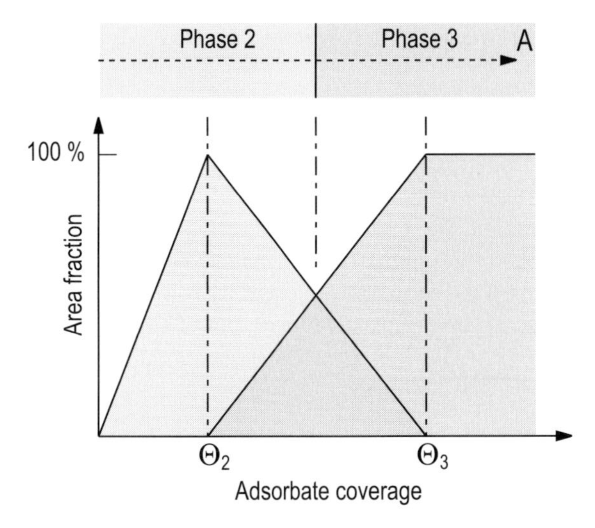

Fig. 9.9. Schematic illustrating the pathway A in Fig. 9.8 in greater detail. The stoichiometric adsorbate coverages of Phase 2 and Phase 3 are indicated by Θ_2 and Θ_3, respectively. With increasing adsorbate coverage from Θ_2 to Θ_3, the area fraction of Phase 2 decreases, while that of Phase 3 increases accordingly. The boundary in the phase diagram corresponds to the adsorbate coverage where the phases occupy about the same area fraction

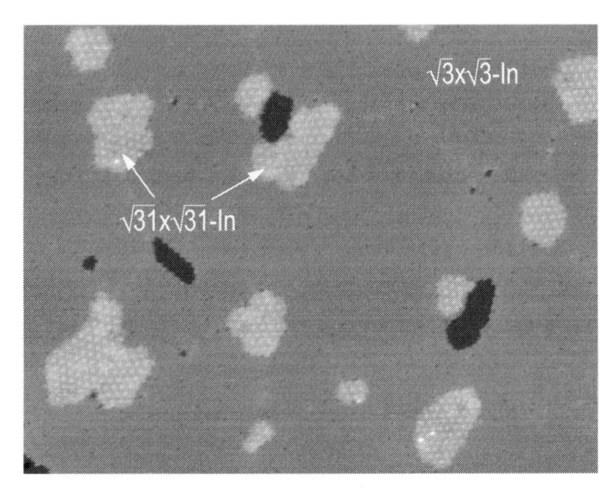

Fig. 9.10. STM image showing the coexistence of two phases: Si(111)$\sqrt{3}\times\sqrt{3}$-In (seen as a uniform gray array) and Si(111)$\sqrt{31}\times\sqrt{31}$-In (seen as bright and dark regions, corresponding to "holes" and "islands"). The stoichiometric In coverage of the $\sqrt{3}\times\sqrt{3}$-In phase is 0.33 ML and it occupies 15% area fraction; In coverage of $\sqrt{31}\times\sqrt{31}$-In is 0.53 ML and it occupies 85% area fraction. The average In coverage in the surface shown is 0.36 ML

Each ordered surface phase has a limited temperature range of stability. At the boundaries of this range, phase transitions occur. These transitions may be

- reversible transitions or
- irreversible transitions.

In the case of a *reversible transition*, the low-temperature phase (Phase 1 in Fig. 9.8) transfers upon heating to the high-temperature phase (Phase 3 in Fig. 9.8), but it restores again upon cooling. In the case of an *irreversible transition*, the high-temperature phase once being formed remains stable upon cooling.

Another subdivision of transitions includes

- order–order transitions and
- order–disorder transitions.

An *order–order transition* means a transition between two ordered surface phase having different structures. The loss of ordering is described by *order–disorder transitions.*

Transitions are also subdivided into

- first-order (discontinuous) transitions and
- second-order (continuous) transitions.

In *first-order transitions*, the system changes abruptly from one distinct surface phase to a second distinct surface phase. Phase coexistence, nucleation and growth of the domains of a new phase are the features of first-order transitions. By contrast, in *second-order transitions*, one phase transfers continuously to a second phase and the competing phases are indistinguishable at the transition temperature.

Some illustrative examples of experimental phase diagrams are shown in Figs. 9.11, 9.12 and 9.13. Figure 9.11 presents the phase diagram for molecular hydrogen (H_2) adsorbed on graphite. This is an example of physisorption (the hydrogen–graphite binding energy is $\simeq$40 meV), which occurs only at low temperatures, 5–35 K, and does not involve any substrate reconstruction. The interaction between physisorbed H_2 molecules results in the formation of a commensurate $\sqrt{3}\times\sqrt{3}$ structure, in which a gas molecule is located above every third carbon hexagon of the graphite surface. The $\sqrt{3}\times\sqrt{3}$ phase is completed at 1/3 ML H_2 coverage. With increasing coverage, an incommensurate close-packed phase is formed. The nearest neighbor distance changes from 4.26 Å in the $\sqrt{3}\times\sqrt{3}$ phase to 3.51 Å in the incommensurate phase. Further adsorption results in the growth of a second molecular layer.

Figure 9.12 shows a phase diagram for the H/Fe(110) system. Hydrogen adsorbs dissociatively with an adsorption energy of $\simeq$1.1 eV, thus showing an example of chemisorption. However, chemisorbed hydrogen does not induce reconstruction of the Fe(110) surface (at most, a slight surface relaxation occurs) and the system is a reasonable realization of a 2D *lattice gas* model.

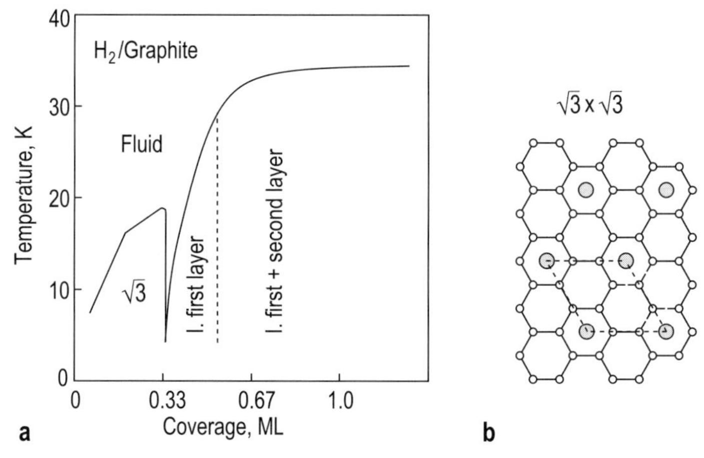

Fig. 9.11. (**a**) Formation phase diagram for H_2 physisorbed on (0001) graphite surface. I stands for incommensurate. (**b**) The model of a commensurate $\sqrt{3}\times\sqrt{3}$ phase (after Nielsen et al. [9.7])

The lattice gas approximation assumes that adatoms can occupy only specific locations on the crystalline substrate. These sites can be empty or at most singly occupied. The interaction between adatoms, which depends only on the lateral distance between them, determines the phase diagram of such a system. In the H/Fe(110) system, the H bonding locations are three-fold sites and the surface phases formed are $c(2\times2)$ at 0.5 ML H coverage and 3×3 at 0.67 ML H coverage. At intermediate coverages, less sharp streaky LEED patterns are seen, which were attributed tentatively to antiphase domain boundaries. The order–disorder transitions were found to occur at 245 K for the $c(2\times2)$ phase and at 265 K for the 3×3 phase.

Figure 9.13 displays the phase diagram for the submonolayer In/Si(111) system. This system furnishes an example of a strong adsorbate–substrate interaction (binding energy is on the order of 2–3 eV), which induces a strong reordering of the substrate surface. Recall that the original Si(111)7×7 substrate surface is already reconstructed into the DAS structure (see Sect. 8.5.2). Upon In deposition onto a heated Si(111) substrate, 7×7 reconstruction is eliminated and the Si(111) substrate surface becomes almost bulk-like terminated. Indium adatoms residing on this surface are arranged into the $\sqrt{3}\times\sqrt{3}$-In structure with a stoichiometric In coverage of 0.33 ML. With increasing In coverage, the $\sqrt{31}\times\sqrt{31}$ and 4×1 phases are formed successively. Formation of both phases involves non-conservative reconstructions of the Si(111) substrate surface: the Si atom coverage was found to be 0.88 ML for the $\sqrt{31}\times\sqrt{31}$-In phase and 0.5 ML for the 4×1-In phase. At high In coverages, $\Theta_{\mathrm{In}} \gtrsim 1$ ML, LEED displays a $\sqrt{7}\times\sqrt{3}$ (or more strictly, $\begin{pmatrix} 3 & 1 \\ 1 & 2 \end{pmatrix}$) pattern, while STM observations show that the surface contains coexisting domains

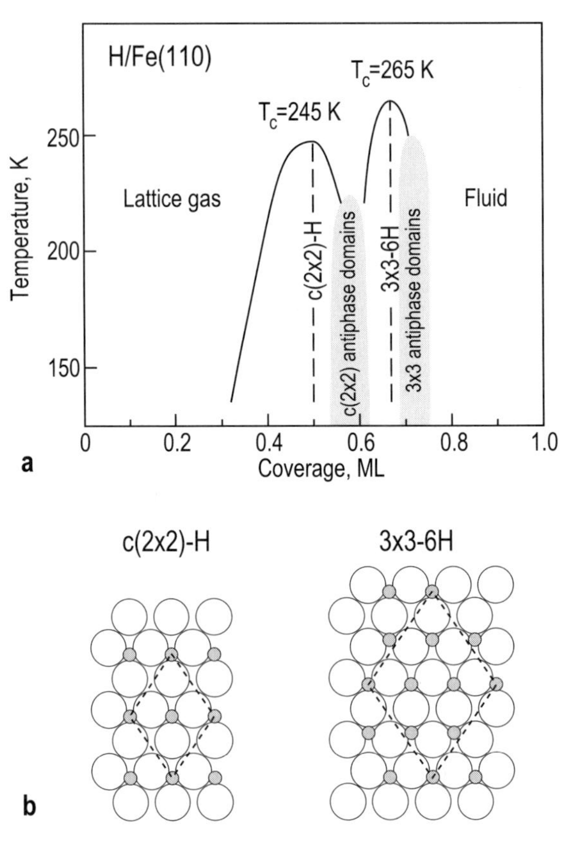

Fig. 9.12. (**a**) Formation phase diagram for H chemisorbed on a Fe(110) surface and (**b**) schematic of the structural models for the forming phases (after Imbihl et al. [9.8])

of quasi-hexagonal (hex-$\sqrt{7}\times\sqrt{3}$-In) and quasi-rectangular (rec-$\sqrt{7}\times\sqrt{3}$-In) phases.

9.4 Metal Surfaces with Adsorbates

Here some examples of the most famous adsorbate/metal surface phases are presented.

9.4.1 Family of $\sqrt{3}\times\sqrt{3}$ Structures on (111) fcc Metal Surfaces

On the (111) surface of fcc metals, adsorption of many adsorbates results in the formation of $\sqrt{3}\times\sqrt{3}$ structures at an adsorbate coverage of 1/3 ML. Depending on the adsorbate position, these $\sqrt{3}\times\sqrt{3}$ structures fall into two main categories:

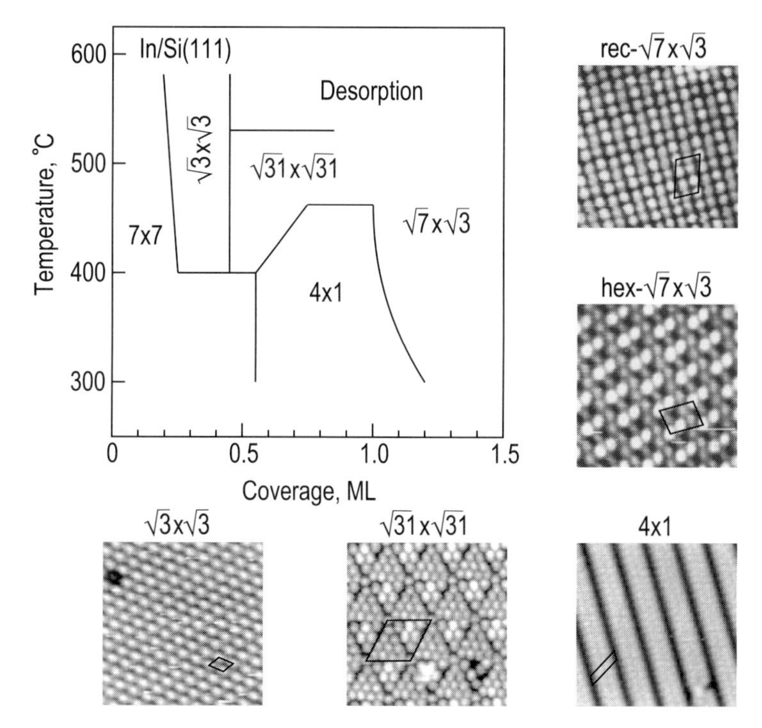

Fig. 9.13. Formation phase diagram of the In/Si(111) system (after Hirayama et al. [9.9]) and high-resolution STM images of the In/Si(111) surface phases (after Saranin et al. [9.10, 9.11] and Kraft et al. [9.12]). The unit cells are outlined in the STM images by solid lines

- substitutional $\sqrt{3}\times\sqrt{3}$ phases (or surface alloy);
- adatom-type $\sqrt{3}\times\sqrt{3}$ phases.

In some adsorbate/substrate systems, both types of $\sqrt{3}\times\sqrt{3}$ occur. For example, in K/Al(111) and Rb/Al(111) systems, $\sqrt{3}\times\sqrt{3}$ of adatom type is formed at 100 K and substitutional $\sqrt{3}\times\sqrt{3}$ at 300 K; upon heating the former transfers irreversibly to the latter.

Substitutional $\sqrt{3}\times\sqrt{3}$ Phases. Figure 9.14 shows a schematic of the substitutional $\sqrt{3}\times\sqrt{3}$ phases at a (111) fcc surface. In this structure, adsorbate atoms occupy *substitutional* sites formed by displacement of the host atoms from the first layer of the substrate. Hence, the substrate coverage in this phase is 2/3 ML. Substitutional $\sqrt{3}\times\sqrt{3}$ phases have been found to form upon annealing of Sn submonolayers on Cu(111), Ni(111), Pt(111) and Rh(111) of about 700°C, Sb submonolayers on Ag(111) at about 300°C and upon RT deposition of alkali metals, Li, K, Na and Rb, on Al(111).

Strictly speaking, adsorbate atoms occupy *quasi-substitutional* sites, as they are shifted outward from the exact substitutional positions. The adsorbate atoms protruding above neighboring substrate atoms make the surface

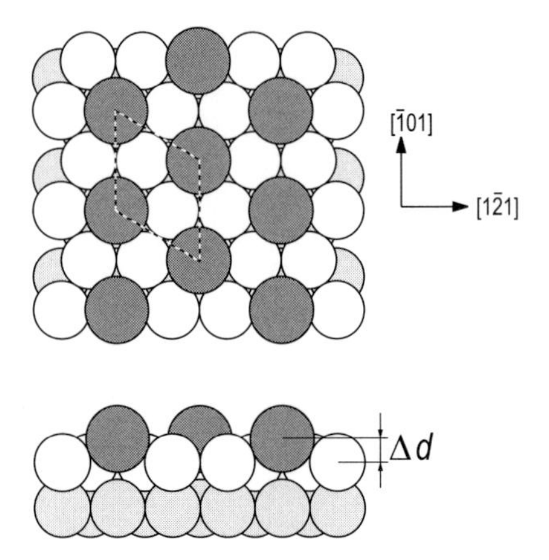

Fig. 9.14. Schematic of substitutional $\sqrt{3}\times\sqrt{3}$ phases forming at (111) fcc metal surfaces. The surface rippling due to outward shifting of the adsorbate atoms is indicated by Δd. Adsorbate atoms are shown by dense gray circles, top-layer substrate atoms by white circles, and second-layer substrate atoms by light gray circles. The $\sqrt{3}\times\sqrt{3}$ unit cell is outlined

rippled. The value of rippling, Δd, increases with the atomic size of adsorbate (for example, it is 0.4 Å for Li at Al(111) and 2.4 Å for Rb at Al(111)) and decreases with the substrate lattice constant (for example, it is 0.46 Å for Sn at Ni(111) ($a_0 = 3.52$ Å) and 0.22 Å for Sn at Pt(111) ($a_0 = 3.92$ Å)).

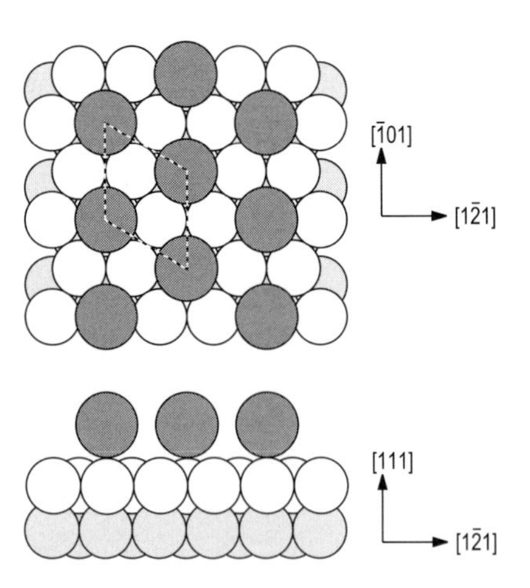

Fig. 9.15. Schematic of the adatom-type $\sqrt{3}\times\sqrt{3}$ phases forming at (111) fcc metal surfaces. Though in the shown example adsorbate adatoms occupy the on-top sites, other adsorption sites are also possible (see Fig. 9.16). Adsorbate atoms are shown by dense gray circles, top-layer substrate atoms by white circles, and second-layer substrate atoms by light gray circles. The $\sqrt{3}\times\sqrt{3}$ unit cell is outlined

Adatom-Type $\sqrt{3}\times\sqrt{3}$ Phases. In $\sqrt{3}\times\sqrt{3}$ phases of this type, adsorbate species occur in the form of adatoms above the (111) substrate surface (Fig. 9.15). Thus, 1 ML coverage of substrate atoms is preserved. There are four different adsorption sites for adatoms on fcc (111) surface (see Fig. 9.16):

- on-top site,
- fcc-hollow site,
- hcp-hollow site,
- bridge site.

An *on-top* site corresponds to the location of adsorbate directly above the surface substrate atoms; *fcc-hollow* and *hcp-hollow* sites are three-fold coordinated sites above the third-layer and the second-layer substrate atoms, respectively, and a *bridge* site is a two-fold coordinated site between two neighboring surface atoms. It appears that, depending on the adsorbate/substrate system, adsorbate might occupy each of the four possible sites. Some examples are listed in Table 9.3.

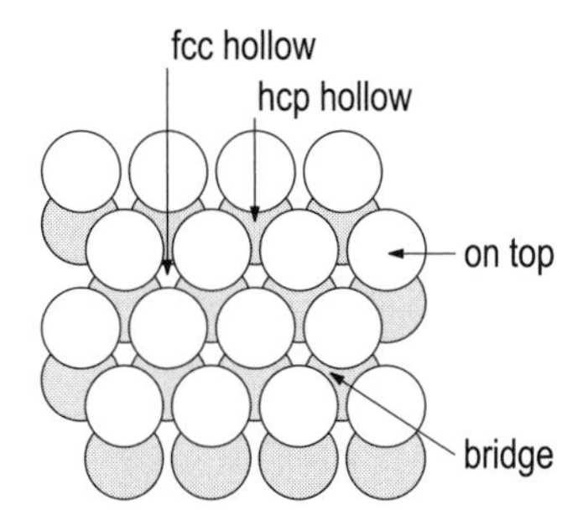

Fig. 9.16. Adsorption sites for adatoms on the (111) fcc surface. Top-layer atoms are shown by white circles and second-layer atoms by light gray circles

9.4.2 Ni(110)2×1-CO

The Ni(110)2×1-CO surface phase is formed by saturated adsorption of 1 ML of CO molecules at 130 K. At room temperature, a saturated 1 ML coverage of CO molecules is achievable only by keeping a relatively high pressure of CO gas ($P_{CO} \simeq 10^{-6}$ Torr) in the UHV chamber. Under these conditions, the Ni(110)2×1-CO surface phase is formed whose atomic structure is shown in Fig. 9.17. CO molecules are bonded in the two-fold "short-bridge" sites via the Ni-C bond. Along the Ni $[\bar{1}10]$ row, the CO molecules alternately tilt away from the normal direction within the (001) plane. The tilt angle was determined to be $19{\pm}2°$. An abundance of various techniques was used for the detailed solution of this structure, for example, LEED for elucidation of the structure symmetry, HREELS and infra-red adsorption spectroscopy to probe

Table 9.3. Adsorption sites for selected elements which form adatom-type $\sqrt{3}\times\sqrt{3}$ phases on (111) fcc metal surfaces

Substrate	Adsorbate	Adsorption site	Substrate	Adsorbate	Adsorption site
Ag(111)	Cl	fcc hollow	Pd(111)	CO	fcc hollow
	I	fcc + hcp hollow		S	fcc hollow
	K	hcp hollow	Pt(111)	S	fcc hollow
	Rb	hcp hollow		Xe	fcc + hcp hollow
Al(111)	Na	on top	Rh(111)	CO	on top
	Rb	on top		Cs	hcp hollow
Cu(111)	Cl	fcc hollow		I	fcc hollow
Ir(111)	S	fcc hollow		K	hcp hollow
Ni(111)	Cl	fcc hollow		Na	hcp hollow
	CO	bridge		Rb	hcp hollow
	S	fcc hollow		S	fcc hollow

the bonds, LEED I–V, XPD, ARUPS, and STM to establish the bonding site and the tilt angle. Figure 9.18 shows a high-resolution STM image of the Ni(110)2×1-CO phase. The protrusions in the image correspond to the oxygen atoms of the chemisorbed molecules. In agreement with the model, they demonstrate a zigzag arrangement.

9.4.3 $n\times1$ Structures in Pb/Cu(110), Bi/Cu(110), Li/Cu(110), and S/Ni(110) Systems

Pb deposition onto a Cu(110) surface, held at about room temperature, results in successive formation of a set of $n\times1$ surface structures, 4×1 at $\Theta_{\mathrm{Pb}} = 0.75$ ML, 9×1 at $\Theta_{\mathrm{Pb}} = 0.778$ ML, and 5×1 at $\Theta_{\mathrm{Pb}} = 0.80$ ML. By means of atomically resolved STM images (Fig. 9.19), the structure of the $n\times1$ structures was elucidated to be formed by substitution of every n-th row of Cu in the [001] direction by Pb atoms. Between these rows of Pb atoms which are located *in* the top Cu layer, there are also Pb atoms in rows aligned along the $[\bar{1}10]$ which are situated *on* the top Cu layer (Fig. 9.20). One can see that due to the difference in atomic size the "ridge" of upper Pb atoms contains

2 Pb atomic rows per 3 underlying Cu rows in the 4×1 phase,
6 Pb atomic rows per 8 Cu rows in the 9×1 phase and
3 Pb atomic rows per 4 Cu rows in the 5×1 phase.

Consequently, the stoichiometric compositions of the phases are

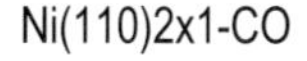

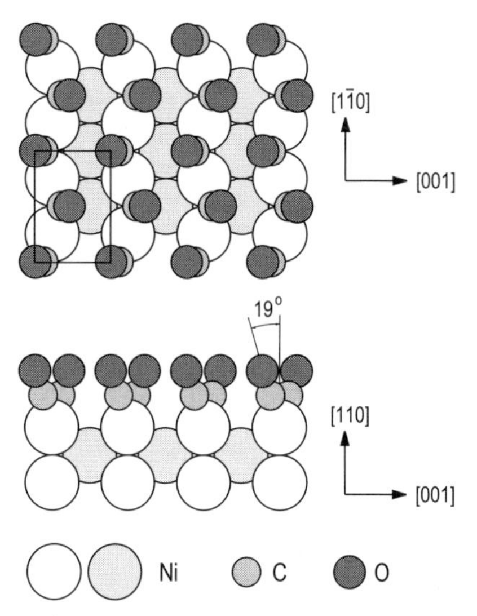

Fig. 9.17. Structural model of the Ni(110)2×1-CO surface phase

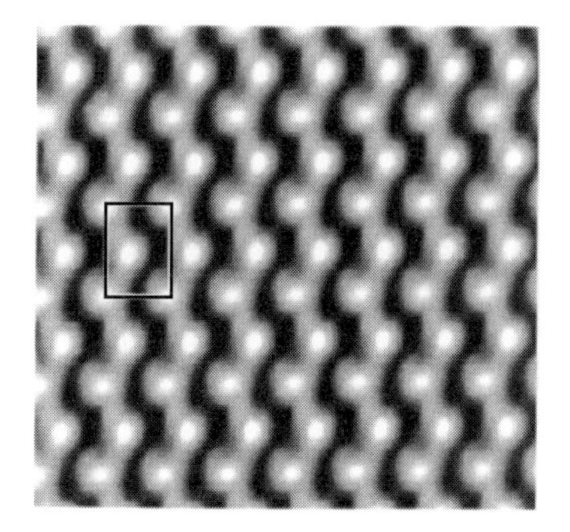

Fig. 9.18. STM image of the Ni(110)2×1-CO surface phase. The 2×1 unit cell is indicated. The zigzag chains of protrusions correspond to surface arrangement of O atoms of CO molecules chemisorbed via C-Ni bonding (after Sprunger et al. [9.13])

3/4 ML of Pb and 3/4 ML of Cu in the 4×1 phase,
7/9 ML of Pb and 8/9 ML of Cu in the 9×1 phase and
4/5 ML of Pb and 4/5 ML of Cu in the 5×1 phase.

A similar atomic structure was established for the Cu(110)4×1-Bi surface phase using XRD [9.15] (Fig. 9.21) and for the Cu(110)4×1-Li phase by means of LEED I–V analysis [9.16]. In contrast, the Ni(110)4×1-S surface was found to have an atomic arrangement, which can be thought of as a "reverse"

Fig. 9.19. Atomically resolved STM image of the Cu(110)5×1-Pb surface. Three atomic rows of Pb forming a "ridge" are clearly seen, the Pb row in the "trough" is not resolved due to the significant height difference. The 5×1 unit cell is outlined (after Nagl et al. [9.14])

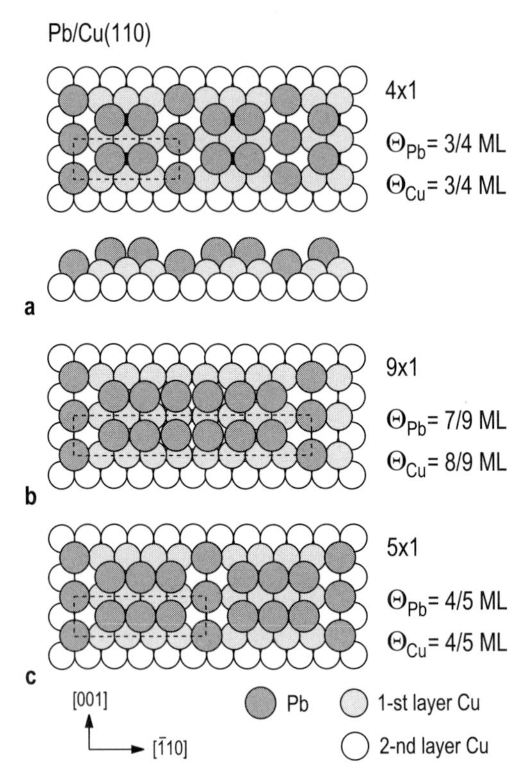

Fig. 9.20. Structural models of the Cu(110)n×1-Pb surface phases: (**a**) 4×1; (**b**) 9×1; and (**c**) 5×1. Pb atoms are shown by dark gray circles, first-layer Cu atoms by light gray circles, and second-layer Cu atoms by white circles. The unit cells are marked by a rectangle (after Nagl et al. [9.14])

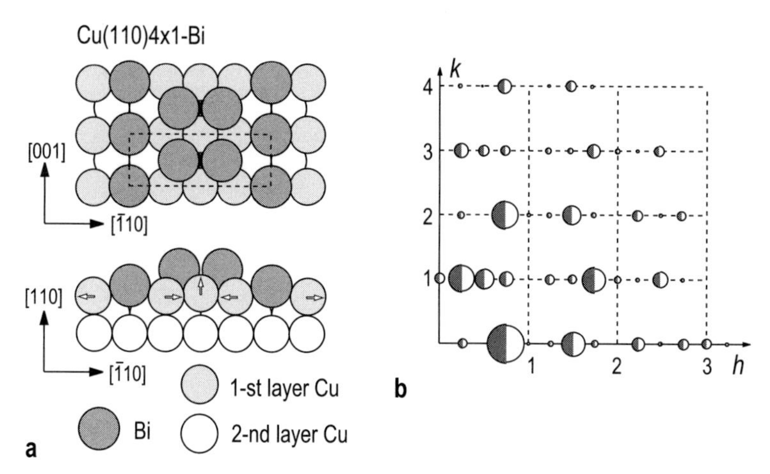

Fig. 9.21. (**a**) Structure of the Cu(110)4×1-Bi surface phase as determined using XRD analysis. Bi atoms are shown by dark gray circles, first-layer Cu atoms by light gray circles, and second-layer Cu atoms by white circles. The directions of the sub-surface displacements of the Cu atoms are indicated by arrows. (**b**) Comparison between the measured (filled semicircles) and calculated (open semicircles) in-plane structure-factor intensities of Cu(110)4×1-Bi for the optimized model. The areas of the semicircles are proportional to the diffraction beam intensities (after Lottermoser et al. [9.15])

structure as compared to that of Cu(110)4×1-Pb. While Cu(110)4×1-Pb and similar structures contain two rows of adsorbate in a "ridge" and one row of adsorbate in a "trough", Ni(110)4×1-S displays the opposite combination, one row in a "ridge" and two rows in a "trough" (Fig. 9.22). Thus, the composition of the latter phase is $\Theta_S = 3/4\,\text{ML}$ and $\Theta_{Ni} = 1/2\,\text{ML}$.

9.5 Semiconductor Surfaces with Adsorbates

We consider here some illustrative examples from the abundance of surface phases formed at silicon and germanium surfaces.

9.5.1 Family of $\sqrt{3}\times\sqrt{3}$ Structures on Si(111) and Ge(111)

Among other superstructures observed at Si(111) and Ge(111) surfaces with metal adsorbates, $\sqrt{3}\times\sqrt{3}$ is of greatest abundance. However, this is not a single structure, but rather a family of quite different phases having the same periodicity.

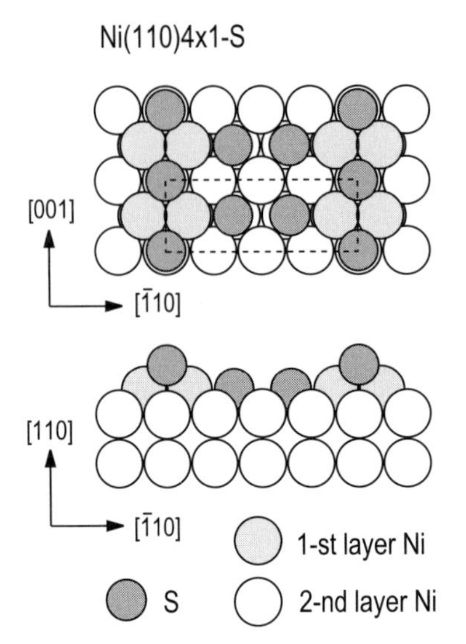

Fig. 9.22. Atomic structure model of Ni(110)4×1-S surface phases as determined on the basis of XRD measurements. S atoms are shown by dark gray circles, first-layer Ni atoms by light gray circles, and second-layer Ni atoms by white circles. The unit cell is marked by a rectangle (after Foss et al. [9.17])

Adatom-Type $\sqrt{3}\times\sqrt{3}$. This is the simplest $\sqrt{3}\times\sqrt{3}$ structure. It contains 1/3 ML of adsorbate adatoms, which reside on the almost bulk-like Si(111) surface (Fig. 9.23). Each adatom is bonded to three top Si atoms. As a result, all silicon dangling bonds are completely terminated. However, this simple adatom model has non-trivial alternatives for the adatom position, as there are two types of three-fold sites at the Si(111) surface, called H_3 and T_4. These sites are shown in Fig. 9.24. The T_4 site is located above the Si atom in the second layer and the H_3 site is above the Si atom in the fourth layer. The correct adsorption site was a subject of long-standing extensive debates. Now it is conclusively established that adsorbate adatoms occupy the T_4 sites.

The results of STM observations shown in Fig. 9.25 provide clear evidence for the T_4 adsorption geometry. In the experiment, the surface was prepared in which domains of the $\sqrt{3}\times\sqrt{3}$-Ga phase coexist with the regions of the original 7×7 surface as seen in Fig. 9.25a. Using as a reference the known 7×7 Si adatom sites (which are T_4 sites), one can establish the location of the Ga adatoms in the $\sqrt{3}\times\sqrt{3}$-Ga surface phase. The corrugation trace of the STM tip spanning the boundary between 7×7 and $\sqrt{3}\times\sqrt{3}$ domains demonstrates that Ga atoms populate the T_4 sites (Fig. 9.25b). The trace that would have resulted from H_3 configuration is indicated to show that STM resolves the distinction comfortably.

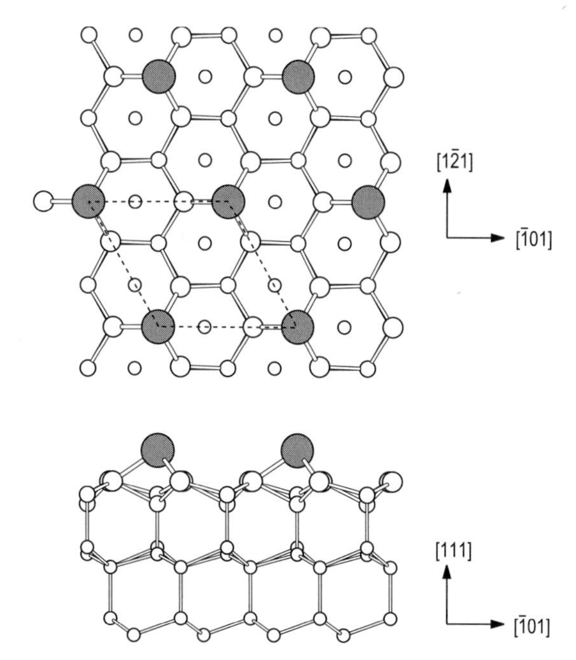

Fig. 9.23. Schematic of the $\sqrt{3}\times\sqrt{3}$ structure built of 1/3 ML of adsorbate adatoms residing in T_4 sites

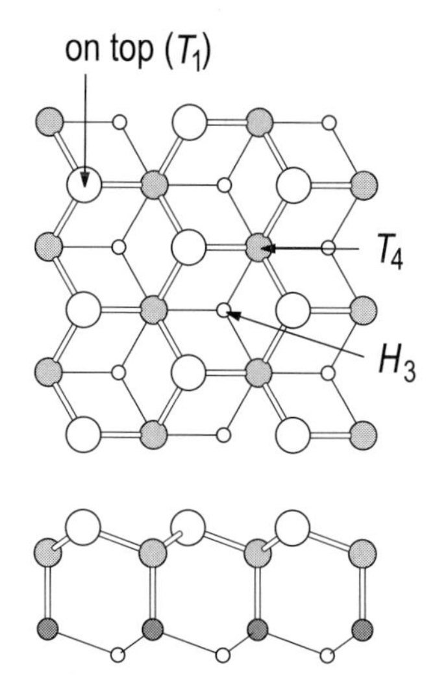

Fig. 9.24. Main adsorption sites on the bulk-like Si(111) and Ge(111) surfaces: T_4 site – the four-fold coordinated site above the second-layer atom; H_3 site – the three-fold coordinated site above the fourth-layer atom; on-top (T_1) site – the site above the first-layer atom

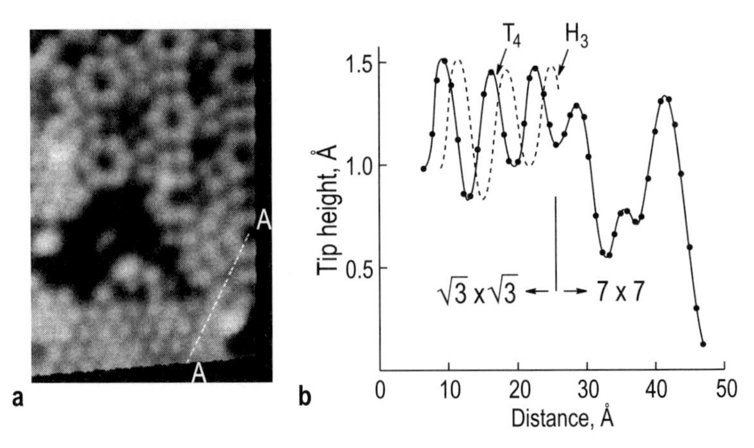

Fig. 9.25. (**a**) Empty state STM image showing coexisting regions of 7×7 and $\sqrt{3}\times\sqrt{3}$-Ga structures on a partially Ga-covered Si(111) surface. (**b**) Line scan along the *A–A* section shown in (**a**). The dotted line shows how the scan would look for the H_3-type geometry (after Zegenhagen et al. [9.18])

The T_4 adatom $\sqrt{3}\times\sqrt{3}$ structure has been found to occur for adsorption of 1/3 ML of Group III atoms, Al, Ga and In, Group IV atoms, Sn and Pb and Group V Bi atoms on Si(111), as well as of Sn, Pb and Bi on Ge(111).

Si(111)$\sqrt{3}\times\sqrt{3}$-B. One might expect boron to form a $\sqrt{3}\times\sqrt{3}$ of adatom type, like that of the other Group III atoms, Al, Ga, and In. However, it appears that, due to the small size of B, the usual T_4 adatom configuration is not stable. B adsorbs as an adatom only at low temperatures, while the lowest-energy Si(111)$\sqrt{3}\times\sqrt{3}$-B structure has been found to be the one shown in Fig. 9.26, in which the B atom substitutes for a second-layer Si atom directly beneath a T_4 Si adatom. This bonding site of B is called conventionally an S_5 site. The electronic consequence of this geometry is the charge transfer from the Si adatom to the B atom, which leaves the Si dangling-bond state empty.

Several techniques are employed to prepare a Si(111)$\sqrt{3}\times\sqrt{3}$-B phase, including adsorption of elemental boron, dissociation of boron-containing compounds (for example, HBO_2, $B_{10}H_{14}$) at the Si(111)7×7 surfaces heated to 600–700°C, or simply by annealing highly B-doped Si(111) wafers to 900–1300°C.

Trimer-Type Si(111)$\sqrt{3}\times\sqrt{3}$. The $\sqrt{3}\times\sqrt{3}$ structure of this type is formed under adsorption of 1 ML of Group V atoms, Sb and Bi, at a Si(111) surface heated to ∼500–700°C. In this structure, which is also referred to as a *"milk-stool"* (Fig. 9.27), adsorbate atoms are located basically at the on-top sites, but are shifted to form trimers. The adsorbate trimers are centered at a T_4 site, the side of the trimer approaches the bond length in metal (for example, ∼2.9 Å for Sb trimers). One can see that each adsorbate atom has

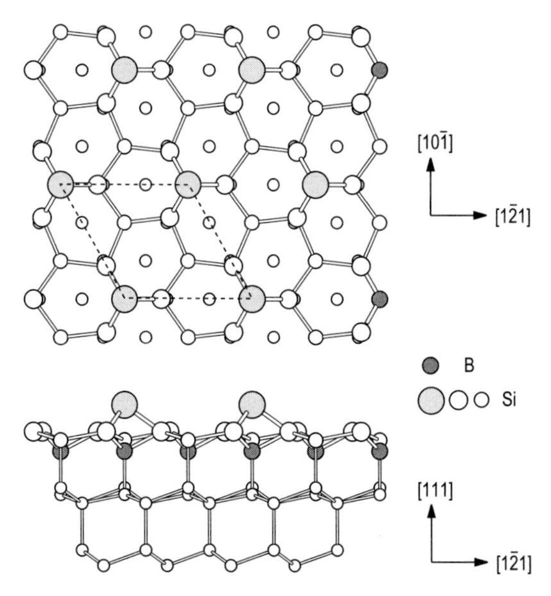

Fig. 9.26. Schematic of the Si(111)$\sqrt{3}\times\sqrt{3}$-B structure built of 1/3 ML of B atoms residing in S_5 sites

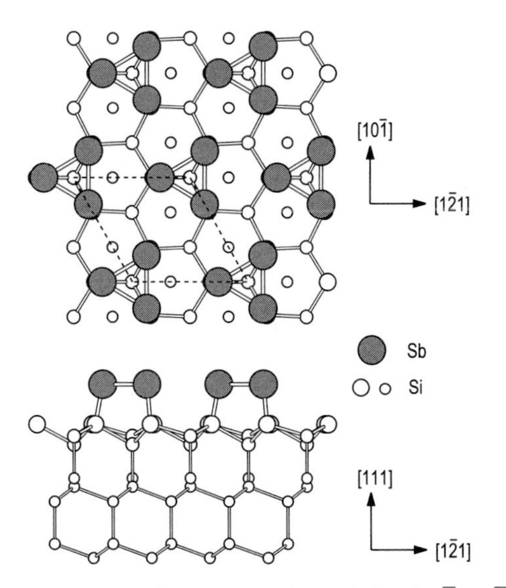

Fig. 9.27. Schematic of the Si(111)$\sqrt{3}\times\sqrt{3}$ structure built of 1 ML of Sb or Bi atoms forming trimers centered in T_4 sites

one bond to a top Si atom and two bonds to the two remaining adsorbate atoms of a trimer. As the adsorbate is a Group III atom, two valence electrons are left and form a lone pair, which gives the dominant contribution to the protrusions associated with the Sb atoms in the filled state STM images (Fig. 9.28).

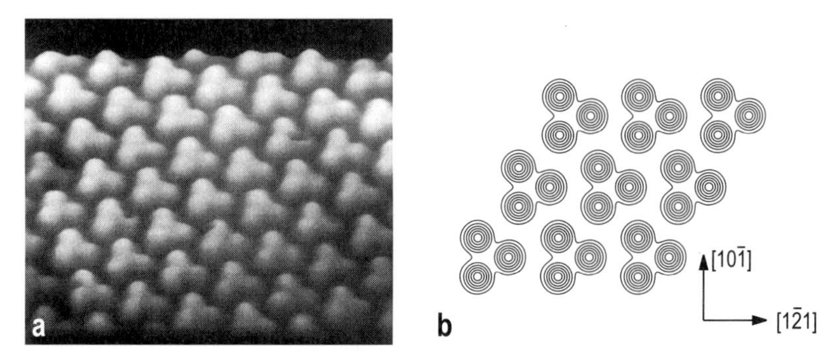

Fig. 9.28. (**a**) Experimental perspective STM image of the Si(111)$\sqrt{3}\times\sqrt{3}$-Sb surface obtained at −2.0 V sample bias (filled states). (**b**) Theoretical integrated charge-density contours corresponding to filled states down to 2 eV below E_F (after Mårtensson et al. [9.19])

Si(111)$\sqrt{3}\times\sqrt{3}$-Ag. The Si(111)$\sqrt{3}\times\sqrt{3}$-Ag phase is formed under saturation adsorption of Ag onto the Si(111)7×7 surface held at about 500°C. The atomic structure of the Si(111)$\sqrt{3}\times\sqrt{3}$-Ag phase is rather complicated and almost 25 years have passed since the first study [9.20] on this surface until the conclusive determination of its atomic arrangement [9.21, 9.22]. During that period, almost all surface-sensitive techniques were employed to analyze its structure and about a dozen structural models were proposed. The final model, called the *honeycomb-chained-trimer (HCT)* model, is shown in Fig. 9.29. According to the model, the topmost layer of the phase is built of Ag atoms with a HCT arrangement, in which the intratrimer Ag–Ag distance is 5.1 Å. At 0.75 Å below the Ag layer, there exists a layer of Si trimers. The composition of the Si(111)$\sqrt{3}\times\sqrt{3}$-Ag phase is $\Theta_{Ag} = 1.0$ ML and $\Theta_{Si} = 1.0$ ML. The Ge(111)$\sqrt{3}\times\sqrt{3}$-Ag phase was established to have a similar HCT structure.

Close-Packed $\sqrt{3}\times\sqrt{3}$-Pb at Si(111) and Ge(111). Upon Pb adsorption on Si(111) and Ge(111), two types of $\sqrt{3}\times\sqrt{3}$ phases occur. The first one, labeled conventionally $\alpha-\sqrt{3}\times\sqrt{3}$-Pb, is formed at 1/3 ML Pb coverage and has a simple adatom structure (see Fig. 9.23). The second one, labeled $\beta-\sqrt{3}\times\sqrt{3}$-Pb, is formed at 4/3 ML Pb coverage and has a structure, which corresponds to the close-packed Pb layer rotated by 30° (see Fig. 9.30). The $\sqrt{3}\times\sqrt{3}$ unit cell contains one Pb atom in an H_3 site and three Pb atoms, which occupy a site along the bond direction between the

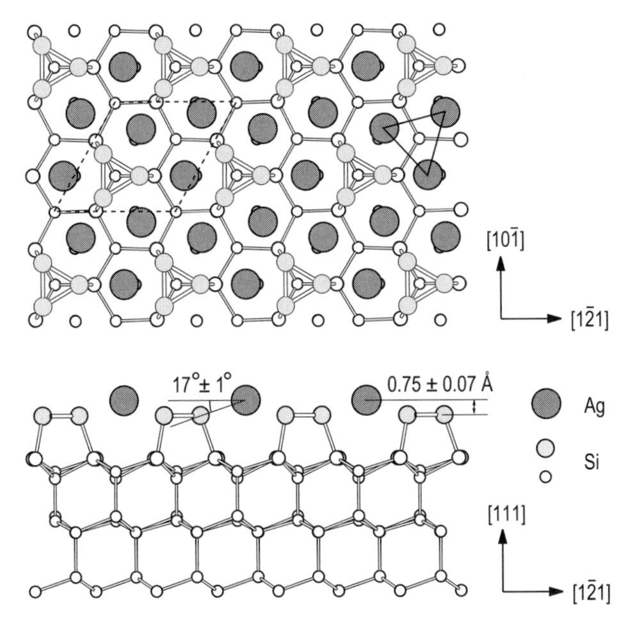

Fig. 9.29. Honeycomb-chained-trimer model of the Si(111)$\sqrt{3}\times\sqrt{3}$-Ag surface as derived in CAICISS experiments (after Katayama et al. [9.21])

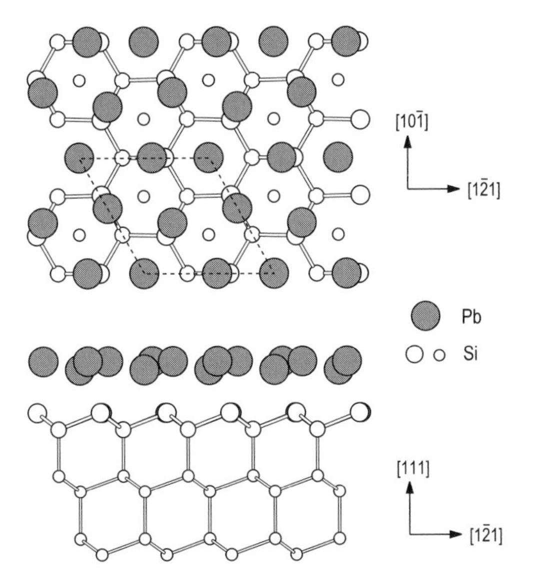

Fig. 9.30. Atomic structure of the close-packed $\beta\sqrt{3}\times\sqrt{3}$-Pb phase on Si(111) and Ge(111) at 4/3 Pb coverage. The $\sqrt{3}\times\sqrt{3}$ unit cell contains one Pb atom in an H_3 site and three Pb atoms displaced from the T_1 site towards the T_4 site (a so-called off-centered T_1 site)

on-top (T_1) and T_4 sites. This site is referred to as an *off-centered* T_1 site. In the case of the $\beta-\sqrt{3}\times\sqrt{3}$-Pb phase on the Ge(111) substrate, the Pb adlayer is compressed by 1% to adopt a commensurate $\sqrt{3}\times\sqrt{3}$-Pb structure. A commensurate $\beta-\sqrt{3}\times\sqrt{3}$-Pb phase on the Si(111) substrate would require a Pb overlayer compression of 5%. This is apparently too much, hence the Si(111)$\beta-\sqrt{3}\times\sqrt{3}$-Pb phase is incommensurate.

9.5.2 2×1, 1×1, and 3×1 Phases in the H/Si(100) System

In the H/Si(100) system, a noticeable interaction can only take place if hydrogen is offered in atomic form; then, however, the interaction is quite strong. The reason is that H_2 molecules do not dissociate easily on Si surfaces. Atomic hydrogen is typically generated by the decomposition of H_2 on a tungsten filament, which is heated to 1500–1800°C and is placed in front of the sample. As a result of atomic H interaction, three distinct H/Si(100) phases are formed, depending on substrate temperature and H exposure. These are Si(100)2×1-H, Si(100)1×1-H and Si(100)3×1-H surface phases.

Monohydride Si(100)2×1-H. At the initial stage of room-temperature adsorption, below ~25 L exposure (or below ~1 ML H coverage), hydrogen atoms adsorb at the Si dangling bond, preserving the Si-Si dimer arrangement. As a result, a monohydride Si(100)2×1-H phase is formed, as shown in Fig. 9.31a. Note that due to H saturation of Si dangling bonds, the Si-Si dimers become symmetric. However, RT H adsorption does not offer a perfect monohydride Si(100)2×1-H surface, as under these conditions the formation of higher hydrides also takes place. The preparation of the well-ordered

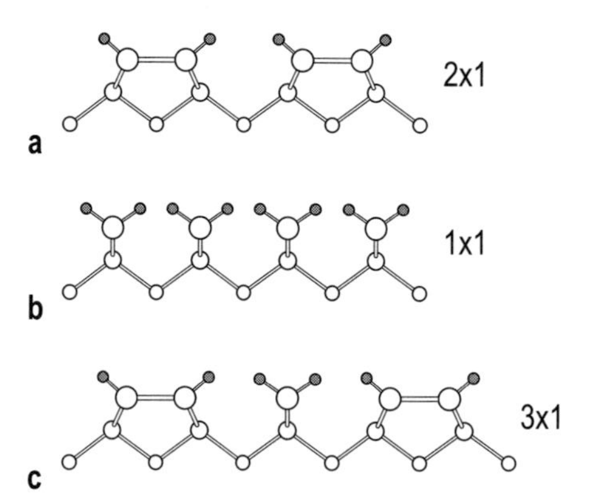

Fig. 9.31. Schematic of the structure of the ordered H/Si(100) surface phases: (**a**) Si(100)2×1-H monohydride; (**b**) Si(100)1×1-H dihydride; (**c**) Si(100)3×1-H monohydride plus dihydride. Silicon atoms are shown by white circles, hydrogen atoms by small hatched circles

monohydride Si(100)2×1-H surface phase is possible at saturation exposure to atomic hydrogen at about 400°C. The higher hydrides are unstable at temperatures above ∼350°C and are not formed under high-temperature conditions. In agreement with the model, the monohydride phase is characterized by a definite saturation coverage of 1 ML (Fig. 9.32).

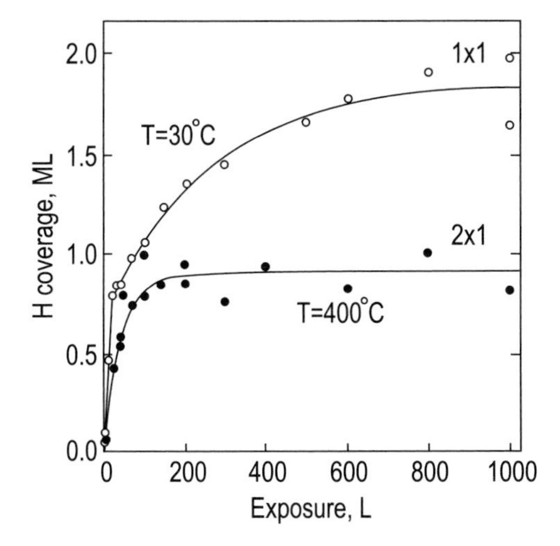

Fig. 9.32. Hydrogen coverage at a Si(100) surface as a function of H_2 exposure for atomic hydrogen adsorption at RT and 400°C (after Oura et al. [9.23])

Dihydride Si(100)1×1-H. The dihydride Si(100)1×1-H surface phase is formed at 2 ML H coverage (Fig. 9.33b) when the arriving H atoms break the Si-Si dimers and saturate the forming dangling bonds. The ideal structure of the dihydride Si(100)1×1-H phase is shown in Fig. 9.31b. In practice, the phase contains a considerable number of defects, which are associated with the breaking of Si-Si back bonds leading to the formation of trihydride species and gaseous silane, i.e., etching of silicon, as well as with remaining monohydride species.

Mixed Si(100)3×1-H. Upon atomic H adsorption at the intermediate temperature of 110±20°C, an ordered Si(100)3×1-H phase is formed. The structure of this phase is recognized to consist of alternating monohydride and dihydride species, as shown in Fig. 9.31c. Thus, the ideal H coverage of the phase is 1.33 ML.

STM images, illustrating the real-space appearance of the Si(100)2×1-H, Si(100)1×1-H, and Si(100)3×1-H surface phases with atomic resolution, are shown in Fig. 9.33.

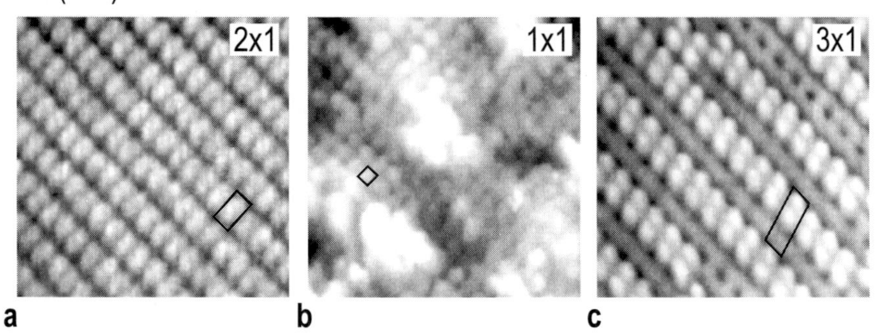

Fig. 9.33. High-resolution STM images of the ordered H/Si(100) surface phases: (**a**) Si(100)2×1-H monohydride; (**b**) Si(100)1×1-H dihydride; (**c**) Si(100)3×1-H monohydride plus dihydride. The corresponding unit meshes are outlined (after Boland [9.24])

Problems

9.1 Consider substitutional adsorption on the Si(111)7×7 surface. Assuming that adsorbate atoms substitute Si adatoms in the DAS structure, determine the composition, Θ_a and Θ_{Si}, (a) in the case when all Si adatoms are substituted and (b) in the case when all Si adatoms in only certain 7×7 unit cell halves (for example, in the faulted halves) are substituted.

9.2 The Si(111)6×1-Ag phase contains 1/3 ML of Ag and is formed upon Ag desorption from a Si(111)$\sqrt{3}\times\sqrt{3}$-Ag phase with 1 ML Ag coverage. Experiment has shown that, at the intermediate stage of the $\sqrt{3}\times\sqrt{3}$-to-6×1 transition, the area fractions occupied by phase domains are as follows:

$\sqrt{3}\times\sqrt{3}$	(upper level)	–	28 %,
$\sqrt{3}\times\sqrt{3}$	(lower level)	–	37 %,
6×1	(upper level)	–	20 %,
6×1	(lower level)	–	15 %

Taking into account that Si coverage is 2.08 ML in Si(111)7×7 and 1.0 ML in Si(111)$\sqrt{3}\times\sqrt{3}$-Ag, determine the Si coverage in Si(111)6×1-Ag.

9.3 Two phases with a $\sqrt{7}\times\sqrt{3}$ (or more strictly, $\begin{pmatrix} 3 & 1 \\ 1 & 2 \end{pmatrix}$) superlattice are formed in the In/Si(111) system, called quasi-hexagonal (hex-$\sqrt{7}\times\sqrt{3}$-In) and quasi-rectangular (rec-$\sqrt{7}\times\sqrt{3}$-In) (see Fig. 9.13). In high-resolution STM images, five protrusions per unit mesh are seen in the case of hex-$\sqrt{7}\times\sqrt{3}$-In and six protrusions in the case of rec-$\sqrt{7}\times\sqrt{3}$-In.

Assuming that each protrusion corresponds to a single In atom, calculate the ideal In coverage of these phases.

Further Reading

1. V.G. Lifshits, A.A. Saranin, A.V. Zotov: *Surface Phases on Silicon* (John Wiley, Chichester 1994) (structures occurring at the adsorbate-covered Si surfaces)
2. G.A. Somorjai: *Introduction to Surface Chemistry and Catalysis* (John Wiley, New York 1994) Chapter 2 (vast list of structures occurring at clean crystal surfaces with adsorbates in tabular form with brief comments)
3. *National Institute of Standards and Technology (NIST) Surface Structure Database (SSD), Version 3.0* `http://www.nist.gov/srd/nist42.htm`. (powerful graphics of SSD allow detailed assessment of atomic-scale structures of surfaces)
4. H.P. Bonzel (Ed.): *Physics of Covered Solid Surfaces. Landolt-Börnstein. Vol. III/42.* (Springer, Berlin, Heidelberg, New York 2001) (structures occurring on adsorbate-covered metal and semiconductor surfaces)

10. Structural Defects at Surfaces

An ideal ordered surface with complete translational symmetry does not occur in reality. Each real surface always contains a certain number of structural defects. The surface defects are usually classified according to their dimensionality into

- zero-dimensional or point defects and
- one-dimensional or line defects.

Zero-dimensional defects include adatoms, vacancies, and dislocation emergence points the terrace; kinks, step adatoms, and step vacancies at steps, as well as anti-site defects in the case of compound surfaces. *One-dimensional defects* involve step edges and domain boundaries.

The concerted behavior of surface defects might change the surface structure drastically. For example, ordered arrays of monatomic steps result in the formation of vicinal surfaces, an ordered network of domain boundaries leads to long-period superstructures, and surface roughening and melting can be thought of as uncorrelated defects involving nearly all surface atoms. But even in relatively small amounts, surface defects might play a predominant role in many processes at a surface, like adsorption, surface diffusion, chemical reactions, and thin film growth.

10.1 General Consideration Using the TSK Model

Most surface defects can be illustrated schematically in terms of a model, called the *terrace–step–kink* or *TSK* model, as shown in Fig. 10.1. It is also refereed to as the *terrace–ledge–kink* or *TLK* model. The TSK model portrays a simple cubic crystal, in which each atom in the lattice is represented by a cube or a sphere. Note that, though a simple cubic crystal is widely used for theoretical considerations, in reality only a single example is known, namely, one of the crystal modifications of the metal polonium.

10.1.1 Point Defects

Using the TSK model, consider first point defects. They can be subdivided into two types:

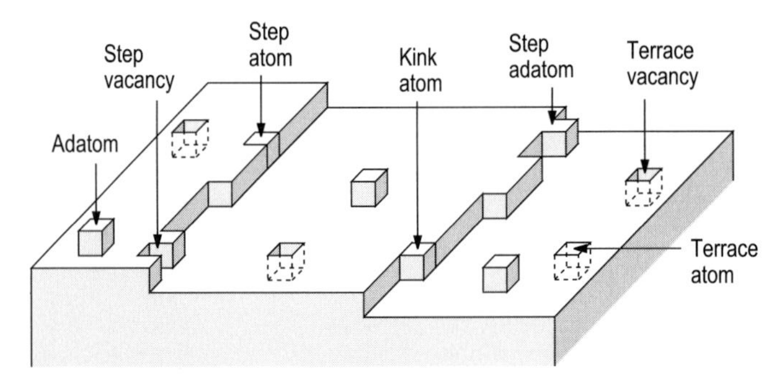

Fig. 10.1. TSK model showing typical atomic sites and defects on a simple cubic (100) surface

- kinetically stable defects;
- thermodynamically stable defects.

The main kinetically stable defects are points of dislocation emergence at the surface. The dislocations are line defects in the crystal bulk. Each dislocation can terminate only at another dislocation or at the crystal surface. At the point of dislocation emergence, the ordering of the surface atoms is disrupted. Figures 10.2a and b illustrate the emergence of edge and screw dislocations, respectively. As one can see, the emergence of a screw dislocation creates a step, which is tied at one end to the dislocation emergence point.

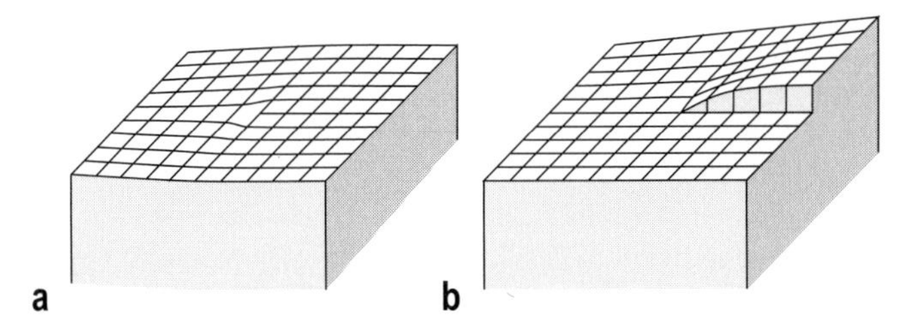

Fig. 10.2. Emergence of (**a**) an edge dislocation and (**b**) a screw dislocation on the (100) surface of the TSK model crystal

Thermodynamically stable point defects are those that are present at equilibrium at any temperature above 0 K. These are adatoms, vacancies, step kinks, step adatoms, and step vacancies. The relative number of defects of each kind depend on their formation energies and on the temperature. The formation energies are determined by the energies of the atoms in the various positions. In turn, these energies are determined primarily by the number of first, second, third, and so on, nearest neighbors.

As a simple example, consider the energetics of the point defects on a cubic (100) surface using the TSK model (see Fig. 10.1). For simplicity, assume that the energy of an atom in a certain crystal site is determined by the number of nearest neighbors only. Then the binding energy, expressed as the number of bonds, is 6 for a bulk atom, 5 for a surface atom, and so on (see Table 10.1).

Table 10.1. Positions of the atoms in a crystal and notation of their energy for a simple cubic lattice

Site	Energy	Numbers of neighbors		
		1st	2nd	3rd
Adatom	ε_{A}	1	4	4
Step adatom	$\varepsilon_{\mathrm{LA}}$	2	6	4
Kink atom	ε_{K}	3	6	4
Step atom	ε_{L}	4	6	4
Surface atom	ε_{T}	5	8	4
Bulk atom	ε_{B}	6	12	8

The formation of each defect can be thought of as moving an atom from a kink site to another site, or vice versa. For example, the formation of a surface adatom can be represented by removing a kink atom and placing it on the terrace. In this process, three bonds are broken to remove an atom from the kink and one bond is reformed when the atom is placed in the adatom site, hence, the net change is minus two bonds. In a more general form, the formation energy of the adatom can be written as

$$\Delta G_{\mathrm{A}} = \varepsilon_{\mathrm{A}} - \varepsilon_{\mathrm{K}} \,, \tag{10.1}$$

where ε_{A} and ε_{K} are the binding energies in the adatom and kink sites, respectively. The equilibrium concentration of surface adatoms (n_{A}) at temperature $T > 0$ is given by

$$n_{\mathrm{A}} = n_0 \exp - \frac{\Delta G_{\mathrm{A}}}{k_{\mathrm{B}} T} \,, \tag{10.2}$$

where n_0 is the total number of surface sites per unit area.

In a similar way, the formation of a surface vacancy can be thought of as removing a surface atoms (breaking of five bonds) and placing this atom at the kink (reformation of three bonds). That is,

$$\Delta G_{\mathrm{V}} = \varepsilon_{\mathrm{K}} - \varepsilon_{\mathrm{T}} \,, \tag{10.3}$$

and

$$n_{\mathrm{V}} = n_0 \exp - \frac{\Delta G_{\mathrm{V}}}{k_{\mathrm{B}} T} \,. \tag{10.4}$$

The reader can easily continue the above consideration for step adatoms and step vacancies.

One might notice that in the above consideration the formation energies have been evaluated with respect to the kink site. This choice is not accidental. Among other positions, the kink is a peculiar site. Irrespective of the crystal type, an atom in the kink position has half the number of bonds of an atom in the crystal bulk. This is true also for the number of second, third, and so on, neighbors. Hence, the other name for the kink position is the *half-crystal position*. The energy of an atom in the kink site is half that of an atom in the bulk. But the sublimation energy of a crystal is also half of the energy of the bulk atom, as the work necessary to detach an atom from the bulk should be divided by two to avoid double counting (one bond involves two atoms). Thus, the change in the internal energy caused by detachment of an atom from the kink is equal to the heat of sublimation, which can be measured experimentally and, thus, provides a scale for the formation energies of other point defects. For more precise evaluation, one should take into account lattice relaxation around the defect and consider also effects associated with second and more distant neighbors in a crystal. Table 10.2 shows the number of neighbors of a bulk atom in the main crystal lattices.

Table 10.2. The number of first, second, and third neighbors of an atom in bulk position for various lattices

Crystal lattice	Numbers of neighbors		
	1st	2nd	3rd
Simple cubic	6	12	8
Face-centered cubic	12	6	24
Body-centered cubic	8	6	12
Hexagonal close packed	12	6	2
Diamond	4	12	12

10.1.2 Steps, Singular and Vicinal Surfaces, Facets

Using the TSK model, consider the effects associated with atomic steps. As a starting point, consider a step-free low-index surface, for example, the (100) plane of a simple cubic crystal, as shown in Fig. 10.3a. These atomically smooth surfaces are so-called *singular surfaces*. (A more detailed consideration of singular surfaces in terms of the surface energy will be given later.) Surfaces that make a small angle θ with a singular plane, are called *vicinal surfaces*. In the rigid TSK model, such surfaces are built of terraces of the

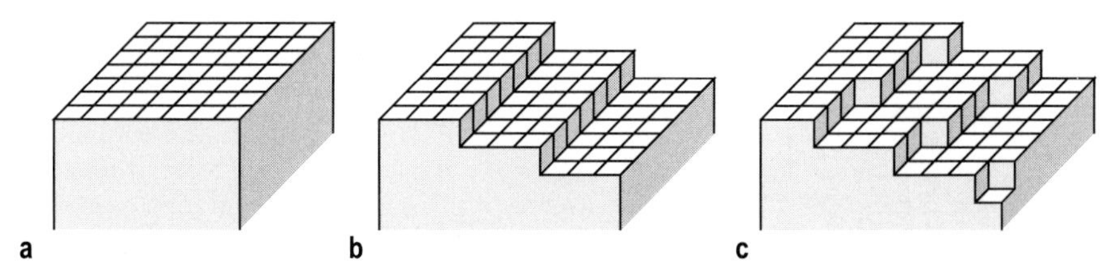

Fig. 10.3. TSK model presentation of (**a**) a singular (100) surface and surfaces vicinal to the (100) surface misoriented in (**b**) one and (**c**) two directions

closest singular plane and monatomic steps (Fig. 10.3b). The terrace width is determined by the inclination angle. When the vicinal surface is misoriented in two directions, the steps contain regular kinks to accomodate the misorientation in the second direction (Fig. 10.3c).

Consider now the energetics of crystal surfaces, including the energy contribution due to the presence of steps. The *surface energy* of a crystal surface, which is also referred to as *surface tension* by analogy with liquids, is the work that must be performed to build a crystal surface of unit area. At $T = 0$, the surface energy equals approximately the half-sum of the energies of all interatomic bonds which have to be broken in order to obtain a given surface. The factor 1/2 is due to the fact that two free surfaces are formed as a result of bond breaking. Here the effects associated with relaxation and reconstruction, which would change the surface energy (up to about 10–20% according to some estimations), are ignored. Similarly, the specific energy of the step is the work that must be done to create a step of unit length, which can be evaluated just by counting the number of broken bonds per unit length of a step.

If $\gamma(0)$ is the surface energy of a singular surface at terraces, and γ_{L} is the surface energy of the step, one can write for the surface energy of a vicinal surface formed by equidistant steps of atomic height a

$$\gamma(\theta) = (\gamma_{\mathrm{L}}/a)\sin\theta + \gamma(0)\cos\theta\,. \tag{10.5}$$

For a vicinal (100) surface in the cubic TSK nearest-neighbor model (see Fig. 10.4), (10.5) becomes

$$\gamma(\theta) = \left(\frac{\Phi}{2a^2}\right)(\sin\theta + \cos\theta) \tag{10.6}$$

or

$$\gamma(\theta) = \left(\frac{\Delta H_{\mathrm{S}}}{6a^2}\right)(\sin\theta + \cos\theta)\,, \tag{10.7}$$

where Φ is the energy per nearest-neighbor bond and ΔH_{S} is the heat of sublimation. In (10.5)–(10.7), the terms responsible for the interstep interactions are omitted, which means that this approximation is better for a large terrace width, i.e., small θ. Note that since $(\sin\theta + \cos\theta) > 1$, the vicinal surfaces always have a higher energy than the corresponding singular surface.

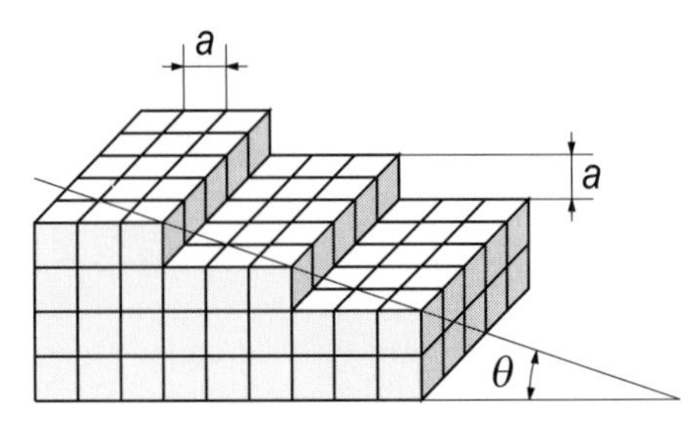

Fig. 10.4. TSK nearest-neighbor model presentation for determining the surface energy of the vicinal surface misoriented from the (100) plane by angle θ

Using (10.5)–(10.7), one can consider the anisotropy of the surface energy in the crystal. Figure 10.5a shows the plot of the surface energy γ versus θ for a model nearest-neighbor cubic crystal. One can see that, at angles corresponding to singular surfaces ($\theta = 0, \pm\pi/2, \pm\pi, ...$), the surface energy plot has sharp minima with discontinuity of the derivative, $\partial\gamma/\partial\theta$, i.e., singular points. Hence, the name *singular surfaces.* Figure 10.5b shows the same function plotted in polar coordinates. Extension of the model to the three-dimensional case produces an outer envelope of eight spherical segments with six sharp singular points corresponding to the $\{100\}$ planes (Fig. 10.5c). The plots of this kind are called *polar plots of* γ or, simply, γ*-plots.*

In the construction of the γ-plot shown in Fig. 10.5, only the first-neighbor interaction was taken into account. A similar consideration can be applied for the second-neighbor interaction, bearing in mind that in the simple cubic crystal the bonds between the second neighbors are oriented at an angle of $\pi/4$ with respect to the first-neighbor bonds and that the second-neighbor bonds are much weaker than the first-neighbor ones. Figure 10.6 shows the γ-plots for contributions that account separately for the first-neighbor (γ_1) and second-neighbor (γ_2) interactions, as well as the resultant plot, which accounts for both types of interactions ($\gamma_1 + \gamma_2$). One can see that new shallow minima appear, corresponding to the $\{110\}$ planes.

Using the γ-plot, one can establish the equilibrium shape of crystal. The procedure, known as the *Wulff construction*, is as follows. Planes perpendicular to the radius vector are drawn in each point of the γ-plot. The inner envelope of the planes yields the crystal shape in equilibrium. In this case, the total surface energy is minimal for a crystal of fixed volume.

One can see that for a model TSK nearest-neighbor crystal the equilibrium shape is simply a cube built of singular planes drawn through the deep cusps. In more realistic cases (Fig. 10.7), less deep cusps can be present. Whether the corresponding faces are stable or not depends on the peculiarities of $\gamma(\theta)$. The criterion is provided by the sign of the value $\gamma(\theta) + \partial^2\gamma(\theta)/\partial\theta^2$, called the *surface stiffness*:

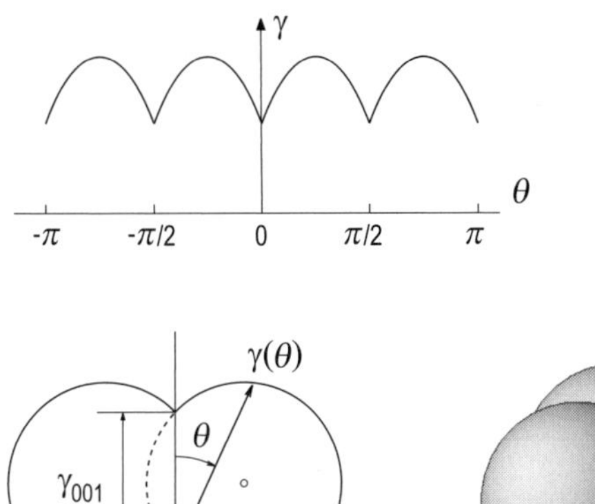

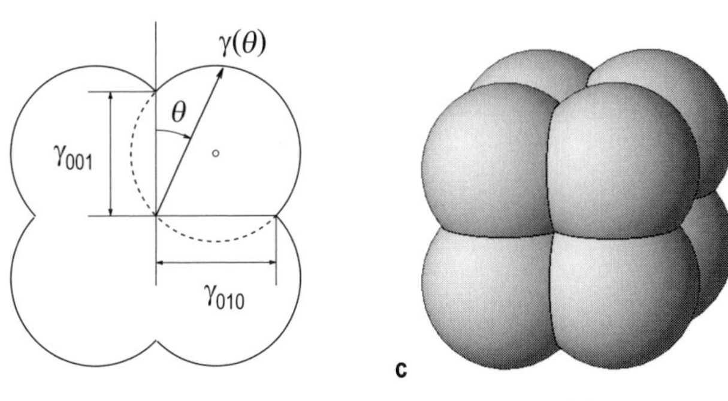

Fig. 10.5. Anisotropy of the surface energy $\gamma(\theta)$ for a simple cubic crystal. The plot of $\gamma(\theta)$ is in (**a**) orthogonal coordinates, (**b**) polar coordinates (two-dimensional representation), and (**c**) spherical coordinates (three-dimensional representation). Only the first-neighbor interactions are taken into account (after Chernov [10.1])

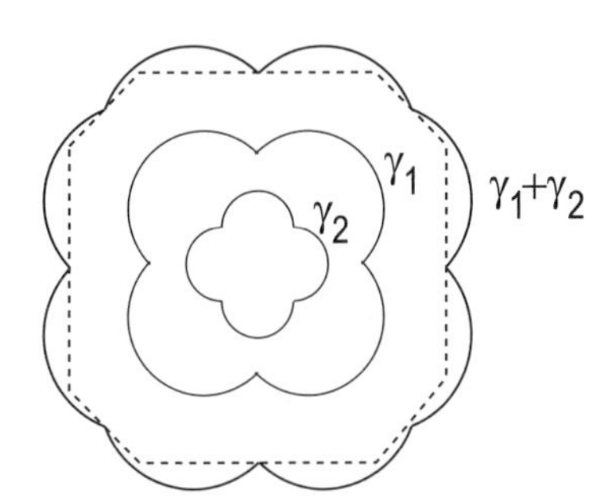

Fig. 10.6. Two-dimensional representation of the γ-plot for a simple cubic crystal, which takes into account both the first- and second-neighbor interactions (outer contour denoted $\gamma_1+\gamma_2$). The contours denoted γ_1 and γ_2 show the γ-plots as calculated by taking into account separately the first- and second-neighbor interactions. The equilibrium shape of the crystal is shown as dashed lines (after Markov [10.2])

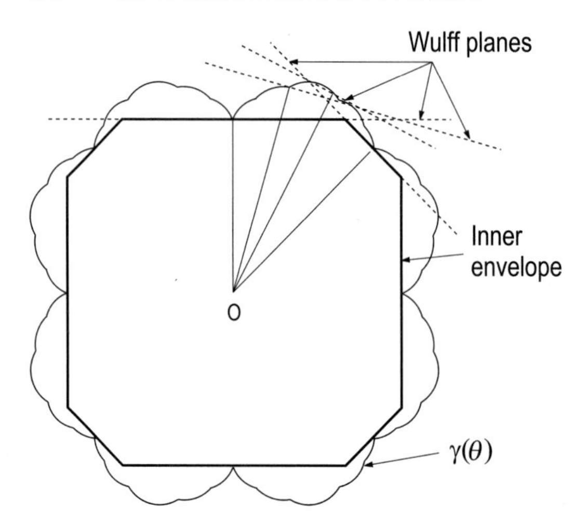

Fig. 10.7. Schematic diagram illustrating the Wulff construction, which yields the equilibrium shape of a crystal (shown by bold solid lines) as the inner envelope of the so-called Wulff planes (shown by dashed lines) which are normals to the radius vectors of the $\gamma(\theta)$ plot (shown in light solid lines) (after Herring [10.3])

$\gamma(\theta) + \partial^2\gamma(\theta)/\partial\theta^2 > 0$, the surface is stable
$\gamma(\theta) + \partial^2\gamma(\theta)/\partial\theta^2 < 0$, the surface is unstable and disintegrates into *facets*.

In the latter case, the increase in the surface area is overcompensated by the gain in surface energy due to replacing the high-energy surface by a set of the low-energy surfaces, i.e., facets (Fig. 10.8).

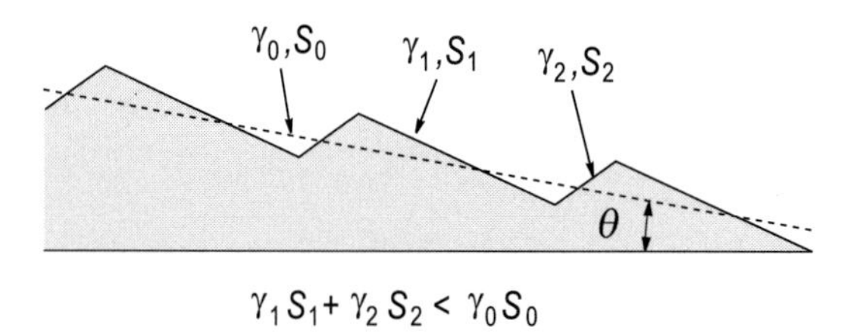

Fig. 10.8. Schematic diagram illustrating facetting of the unstable vicinal surface. Though the total surface area increases, $S_1 + S_2 > S_0$, the total surface energy decreases, $\gamma_1 S_1 + \gamma_2 S_2 < \gamma_0 S_0$

10.2 Selected Realistic Examples

10.2.1 Adatoms

Surface adatoms play a crucial role in a variety of dynamic phenomena which rely on surface diffusion (for example, epitaxial growth, nucleation and decay of non-equilibrium corrugations on the surface, transitions between phases having different atomic densities, etc.). Depending on their origin, the adatoms might be *external* or *thermal*. The external adatoms are those created in non-equilibrium conditions, say, by deposition from the vapor phase. Experiments with such an adatoms allow one to gain information on the adatom diffusion paths and energies. However, the equilibrium concentration and formation energies of the adatoms can be evaluated only in experiments with a two-dimensional gas of adatoms in equilibrium with the atomic step edges, i.e., thermal adatoms. Study of adatoms in equilibrium is generally associated with serious experimental difficulties, as at temperatures when they are sufficiently plentiful, adatoms are also very mobile and are not suitable for direct observation even with an atomic resolution probe such as STM. Thus, a special experimental design is required to study thermal adatoms. Two examples are shown below.

Si Ad-dimers on a Si(100) Surface. In the experiment in [10.4], Si(100) wafers with very large step-free regions were heated to 750–1050°C and then rapidly quenched to room temperature. In situ surface observations were conducted with LEEM using the (1/2,0) diffracted beam. As a consequence, alternating terraces, 1×2 and 2×1, are imaged in an alternating dark-field contrast. Figure 10.9a shows a LEEM image of the Si(100) terrace at high temperature. While the area contains numerous thermal adatoms, they are not resolved by LEEM. When the temperature is rapidly decreases, the 2D adatom gas becomes supersaturated, which results in nucleation and growth of islands. Figure 10.9b shows the same area as in Fig. 10.9a, after quenching. Numerous atomic-layer-high 2D islands (white patches) are seen on the terrace (dark area). The adatom concentration at the original temperature is estimated from the amount of Si accumulated in 2D islands.

Figure 10.10 summarizes the results of experiments in the plot of adatom concentration as a function of temperature. One can see that adatom concentration varies from ~0.02 (2% of a monolayer) to ~0.04 at about 1000°C. Comparison of the experimental data with the adatom concentrations, calculated using (10.2), yields the value of adatom formation energy $\Delta G_{\mathrm{A}} = 0.35 \pm 0.05\,\mathrm{eV}$. Theory [10.5] tells us that this energy value corresponds to the formation energy of an ad-dimer, rather than that of an adatom (in the latter case, the formation energy would be greater). Thus, in the case of the Si(100) surface, the atomic-scale species formed thermally at high temperatures are plausibly ad-dimers, not adatoms. As a final note, one should bear in mind that some adatoms might be "lost" in the experiment due to annihilation with vacancies or due to agglomeration into islands smaller than

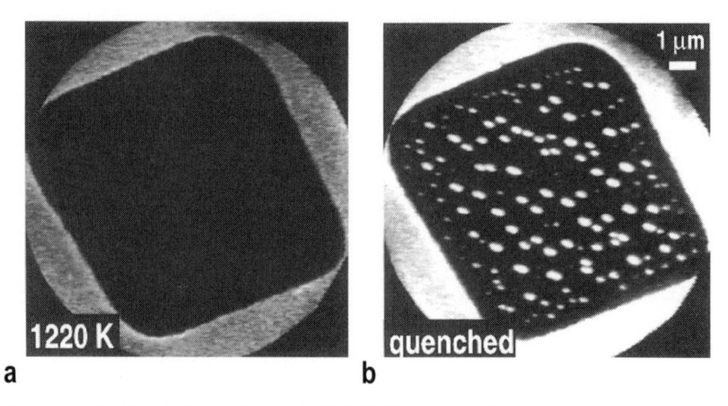

Fig. 10.9. Dark-field (1/2,0) LEEM images of a large atomically flat Si(100) terrace acquired (**a**) at 950°C and (**b**) after quenching (after Tromp and Mankos [10.4])

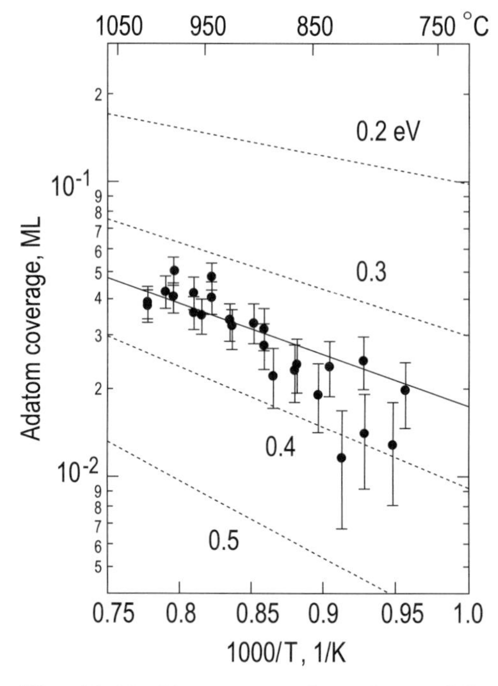

Fig. 10.10. Temperature dependence of Si adatom concentration on Si(100) as determined from quenching experiments. The dashed lines are calculated adatom concentrations using (10.2) with formation energies shown. The best-fit formation energy ($0.35 \pm 0.05\,\mathrm{eV}$) agrees well with the calculated formation energy of an addimer, rather than an adatom (after Tromp and Mankos [10.4])

the LEEM resolution (∼100 Å). Hence, one may view the ad-dimer formation energy of 0.35 eV as an upper bound.

Ag Adatoms on Si(111)$\sqrt{3}\times\sqrt{3}$-Ag. Using STM, "frozen" thermal adatoms can be observed directly, as in the example with Si(111)$\sqrt{3}\times\sqrt{3}$-Ag [10.6]. In order to observe the Ag adatoms, the following procedure was employed. First, about 1.7 ML of Ag was deposited onto the Si(111)7×7 clean surface at 500°C to form the Si(111)$\sqrt{3}\times\sqrt{3}$-Ag surface. Since Ag coverage of the the Si(111)$\sqrt{3}\times\sqrt{3}$-Ag phase is 1.0 ML, the excess Ag (0.7 ML) makes 3D Ag islands in accordance with the Stranski–Krastanov growth mode. At any constant temperature, there is a certain concentration of Ag atoms forming the 2D gas of adatoms in equilibrium with the 3D Ag islands. The sample was kept at RT for 1 h and then transferred to the STM stage at 6 K. The "frozen" Ag adatoms are seen in Fig. 10.11 as random bright "stars" (labeled A) and "propellers" (labeled B). The "stars" were interpreted as corresponding to single Ag adatoms and "propellers" to a group of several, plausibly three, Ag adatoms. On the basis of these STM observations, the equilibrium concentration of the Ag adatoms at RT was estimated to be 0.003 ML.

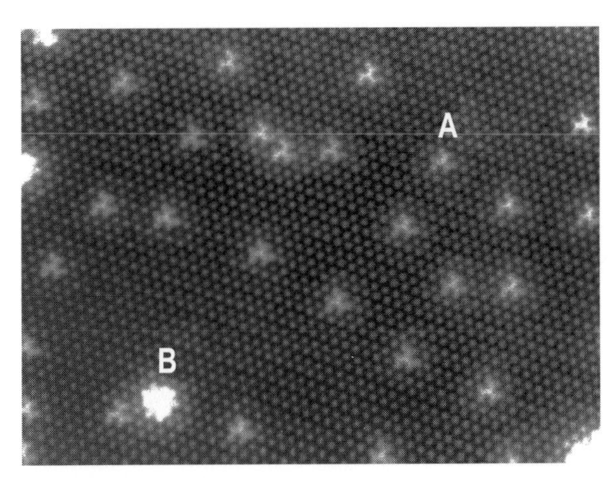

Fig. 10.11. 320×280 Å^2 empty-state STM image of the Ag adatoms "frozen" on the Si(111)$\sqrt{3}\times\sqrt{3}$-Ag surface at 6 K. Features A ("stars") are plausibly single Ag adatoms; feature B ("propeller") is a group of several, plausibly three, Ag adatoms (after Sato et al. [10.6])

10.2.2 Vacancies

Missing-Dimer Defects on Si(100)2×1. High-resolution STM observations have revealed that the major structural defects of a clean Si(100)2×1 surface are of the vacancy type. These are so-called *missing-dimer defects*. Following the notation of the pioneering work by Hamers and Köhler [10.7],

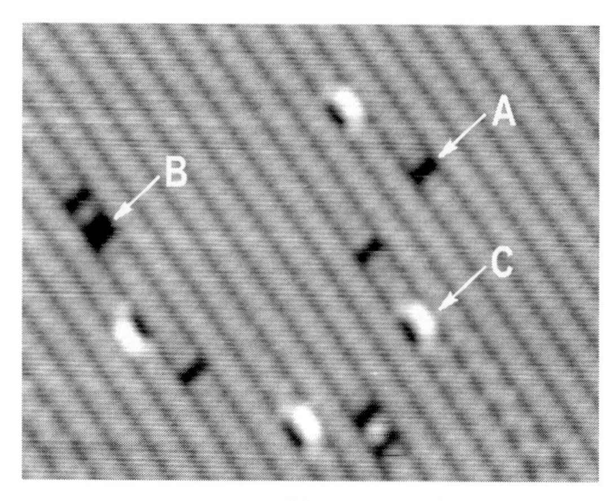

Fig. 10.12. 150×110 Å^2 filled-state STM image of the Si(100)2×1 surface region with typical missing-dimer defects. According to the pioneering work by Hamers and Köhler [10.7], the defects are labeled as type-A, type-B, and type-C defects

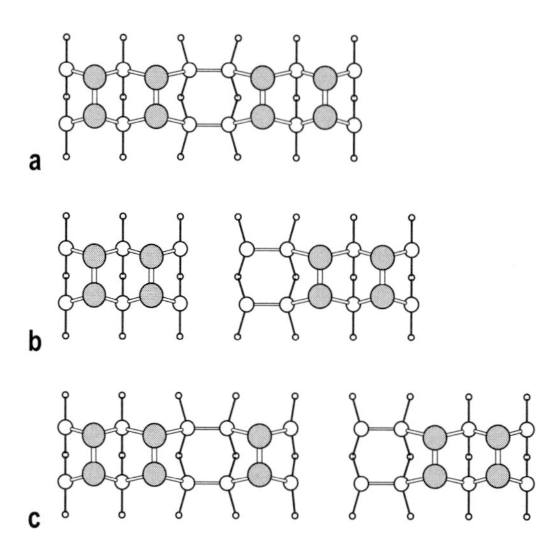

Fig. 10.13. Schematic of the atomic structure of typical missing-dimer defects on Si(100)2×1. (**a**) Single-dimer vacancy (A-type defect). (**b**) Double-dimer vacancy (B-type defect). (**c**) Complexes of single-dimer vacancy and double-dimer vacancy separated by a "split-off" dimer

typical defects can be classified into three types, namely, type A, type B, and type C. Defects of all three types are seen in the Si(100)2×1 surface region shown in Fig. 10.12. Type-A and type-B defects are well established to be a single dimer vacancy and a double dimer vacancy, respectively (Fig. 10.13a and b). In contrast, the atomic structure of the type-C defect remains a debated subject.

The surface density of the missing-dimer defects is very sensitive to the sample preparation process and varies from below 1% to 10% or more. In particular, the presence of minor amounts of Ni or some other metals in the subsurface region might increase greatly the defect density. With increasing density, the missing-dimer defects tend to congregate into complexes (for example, complexes of a single-dimer vacancy and double-dimer vacancy separated by a "split-off" dimer as shown in Fig. 10.13c) and to form line trenches running perpendicular to the Si dimer rows (Fig. 10.14a). In the limiting case of high defect density, the ordering of dimer vacancies results in the occurrence of the Si(100)2×n superstructure with $6 \leq n \leq 10$ (Fig. 10.14b).

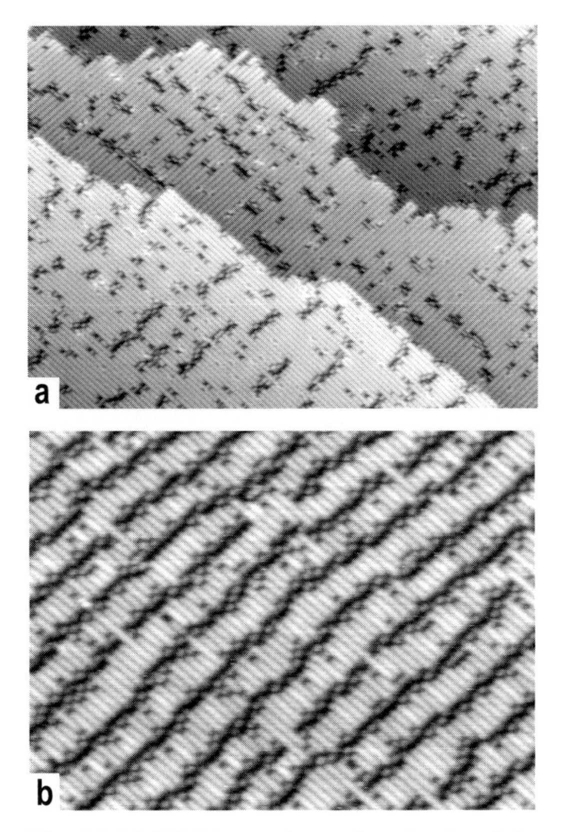

Fig. 10.14. With increasing surface density of the missing-dimer defects on Si(100), they become arranged in line trenches running perpendicular to the dimer rows (**a**) and finally form a Si(100)2×n superstructure (**b**). Scale of the STM images: (**a**) 700×500 Å^2 and (**b**) 400×280 Å^2

Phosphorus Vacancies on InP(110). The presence of vacancies might affect greatly the electronic properties of III-V semiconductors, since they may induce deep-lying electrical states in the bad gap and, at high concentration, can counteract the desired properties obtained by dopant atoms. Furthermore, a vacancy may have a net charge. In turn, the formation energy of the charge vacancies depends on the position of the Fermi level within the band gap. As a result, a certain type of vacancy prevails for a given doping type. In particular, in the case of *p*-doped InP(110), positively charged P (phosphorus) vacancies are formed.

The atomic structure of the P vacancy on GaP(110) is shown schematically in Fig. 10.15a, while Figs. 10.15b and c illustrate the STM appearance in filled and empty states, respectively. Recall that in a defect-free (110) surface of III-V semiconductors due to a charge transfer the empty states are localized at the cation (Group III) and the filled states at the anion (Group V) atoms. Hence, in STM images of the P (anion) vacancy, one filled-state protrusion is missing (Fig. 10.15b), while no protrusion is missing in empty states (Fig. 10.15c). This particular STM appearance is a fingerprint of an anion vacancy. Note that two neighboring P vacancies are shown in Figs. 10.15b and c.

One can see that in the filled-state STM image (Fig. 10.15b) the defects are surrounded by a wide depression about 40–50 Å in diameter, while in the empty-state image (Figs. 10.15c) this region is seen as elevated. This behavior is a consequence of local downward band bending induced by the positive charge of the P vacancies.

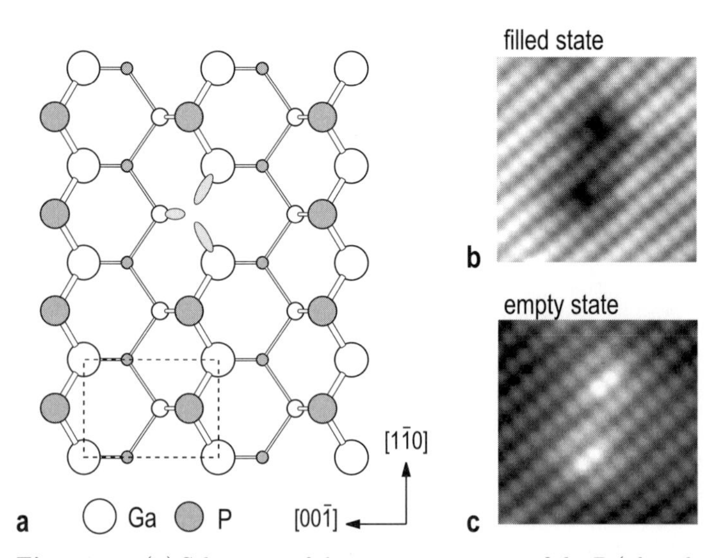

Fig. 10.15. (**a**) Schematic of the atomic structure of the P (phosphorus) vacancy on the GaP(110) surface. Two P vacancies on GaP(110), as seen in the (**b**) filled-state and (**c**) empty-state STM images (after Ebert [10.8])

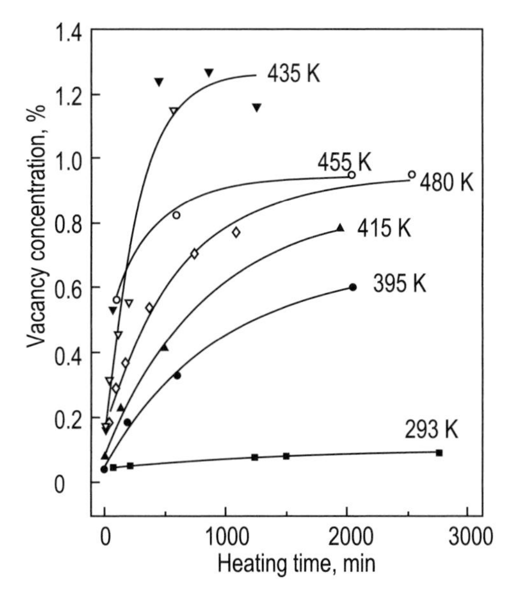

Fig. 10.16. Vacancy concentration on InP(110) (fraction of empty P lattice sites) as a function of heating time for different temperatures between 293 and 480 K. The lines serve only as guidance of the eye (after Ebert [10.8])

The dependence of the vacancy formation rate and concentration on the temperature and annealing time (Fig. 10.16) indicates the thermal mechanism of vacancy formation. It was suggested that the mechanism involves the following successive stages. At the first stage, a vacancy–adatom pair is generated at the terrace. (Note that though adatom formation is usually more favorable energetically at the kink sites of steps (see (10.3)), this formation mechanism is not efficient in the present case, as after initial detachment of the outermost row of P atoms the steps become In-terminated and are thus passivated preventing the further generation of P adatoms.) The forming P adatoms migrate over the surface and, if not annihilate with occurring vacancies, can form P_2 molecules and desorb from the surface. The vacancy saturation concentration increases with temperature, reaching the maximal value of about 1.2% at 435 K. However, at higher temperatures the outdiffusion of phosphorus from the bulk becomes significant, reducing the concentration of P vacancies.

10.2.3 Anti-Site Defects

Anti-Site Defects in GaP(110). In compound materials, *anti-site defects* might occur, i.e., some sites, normally occupied by the atoms of one component, may be occupied by the atoms of the other component. Say, in a GaP crystal, a P atom occupies a Ga site (P_{Ga} anti-site defect) or vice versa (Ga_P anti-site defect). Figure 10.17a shows a filled-state STM image of the GaP(110) surface with four P_{Ga} anti-site defects. Since the occupied states on the GaP(110) surface are known to be localized above the P atoms, the P_{Ga} defects manifest themselves as additional maxima between the rows (Fig. 10.17b).

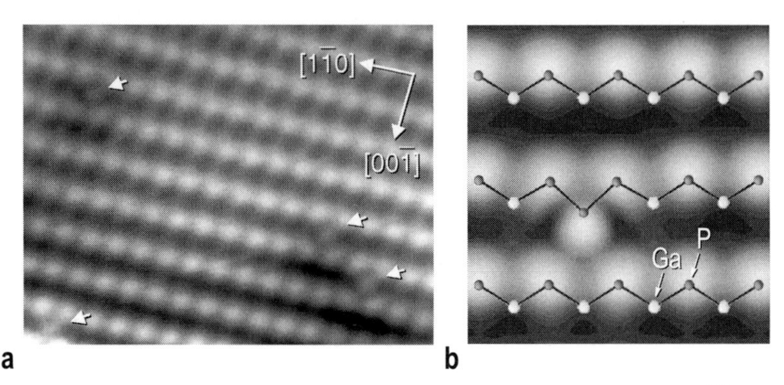

Fig. 10.17. (**a**) Filled state (−2.7 V) STM image of the 55×45 Å^2 area of the n-doped GaP(110) cleavage surface with four P_{Ga} anti-site defects (marked by arrows). (**b**) Calculated filled state STM image for anion anti-site defect superposed with the surface atomic structure (after Ebert et al. [10.9])

Si Substitutional Defects in Group III/Si(111)$\sqrt{3}\times\sqrt{3}$ Phases. The adsorption of Group III metals, Al, Ga, In, on Si(111) results in the formation of the Si(111)$\sqrt{3}\times\sqrt{3}$ surface phase. This surface phase is built of metal adatoms residing in T_4 sites on the almost bulk-like Si(111) surface (see Sect. 9.5.1). In STM images, such surfaces are seen as an ordered array of round protrusions, which correspond to the Group III adatoms. Two types of point defects can be distinguished on the surface. Defects of the first type (marked "V" in Fig. 10.18) appear as deep depressions at both polarities and are associated with *vacancies*. Defects of the second type (marked "S" in Fig. 10.18) are seen darker than the ideal protrusions in the empty states, but brighter in the filled states. These defects were first observed in Si(111)$\sqrt{3}\times\sqrt{3}$-Al [10.10], but later were also found for In and Ga adsorbates. From coverage-dependent measurements, it was concluded that S-type defects are *Si adatoms* which substitute for metal adatoms in the $\sqrt{3}\times\sqrt{3}$ overlayer, hence might essentially be regarded as anti-site defects.

The Si adatoms imbedded in the $\sqrt{3}\times\sqrt{3}$ layer are found to become mobile at temperatures above ~200°C. They migrate through exchanging sites with neighboring metal adatoms. The activation energies of the exchange rate for Al, Ga, and In adatoms are in the range 1.4–1.7 eV, the prefactors are within the range 10^{10}–10^{13} s^{-1} [10.11].

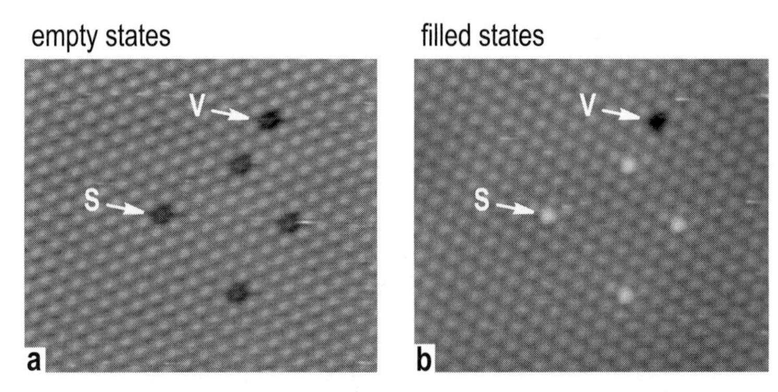

Fig. 10.18. STM images of the same area of Si(111)$\sqrt{3}\times\sqrt{3}$-In phase acquired at (**a**) positive (+2.3 eV) and (**b**) negative (−2.0 eV) sample bias. The point defects, namely, vacancy and Si substitutional defect are labeled "V" and "S", respectively

10.2.4 Dislocations

Emergence of Screw Dislocations on NaCl(100). The presence of screw dislocations greatly affects the crystal growth from a vapor, as well as crystal etching by thermal evaporation. The growth (etching) proceeds through atom attachment to (detachment from) the spiral atomic step generated around the emergence point of a screw dislocation. As an example, Fig. 10.19 illustrates schematically the crystal growth process. The etching process can clearly be visualized in a similar way.

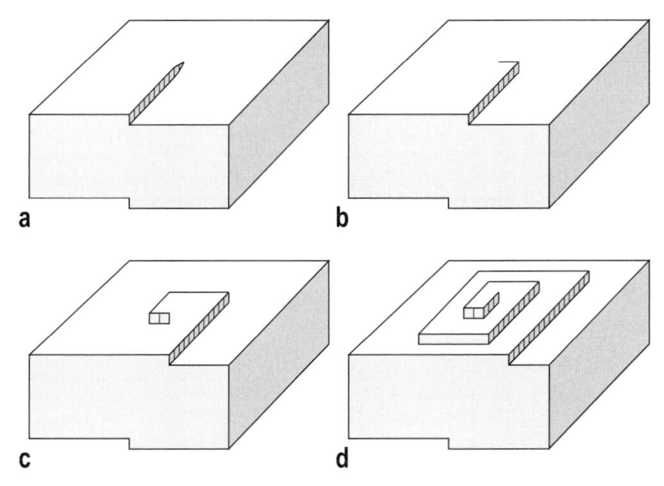

Fig. 10.19. Schematic diagram illustrating consecutive stages from (**a**) to (**d**) of the formation of a growth pyramid around the emergence point of a screw dislocation

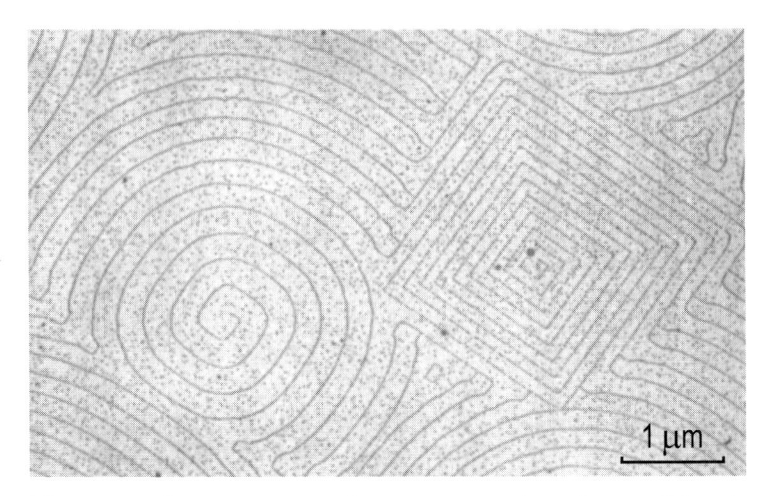

Fig. 10.20. Step structure at the emergence points of screw dislocations produced by sublimation of an NaCl crystal at 400°C for 90 min in a high vacuum (after Krohn and Bethge [10.12])

An experimental example is represented in Fig. 10.20, which shows the step structure at the emergence points of two dislocations produced by evaporation of a NaCl(100) crystal. The image is a TEM micrograph of the NaCl(100) surface decorated by gold (see Sect. 7.3). The step height equals the normal component of the Burgers vector. The left dislocation with a round step spiral has the normal component of Burgers vector equal to a single lattice plane distance ($a/2 = 2.8$ Å), while that of the right dislocation with a square step spiral is twice that distance, i.e., 5.6 Å. The doubling of the step height is clearly seen at the points where two step spirals intersect.

Surface Dislocations in a Reconstructed Au(100) Layer. Like the Pt(100) surface described in Sect. 8.3.1, the reconstructed Au(100) surface consists of a close-packed quasi-hexagonal top-atom layer residing above the lower Au(100) layers, in which atoms are arranged into a square lattice. The reconstructed surface has a $5\times n$ ($n \simeq 25$) periodicity and shows up in STM images as an array of ridges, which are due to the short periodicity (14.4 Å) of the reconstruction. Reconstruction ridges run along the $\langle 011 \rangle$ directions of the substrate. The top layer is known to have a "floating" nature, i.e., it is relatively uncoupled to the lower layers.

If such a Au(100) surface is subjected to a low-dose (~0.05 ML) bombardment with 600 eV Ar^+ ions at 300 K, defects of two types are formed (Fig. 10.21). These are elongated depressions (labeled A in Fig. 10.21) and forklike features appeared generally in pairs (labeled B in Fig. 10.21). The B features are the Moiré-enhanced footprints of the individual 2D dislocations, while depressions are dislocation dipoles of opposite Burgers vector [10.13].

These defects represent one of the few examples of the dislocation structures in quasi-2D systems. Note that the 2D dislocations discussed above are

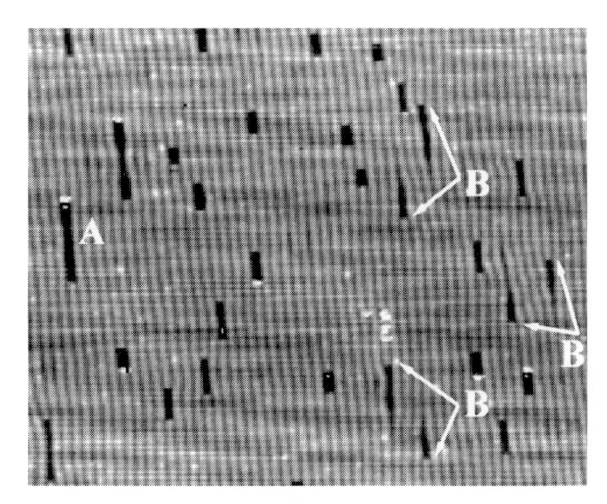

Fig. 10.21. 830×830 Å^2 STM image of the Au(100) surface after room temperature bombardment with 0.05 ML$^+$ of 600 eV Ar$^+$. Two types of defects are seen: depressions (A) and individual 2D dislocations. Note that the latter appear in pairs as shown by arrows (after Rodríguez de la Fuente et al. [10.13])

constrained to the top atomic layer which ends up at the boundary between the reconstructed layer and the crystal bulk. The 2D dislocations have an extra atomic row instead of an extra atomic plane as in case of the 3D dislocations. Hence, in contrast to 3D dislocations, which are line defects, 2D dislocations are point defects.

10.2.5 Domain Boundaries

A continuous surface region occupied by a given surface phase, i.e., a surface phase domain, always has a finite size. Each domain is, at least, limited by the width of the terrace bounded by atomic steps. Even within a single terrace, usually there are several domains separated from each other by frontiers, called, *domain boundaries* or *domain walls.* This occurs also in the case when the domains from both sides of the boundary have the same atomic structure, i.e., refer to the same surface phase. The multidomain structure is a natural consequence of the fact that in most cases nucleation of the new phase takes place independently at several points at the surface. When the growing domains come into contact, they form a single large domain only when their structures are "in phase". Otherwise these domains are called *antiphase domains* and an *antiphase domain boundary* is formed between them.

An illustrative example of antiphase Si(111)$\sqrt{3}\times\sqrt{3}$-In domains is shown in Fig. 10.22. As one can see, an In adatom in a given $\sqrt{3}\times\sqrt{3}$ domain occupies one of three equivalent T_4 sites in a unit cell (labeled A, B, and C in Fig. 10.22b). Thus, the number of equivalent domains is three. All of them are present in the STM image in Fig. 10.22a.

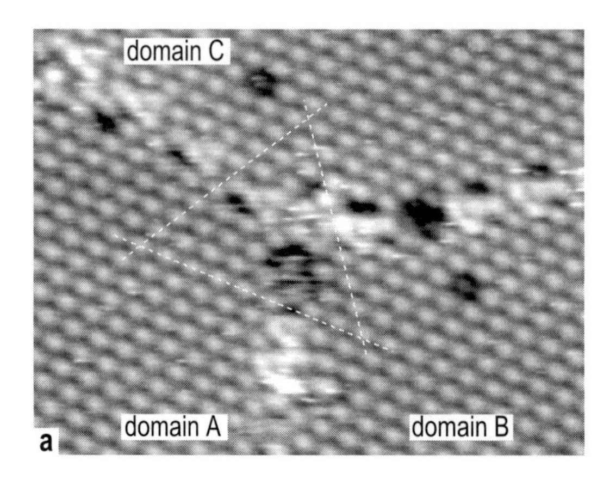

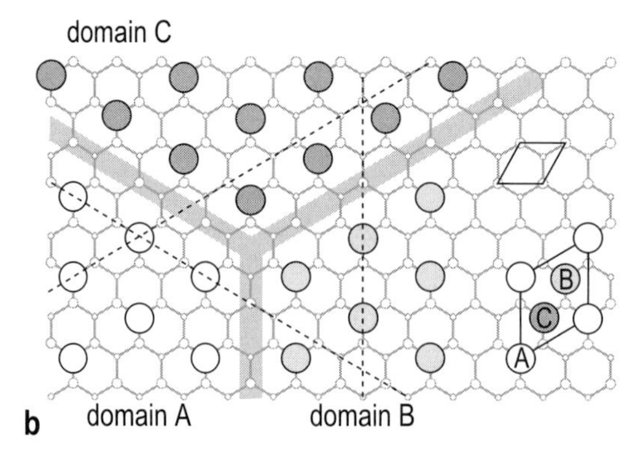

Fig. 10.22. (**a**) STM image showing the boundary between three antiphase domains of the Si(111)$\sqrt{3}\times\sqrt{3}$-In phase. (**b**) Schematic diagram illustrating the occurrence of the three antiphase domains due to the existence of three equivalent T_4 sites within the $\sqrt{3}\times\sqrt{3}$ unit cell, which the In adatom could occupy. These sites are labeled A, B, and C and the corresponding $\sqrt{3}\times\sqrt{3}$ domains are labeled domain A, domain B, and domain C. The domain boundaries are hatched; the 1×1 and $\sqrt{3}\times\sqrt{3}$ unit cells are outlined. The dashed lines crossing the boundaries guide the eye to follow the correspondence between the STM image and the schematic diagram

The formation of multidomain surfaces also takes place when the symmetry of the surface phase is lower than the symmetry of the underlying substrate. For example, the chain-like Si(111)4×1-In phase occurs in three equivalent *orientation domains*, due to the threefold rotational symmetry of the Si(111) substrate (see Fig. 10.23).

The above examples for the antiphase domains and rotation domains concern mostly phases with a commensurate structure. In these cases, the size of the domains might vary greatly depending on the preparation conditions (for example, substrate temperature or adsorbate deposition rate) and there

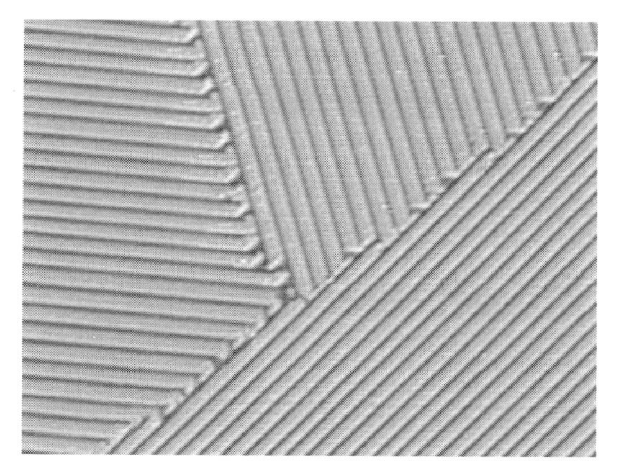

Fig. 10.23. STM image (500×350 Å^2) showing three equivalent rotation domains of the Si(111)4×1-In surface phase

are no apparent reasons for the ordering of the domain boundaries. However, this is not the case for the phases with an incommensurate structure. The latter are mostly characterized by a more or less ordered network of domain boundaries. The periodicity of the network (i.e., the size of the domains) is controlled by the misfit between the lattice constants of the substrate and overlayer.

The origin of the domain boundaries due to the layer–substrate misfit could be qualitatively understood in terms of the one-dimensional model which was proposed by Frenkel and Kontorova already in 1939 and analyzed quite conclusively by Frank and van der Merwe in 1949. Consider a chain of adsorbate atoms bound together by elastic springs and residing atop a periodic potential (Fig. 10.24). The springs stand for the adsorbate–adsorbate interaction, while the corrugated potential simulates the adsorbate–substrate interaction. The substrate potential is assumed to be rigid, which is realistic for films of about monolayer thickness. The natural length of the springs is b, the potential periodicity is a. The springs tend to preserve the interatomic spacing b, while the substrate prefers the atoms to occupy the potential troughs with a separation a. The competition between these tendencies results in the compromise interatomic spacing $\bar{b}$.

Two extreme cases are possible. If $\bar{b} = a$, all atoms occupy the potential minima, the phase is commensurate, and the misfit is accommodated by a homogeneous strain. If $\bar{b} = b$, atoms preserve on average their own spacing and the misfit is accommodated entirely by misfit dislocations. The distance between dislocations is given by

$$s = \frac{ab}{\mid b - a \mid} \,. \tag{10.8}$$

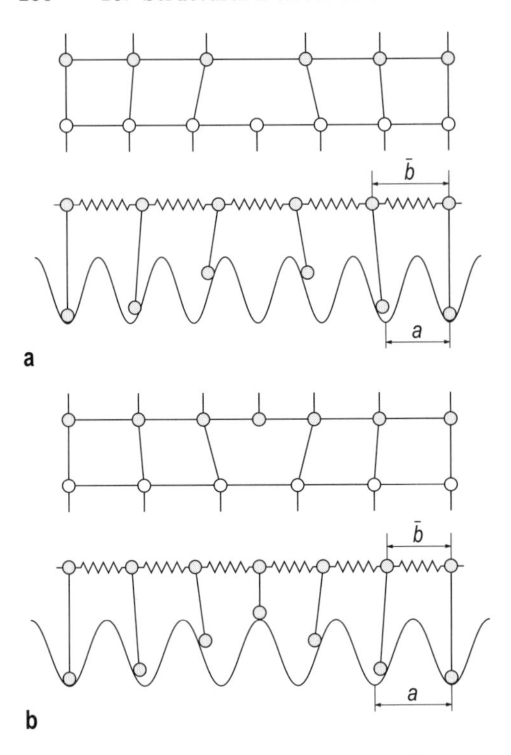

Fig. 10.24. The Frenkel–Kontorova/Frank–van der Merwe model of misfit dislocations at (**a**) positive ($b > a$) and (**b**) negative ($b < a$) misfits. In (**a**) the misfit dislocation represents an empty potential trough, i.e., forms a light wall. In (**b**) the misfit dislocation represents two adatoms in one trough, i.e., forms a heavy wall (after Markov [10.2])

In the general case, $\bar{b}$ is between a and b and the misfit in the first approximation is the sum of the homogeneous strain and the periodic strain due to misfit dislocations.

More elaborate consideration shows that, for the same average atom separation, the system achieves lower energy if the adsorbate atoms form domains, within which they reside close to the potential well minima, and the misfit is accommodated mainly by the compression or expansion of the interatomic spacings at the domain boundaries. The strain built up in the domains is released at the domain boundaries. For example, if the adsorbate is compressed within the domains, it expands in the domain boundaries. Such boundaries, with the adsorbate density lower than that within the domain, are called *light walls*. The opposite case is characterized by the term *heavy walls*.

The surface phases which are built of domains of almost commensurate structure interrupted by a periodic network of domain boundaries are sometimes referred to as *"weakly" incommensurate* or *discommensurate* phases.

Experimental examples are presented in Fig. 10.25 and Fig. 10.26. Figure 10.25 shows the structure of a hexagonal close-packed Pb overlayer on a Cu(111) surface. The ratio of the bulk lattice constants of Pb and Cu is 4.11/3. The lattice constant of the Pb overlayer depends on the Pb coverage and at about 0.6 ML of Pb (the ideal coverage is $9/16 = 0.5625$ ML) corresponds almost exactly to 4/3 of the Cu lattice constant. This results in large areas that show perfect 4×4 periodicity corresponding to the misfit-dislocation network. Thus, most of the 37%-misfit between Pb and Cu is accommodated by misfit dislocations, while only a few percent is accommodated by the compressive strain in the Pb overlayer.

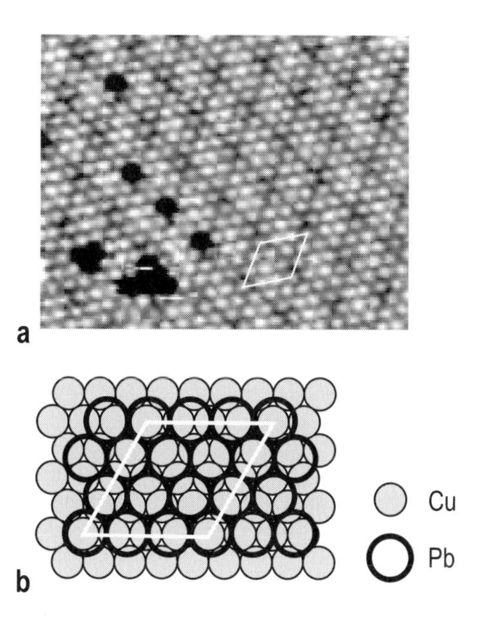

Fig. 10.25. (**a**) STM image of the hexagonal close-packed Pb overlayer on the Cu(111) surface. The Moiré pattern shows a 4×4 periodicity in each direction, where three Pb atoms are placed on four Cu atoms, thus forming a misfit dislocation. (**b**) Atomic model of the Cu(111)4×4-Pb phase. The 4×4 unit cell is outlined in white lines (after Nagl et al. [10.14])

Figure 10.26 presents the discommensurate structure of the so-called Ga/Ge(111) γ-phase formed at saturation Ga coverage of about 0.7 ML. The γ-phase shows up as a quasi-periodic superlattice of domains with an average size of about $7.4a$ ($a = 4.0$ Å for Ge(111)1×1) (Fig. 10.26a). The interior of the domains, resolved in the high-resolution empty-state STM image (Fig. 10.26b), comprises a hexagonal arrangement of the substitutional Ga atoms but with the lattice constant increased by about 10% compared to the Ge substrate lattice. Within the center of each domain, the Ga atoms are laterally almost in registry with ideal bulk positions. But towards the domain

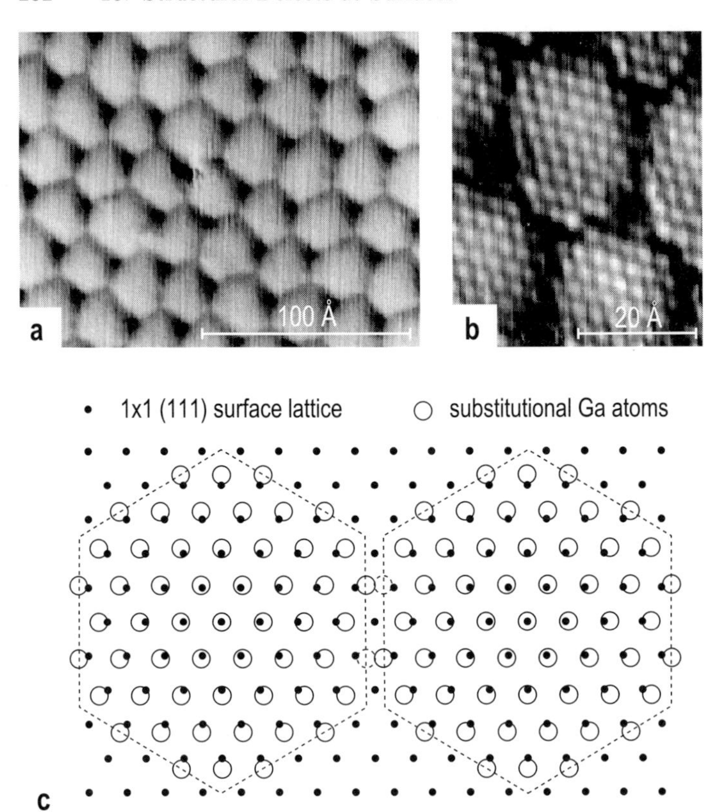

Fig. 10.26. Discommensurate Ga/Ge(111) γ-phase. (**a**) and (**b**) Empty-state STM images of the Ga/Ge(111) γ-phase with different magnifications. (**c**) Schematic model of the atomic structure of the Ga/Ge(111) γ-phase. Ga atoms (shown by open circles) substitute for the top Ge atoms leading to a surface layer with GeGa stoichiometry. The surface strain due to $\sim$10% mismatch leads to the appearance of a domain structure with light walls. The two domains shown have a size of 8×8. In fact, the size of domains fluctuates between 7×7 and 8×8 with an average of 7.4×7.4 (after Zegenhagen et al. [10.15])

periphery, the Ga atoms are progressively more and more out of registry. The compressive strain within the domain is released in the domain boundary depleted of Ga (light domain wall). Figure 10.26c illustrates this schematically for an idealized hexagonally shaped domain [10.15].

10.2.6 Steps

Atomic Steps at a Vicinal Si(100) Surface. Consider the vicinal Si(100) surface tilted towards the [011] direction (Fig. 10.27). In the ideal case, the surface comprises an array of equidistant parallel steps. The single-step height

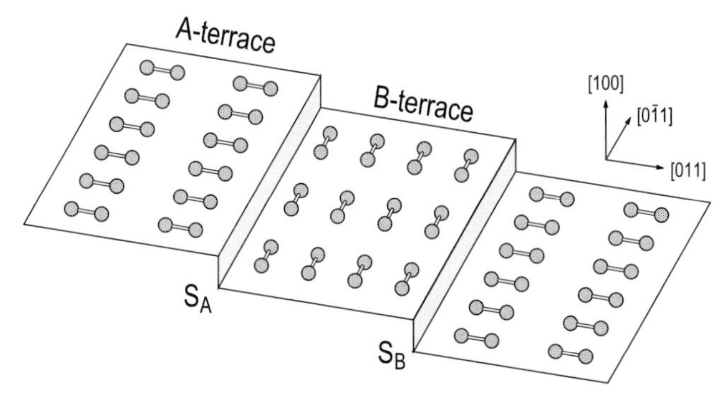

Fig. 10.27. Schematic of the vicinal Si(100) surface tilted towards the [011] direction. The dimer rows are rotated by 90° on every next terrace: On the A terrace, the dimer rows are parallel to the lower step, called the S_A step. On the B terrace, the dimer rows are perpendicular to the lower step, called the S_B step

is one atomic layer, i.e., $a_0/4$, where $a_0 = 5.43$ Å, the lattice constant of Si. The step spacing, i.e., the terrace width, is given by $a_0/(4\tan\alpha)$, where α is the tilt angle. As one can see, the direction of Si dimer rows rotates by 90° across a single step and, thus, the steps are oriented alternately parallel and normal to the dimer rows. According to the notation proposed by Chadi [10.16], the steps are divided into two classes, type A and type B, depending on the step orientation relative to the dimer rows on the upper terrace:

- Above a *type A* single step (labeled S_A), the terrace (known as an *A terrace*) has the dimer rows running *parallel* to the step edge.
- Above a *type B* single step (labeled S_A), the terrace (known as a *B terrace*) has the dimer rows running *perpendicular* to the step edge.

The difference in formation energies of the S_A and S_B steps can be understood qualitatively from the following simple consideration. The formation of the S_A step can be visualized as an imaginary cleaving of the top lattice plane between two dimer rows and shifting the half-plane to infinity (Fig. 10.28). During this process, no first-neighbor bonds are broken and no extra dangling bonds are created. Hence, the formation energy of the S_A step should be low. To form the S_B step, one has to cleave the top lattice plane between two neighboring dimers and remove the half-plane (Fig. 10.29). During this process, one first-neighbor bond per edge atom is broken and consequently an extra dangling bond per atom is created. Thus, the formation energy of the S_B step should be high.

The step shown in Fig. 10.29b is called a *non-bonded* S_B step. One can see that the dangling bond can be eliminated by the rebonding process as illustrated by Figs. 10.29b and c. The forming step is called a *rebonded* S_B

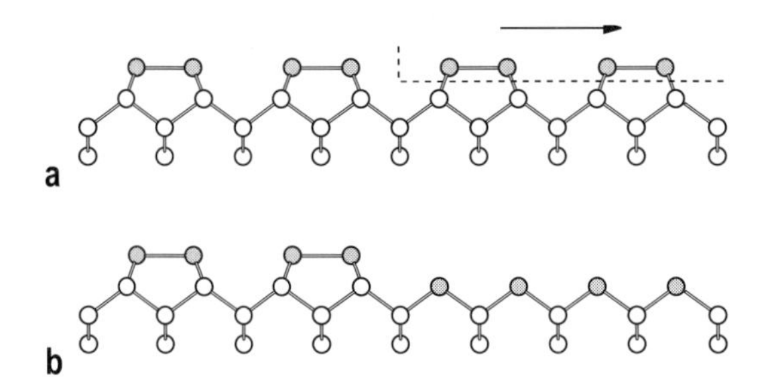

Fig. 10.28. (**a**) Schematic of the imaginary process of S_A step formation, including the cleaving of the top atomic plane between two dimer rows and removing the half-plane to infinity. (**b**) The formed S_A step. The dimerized atoms are shown by gray circles

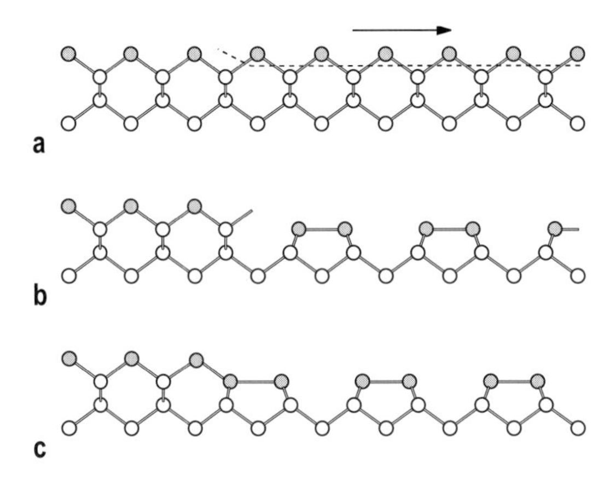

Fig. 10.29. (**a**) Schematic of the imaginary process of S_B step formation, including the cleaving of the top atomic plane between two adjacent dimers and removing the half-plane to infinity. (**b**) The formed *non-bonded* S_B step. (**c**) The *rebonded* S_B step formed from the non-bonded S_A step as a result of the rebonding process. The dimerized atoms are shown by gray circles

step. The energy of the rebonded S_B step is obviously lower than that of the non-bonded S_B step.

In his classical work [10.16], Chadi proposed the atomic structure of the single S_A and S_B steps and the double D_A and D_B steps (Fig. 10.30) and calculated the corresponding formation energies as follows:

$$\begin{aligned}
\varepsilon(S_A) &= 0.01 \pm 0.01 \ \text{eV/atom} \ , \\
\varepsilon(S_B) &= 0.15 \pm 0.03 \ \text{eV/atom} \ , \\
\varepsilon(D_A) &= 0.54 \pm 0.10 \ \text{eV/atom} \ , \\
\varepsilon(D_B) &= 0.05 \pm 0.02 \ \text{eV/atom} \ .
\end{aligned}$$

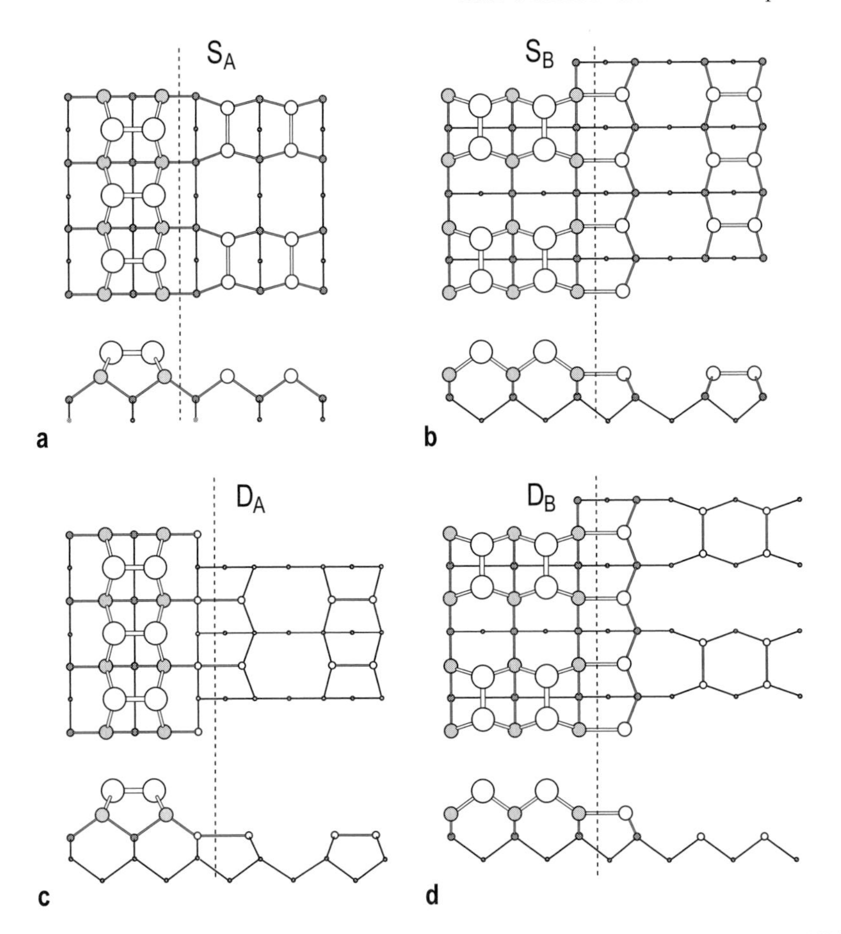

Fig. 10.30. Atomic arrangement of the various types of steps on the vicinal Si(100) surface: (**a**) single S_A step; (**b**) single S_B step; (**c**) double D_A step; (**c**) double D_B step (after Chadi [10.16])

Note that the $\varepsilon(S_B)$ energy refers to the rebonded S_B step (the non-bonded S_B step was estimated to be at least 0.16 ± 0.02 eV/atom more energetic than the rebonded configuration).

The values of $\varepsilon(S_A)$ and $\varepsilon(S_B)$ extracted from the STM and LEEM data coincide by an order of magnitude with the energies calculated by Chadi. However the experimental values for $\varepsilon(S_A)$ (for example, 0.028 ± 0.002 eV/atom [10.17], 0.023 ± 0.001 eV/atom [10.18], 0.025 ± 0.003 eV/atom [10.19]) are systematically higher, while those for $\varepsilon(S_B)$ (for example, 0.09 ± 0.01 eV/atom [10.17], 0.065 ± 0.005 eV/atom [10.18], 0.053 ± 0.006 eV/atom [10.19]) are systematically lower than the corresponding values obtained by Chadi. The results of the more recent calculations (for example, $\varepsilon(S_A) =$

0.019 ± 0.005 eV/atom; $\varepsilon(S_B) = 0.08 \pm 0.01$ eV/atom [10.18]) are in better agreement with the experimental results.

As one can see, the kinks on one type of step are made up of segments of the other type of step, i.e., on the S_B step the kink contains a $2a$ fragment of the S_A step and vice versa. The kink formation energy is essentially the work required to form a $2a$ fragment of the step of the other type. Hence, the formation energy of the kink is greater for the step with lower formation energy. This means, in turn, that S_A steps should be smoother than S_B steps. STM observations clearly confirm this conclusion (Fig. 10.31).

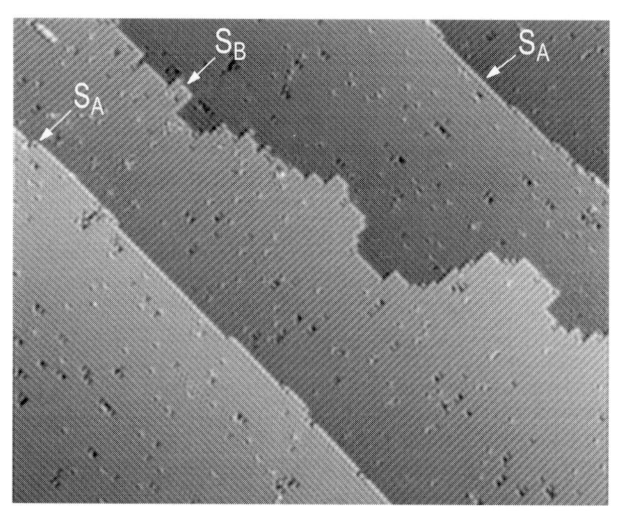

Fig. 10.31. 900×600 Å^2 STM image of the vicinal Si(100) surface illustrating the step structure. Due to the difference in the kink formation energies, S_A steps are smooth and S_B steps are rough

Back to (10.9), one can notice that the formation energy of the double D_B step is lower than the sum of the formation energies of the S_A and S_B steps, $\varepsilon(D_B) < \varepsilon(S_A) + \varepsilon(S_B)$. Thus, one can expect replacement of the monatomic steps by biatomic steps under appropriate conditions. Indeed, while at low tilt angles (below $\sim 2°$), most steps are monatomic, at tilt angles above $\sim 3°$ the vicinal Si(100) surfaces demonstrate a strong tendency towards the formation of biatomic steps. In the ideal case when all steps are biatomic, only a single orientation of the dimer rows is preserved and such a surface is referred to as *single-domain Si(100)2×1*.

As a result of external action, the Si(100) surface with an equal population of type A and type B terraces (i.e., 2×1 and 1×2 domains) might change to a surface with preferential population of one type of domain. For example, bending of a nominally flat Si(100) sample held at 550–800°C produces an unequal population of the 2×1 and 1×2 domains with the domain compressed

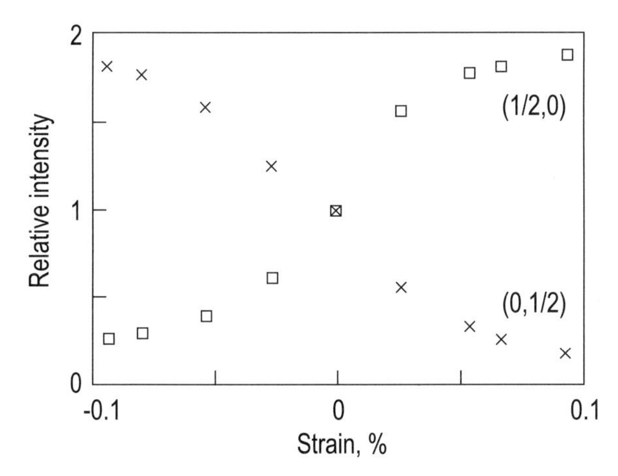

Fig. 10.32. LEED intensity of the 1×2 (crosses) and 2×1 (squares) domains on the Si(100) surface as a function of the surface strain. The domain compressed along the dimer bond is favored (after Men et al. [10.20])

along the dimer bond being favored. The effect was found to saturate at a strain of 0.1% when about 90% of the surface area is occupied by the favorable domain (Fig. 10.32).

Another example is the step rearrangement induced by direct current heating. It appears that the direction of the current affects the step motion at sublimation and modifies the population of the 2×1 and 1×2 domains. In the case of the step-up current direction, the A terraces prevail, while with the step-down current direction the B terraces become favored. The change-over from one type of surface to the other with changing current direction was found to be reversible. Figure 10.33 illustrates this phenomenon schematically; for experimental REM observations see Fig. 7.9.

10.2.7 Facetting

Facetting of the Vicinal Si(111) Surface. The vicinal Si(111) surface misoriented 10° towards the $[11\bar{2}]$ direction is covered with single steps at temperatures above the "1×1"-to-7×7 transition. Upon cooling, 7×7-reconstructed (111) facets appear at about 775°C, that is about 90°C lower than the phase transition temperature on a nominally flat Si(111) surface. As the temperature is lowered, the Si(111)7×7 facets expand and single steps are arranged into step bunches. The inclination angle of the step bunches increases continuously and, when reaching about 15° at a temperature below 700°C, the step bunch is transformed into a reconstructed (331) facet inclined at 22°. At 620°C, no step bunches are left and the surface is built of (331) and (111) facets only. The wide (111) facets are always covered by the 7×7 reconstruction, while 5×5 reconstruction occurs only at the narrow (111)

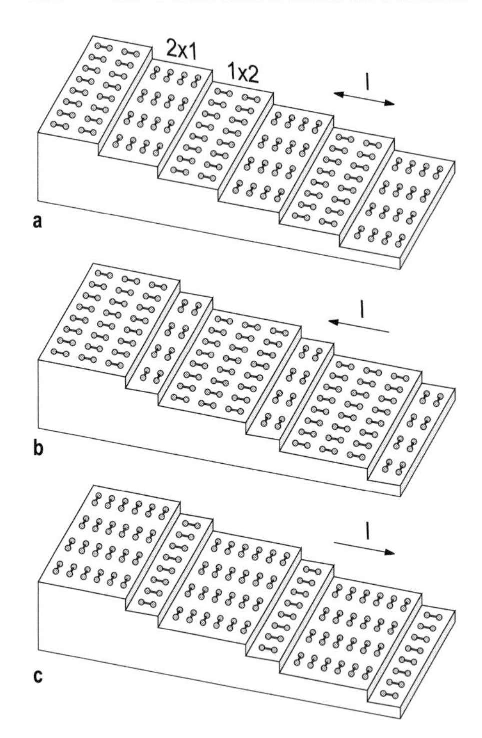

Fig. 10.33. Effect of the current direction on the step structure of the Si(100) surface at sublimation. (**a**) With an alternating current the populations of the 2×1 (A terraces) and 1×2 (B terraces) domains are equivalent. (**b**) With step-up current direction the A terraces dominate. (**c**) With step-down current direction the B terraces dominate (after Latyshev et al. [10.21])

facets adjacent to the (331) facets (Fig. 10.34). As one can see, the facetting of the vicinal Si(111) surface is induced by the "1×1"-to-7×7 transition and, hence, furnishes an example of *reconstruction-driven facetting.*

Au-induced Facetting of the Vicinal Si(100) Surface. As an example of *adsorbate-induced facetting*, consider the effect of Au adsorption on the surface morphology of the vicinal Si(100) surface tilted 4° towards the [011] direction. Before Au deposition, the surface comprises a single-domain 1×2 structure on the B terraces bounded by double D_B steps (Fig. 10.35a and c). Upon Au adsorption conducted at 800°C, the incommensurate "5×3.2"-Au structure develops on the terraces. The forming reconstruction changes the surface energy and the initial regular step array transforms into a "hill-and-valley" structure composed of wide (100) terraces and step bunches. Eventually, the step bunches transform to the well-ordered (911) facets with a "8×2"-Au structure (Fig. 10.35b and d). It is interesting that the LEED pattern reveals a single-domain "5×3.2"-Au structure with a fivefold direction

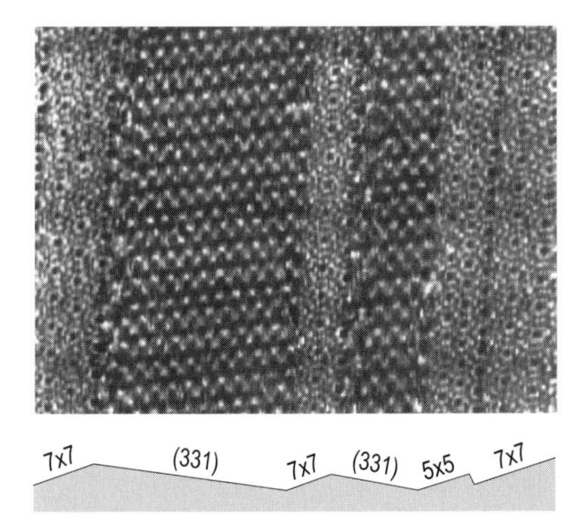

Fig. 10.34. 600×420 Å^2 STM image of the vicinal Si(111) surface misoriented 10° towards the $[11\bar{2}]$ direction at 620°C (after Hibino et al. [10.22])

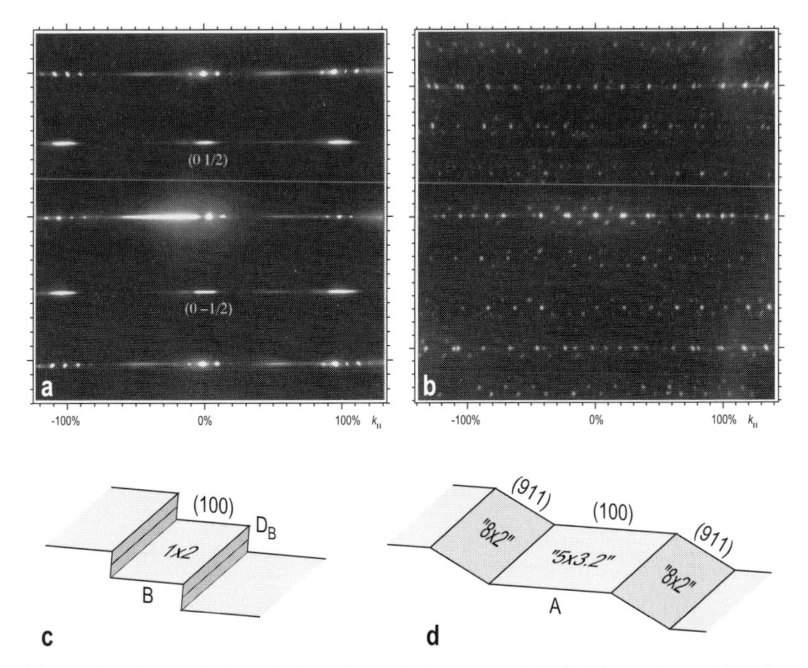

Fig. 10.35. Au-induced facetting of the vicinal Si(100) surface. LEED patterns are taken (**a**) before and (**b**) after Au-induced facetting at 800°C. (**c**) and (**d**) show schematically the surface structure and morphology before and after the transition (after Minoda et al. [10.23])

normal to the step edges. This means that all the (100) terraces are of A type, though the initial Si(100)1×2 surface contained only B-type terraces [10.23, 10.24].

Problems

10.1 Using the TSK nearest-neighbor model, write expressions for the formation energy of step adatom and step vacancy.

10.2 Using the Wulff construction, determine the equilibrium 3D shape of the simple cubic TSK crystal taking into account
(a) only the first-neighbor interactions;
(b) the first-neighbor and second-neighbor interactions.

10.3 At high surface concentrations, the dimer vacancies on the Si(100) surface are arranged into an ordered Si(100)2×n superstructure with $6 \leq n \leq 10$ (Fig. 10.14b). The superstructure is built of dimer-row segments bounded by dimer-vacancy troughs. Estimate the range of the vacancy concentration at which the 2×n superstructure occurs. Assume the troughs to be built of
(a) double dimer vacancies;
(b) vacancy complexes of single and double vacancies separated by a "split-off" dimer (see Fig. 10.13).

Further Reading

1. Ph. Ebert: *Nano-Scale Properties of Defects in Compound Semiconductor surfaces*. Surf. Sci. Rep. **33**, 121–303 (1999) (review of atomic-scale properties of point defects, vacancies, and dopant atoms at (110) surfaces of III-V and II-VI semiconductors)
2. H.-C. Jeong, E.D. Williams: *Steps on Surfaces: Experiment and Theory*. Surf. Sci. Rep. **34**, 171–294 (1999)
3. M. Henzler, W. Ranke: 'Structural defects at surfaces'. In: *Physics of Solid Surfaces. Landolt Börnstein. Vol. III/24a*. ed. by G. Chiarotti (Springer, Berlin 1993) Chapter 2.3 (detailed data on structural defects at clean surfaces)
4. A.A. Chernov: *Modern Crystallography III. Crystal Growth* (Springer, Berlin 1984) Chapter 1 (TSK model and related subjects)
5. I.V. Markov: *Crystal Growth for Beginners* (World Scientific, Singapore 1995) Chapter 1 (TSK model and related subjects)

11. Electronic Structure of Surfaces

By breaking the 3D periodicity of the crystal bulk, a surface modifies strongly the electronic structure in its vicinity. Additional modification is introduced by surface reconstruction. The modification concerns the charge density redistribution in the near-surface region and the formation of specific electronic states, called surface states. The electronic structure manifests itself in the surface properties, like the surface conductivity and work function.

11.1 Basics of Density Functional Theory

In order to characterize the electronic structure of a solid, one has to solve the Schrödinger equation for all electrons in a solid. This is clearly an unrealistic task, as a solid contains of the order of 10^{23} electrons per cm^3, while exact solutions of the Schrödinger equation are only possible for very small numbers of electrons (of the order of one). Hence, drastic simplifying approximations are required. On the other hand, one would like to preserve sufficient accuracy and reliability of the theoretical predictions. *Density functional theory (DFT)* has proved to be particularly successful in reaching the goal. The achievements of this approach were recognized in 1998 by the award of the Nobel Prize in chemistry to Walter Kohn for his development of density functional theory. The alternative approach, namely, *Hartree–Fock theory*, allows the calculation of the electronic structure of molecules or small clusters but appears to be clumsy for extended systems like solid surfaces and it will not be discussed here.

DFT is based on the theorem formulated by Hohenberg and Kohn [11.1], which states that the total energy of a system (for example, a bulk solid or a surface) is completely specified by the electron density, $n(\boldsymbol{r})$, of its ground state. Furthermore, it is possible to define an energy functional

$$E = E[n(\boldsymbol{r})] \tag{11.1}$$

with the property that it has a minimum when $n(\boldsymbol{r})$ corresponds to the ground state density. Typically, $E[n(\boldsymbol{r})]$ is decomposed into three parts, namely, kinetic energy T, electrostatic (or Coulomb) energy U, and an exchange–correlation term E_{xc}:

$$E[n(\boldsymbol{r})] = T + U + E_{\mathrm{xc}} \,. \tag{11.2}$$

The kinetic energy term stands for the kinetic energy of a non-interacting inhomogeneous electron gas in its ground state. The Coulomb term is purely classical and describes the electrostatic energy arises from the Coulomb attraction between the valence electrons and the ion cores, the repulsion between the electrons and the repulsion between the ion cores:

$$U = U_{\mathrm{ec}} + U_{\mathrm{ee}} + U_{\mathrm{cc}} \tag{11.3}$$

with

$$U_{\mathrm{ec}} = -e^2 \sum_{\boldsymbol{R}} Z \int \frac{n(\boldsymbol{r})}{|\boldsymbol{r} - \boldsymbol{R}|} \, \mathrm{d}\boldsymbol{r}, \tag{11.4a}$$

$$U_{\mathrm{ee}} = \frac{1}{2} e^2 \iint \frac{n(\boldsymbol{r}) n(\boldsymbol{r}')}{|\boldsymbol{r} - \boldsymbol{r}'|} \, \mathrm{d}\boldsymbol{r} \mathrm{d}\boldsymbol{r}', \tag{11.4b}$$

$$U_{\mathrm{cc}} = e^2 \sum_{\boldsymbol{R}\boldsymbol{R}'} \frac{ZZ'}{|\boldsymbol{R} - \boldsymbol{R}'|}, \tag{11.4c}$$

where $\boldsymbol{R}$ denotes the position of atoms (ion cores) and Z is the atomic number of the atom. Summations extend over all atoms and the integration over the entire space.

The exchange–correlation term E_{xc} lumps together the remaining contributions due to many-body quantum mechanical effects. The most important of these contributions is the exchange term, which is associated with the action of the Pauli exclusion principle: in real space electrons of the same spin tend to avoid each other, hence the electron–electron Coulomb repulsion is reduced. The corresponding energy gain is called the exchange energy. The additional terms describing the interaction between electrons of opposite spin are defined as the correlation energy. Typically, the kinetic energy and the Coulomb terms are of similar magnitude, while the exchange–correlation term is about 10% of this value with the exchange energy being greater than the correlation energy.

The electron density which minimizes the energy functional $E[n(\boldsymbol{r})]$ (11.2) is found by a self-consistent solution of a set of Schrödinger-like one-electron equations (called the Kohn–Sham equations [11.2]):

$$-\frac{\hbar^2}{2m} \nabla^2 \psi_i(\boldsymbol{r}) + v_{\mathrm{eff}}(\boldsymbol{r}) \psi_i(\boldsymbol{r}) = \varepsilon_i \psi_i(\boldsymbol{r}) \tag{11.5}$$

with the effective one-electron potential $v_{\mathrm{eff}}(\boldsymbol{r})$ defined as

$$v_{\mathrm{eff}}(\boldsymbol{r}) = -e^2 \sum_{\boldsymbol{R}} \frac{Z}{|\boldsymbol{r} - \boldsymbol{R}|} + e^2 \int \frac{n(\boldsymbol{r}')}{|\boldsymbol{r} - \boldsymbol{r}'|} \, \mathrm{d}\boldsymbol{r}' + v_{\mathrm{xc}}[n(\boldsymbol{r})] \,. \tag{11.6}$$

The required electron density is generated from one-electron wave functions as

$$n(\boldsymbol{r}) = \sum |\psi_i(\boldsymbol{r})|^2 \,. \tag{11.7}$$

The result is formally exact, but in practice the exchange–correlation energy E_{xc} (and, consequently, the exchange–correlation potential $v_{\mathrm{xc}}[n(\boldsymbol{r})] = \delta E_{\mathrm{xc}}/\delta n(\boldsymbol{r})$) is unknown and an appropriate approximation is required. The so-called *local density approximation (LDA)* provides a simple, but surprisingly successful approach. Here, the exchange–correlation energy density at the *local* electron density $n(\boldsymbol{r})$ of the *inhomogeneous* electron distribution is approximated by the value for the *homogeneous* electron gas with the same electron density and, thus, the exchange–correlation energy is given by

$$E_{\mathrm{xc}}[n(\boldsymbol{r})] = \int n(\boldsymbol{r})\varepsilon_{\mathrm{xc}}[n(\boldsymbol{r})]\mathrm{d}\boldsymbol{r}' , \tag{11.8}$$

where $\varepsilon_{\mathrm{xc}}[n(\boldsymbol{r})]$ is the exchange–correlation energy per electron of the homogeneous electron gas. It is accurately known for all densities of physical interest. For practical calculations, $\varepsilon_{\mathrm{xc}}[n(\boldsymbol{r})]$ is expressed as an analytical function of the electron density.

11.2 Jellium Model

In the *jellium model*, the ion cores are replaced by a uniform background of positive charge with density equal to the spatial average of the ion charge distribution. This is the simplest model to give a reasonably accurate description of simple metals, like Na, Mg or Al, where the conduction band comprises only s- and p-electrons.

Let us apply the Jellium model to the surface problem. For a semi-infinite substrate with z direction normal to the surface, the positive charge distribution $n^{+}(\boldsymbol{r})$ is given by a step at $z = 0$, i.e.,

$$n^{+}(\boldsymbol{r}) = \begin{cases} \overline{n}, & z \leqslant 0 \\ 0, & z > 0. \end{cases} \tag{11.9}$$

It is common practice to express the positive background charge density $\overline{n}$ in terms of a dimensionless measure r_{s} of the average distance between the electrons and the Bohr radius $a_0 = \hbar^2/me^2 = 0.529\,\text{Å}$. By definition, $r_{\mathrm{s}}a_0$ is the radius of a sphere containing exactly one electron, i.e.,

$$\frac{4\pi}{3}(r_{\mathrm{s}}a_0)^3 = \frac{1}{\overline{n}} . \tag{11.10}$$

The associated electron density will obey the conditions

$$\lim n(\boldsymbol{r}) = \begin{cases} \overline{n}, & z \to -\infty \\ 0, & z \to +\infty. \end{cases} \tag{11.11}$$

The ground state electron density profile for the semi-infinite jellium model was calculated in the seminal work of Lang and Kohn [11.3] using the DFT-LDA approach and is shown in Fig. 11.1. Two main features are clearly seen.

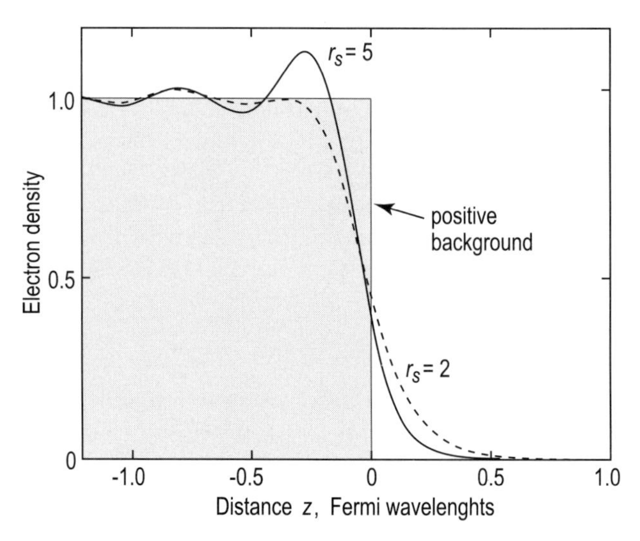

Fig. 11.1. Electron density profile at a jellium surface for two background densities: $r_\mathrm{s} = 2$ (solid line) approximates Al and $r_\mathrm{s} = 5$ (dashed line) approximates Cs. The distance is given in units of the Fermi wavelength, $2\pi/k_\mathrm{F}$, which is 3.45 Å for $r_\mathrm{s} = 2$ and 8.65 Å for $r_\mathrm{s} = 5$ (after Lang and Kohn [11.3])

- First, the electron distribution leaks out of the surface into the vacuum, dropping to zero at 1–3 Å away from the surface. The spill-over generates an imbalance in positive and negative charge, which creates an electrostatic dipole layer at the surface with the negative end of the dipole outmost.
- Second, towards the bulk the electron density oscillates while reaching asymptotically the bulk density value $\overline{n}$. These are *Friedel oscillations* having a period of π/k_F, where the Fermi wave vector is $k_\mathrm{F} = (3\pi\overline{n})^{1/3}$. The Friedel oscillations arise as a response of the electron gas to the abrupt variation in the positive background charge distribution. They are a feature of charged defects in metals in general, not just surfaces. Moreover, while it has proven difficult to access Friedel oscillations in the bulk, the electron density oscillations induced by steps and charged defects at the surface are clearly resolved by STM (Fig. 11.2).

The presence of a surface dipole layer leads to the variation of the electrostatic potential $v(z)$ as shown in Fig. 11.3 by a dashed line. Variation of the total effective potential, which incorporates also the contribution from exchange–correlation effects, $v_\mathrm{eff}(z) = v(z) + v_\mathrm{xc}(z)$, is shown in Fig. 11.3 by a solid line.

Knowledge of the magnitude of potentials allows the determination of the work function, which is the minimum energy required to remove an electron from the crystal bulk to a point a macroscopic distance outside the surface. Taking into account that the most energetic electrons in the bulk have a

Fig. 11.2. Constant-current 500×500 Å^2 STM image of a Cu(111) surface acquired at a sample bias of 0.1 V. Friedel oscillations of the surface electron density near steps and point defects are clearly evident. The vertical scale has been greatly exaggerated to display the Friedel oscillations more clearly (after Crommie et al. [11.4])

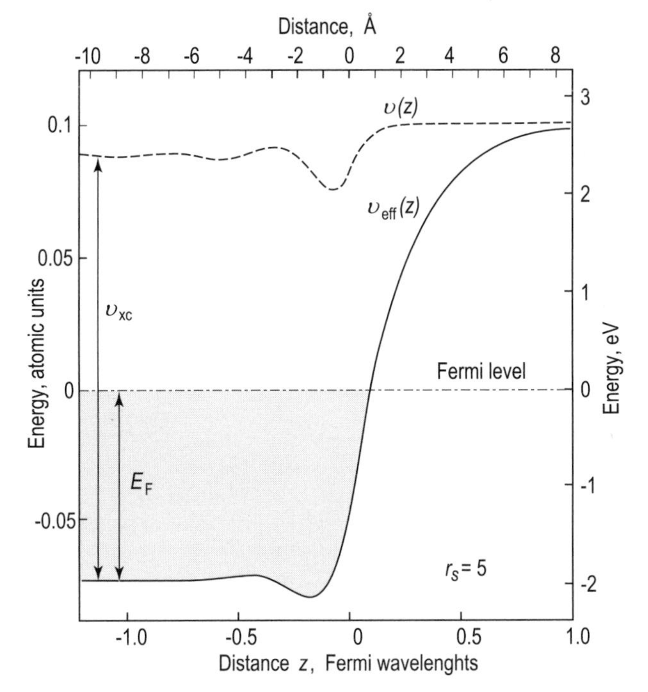

Fig. 11.3. Spatial variation of the electrostatic potential $v(z)$ (dashed line) and total effective one-electron potential $v_{\rm eff}$ (solid line) near the jellium surface for $r_{\rm s} = 5$ (after Lang and Kohn [11.3])

kinetic energy equal to the Fermi energy $E_F = (\hbar^2/2m)(3\pi^2\overline{n})^{2/3}$, one can see from Fig. 11.3 that the work function ϕ is given by

$$\phi = v_{\text{eff}}(+\infty) - v_{\text{eff}}(-\infty) - E_F \ . \tag{11.12}$$

The values of the work functions calculated for simple metals within a framework of the jellium model demonstrate reasonably good agreement with the experimental results (Fig. 11.4).

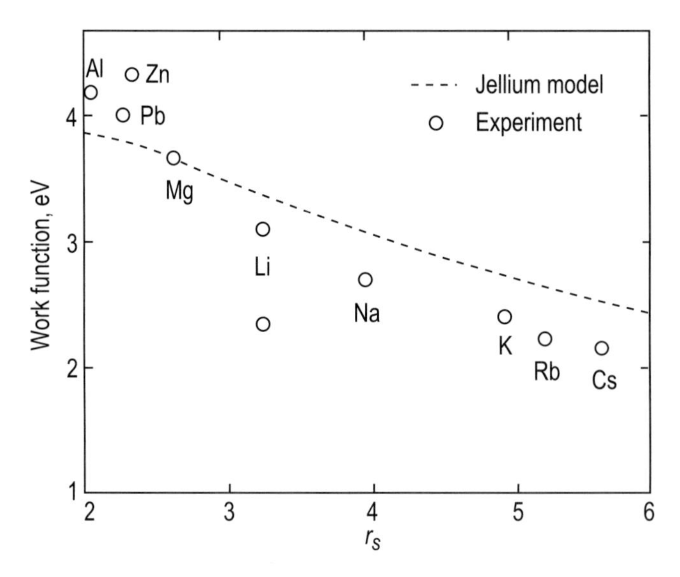

Fig. 11.4. Experimentally determined work functions (open circles) compared with the results of the Jellium model calculation shown by the dashed line (after Lang and Kohn [11.5])

11.3 Surface States

The main conclusion of the electron band theory of solids resides in the existence of successive bands of allowed energies, which are always separated by forbidden energy gaps. The solution of the Schrödinger equation for the periodic potential associated with an infinite crystalline lattice is known to reproduce these trends for a solid bulk. The presence of a free surface breaks the periodicity and, hence, alters the boundary condition for the Schrödinger equation. For the one-dimensional problem, the potential can be taken in the simplified form shown in Fig. 11.5a. Two types of solution are available in this case.

- The first type of solution (Fig. 11.5b) corresponds to *bulk states* with wave functions extended into the bulk and decaying exponentially into the vacuum.
- The second type of solution (Fig. 11.5c) corresponds to *surface states*, whose wave functions are localized in the surface region, decaying exponentially both into the bulk and vacuum.

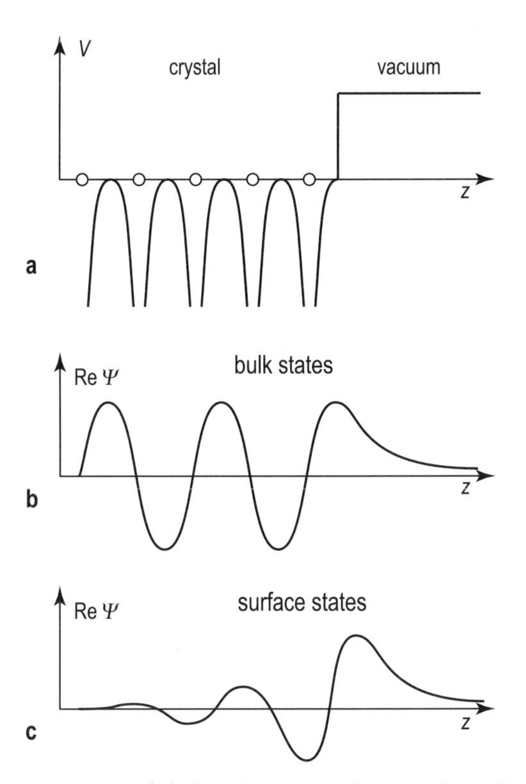

Fig. 11.5. (**a**) One-dimensional semi-infinite lattice model potential. Two types of wave functions of a semi-infinite crystal: (**b**) bulk states and (**c**) surface states

In the discussion of surface states it is customary to distinguish Shockley states from Tamm states and "true" surface states from surface resonance.

- *Shockley states* arise as a solution of the Schrödinger equation in the framework of the nearly free electron model. In this approximation, the electron–electron interaction is neglected and the solution is sought in the form of plane wave functions. The occurrence of Shockley states is due only to crystal termination and does not require any other deviation from the bulk parameters. This approach is appropriate primarily to normal metals and some narrow-gap semiconductors.

- *Tamm states* are obtained using the tight-binding model, which focuses on the wave functions constructed from atomic-like orbitals. The approximation applies to fairly localized electrons. The occurrence of the Tamm states implies a considerable perturbation of the potential in the surface region (for example, due to surface reconstruction or to the presence of dangling bonds). The approach is suitable for d electrons of the transition metals and also for semiconductors and insulators.

In the one-dimensional representation of $E(k_\perp)$, the surface states always appear in the bulk zone-boundary gap (Fig. 11.6a). In the three-dimensional picture, one has to consider the dispersion of the surface state $E(k_\|)$ together with the projection of the bulk bands onto the surface Brillouin zone (Fig. 11.6b). When the surface states are located within the gap of the projected bulk bands, one speaks about *"true" surface states* (shown as a solid line in Fig. 11.6b). The states that cross the projection of the bulk bands are referred to as *surface resonances* (shown as a dashed line in Fig. 11.6b). Reso-

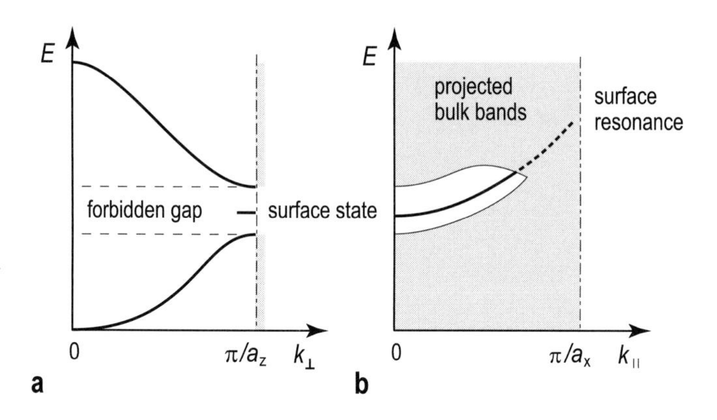

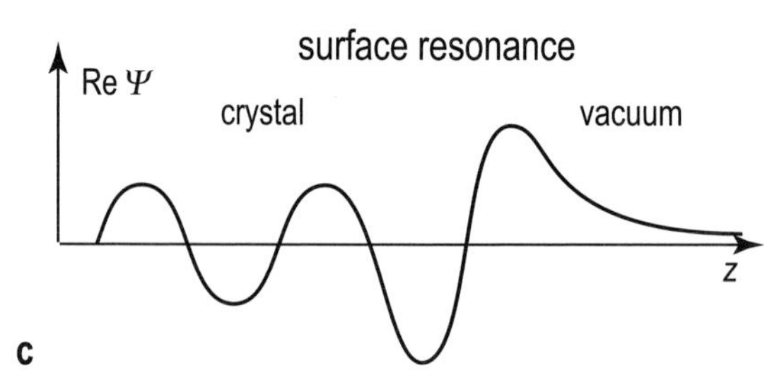

Fig. 11.6. (**a**) One-dimensional representation of the bulk band structure in the direction normal to the surface, $E(k_\perp)$. The projection of the bulk bands to the surface Brillouin zone is shown in gray bars. The surface state appears in the forbidden energy gap. (**b**) Dispersion of the surface state along the direction parallel to the surface $E(k_\|)$ superposed on the projection of the bulk bands (gray area). The "true" surface state band is shown as a solid line; the surface resonance band as a dashed line. (**c**) Wave function of the surface resonance state

nances are energetically degenerate with the bulk bands. The resonance wave function is bulk-like but with enhanced amplitude at the surface (Fig. 11.6c).

The type of surface states discussed above is commonly termed *intrinsic surface states* or *"crystal-induced" surface states* to denote that these surface states characterize the electronic structure of the well-ordered crystal surface in its ground state. In addition to the intrinsic states, there are other types of surface states that originate from perturbation of the ideal surface potential, namely, extrinsic surface states and image-potential surface states.

- *Extrinsic surface states* are related to surface structure imperfections, like vacancies, foreign atoms, steps. Generally, defect-induced surface states do not display 2D translational symmetry along the surface and their wave functions are localized near the defects.
- *Image-potential induced surface states* (or in brief, *image-potential surface states*) refer to surface states due to the electrostatic interaction of an electron with its image charge in a surface. The presence of an electron in front of a metal surface redistributes the charge in the solid so that an attractive potential is induced. The attraction can be visualized as the Coulomb interaction between a real electron outside and its positive image charge in the solid. In the simplest case, the attractive image potential is $v_{\mathrm{image}} = -e^2/4z$, where z is the distance from surface. If the bulk-band structure projected onto the surface presents a gap near the vacuum level, an electron in front of the surface cannot penetrate to the solid. As a result, the electron can be trapped in a bound state. These image-potential states are described by a Rydberg-like series and their wave functions are localized in vacuum several Å apart from the topmost atomic layer.

Several experimental techniques exist to characterize the surface states, of which the most important ones are

- angle-resolved ultraviolet photoemission,
- $\boldsymbol{k}_{\|}$-resolved inverse photoemission,
- scanning tunneling microscopy/spectroscopy.

Angle-resolved ultraviolet photoemission (ARUPS) is the main technique to probe the *filled (occupied)* states. The basic principle of ARUPS is that the measured kinetic energy and wave function of the electrons photoemitted from a solid allow the restoration of the dispersion of the surface states. A detailed description of the technique is given in Sect. 5.4

$\boldsymbol{k}_{\|}$-resolved inverse photoemission (KRIPS) provides information on the *empty (unoccupied)* states. In the inverse photoemission process (Fig. 11.7a), an electron of energy eU is sent into the solid, enters an excited state $E_1 = eU$, and is then deexcited into a lower empty state E_2. The corresponding deexitation energy is released as a photon of $\hbar w = E_1 - E_2$. There are two methods to acquire the empty state spectrum:

- *Isochromate spectroscopy*, in which the primary electron energy is varied, while the yield of photons with a fixed energy is detected (Fig. 11.7b).

- *Bremsstrahlen spectroscopy*, which means braking radiation spectroscopy. Here the primary electron energy is fixed and the spectrum of the emitted photons is recorded (Fig. 11.7c).

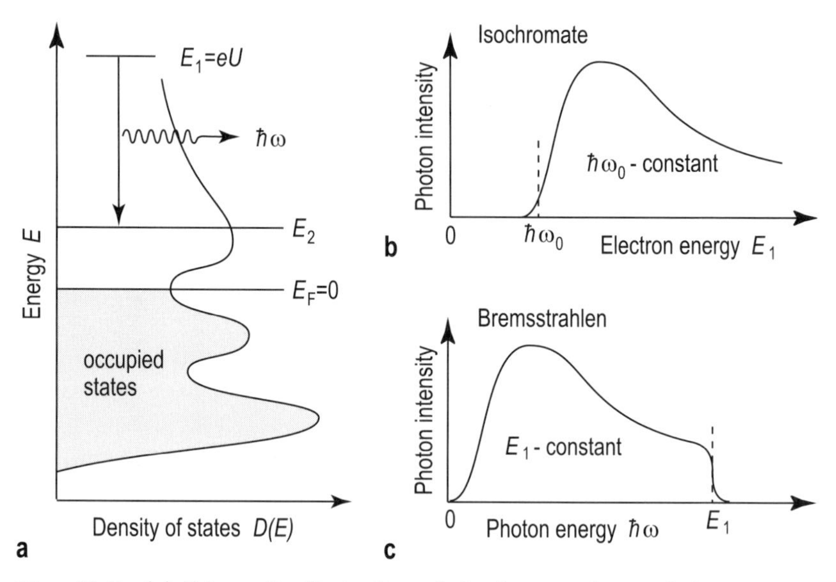

Fig. 11.7. (**a**) Schematic illustration of the inverse photoemission process. (**b**) Schematic isochromate spectrum. (**c**) Schematic Bremsstrahlen spectrum

The angle resolution, which is required to determine the dispersion $E(\boldsymbol{k})$ of the states, is achieved by changing the direction of incidence of the injected electrons and using a tunable photon detector.

Another method to probe the empty states is *two-photon photoemission spectroscopy.* The principle of the technique is to populate the empty states using photoexcitation with a laser source and then to probe them by photoemission.

Scanning tunneling microscopy/spectroscopy (STM/STS) (see Sect. 7.7) probes either *filled* or *empty states*, depending on the polarity of the bias voltage. In STS mode, the density of states (DOS) in the local area at the atomic scale is acquired. The STM mode allows one to obtain a real space image of the spatial distribution of the specific DOS. Access to the dispersion $E(\boldsymbol{k}_{\|})$ of the surface states by means of STM is achieved, when one measures the *periodicity* of the surface Friedel oscillations (Fig. 11.2), which gives directly the parallel wave vector $\boldsymbol{k}_{\|}$ at an energy E related to the tip bias voltage (Fig. 11.8).

Figure 11.9 shows an example of the experimental evaluation of surface state dispersion by a set of techniques, ARUPS, KRIPS, and STM, for the case of the Cu(111) surface. The dispersion of the surface state below E_{F}

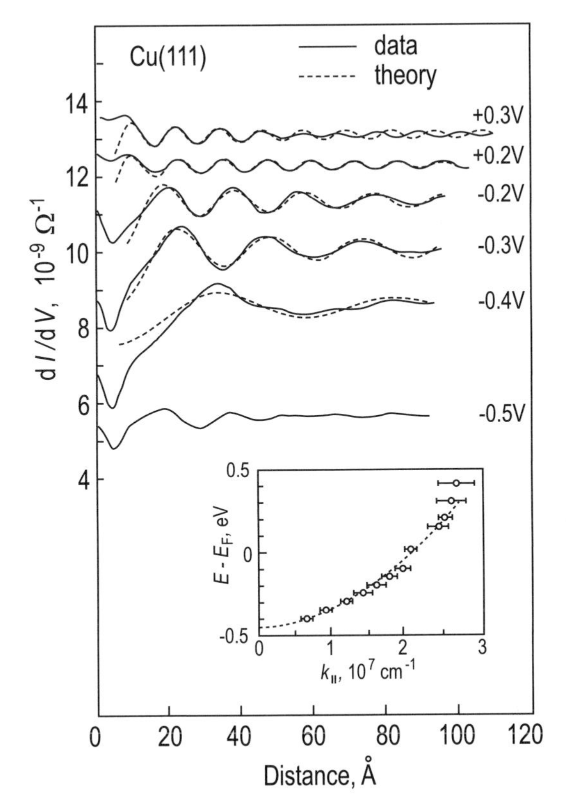

Fig. 11.8. Evaluation of the surface state dispersion $E(\boldsymbol{k}_{\|})$ from the periodicity of the Friedel oscillation of the surface electron density near the step edges (see Fig. 11.2). The spatial distribution of the density of states is given by $\mathrm{d}I/DV$, measured as a function of distance (along the upper terrace) from the step edge at different bias voltages. Changing the sample bias, one probes the density of states at different energies and, hence, for different parallel wave vectors $|\boldsymbol{k}_{\|}| = (2m^* E/\hbar)^{1/2}$, where E is the probed energy and m^* the effective mass of the surface-state electron. The oscillation periodicity is $\pi/k_{\|}$. The evaluated surface state dispersion is shown in the inset (after Crommie et al. [11.4])

has been determined by ARUPS and appears to have a parabolic shape with an effective electron mass m^* of 0.46 times the free electron mass m_e and a band minimum 0.39 eV below E_F. The dispersion above E_F, determined using KRIPS, connects smoothly to the ARUPS data. The STM results (Fig. 11.8) cover the energy range both below and above E_F and yield the parabolic dispersion with band origin 0.44 eV below E_F and $m^* = 0.38 m_\mathrm{e}$.

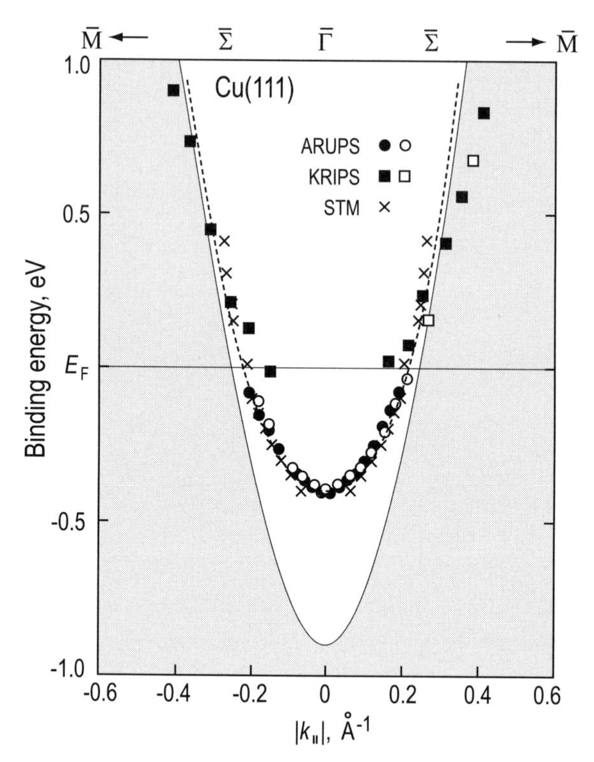

Fig. 11.9. Experimental dispersion relation for the Cu(111) zone center surface band near E_F determined by a combination of probing techniques, including ARUPS (open and closed circles for a photon energy of 16.8 eV and 11.8 eV, respectively, [11.6]), KRIPS (open and closed squares for a photon energy of 11.0 and 10.2 eV, respectively [11.7]) and STM (crosses [11.4]). The shaded region is the projected bulk continuum

11.4 Electronic Structure of Selected Surfaces

Here the electronic structure of several surfaces is considered. All examples refer to the Si(111) surface, which allows us to trace the effects of reconstruction (Si(111)2×1 versus Si(111)7×7) and adsorbates (Si(111)2×1 and Si(111)7×7 versus Si(111)1×1-As and Si(111)$\sqrt{3}\times\sqrt{3}$-In) on the electronic surface structure. The choice was also dictated by the fact that the atomic structure of these surfaces is well known, which allows us to establish a relationship between the observed electronic states and the features in the surface atomic structure.

11.4.1 Si(111)2×1

The cleaved Si(111)2×1 surface is known to have a structure described by the π-bonded chain model of Pandey (see Fig. 8.15b). The basic feature of the model is the zigzag chain of surface atoms forming π bonds with each other just as in organic materials. Formation of the π bonds results in the appearance of two surface state bands, of which the filled state band is due to the bonding π-orbitals and the empty state band is due to the antibonding π^*-orbitals. The dispersion of these surface state bands has been determined in ARUPS and KRIPS experiments and has been obtained in calculations based on the π-bonded chain model (Fig. 11.10). The bands are strongly dispersive along the chain direction ($\overline{\Gamma \mathrm{J}}$) and weakly dispersive perpendicular to the chain direction ($\overline{\mathrm{JK}}$). The π and π^* bands are separated at the $\overline{\mathrm{J}}$ point by a band gap of ~0.5 eV.

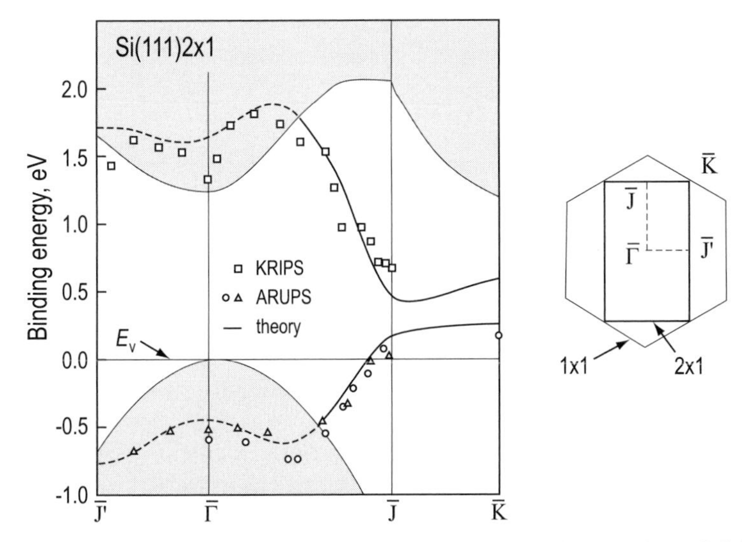

Fig. 11.10. Dispersion of the surface state bands for the cleaved Si(111)2×1 surface. The ARUPS data are shown by circles [11.8] and triangles [11.9]; the KRIPS data by squares [11.10]. The results of calculations based on Pandey's π-bonded chain model [11.11] are shown as a solid line for the "true" surface states and as a dashed line for surface resonance states. The shaded region is the projected bulk band structure. The right panel shows the surface Brillouin zones for the Si(111)1×1 and Si(111)2×1 surfaces. The corresponding symmetry points for the 2×1 Brillouin zone are indicated

11.4.2 Si(111)7×7

The most stable reconstruction of the Si(111) surface is well established to have the complex dimer–adatom–stacking fault (DAS) structure shown in Fig. 8.18. ARUPS investigations on the Si(111)7×7 surface reveal the presence of three distinct surface states at ~ 0.2 eV, ~ 0.8 eV, and ~ 1.7 eV below the Fermi level. Conventionally, these states are labeled S_1, S_2, and S_3, respectively. KRIPS results show the existence of an unoccupied state U_1 at

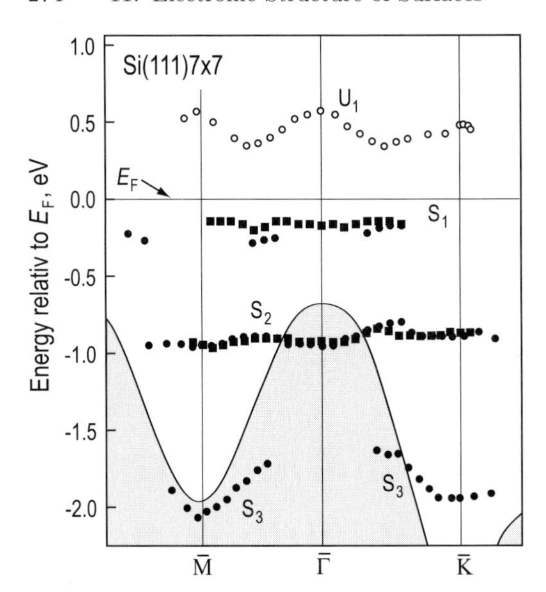

Fig. 11.11. Surface band structure of the Si(111)7×7 surface (referring to the Brillouin zone for the Si(111)1×1 surface) as resulting from ARUPS (closed circles [11.12] and closed squares [11.13]) and KRIPES (open circles [11.12]) measurements. The shaded region is the projected bulk band structure

~ 0.5 eV above the Fermi level. The data of both techniques are compiled in Fig. 11.11.

The origin of each of the surface states was clarified in the STM/STS study [11.14], where the spatial distribution of the surface states within the 7×7 unit cell was mapped out (Fig. 11.12). It appears that the S_1 state, as well as the empty U_1 state, is localized at the *adatoms*. The S_2 state was found to be localized at the *rest atom*. The S_3 state was assigned to the *adatom backbonds*.

Note that Fig. 11.11 gives but a rough idea of the surface state dispersion, as the data are plotted for the 1×1 surface Brillouin zone, but not for the 7×7 Brillouin zone. The size of the 7×7 Brillouin zone is seven times smaller than that of the 1×1 zone and the angular resolution of the conventional ARUPS set-up (typically, a few degrees) is not sufficient to resolve the fine structure within the small 7×7 Brilloiun zone. The goal was reached only with achieving the extraordinary angular resolution of $\sim 0.3°$ [11.18]. The measured dispersion of the adatom surface state band within the 7×7 Brillouin zone is shown in Fig. 11.13.

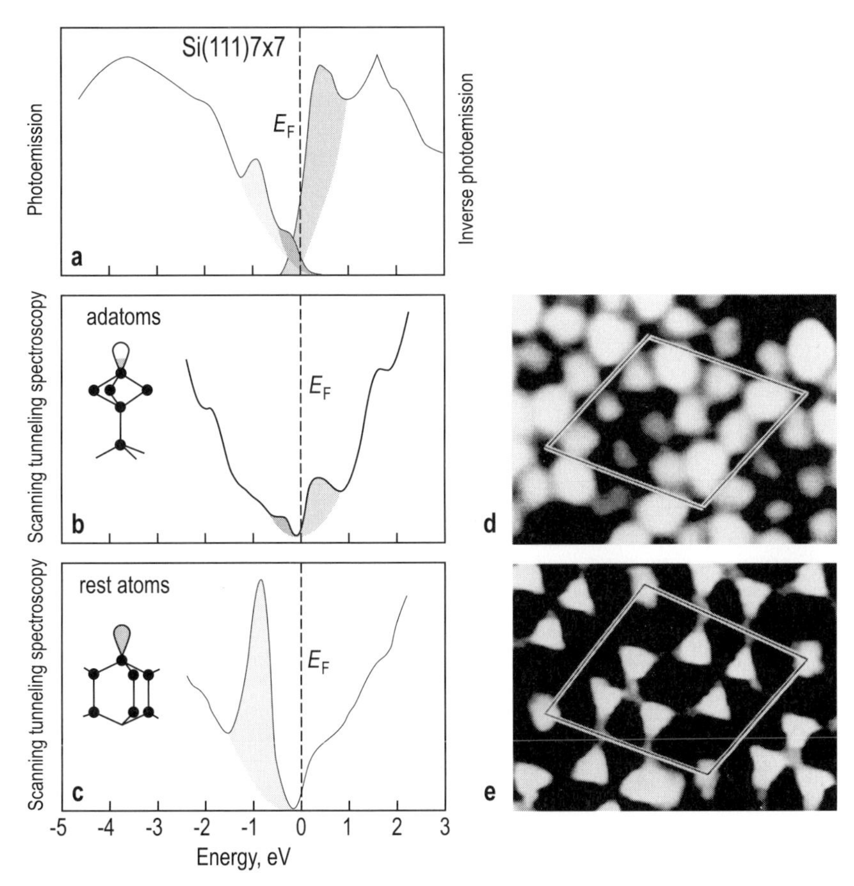

Fig. 11.12. Surface states on the Si(111)7×7 surface detected with (**a**) photoemission, inverse photoemission [11.15] and **b**, (**c**) scanning tunneling spectroscopy [11.16]. Two types of dangling bond states occur; a lone pair at the rest atom sites and a partially filled state at the adatom sites. The surface localization of the states is revealed in STM/STS experiments by using the bias voltage corresponding to the energy position of a given state, (**d**) at −0.35 V for the adatom state and (**e**) at −0.8 V for the rest atom state [11.14] (after Himpsel [11.17])

11.4.3 Si(111)1×1-As

The Si(111)1×1-As surface phase is formed by annealing the Si(111)7×7 surface while exposing it to an As_4 flux. Adsorption of As removes the 7×7 reconstruction and results in the bulk-like Si(111)1×1 surface, in which the outermost Si atoms are replaced by As atoms (Fig. 11.14a). Being incorporated into the 1×1 structure, the As atom donates three electrons to the bonds with the neighboring Si atoms, while the two electrons left form a lone pair. The photoelectron spectra from the Si(111)1×1-As surface were found to be dominated by a peak, which corresponds to the As lone pair surface

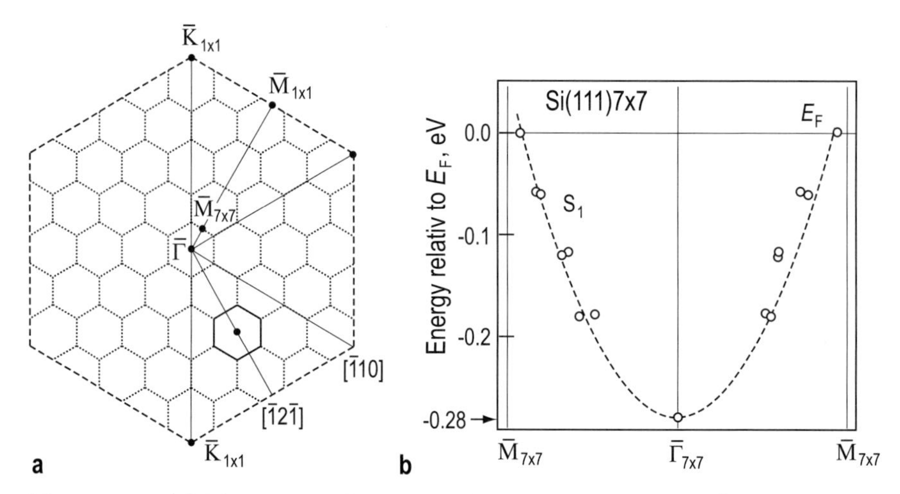

Fig. 11.13. (**a**) The 1×1 Brillouin zone (shown as a dashed line) with the 7×7 Brillouin zones (shown as a dotted line) incorporated inside. The symmetry points are indicated. (**b**) The experimental dispersion of the adatom surface state band resolved within the 7×7 Brillouin zone outlined by a solid line in (**a**) (After Losio et al. [11.18])

state. The dispersion of this surface state band is shown in Fig. 11.14b. The lone pair state is a surface resonance, overlapping with the projected bulk bands, except for the $\boldsymbol{k}_{\|}$ near the $\overline{\mathrm{K}}$ point, where it falls in the projected bulk band gap.

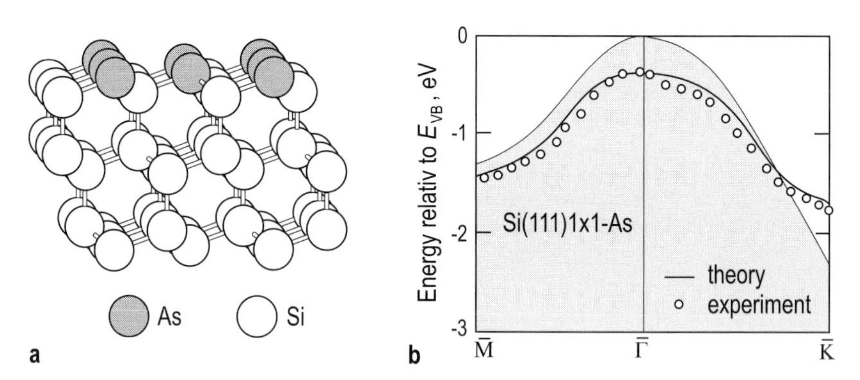

Fig. 11.14. (**a**) Model of the atomic structure of the Si(111)1×1-As surface with As atoms (hatched) substituting for Si atoms in the outermost layer. (**b**) Dispersion of the occupied surface state band, associated with the As lone pairs, as determined using ARUPS measurements [11.19] (open circles) and theoretical calculations [11.20] (solid line). The hatched area shows the projection of the bulk bands

11.4.4 Si(111)$\sqrt{3}\times\sqrt{3}$-In

The Si(111)$\sqrt{3}\times\sqrt{3}$-In surface is a member of a family of $\sqrt{3}\times\sqrt{3}$ reconstructions induced by Group III metal atoms. All of them have a similar atomic structure, in which metal adatoms, residing at the T_4 sites, saturate all dangling bonds of the top Si(111) layer (see Fig. 9.23). In this structure, Si atoms contribute three dangling-bond electrons per $\sqrt{3}\times\sqrt{3}$ unit cell and the In atom contributes one $3p$ electron per unit cell. These four electrons fill two surface state bands, labeled S_2 and S_3 in Fig. 11.15, showing the surface band structure of the Si(111)$\sqrt{3}\times\sqrt{3}$-In surface determined in ARUPS and KRIPES experiments. Besides S_2 and S_3, a third state S_1 exists close to the Fermi level, which show a close resemblance to the S_1 adatom state on the Si(111)7×7 surface. The intensity of this state was found to vary with sample preparation. It was suggested that the extra state is an *extrinsic state* due to the special type of defect which was identified as Si adatoms substituting for Al in the adatom layer (see Fig. 10.18).

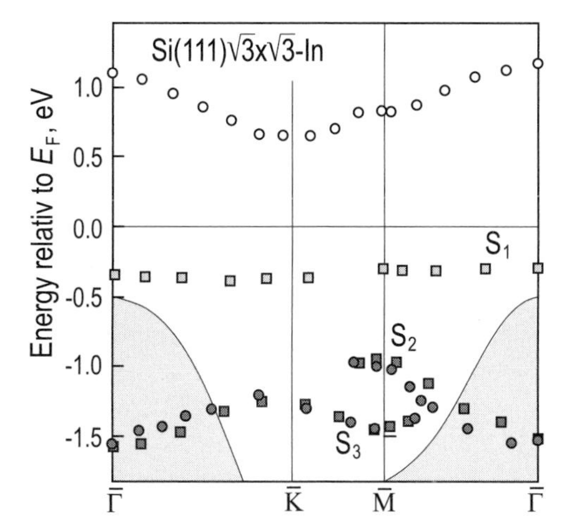

Fig. 11.15. Dispersion of the surface state bands of the Si(111)$\sqrt{3}\times\sqrt{3}$-In surface as determined using ARUPS (solid and hatched squares [11.21] and solid circles [11.22] for the filled states) and KRIPES (open circles [11.23] for the empty state)

11.5 Surface Conductivity

By analogy with bulk materials, surfaces might be separated into the metallic and semiconducting (insulating). A clear distinction is based on the surface state band structure. A *metallic surface* has appreciable density of states at the Fermi level or, in other words, there is a surface state with a dispersion

$E(\boldsymbol{k}_{\parallel})$ crossing the Fermi level. Hence, this surface state band is partially occupied. A *semiconducting surface* has a gap at the Fermi level and, in general, has surface state bands with smaller dispersion (recall that the electron effective mass is $m^* = \hbar^2(\mathrm{d}^2E/\mathrm{d}k^2)^{-1}$, i.e., a band with small dispersion means a large value of m^*). Depending on the gap width, one can loosely distinguish between semiconductor and insulator (a semiconductor has a smaller gap), but actually there is no principal difference between them. Back to the example surfaces discussed above, one can see that the Si(111)2×1 (Fig. 11.10), Si(111)1×1-As (Fig. 11.14), and intrinsic Si(111)$\sqrt{3}\times\sqrt{3}$-In (Fig. 11.15) surfaces are semiconducting, while the Cu(111)1×1 (Fig. 11.9) and Si(111)7×7 surfaces (Figs. 11.11, 11.12, and 11.13) are metallic. The latter example illustrates that a semiconductor crystal might have surfaces with metallic properties.

The presence of electronic surface states at the surface of a semiconductor disturbs the electronic structure inside the semiconductor material, leading to near-surface *band bending*. This phenomenon can be visualized as follows. Consider first an n-type semiconductor (Fig. 11.16a). Deep in the bulk, the position of the Fermi level is shifted by doping from the mid-gap towards the conduction band minimum. At the surface, the energetic position of the surface state band within the gap is fixed. Upon "contact" between the surface and bulk, the filled bulk donor levels appear to be well above the empty surface states. This energetically unfavorable situation cannot be stable and electrons from the bulk donor level transfer to the empty surface states until the Fermi levels in the bulk and at the surface become equilibrated (Fig. 11.16b). As a result of charge transfer, an additional negative charge is accumulated at the surface. The near-surface region is depleted of electrons, hence, positively charged. This region is referred to as the *space charge layer*. The non-uniform charge distribution generates an electrostatic potential, which is called the *built-in potential*. Under simplifying assumption that the concentration of ionized bulk donors, N_{d}, is constant throughout the charge density layer of thickness d, the built-in potential, $v(z)$, within the space charge layer is given by

$$v(z) = v_{\mathrm{bulk}} - \frac{2\pi N_{\mathrm{d}}}{\varepsilon}(z-d)^2 \,, \tag{11.13}$$

where ε is the dielectric constant of the semiconductor and z is the distance from the surface. The simplest way to visualize band bending associated with this parabolic built-in potential is to recall that the surface negative charge repels the electrons of the bulk, i.e., work has to be done to move an electron to the surface: this is an uphill process. Similar considerations for the case of a p-type semiconductor shows that the surface accumulates positive charge and downward band bending takes place (Fig. 11.16d).

The examples shown in Fig. 11.16 correspond to the case when the space charge layer is a *depletion layer* (i.e., the carriers are depleted from the near-surface region, hence the conductivity of the space charge layer is low). When

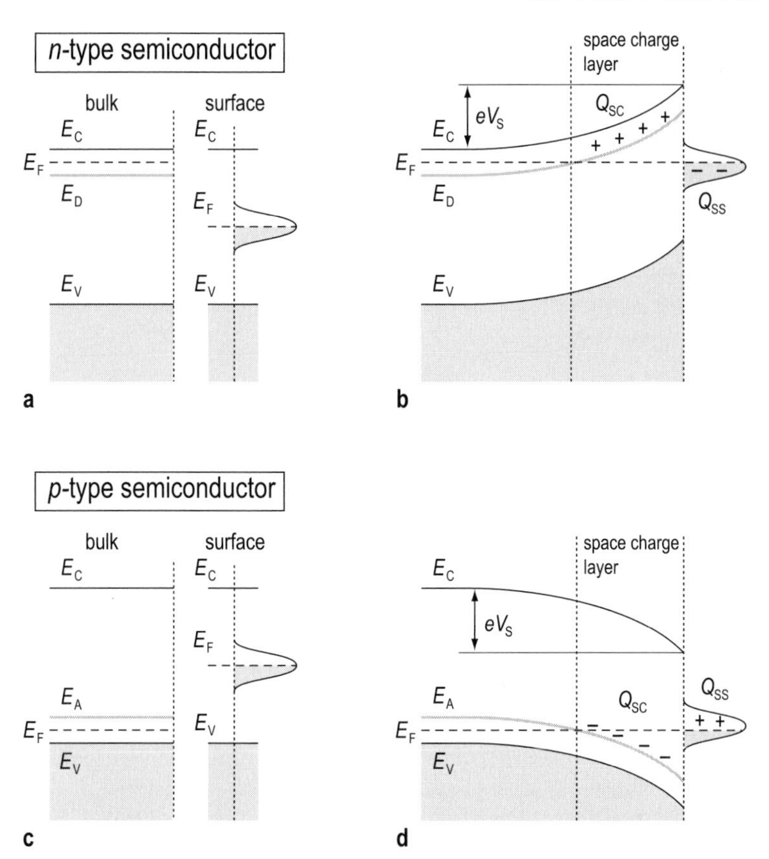

Fig. 11.16. Schematic illustration of band bending near the surface of **a**, (**b**) n-type and **c**, (**d**) p-type semiconductors. (**a**) and (**c**) illustrate the disequilibrium between the bulk and the surface. (**b**) and (**d**) show band bending at equilibrium. E_C and E_V are the conduction- and valence-band edges, E_F the Fermi energy, E_D and E_A the energy of the bulk donor and acceptor levels. $Q_{SS} = -Q_{SC}$ are the charges accumulated at the surface and in the space charge layer. $eV_S = ev\,(z = 0)$ denotes the band bending

the surface state band is located close to the edge of the bulk gap, band bending may result in the formation of an *accumulation layer*, which is a hole-accumulation layer when the surface state band is near the bulk valence-band maximum and an electron-accumulation layer when the surface band is near the bulk conduction-band minimum. Accumulation of free carriers in the near-surface region leads to an enhancement of the surface conductivity. In the case of strong band bending, the minority carriers might dominate in the space charge layer, i.e., a surface *inversion layer* is formed.

Consider now the surface conductivity. Correlation between the structural transformations at the surface and the changes in the surface conductivity was established as early as in 1970s [11.24]. As an example, Fig. 11.17

shows the results of an experiment in which surface conductivity measurements and LEED observations were conducted simultaneously in the course of isochronal annealing of a cleaved Si(111)2×1 sample. One can see that the surface conductivity decreases with the disappearance of the 2×1 superstructure, passes through a distinct minimum at the conversion temperature of 370°C and increases again with the development of the 7×7 structure. It should be emphasized, however, that the interpretation of such results is not straightforward even now, as the current flowing through the sample is actually a total of three main contributions. These contributions are associated with three parallel channels:

- surface state bands;
- space charge layer;
- sample bulk.

Thus, one can write for the electrical sheet conductance g of a semiconductor of a square piece of arbitrary length and thickness d:

$$g = g_0 + \Delta\sigma_{SC} + \Delta\sigma_{SS} . \tag{11.14}$$

Here g_0 is the bulk contribution to the sheet conductivity. It is expressed in units of [S/□] and is related to the bulk conductivity σ_B [S/cm] as $g_0 = \sigma_B d$. $\Delta\sigma_{SC}$ and $\Delta\sigma_{SS}$ stand for the contributions due to the space charge layer and surface states, respectively.

In general it is very difficult to separate the contributions from each channel. Moreover, the bulk contribution in most cases far exceeds the surface contributions. To minimize the bulk contribution, special experimental efforts are required.

Figure 11.18 illustrates how the surface contribution can be enhanced by miniaturizing the four-point probe. Recall that the four probe method is the most common technique to measure electrical conductivity. Here a constant current I is injected through the outer pair of probes and the voltage drop V is measured across the inner pair of probes (see top inset in Fig. 11.18). The four-probe resistance is then $R = V/I$ (with geometrical correction factors depending on the sample shape and probe spacing). The data points in Fig. 11.18 show the resistance of a silicon crystal (n-type, resistivity of 5–15 Ω cm, 4×15×0.4 mm^3 in size) having a Si(111)7×7 surface on the front face, measured using the four-point probe method as a function of the probe spacing, d. For comparison, the gray straight band shows the calculated dependence for a *semi-infinite homogeneous* sample, whose resistance is given by $R = \rho/2\pi d$, where ρ is the resistivity of the crystal. The $R \propto 1/d$ dependence can be understood if one supposes that, when the probe spacing is d, the current distribution is confined roughly to the crystal volume of length, width, and depth, each on the order of d.

In Fig. 11.18, one can see that experimental data are consistent with the calculated values only at intermediate probe spacings of 10–100 μm, but

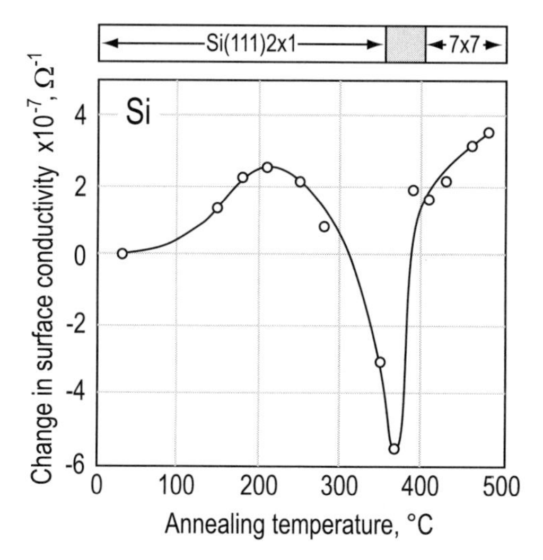

Fig. 11.17. Change of surface conductivity and LEED pattern as measured at 300 K after isochronal annealing as a function of annealing temperature (after Bäuerle et al. [11.24])

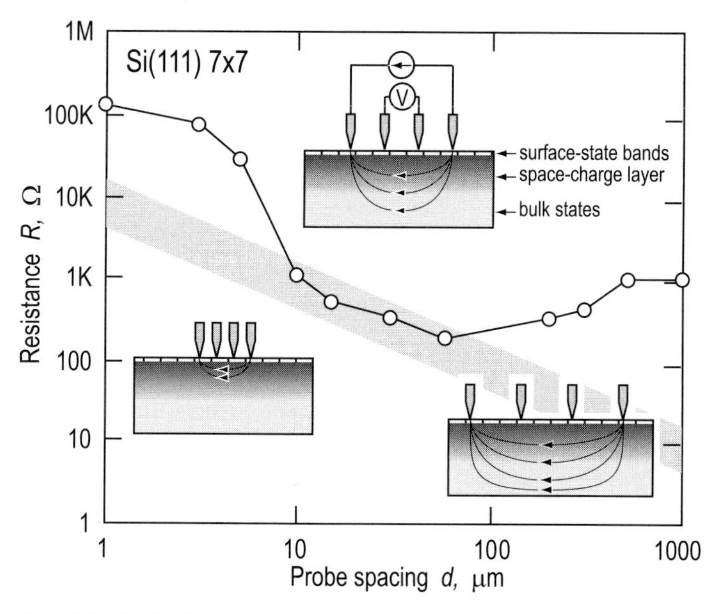

Fig. 11.18. Resistance of a silicon sample with a Si(111)7×7 surface measured as a function of the probe spacing. Insets show schematically the current flow distribution in the sample at four-point probe measurements with different probe spacings. The gray straight band shows the calculated dependence for the semi-infinite sample, $R = \rho/2\pi d$, with resistivity ρ =5–15 Ω cm (after Shiraki et al. [11.25])

deviate upward both at larger and smaller values of d. Taking into account that the thickness of the sample is 0.4 mm, the semi-infinite sample is a good approximation for the case when d is 10–100 μm, but not for larger d. When d exceeds the thickness of the sample, the current distribution is limited by the sample thickness and the $R \propto 1/d$ dependence no longer holds. At $d < 10$ μm, the current flows only near the surface and the data points indicate that the resistance of the near-surface layer is larger than that of the bulk. This result is consistent with the fact that the space charge layer beneath the Si(111)7×7 surface is depleted of mobile charge carriers.

The above results demonstrate that reducing the probe spacing makes the measurements more surface sensitive. This can also be illustrated by comparison of the results obtained with micro-probes and macro-probes for the Si(111)7×7 and Si(111)$\sqrt{3}\times\sqrt{3}$-Ag surfaces. The resistance measured with micro-probes on the $\sqrt{3}\times\sqrt{3}$-Ag surface is smaller than that for a clean 7×7 surface by about two orders of magnitude, while with macro-probes the difference of resistance between the two surfaces does not exceed 10% [11.26].

When dealing with the surface conductivity, one has to take into account the presence of defects on the surface. Atomic steps and point defects act as barriers for surface state electrons, as evidenced by observation of the Friedel oscillations near the step edges and defects (see Fig. 11.2). This should cause additional resistance at steps, which is clearly resolved in measurements with the micro-four-point probe. The resistance measured across a step bunch is much larger than that measured on a step-free terrace (Fig. 11.19).

11.6 Work Function

11.6.1 Work Function of Metals

The work function of a metal is often defined as the minimum energy required to extract one electron from a metal. In this definition, the work function is the energy difference between two states of the whole crystal. In the initial state, the neutral crystal of N electrons is in its ground state with energy E_N. In the final state, one electron is removed from the crystal and possesses only electrostatic energy, described by the vacuum level E_{vac}. The crystal with the remaining $N-1$ electrons enters a new ground state with energy E_{N-1}. Thus, the work function is written as

$$\phi = E_{N-1} + E_{\mathrm{vac}} - E_N \ . \tag{11.15}$$

For the thermodynamic change of state, the difference $E_{N-1} - E_N$ can be expressed as the derivative of the free energy F with respect to the electron number N, the temperature T and volume V being constant. This derivative $(\partial F/\partial N)_{T,V}$ is the electrochemical potential μ of the electrons (or Fermi energy E_{F} at non-zero temperature):

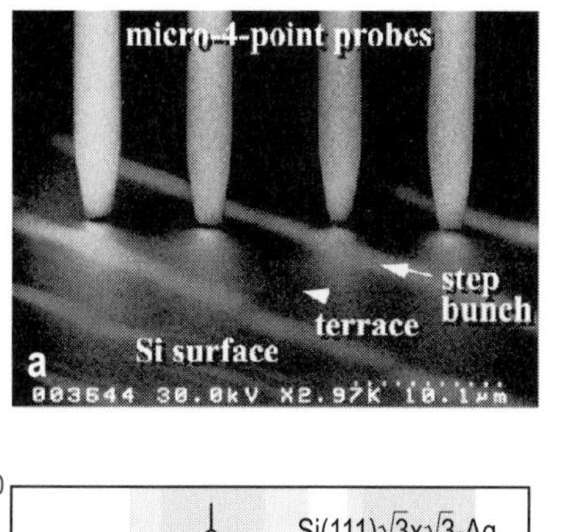

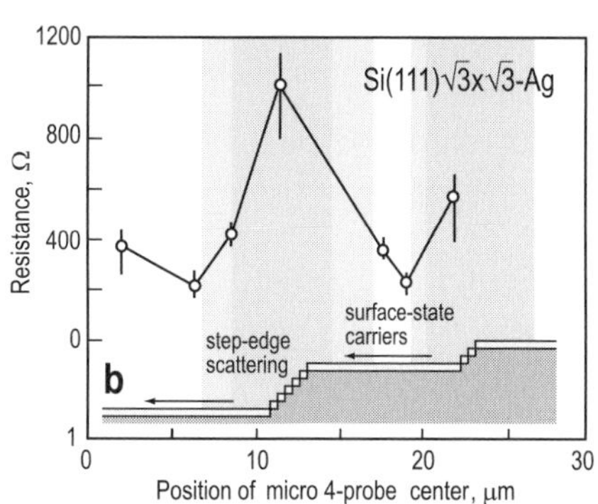

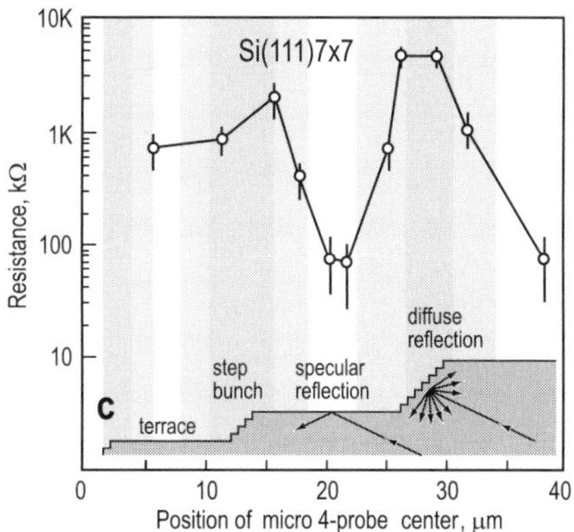

Fig. 11.19. (**a**) Scanning electron micrograph showing the micro-four-point probe contacting a silicon surface for measuring the conductivity in UHV. Slightly brighter bands on the sample surface are step bunches and wider darker bands are terraces. "Line profiles" of resistance of (**b**) Si(111)$\sqrt{3}\times\sqrt{3}$-Ag and (**c**) Si(111)7×7 surfaces, measured by shifting the micro-four-point probes (8 μm probe spacing) along a line across step bunches on the surfaces. The morphology of the sample surfaces, as determined by SEM, is schematically drawn at the bottom of graph. Areas of dark shadow show the situation that the voltage drop is measured by the inner pair of probes across a step bunch (like in **a**). Areas of light shadow show the situation that both the inner probes contact the same terrace (i.e., no step bunch runs between the inner probes). Areas of intermediate shadow show the situation when one of the inner probes is on a terrace and the other probe is on a step-bunch region. The resistance measured across a step bunch is much larger than that measured on a step-free terrace (after Hasegawa et al. [11.27, 11.28])

$$E_{N-1} - E_N \rightarrow \left(\frac{\partial F}{\partial N}\right)_{T,V} = \mu \tag{11.16}$$

and hence the generalized expression for the work function is

$$\phi = E_{\mathrm{vac}} - E_{\mathrm{F}} \; . \tag{11.17}$$

When dealing with a crystal of finite size, the final position of the electron needs to be specified, as it controls the value of E_{vac}. For a clean single-crystalline face, it is accepted that the distance of the electron from the face should be large enough, so that the image force is already negligible (typically, ∼1 μm), but it should be small compared with the distance from another crystal face with a different work function.

For convenience, the work function is often expressed as a sum of two contributions. The first contribution is associated with the bulk properties of a crystal and is given by the electrostatic potential deep in the bulk (with Fermi level taker as zero). In terms of the potential diagram, shown in Fig. 11.3, this is $v(-\infty)$. The second contribution is a surface-sensitive term given by the electrostatic potential energy difference:

$$\Delta v = v(+\infty) - v(-\infty) = 4\pi e \int_{-\infty}^{\infty} z\,[n(z) - n_+(z)]\,\mathrm{d}z\,, \tag{11.18}$$

where $n(z)$ is the distribution of the electron charge density and $n_+(z)$ the density of the positive background charge. This term is called *electrostatic dipole barrier*, as it stands for the work to be done to move an electron across the surface dipole layer.

The surface dipole moment is a characteristic feature of a surface and it varies from one surface to another. For the same metal, the closely packed (atomically smooth) faces have generally a greater dipole moment than that of loosely packed (atomically rough) faces. This can be understood qualitatively if one visualizes the rough surface as having a fraction of positive ion cores, which protrude into the negatively charged layer, hence reducing the dipole moment and, consequently, the work function. Indeed, many metals follow this empirical rule that the work function is lower for loosely packed faces (see Table 11.1).

Table 11.1. Experimental values of the work function for selected metals [11.29]

Metal	Structure	Work function ϕ, eV			
		Plane			Polycrystalline
		(110)	(100)	(111)	
Cs	bcc				2.9
Li	bcc				2.30
Ag	fcc		4.42	4.56	4.3
Cu	fcc	4.48	4.63	4.88	4.65
Pt	fcc		5.84	5.82	5.65
Ir	fcc	5.42	5.67	5.76	5.3

Adsorption of atoms or molecules changes the surface dipole moment and, thus, changes the work function. As an example, Fig. 11.20 shows the work function changes due to chlorine and cesium adsorption on a Cu(111) surface. Since the *electronegative* Cl accepts electrons from the metal, the total surface dipole moment increases, leading to an increase of the work function (Fig. 11.20a). In contrast, *electropositive* Cs donates electrons to the metal, where they remain in the immediate vicinity to screen the adsorbed ions. The adsorbate-induced dipole moment opposes the original dipole moment of the clean Cu(111) surface with a concomitant decrease in the work function (Fig. 11.20b). The shape of the dependence for Cs is typical of alkali metal adsorption on metal surfaces in that there is a rapid (almost linear) initial decrease in the work function followed by a minimum and a small rise near a saturated monolayer. This behavior can be understood simply as follows. Initially, each adsorbed ion contributes to the work function change individually. Thus, the dipole moment of a single atom can be evaluated from the initial slope of the measured $\Delta\phi$ versus coverage dependence. However, with the growth of coverage the dipoles begin to feel the fields of the neighboring

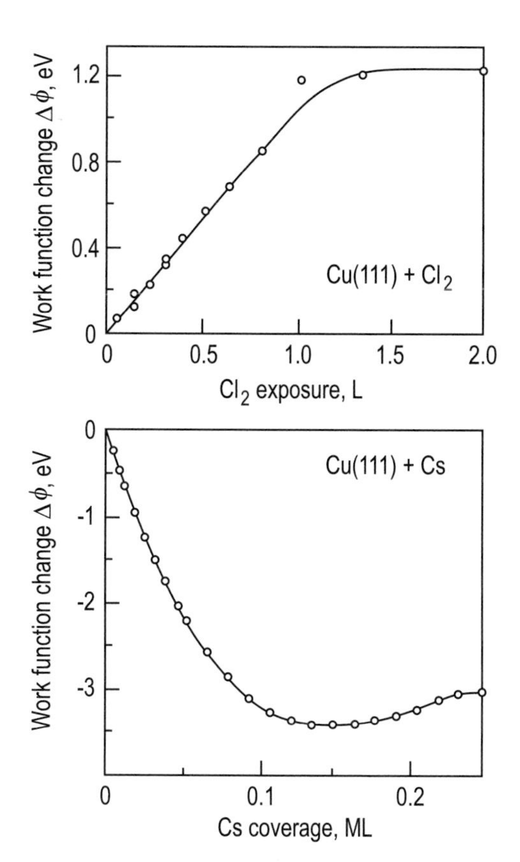

Fig. 11.20. Work function changes induced by adsorption of (**a**) chlorine and (**b**) cesium on a Cu(111) surface. The work function of the clean Cu(111) surface is 4.88 eV (after Goddard and Lambert [11.30] and Lindgren and Waldén [11.31])

dipoles. The result of the interdipole interaction is a *depolarization* of the closely packed dipoles.

11.6.2 Work Function of Semiconductors

For semiconductors, an additional effect of the band bending arises and the total work function of the semiconductor is conventionally described as a sum of three terms (Fig. 11.21):

$$\phi = \chi + eV_{\mathrm{S}} + (E_{\mathrm{C}} - E_{\mathrm{F}}) \,. \tag{11.19}$$

Here χ is the electron affinity, eV_{S} stands for the band bending, and the term $(E_{\mathrm{C}} - E_{\mathrm{F}})$ denotes the energy difference between the Fermi level and the conduction band minimum in the bulk. Adsorbate-induced dipoles affect the two first terms, while semiconductor doping changes the third term.

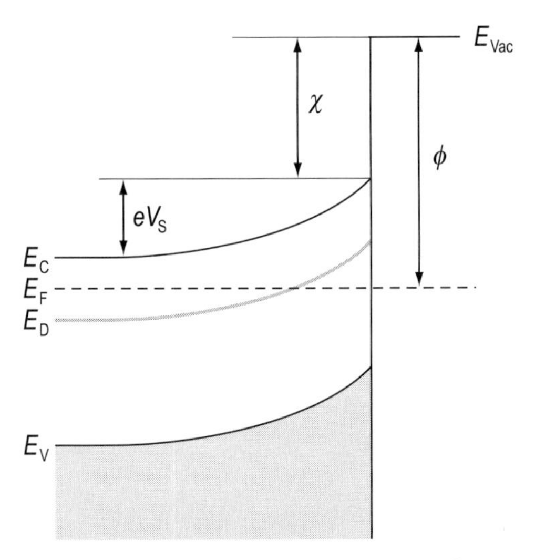

Fig. 11.21. Schematic band diagram for a semiconductor surface. ϕ is the work function, χ the electron affinity, eV_{S} the band bending, E_{V} the valence band maximum, E_{C} the conduction band minimum, and E_{F} the Fermi level

11.6.3 Work Function Measurements

There are a number of experimental techniques that are used to measure the work function. They are divided into two groups:

- absolute measurements and
- relative measurements.

Absolute measurements are based on electron emission phenomena, namely,

- field emission (electron emission stimulated by a high electric field);
- thermionic emission (electron emission stimulated by high temperature);
- photoemission (electron emission stimulated by photon irradiation).

Field Emission. The basics of field emission is illustrated in Fig. 11.22, which shows the schematic potential diagram for the case when an external electric field F is present near the surface. The original barrier (approximated here by an image potential $\sim -e^2/4z$) is deformed by the applied field potential $-Fez$ and obtains a shape represented by the "total potential" curve. This effective barrier has a reduced height and a finite thickness. The latter is $\sim \phi/eF$ at the Fermi level and for electric fields on the order of 1 V/Å (accessible by applying ~1000 V to a sharp tip with a radius of curvature ~1000 Å) equals several Å. In this case, an electron can escape from the metal by quantum mechanical tunneling through the barrier. The current density j for this process is given by the *Fowler–Nordheim equation* as

$$j = \frac{1.54 \times 10^{-6} F^2}{\phi t^2(\xi)} \exp\left[\frac{-(6.83 \times 10^7 \phi^{3/2} f(\xi))}{F}\right] \quad [\mathrm{A/cm^2}] \,, \tag{11.20}$$

where F is the applied field in V/cm, ϕ is the metal work function in eV, and $t(\xi)$ and $f(\xi)$ are slowly varied functions of the dimensionless parameter ξ. The latter are available in tabular form. An experimental plot of $\ln(j/F^2)$ versus $(1/F)$ is used for the determination of the work function ϕ, as illustrated in Fig. 11.23. The field emission is also the basis of the field emission microscope (FEM), which is discussed in some detail in Sect. 7.1.

Thermionic Emission. At finite temperature, electrons have a finite probability of surmounting the surface energy barrier and appear as thermally emitted electrons. The density of the thermionic current j from a homogeneous metal surface at temperature T is given by the *Richardson–Dushman equation*:

$$j = AT^2 \exp(-\phi/k_\mathrm{B}T) \,, \tag{11.21}$$

where

$$A = \frac{4\pi m k_\mathrm{B}^2 e}{h^3} \approx 120 \ \mathrm{A\,cm^{-2}K^{-2}} \,, \tag{11.22}$$

with m and e being the electron mass and charge respectively, h Planck's constant, and k_B Boltzmann's constant. In thermionic experiments, the work function is obtained from the slope of the measured $\ln(j/T^2)$ versus $(1/T)$ curves, as shown in Fig. 11.24 for several faces of a W crystal. Table 11.2 combines the data on work function of tungsten measured in field emission and thermionic emission experiments.

It is worth noting that in practice the measurement of the saturation thermionic current requires the application of an accelerating field, which

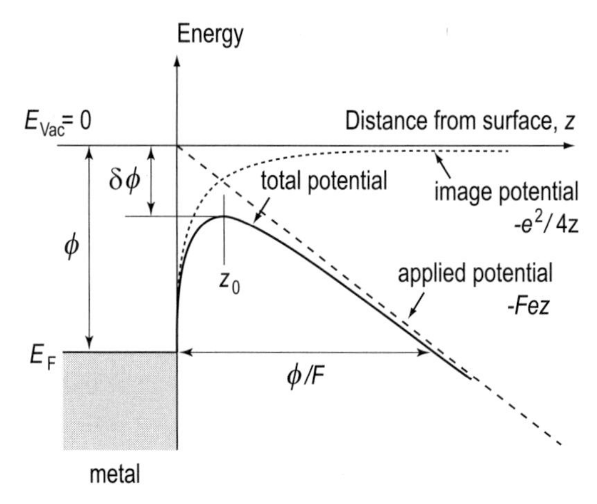

Fig. 11.22. Potential energy diagram for an electron near a metal surface in the presence of an applied external electric field of strength F. The total potential (shown as a solid line) is the sum of the image potential (shown as a dotted line) and the applied potential (shown as a dashed line). ϕ is the work function in the absence of the applied field. The lowering of the potential barrier by $\delta\phi$ due to the Schottky effect is indicated. z_0 denotes the position of the maximum of the total potential

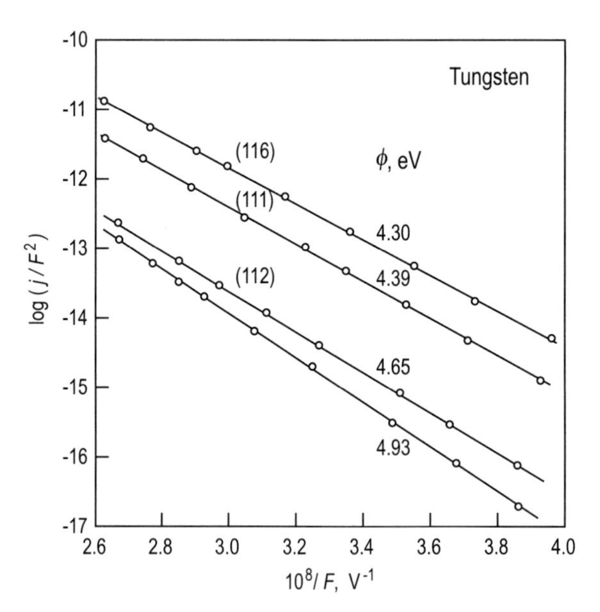

Fig. 11.23. Fowler–Nordheim plots of field emission from various faces of tungsten (after Müller [11.32])

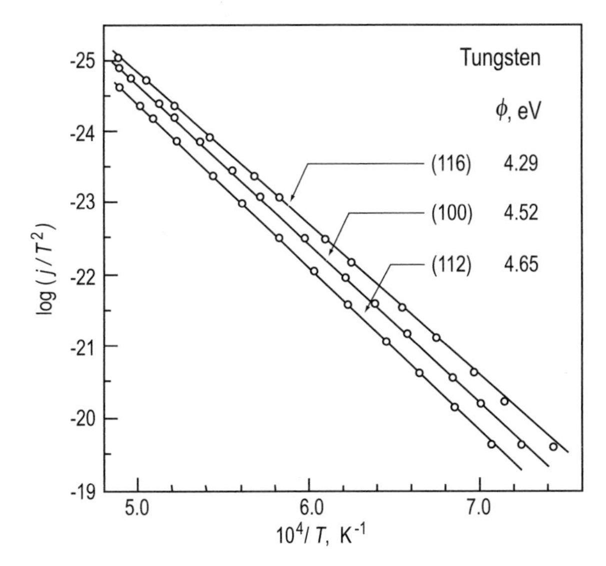

Fig. 11.24. Richardson–Dushman plots of thermionic emission from various faces of tungsten (after Smith [11.33])

Table 11.2. Values of the work function of tungsten crystal faces, as determined in field emission [11.32] and thermionic emission [11.33] experiments

Plane	Work function ϕ, eV	
	Field emission	Thermionic emission
(116)	4.30	4.29
(111)	4.39	4.39
(112)	4.65–4.93	4.65
(116)	5.70-5.99	~5.26

leads to a lowering of the potential barrier at the surface. This phenomenon is known as the *Schottky effect.* As one can see in Fig. 11.22, the lowering of the work function can be estimated from the condition for searching for a maximum in the total potential curve:

$$\frac{\mathrm{d}}{\mathrm{d}z}\left(-\frac{e^2}{4z} - Fez\right) = 0\,. \tag{11.23}$$

The estimation yields the position z_0 of the maximum in the potential barrier as

$$z_0 = \frac{1}{2}e^{1/2}F^{-1/2} \tag{11.24}$$

and the variation of the work function $\delta\phi$ versus accelerating field as

$$\delta\phi = e^{3/2}F^{1/2}\,. \tag{11.25}$$

Photoelectron Emission. When a solid is irradiated by photons, some electrons absorb the photon energy and may escape from the solid. This process is called *photoelectron emission*, or simply *photoemission* (for more details see Sect. 5.4). The ejected electrons are called *photoelectrons*. For metal at zero temperature, the minimal photon energy to generate a photoelectron (*photoemission threshold*) equals the work function:

$$h\nu_0 = \phi \, . \tag{11.26}$$

At a finite temperature, some electrons occupy states above the Fermi level and photons of energy lower than $h\nu_0$ can cause photoemission. The saturation photocurrent from a homogeneous metal at temperature T is given by the *Fowler expression*:

$$j = B(k_\mathrm{B}T)^2 f\left(\frac{h\nu - \phi}{k_\mathrm{B}T}\right) , \tag{11.27}$$

where B is a parameter which depends on the material and f is the universal Fowler function [11.34]. For $h\nu - \phi > 2k_\mathrm{B}T$, (11.27) can be approximated by $j \propto (h\nu - \phi)^2$. In practice, in order to determine the work function, one plots $j^{1/2}$ versus $h\nu$. Except for a slight rounding off near the threshold due to the finite-temperature tail of the Fermi–Dirac distribution, such plots yield straight lines, which intercept the $h\nu$ axis at $h\nu_0 = \phi$. As an example, Fig. 11.25 illustrates the evaluation of the work function for several faces of copper by means of the photoelectric method.

Besides the absolute techniques discussed so far, a number of *relative methods* of work function measurements have been developed. All relative methods are based on the fact that between two metal samples with an external electric connection there exists a *contact potential difference (CPD)*, which equals their work function difference (divided by the electron charge). Variation in the work function of a sample surface as a result of certain physical processes (for example, adsorption) can be monitored relative to the reference work function of the probe by monitoring the CPD between the two surfaces. If the work function of the probe is accurately known, the relative work function variation can be transferred to the absolute value. The most popular *relative methods* are

- vibrating capacitor (or Kelvin probe) method and
- diode method.

Vibrating Capacitor Method. The principle of the method is illustrated in Fig. 11.26. The probe electrode and the sample surface form a capacitor of variable capacitance. In the external circuit, the two electrodes are connected through an AC ammeter and a variable voltage source. The voltage between the sample and the probe is

$$U = \Delta\phi/e + U_\mathrm{comp} \, , \tag{11.28}$$

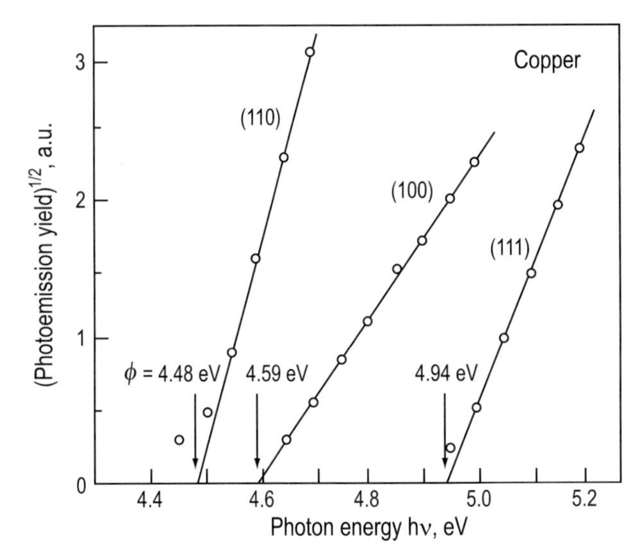

Fig. 11.25. Square root of photoelectric yield as a function of photon energy for selected faces of a Cu crystal. The intercept of the extrapolated dependence with the abscissa yields the value of the work function with an estimated accuracy of $\pm 0.03\,\mathrm{eV}$ (after Gartland et al. [11.35])

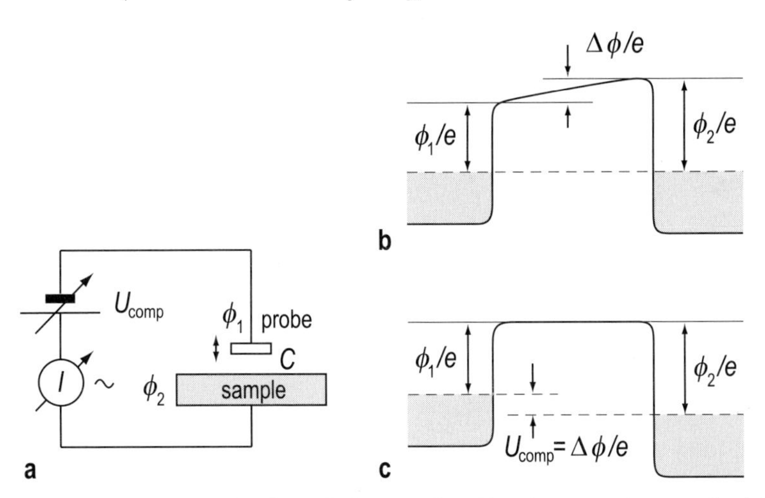

Fig. 11.26. Principle of the Kelvin probe vibrating capacitor method for relative measurements of the work function. (**a**) Schematic of the circuit including a capacitor of variable capacitance formed by the vibrating probe electrode and the sample surface, variable voltage source, and AC ammeter. The AC current generated by the vibrating probe is compensated by adjusting the voltage U_{comp} of the voltage source. (**b**) Variation of potential between the probe and the surface, when $U_{\mathrm{comp}} = 0$. (**c**) Variation of potential between the probe and the surface, when $U_{\mathrm{comp}} = -\Delta\phi/e$

where $\Delta\phi/e = (\phi_2 - \phi_1)/e$ is the contact potential difference and U_{comp} the voltage of the variable source. Thus, the capacitor formed by the sample and probe and having capacitance C carries a charge

$$Q = CU = C(\Delta\phi/e + U_{\mathrm{comp}}) \,. \tag{11.29}$$

If the probe is vibrating at some constant frequency, the capacitance is changed periodically, giving rise to an AC current in the external circuit of magnitude

$$I = \frac{\mathrm{d}C}{\mathrm{d}t}U = \frac{\mathrm{d}C}{\mathrm{d}t}(\Delta\phi/e + U_{\mathrm{comp}}) \,. \tag{11.30}$$

By adjusting the voltage U_{comp} of the variable source, the current is set to zero, though the capacitance is changed. As one can see from (11.30), this takes place at the value of U_{comp} equal to CPD:

$$U_{\mathrm{comp}} = -\Delta\phi/e \,. \tag{11.31}$$

Thus, the work function of the sample relative to the reference probe is obtained.

Diode Method. In this method, a diode scheme is used with the cathode as a reference electrode and the sample under investigation as the anode. The measuring technique is based on the fact that variation of the anode work function by $\Delta\phi_{\mathrm{a}}$ shifts the characteristic curve of the diode by $\Delta U_{\mathrm{a}} = \Delta\phi_{\mathrm{a}}$, as shown in Fig. 11.27. In practice, the work function variation is monitored by maintaining the anode current at a constant value by readjustment of the anode potential.

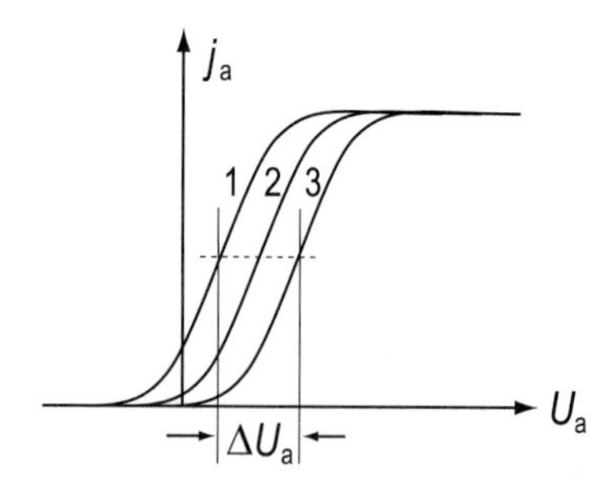

Fig. 11.27. Principle of the diode method for determination of the work function change: the characteristic curve of the diode, I_{a} versus U_{a}, shifts following the variation of the work function ϕ of the anode. Here $\phi_1 < \phi_2 < \phi_3$, the displacement $\Delta U_{\mathrm{a}} = (\phi_3 - \phi_1)/e$

Problems

11.1 Within the framework of the jellium model, calculate the period of the Friedel oscillations for Na, which is an alkali bcc metal with a lattice constant of 4.23 Å.

11.2 From the plot of the work function variation induced by adsorption of cesium on a Cu(111) surface, shown in Fig. 11.20b, estimate the dipole moment of a single Cs atom on a Cu(111) surface. The surface atom density of the Cu(111) surface is $1.77 \cdot 10^{15}$ cm^{-2}. Express the obtained result in Debye units (1 Debye = 10^{-18} esu cm).

11.3 Using the Richardson–Dushman equation (11.21), estimate the temperature for the different cathode materials listed in the following Table at which the thermionic current density is 1 A/cm^2.

Cathode	ϕ, eV	j, A/cm^2	T, °C
W	4.6	1	
Th-W	3.2	1	
LaB_6	2.8	1	
BaO	1.5	1	

Further Reading

1. S.G. Davison, M. Stęślicka: *Basic Theory of Surface States* (Oxford University Press, Oxford 1992) (theory of Tamm states in detail)
2. K. Horn, M. Scheffler (Ed.): *Electronic Structure. Handbook of Surface Science. Vol. 2.* (Elsevier, Amsterdam 2000) (set of reviews on various aspects of surface electronic structure of metals and semiconductors)
3. S. Hasegawa, F. Grey: *Electronic Transport at Semiconductor Surfaces–From Point-Contact Transistor to Micro-Four-Point Probes.* Surf. Sci. **500**, 84–104 (2002) (discussion on the measurements of surface conductivity)
4. J. Hölzl, F.K. Schulte: 'Work Function of Metals'. In: *Solid State Physics. Springer Tracts in Modern Physics. Vol.85* (Springer, Berlin, Heidelberg, New York 1979)

12. Elementary Processes at Surfaces I. Adsorption and Desorption

Adsorption/desorption phenomena have already been treated in Chap. 9 devoted to the structure of surfaces with adsorbates. In particular, the basic terms (for example, adsorbate, substrate, physisorption, chemisorption, coverage) have been introduced. However, there the final result of the adsorption/desorption process has been discussed, not the process itself. The latter is the subject of the present chapter, in which adsorption/desorption phenomena are treated within the kinetic approach. It resides in establishing the relationship between the rates of adsorption and desorption of atoms or molecules as a function of the external macroscopic variables, such as vapor pressure, temperature of the substrate and vapor, substrate surface structure, etc. Kinetics is an external manifestation of the atomic-scale dynamic processes occurring at the surface. Analysis of the kinetic data gives an insight into the atomic mechanisms involved in the adsorption and desorption. In particular, adsorption/desorption energetics (the height of energy barriers and the depth of energy wells) are commonly extracted from the experimental kinetic data.

12.1 Adsorption Kinetics

Consider the kinetic approach for the case of a uniform solid surface exposed to an adsorbing gas. According to the kinetic theory of gases, the flux I of the gas molecules (atoms) impinging on the surface from the gas phase is given by

$$I = \frac{p}{\sqrt{2\pi m k_{\mathrm{B}} T}} , \qquad (12.1)$$

where p is the partial pressure of the adsorbing gas, m is the mass of the gas molecule (atom), k_{B} is Boltzmann's constant, and T is the temperature.

However, not all of the impinging molecules become adsorbed at the surface, i.e., contribute to the adsorption rate. The ratio of the adsorption rate to the impingement rate is defined as the *sticking coefficient* (or *sticking probability*), s. Thus, the adsorption rate r_{a} is

$$r_{\mathrm{a}} = sI .$$

The general expression for the sticking coefficient in the case of activated adsorption is written as

$$s = \sigma f(\Theta) \exp(-E_{\mathrm{act}}/k_{\mathrm{B}}T) \,. \tag{12.3}$$

It contains the following terms:

σ, called the *condensation coefficient*, is responsible for the effects of the orientation (steric factor) and the energy accommodation of the adsorbed molecules.

$f(\Theta)$ is a coverage dependent function which describes the probability of finding an adsorption site. The type of the adsorption reaction (i.e., dissociative or non-dissociative), the statistics of site occupation, and the mobility of molecules in a precursor adsorption state (if available) affect the $f(\Theta)$ term.

The temperature-dependent Boltzmann term, $\exp(-E_{\mathrm{act}}/k_{\mathrm{B}}T)$, is associated with energetics of the activated adsorption.

We now consider the contribution of the above factors in greater detail.

12.1.1 Coverage Dependence

Langmuir Adsorption Model. As a starting point, consider the simplest case, referred to as the *Langmuir adsorption model.* The Langmuir model is based on the following assumptions:

- Adsorption is limited by monolayer coverage.
- All adsorption sites are equivalent.
- Only one molecule can reside in the adsorption site.

Consider non-dissociative and dissociative adsorption in terms of the Langmuir model (Fig. 12.1)

Non-Dissociative Langmuir Adsorption. For non-dissociative adsorption, the free sites are readily accessible for impinging molecules and $f(\Theta)$ is simply

$$f(\Theta) = 1 - \Theta \tag{12.4}$$

and the Langmuir adsorption kinetics is given by

$$\mathrm{d}\Theta/\mathrm{d}t = s_0 I(1 - \Theta) \,, \tag{12.5}$$

where s_0 is the sticking probability at zero coverage.

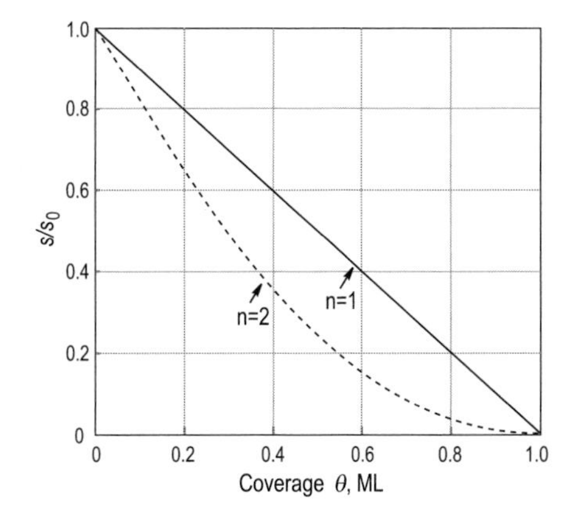

Langmuir adsorption

Sticking coefficient

$$s = s_0(1-\Theta)^n$$

$n = 1$ – non-dissociative,
$n = 2$ – dissociative.

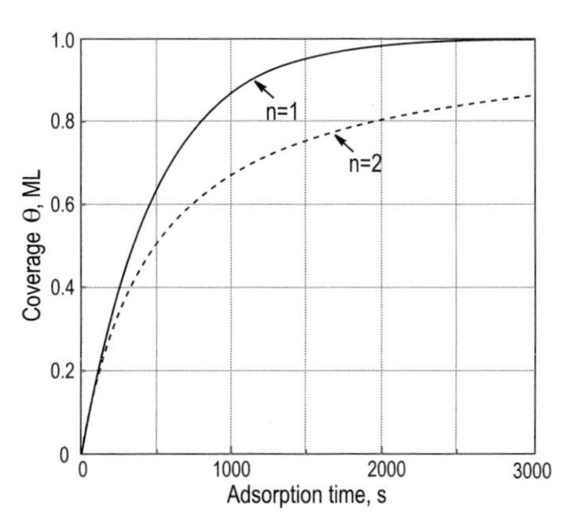

Adsorption kinetics

non-dissociative ($n = 1$):

$$\Theta = 1 - \exp(-s_0 I \cdot t)$$

dissociative ($n = 2$):

$$\Theta = \frac{s_0 I \cdot t}{1 + s_0 I \cdot t}$$

Fig. 12.1. Sticking coefficient and adsorption kinetics for non-dissociative first-order ($n = 1$) and dissociative second-order ($n = 2$) Langmuir adsorption. The kinetic curves are calculated for an adsorbate flux of 0.002 ML/s.

Dissociative Langmuir Adsorption. For the dissociative adsorption of diatomic molecules, where the impinging molecule dissociates into two atoms that are then trapped in the adsorption sites,

$$f(\Theta) = (1-\Theta)^2 \,, \tag{12.6}$$

if the dissociative products are mobile, and

$$f(\Theta) = \frac{z}{z-\Theta}(1-\Theta)^2 \,, \tag{12.7}$$

if the dissociative products are immobile. Here z is the number of nearest-neighbor sites. Actually, the difference between (12.6) and (12.7) is not so great and becomes almost negligible at relatively low Θ.

Note that if the adsorbing molecule dissociates into n species, the exponent in (12.6) is changed from 2 to n (for example, for dissociation of an Sb_4 molecule into four atoms, $f(\Theta) = (1 - \Theta)^4$. Thus, the Langmuir kinetics can be expressed in general form as

$$f(\Theta) = (1 - \Theta)^n \,, \tag{12.8}$$

where n gives the *order of the kinetics.*

Simple Langmuir kinetics appear to describe the adsorption process in some real chemisorption systems. As an example, Fig. 12.2 shows the data for adsorption of energetic O_2 molecules on the Rh(111) surface. As one can see, the experimental $s(\Theta)$ curve can be decomposed into two curves corresponding to first- and second-order Langmuir kinetics, which indicates that oxygen is adsorbed initially dissociatively, but then in molecular form.

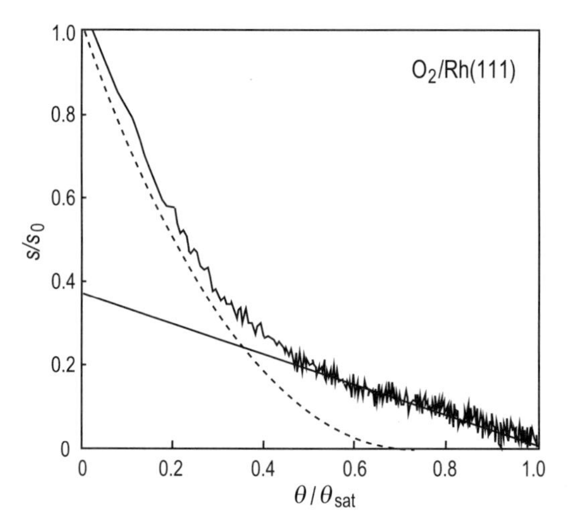

Fig. 12.2. Time evolution of the sticking coefficient of O_2 on Rh(111). The incident beam energy is 490 meV, and the substrate temperature is 110 K. The dashed and solid lines are the fits using, respectively, first- and second-order Langmuir kinetics (after Brault et al. [12.1])

Precursor-Mediated Adsorption. Often the chemisorption of the molecule is not direct but involves a transient, so-called, *precursor state*, in which the molecule is first trapped and from which it attempts to dissociate and enter the final adsorption state. The existence of a precursor state is possible also in the case of non-dissociative adsorption, in which case it means that the lightly bonded molecule can migrate around the surface until becoming trapped eventually in a tightly bonded state.

When adsorption proceeds via a precursor state, the mechanism of adsorption can be represented schematically by

$$(\mathrm{A})_\mathrm{g} \underset{k_\mathrm{d}}{\overset{s_\mathrm{p} I}{\rightleftarrows}} (\mathrm{A})_\mathrm{p} \overset{k_\mathrm{a}}{\rightarrow} (\mathrm{A})_\mathrm{a} \qquad \text{(non-dissociative adsorption)} \tag{12.9}$$

$$(\mathrm{A}_2)_\mathrm{g} \underset{k_\mathrm{d}}{\overset{s_\mathrm{p} I}{\rightleftarrows}} (\mathrm{A}_2)_\mathrm{p} \overset{k_\mathrm{a}}{\rightarrow} 2(\mathrm{A})_\mathrm{a} \qquad \text{(dissociative adsorption, } n = 2\text{)}. \tag{12.10}$$

Here $(\mathrm{A})_\mathrm{g}$ and $(\mathrm{A}_2)_\mathrm{g}$ denote the molecules in the gas phase, $(\mathrm{A})_\mathrm{p}$ and $(\mathrm{A}_2)_\mathrm{p}$ denote the molecules in the precursor state, $(\mathrm{A})_\mathrm{a}$ stands for the chemisorbed species, s_p is the trapping probability from the gas phase into the precursor state, k_a the rate constant for adsorption from the precursor state into the final chemisorbed state, and k_d the rate constant for desorption from the precursor state. The precursor state can be located spatially over an empty site (an *intrinsic precursor*) or over an occupied site (an *extrinsic precursor*). The resultant adsorption kinetics depend on whether the intrinsic and extrinsic precursors are energetically equivalent or not.

Precursor-Mediated Adsorption for Non-Interacting Adsorbates. If one assumes the intrinsic and extrinsic precursors to be energetically equivalent, the adsorption rate can be written as

$$r_\mathrm{ads} = \frac{s_\mathrm{p} I k_\mathrm{a} (1 - \Theta)^n}{k_\mathrm{d} + k_\mathrm{a}(1 - \Theta)^n} \tag{12.11}$$

and, consequently, the overall sticking coefficient can be expressed as

$$f(\Theta) = \frac{(1 + K)(1 - \Theta)^n}{1 + K(1 - \Theta)^n}\ , \tag{12.12}$$

where $K = k_\mathrm{a}/k_\mathrm{d}$ and $n = 1, 2, \ldots$ denotes the order of the kinetics.

Figure 12.3a shows a plot of (12.12) for first-order non-dissociative kinetics ($n = 1$) with various values of the parameter K. Figure 12.3b is the same but for fourth-order dissociative kinetics ($n = 4$). When $K \to \infty$ (the case of low desorption rate from the precursor state), $s \to 1$. When $K \to 0$ (the case of high desorption rate from the precursor state), $s \to 1 - \Theta$, i.e., Langmuir-type kinetics. Note that K is temperature dependent and, hence, the adsorption kinetics might vary with temperature.

Figure 12.4 shows an experimental example of the dissociative adsorption of Sb_4 molecules on the Si(100) surface. As one can see, the experimental data are well described by the curves calculated using (12.12) for $n = 4$. Fourth-order kinetics is a natural consequence of the dissociation of an Sb_4 molecule into four Sb atoms.

Precursor-Mediated Adsorption for Interacting Adsorbates. Here, the intrinsic and extrinsic precursors are considered to be energetically inequivalent. Thus,

k_d denotes the desorption rate constant for the intrinsic precursor site (situated over the empty chemisorption site) and

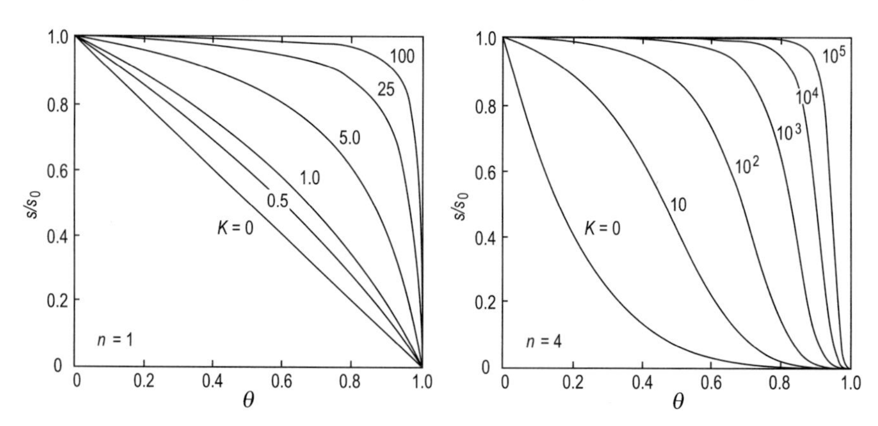

Fig. 12.3. Coverage dependent sticking probability for precursor-mediated chemisorption of non-interacting adsorbates as calculated using (12.12) (**a**) for first-order kinetics ($n = 1$) and (**b**) for fourth-order kinetics ($n = 4$). The precursor parameter is $K = k_a/k_d$ (after Lombardo and Bell [12.2])

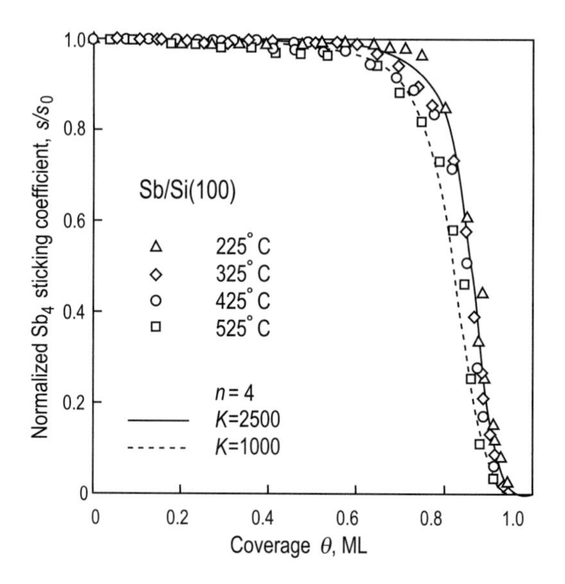

Fig. 12.4. Coverage dependence of the sticking coefficient of Sb on Si(100). The curves are calculated from (12.12) for $n = 4$ and $K = 2500$ (solid line) and $K = 1000$ (dashed line) (after Barnett et al. [12.3])

k'_d denotes the desorption rate constant for the extrinsic precursor site (situated over the occupied chemisorption site).

A statistical analysis yields the following expression for dissociative adsorption under these conditions:

$$f(\Theta) = [1 + K'(1/P_{VV} - 1)]^{-1} , \tag{12.13}$$

where $K' = k'_d/(k_a + k_d)$ and P_{VV} is the probability that two nearest-neighbor sites are vacant. P_{VV} depends on Θ and the short-range order parameter B, which describes the possibility of ordering within the chemisorbed layer: $B = 1$ corresponds to complete order and $B = 0$ to complete disorder.

A family of curves calculated using (12.13) is plotted in Fig. 12.5 for (a) fixed B and variable K' and (b) vice versa. For comparison, Fig. 12.6 shows the experimental curves for N_2 adsorption on W(100). As one can see, the general shape of the experimental curves is similar to the calculated curves with values of B close to unity. This is in agreement with LEED observations which reveal the formation of the ordered W(100)$c(2\times2)$-N structure.

12.1.2 Temperature Dependence

The temperature dependence of the sticking coefficient is intimately related to the energetics of adsorption. To visualize the general trends of this relationship, consider the one-dimensional Lennard-Jones potential of gas–surface interaction. The case of a single chemisorption well (Fig. 12.7) corresponds to simple adsorption without any precursor state. Since for most chemisorption systems $E_{ads} \gg k_B T$, the effect of adsorbate re-evaporation is negligible and, hence, the sticking coefficient for simple adsorption is almost temperature independent.

In the case of precursor-mediated chemisorption, the potential diagram contains two wells: a shallow well for the physisorption precursor state and a deep well for the final chemisorption state (Fig. 12.8). For example, this behavior is typical for the dissociative adsorption of molecules, where the adsorption process involves physisorption of a molecule followed by its dissociation into atomic species which form strong chemical bonds with the substrate. The two wells are separated by an activation barrier whose value controls the temperature dependence of the sticking coefficient.

Consider a molecule trapped in the precursor state. It can either desorb back to the gas phase or become adsorbed at the chemisorption state. The rates of desorption and adsorption from the precursor state can be expressed respectively as

$$k_d = \Theta_p \nu_d \exp(-\varepsilon_d / k_B T) \tag{12.14}$$

and

$$k_a = \Theta_p \nu_a \exp(-\varepsilon_a / k_B T) , \tag{12.15}$$

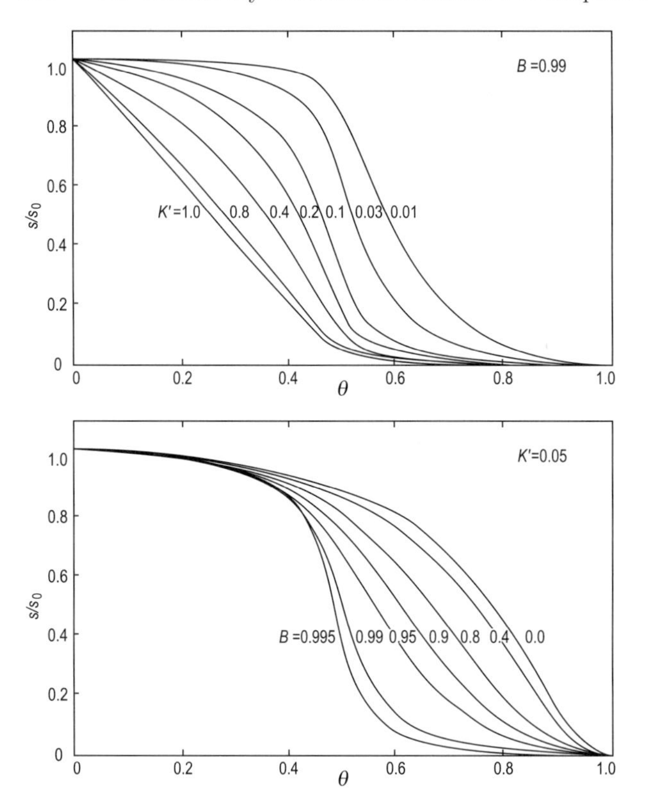

Fig. 12.5. Computed coverage dependent sticking probability for precursor-mediated chemisorption of interacting adsorbates. (**a**) Fixed degree of the short-range order parameter ($B = 0.99$) and various values of $K' = k'_\mathrm{d}/(k_\mathrm{a} + k_\mathrm{d})$. (**b**) Fixed $K' = 0.05$ and variable B (after King and Wells [12.4])

where ν_d and ν_a are the rate constants and Θ_p the coverage in the precursor state. Consequently, the initial sticking coefficient can be written as

$$s_0 = \frac{k_\mathrm{a}}{k_\mathrm{a} + k_\mathrm{d}} = \left[1 + \frac{\nu_\mathrm{d}}{\nu_\mathrm{a}} \exp\left(-\frac{\varepsilon_\mathrm{d} - \varepsilon_\mathrm{a}}{k_\mathrm{B}T}\right)\right]^{-1} . \tag{12.16}$$

One can see that s_0 exhibits different types of temperature dependence for various relative values of ε_d and ε_a:

- If $\varepsilon_\mathrm{d} < \varepsilon_\mathrm{a}$ (i.e., in the case of activated adsorption, see Fig. 12.8a), s_0 increases as the substrate temperature increases.
- If $\varepsilon_\mathrm{d} > \varepsilon_\mathrm{a}$ (i.e., in the case of non-activated adsorption, see Fig. 12.8b), s_0 decreases as the substrate temperature increases.

The value of $(\varepsilon_\mathrm{d} - \varepsilon_\mathrm{a})$ can be extracted from the slope of the experimental Arrhenius plot of $(1/s_0 - 1)$ versus $1/T$, as shown in Fig. 12.9 for N_2 adsorp-

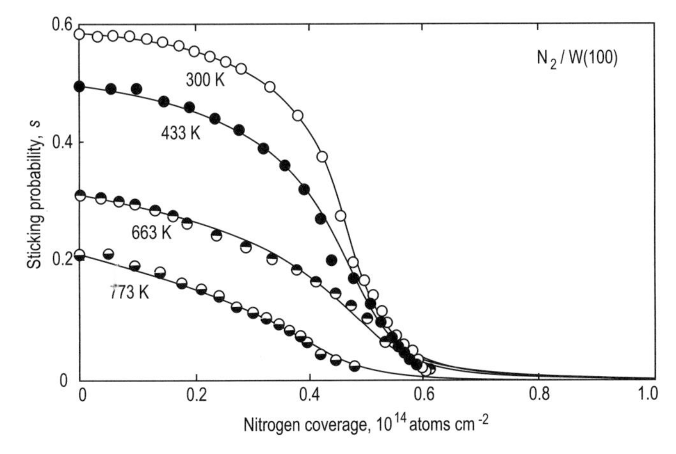

Fig. 12.6. Experimental coverage dependent sticking probability for N_2 on W(100) at 300 K, 433 K, 663 K, and 773 K compared with computer best-fit curves from (12.13) (after King and Wells [12.4])

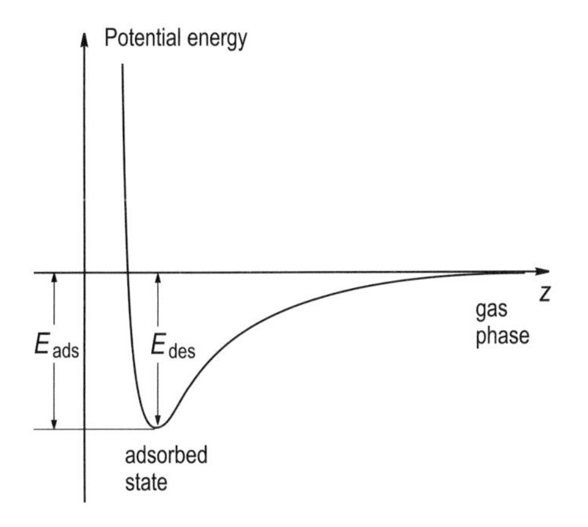

Fig. 12.7. Schematic of the chemisorption potential as a function of the distance of the adsorbed molecule or atom from the surface: the case of simple non-activated adsorption

tion on W(100). The dependence shows that the sticking coefficient decreases with temperature and the plot slope yields the value of $(\varepsilon_d - \varepsilon_a) = 0.19\,\text{eV}$.

Another example is presented in Fig. 12.10 which shows the data for dissociative H_2 and D_2 adsorption on Si(100). As here $s_0 \ll 1$, the plot of s_0 versus $1/k_B T$ is built. The sticking probability increases with temperature revealing activated adsorption with activation barrier $E_{act} = \varepsilon_a - \epsilon_d$ of 0.87 eV.

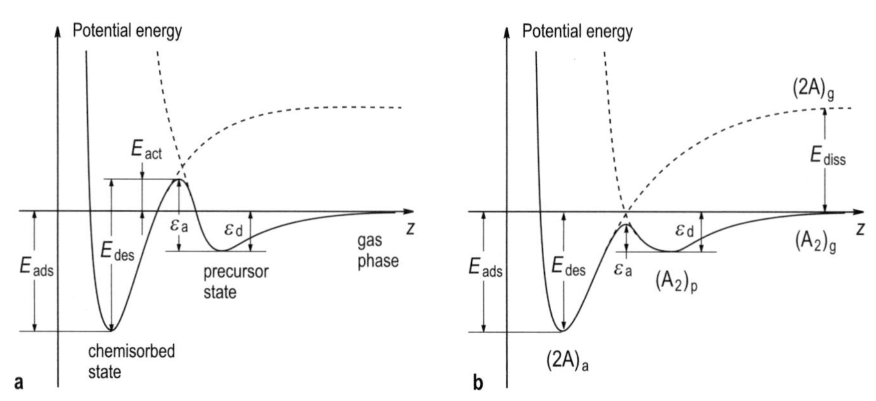

Fig. 12.8. Schematic of the potential energy curves for precursor-mediated chemisorption: (**a**) the case of activated chemisorption with an activation barrier of $E_{act} = \varepsilon_a - \varepsilon_d$; (**b**) the case of non-activated chemisorption with $\varepsilon_a < \varepsilon_d$. The pathway for precursor-mediated dissociative adsorption, $(A_2)_g \rightarrow (A_2)_p \rightarrow (2A)_a$ is shown in (**b**). E_{ads} is the binding energy in the chemisorbed state, E_{des} the barrier for desorption from the chemisorbed state, ε_a and ε_d denote respectively the barriers for adsorption and desorption from the precursor state well, $E_{act} = \varepsilon_a - \epsilon_d$ is the activation barrier for chemisorption, and E_{diss} is the dissociation energy of the molecule in the gas phase

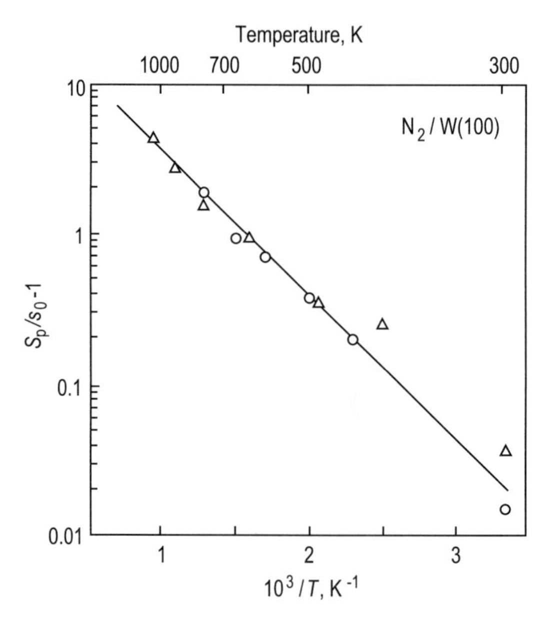

Fig. 12.9. Arrhenius plot of $(1/s_0 - 1)$ against $1/T$ for N_2 adsorption on W(100) (see Fig. 12.6) (after King and Wells [12.4])

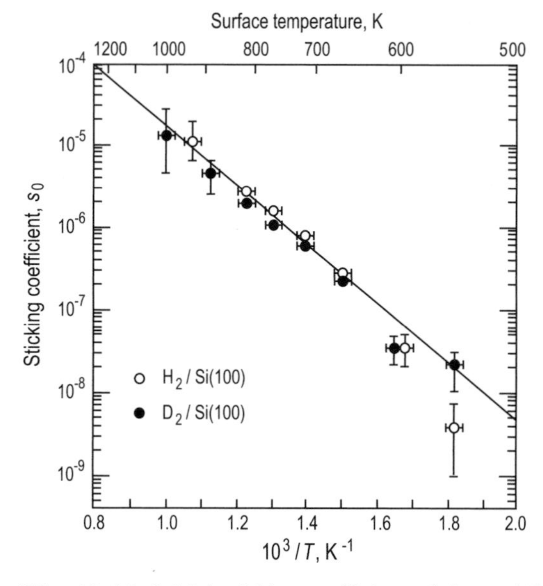

Fig. 12.10. Initial sticking coefficient of thermal H_2 and D_2 gases at $T_{gas} = 300$ K on a Si(100) surface as a function of its reciprocal temperature (after Bratu et al. [12.5])

12.1.3 Angular and Kinetic Energy Dependence

In the case of activated adsorption, the impinging molecules with sufficiently high kinetic energy can surmount the activation barrier directly, without trapping in the precursor state. Recalling that the potential of the molecule–surface interaction is a function of the coordinate z normal to the surface, one would expect that the normal component of molecular velocity is the most important for interaction. Hence, the sticking probability can be scaled with the quantity $E_\perp = E_i \cos^2 \vartheta_i$, the so-called "*normal kinetic energy.*" This behavior is illustrated by the experimental dependence of the sticking coefficient on the normal kinetic energy for the dissociative chemisorption of O_2 on W(110) (Fig. 12.11). The sticking coefficient is $\sim$40% at $E_\perp = 0.1$ eV and rises to essentially unity above $E_\perp = 0.4$ eV, thus indicating the order of magnitude of the activation barrier.

12.2 Thermal Desorption

12.2.1 Desorption Kinetics

The process in which adsorbate species gain enough energy from the thermal vibrations of surface atoms to escape from the adsorption well and leave the surface is called *thermal desorption.* In the kinetic approach, the desorption is

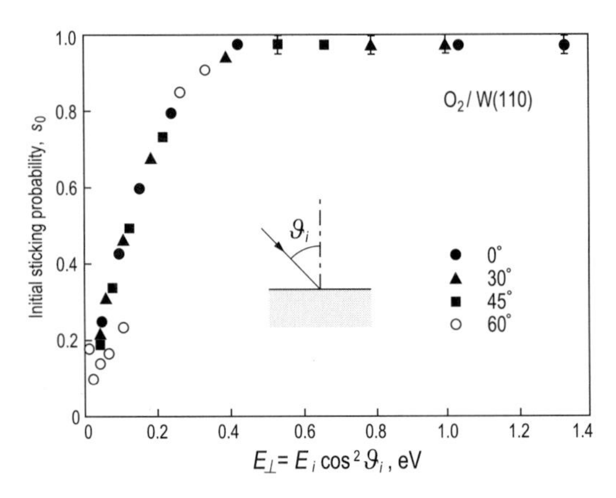

Fig. 12.11. Initial dissociative chemisorption probability of O_2 on W(110) at a surface temperature of 800 K as a function of the "normal kinetic energy" of the incident molecules for angles ϑ_i between 0° and 60° (after Rettner et al. [12.6])

described in terms of a *desorption rate*, r_{des}, which is the number of particles desorbing from unit surface area per unit time. In the most general form, the desorption rate can be written as

$$r_{des} = \sigma^* f^*(\Theta) \exp(-E_{des}/k_B T) , \tag{12.17}$$

where $f^*(\Theta)$ describes the coverage dependence and σ^* is the desorption coefficient standing for steric and mobility factors.

Polanyi–Wigner Equation. Under the assumption that all adsorbed atoms or molecules occupy identical sites and do not interact with each other, the desorption rate is expressed by the *Polanyi–Wigner equation*:

$$r_{des} = -d\Theta/dt = k_n \Theta^n = k_n^0 \Theta^n \exp(-E_{des}/k_B T) , \tag{12.18}$$

where E_{des} is the activation energy of desorption, n the order of the desorption kinetics, and k_n the desorption rate constant.

Kinetic Order. The kinetic order of desorption is given by the value of the exponent n in (12.18). The most general cases are illustrated in Fig. 12.12.

- In *zero-order* kinetics ($n = 0$), the desorption rate is coverage independent, i.e., constant at a given temperature. Zero-order kinetics occurs in the case of the quasi-equilibrium coexistence of a 2D diluted gas of adatoms and a 2D solid phase. It also takes place at the desorption of a homogeneous multilayer film.
- In *first-order* kinetics ($n = 1$), the desorption rate is proportional to Θ. It corresponds to the simplest case, when single atoms desorb directly and independently from their sites. The first-order rate constant k_1^0 is in units of frequency, s^{-1}. This frequency, called the *attempt frequency* ν_0, is roughly on the order of the atomic frequency of the crystal lattice ($\sim 10^{13}$ s^{-1}).

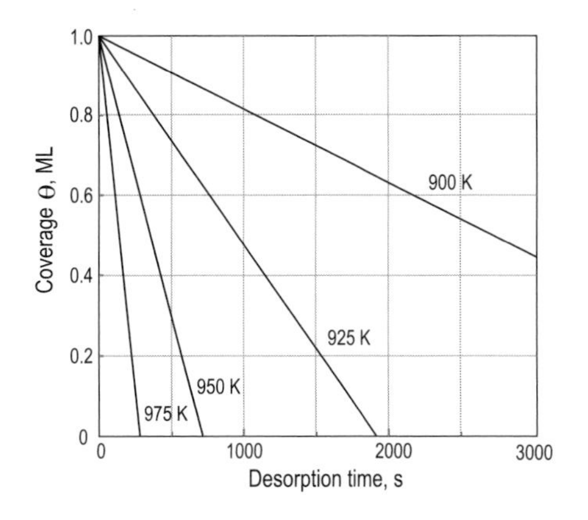

Zero-order $(n = 0)$

$$\Theta = \Theta_0 \left(1 - \frac{k_0 \cdot t}{\Theta_0}\right),$$

where

$$k_0 = k_0^0 \exp\left(\frac{-E_{\text{des}}}{k_B T}\right),$$

$$[k_0] = \left[\frac{\text{ML}}{\text{s}}\right].$$

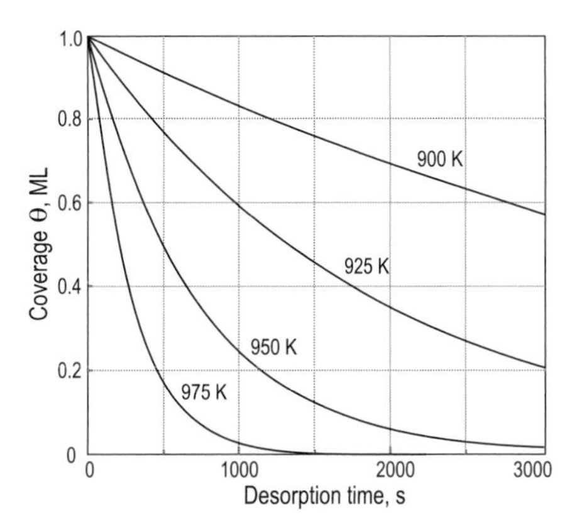

First-order $(n = 1)$

$$\Theta = \Theta_0 \exp(-k_1 \cdot t),$$

where

$$k_1 = k_1^0 \exp\left(\frac{-E_{\text{des}}}{k_B T}\right),$$

$$[k_1] = \left[\frac{1}{\text{s}}\right].$$

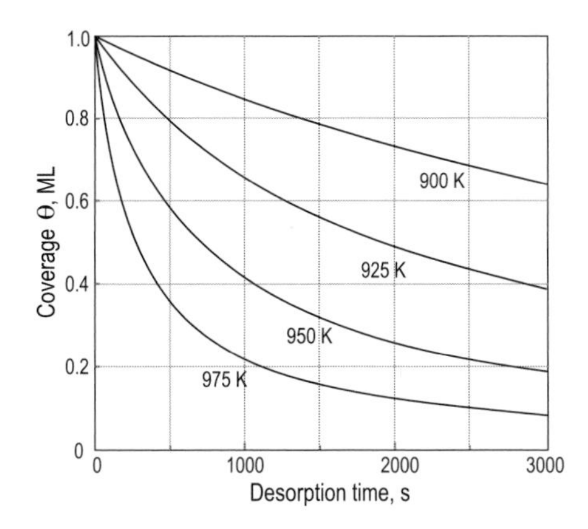

Second-order $(n = 2)$

$$\Theta = \frac{\Theta_0}{1 + k_2 \Theta_0 \cdot t},$$

where

$$k_2 = k_2^0 \exp\left(\frac{-E_{\text{des}}}{k_B T}\right),$$

$$[k_2] = \left[\frac{1}{\text{ML} \cdot \text{s}}\right].$$

Fig. 12.12. Ideal time dependence of adsorbate surface coverage calculated for zero-, first-, and second-order desorption kinetics and various temperatures. All rate constants k_n are assumed to be 10^{13}, the activation energy for desorption is $E_{\text{des}} = 3\,\text{eV}$, and $\Theta_0 = 1\,\text{ML}$.

Often the reciprocal value $\tau = 1/\nu$ is in use, called the *mean stay time for adsorption* or *mean surface lifetime.*
- In *second-order* kinetics ($n = 2$), the desorption rate is proportional to Θ^2. This is the case of *associative molecular desorption*, where the desorbing molecule originates from two radicals residing initially at separate sites.

It should be noted that, in reality, the complicated kinetics might result in other order exponents of desorption, including fractional ones.

Desorption Energy. In order to leave the surface, the adsorbate species has to surmount the activation barrier for desorption, called the *desorption energy* E_{des}. One can see in Fig. 12.8 that

- In the case of *activated* chemisorption (Fig. 12.8a), the desorption energy is the sum of the binding energy in the chemisorbed state and the activation energy for adsorption, $E_{\mathrm{des}} = E_{\mathrm{ads}} + E_{\mathrm{act}}$.
- In the case of *non-activated* chemisorption (Fig. 12.8b), the desorption energy is simply the binding energy in the chemisorbed state, $E_{\mathrm{des}} = E_{\mathrm{ads}}$.

Note that, in the general case, the desorption energy may depend on the adsorbate coverage, leading to relationships for desorption rates more complicated than the relatively simple Polanyi–Wigner equation. For example, one would expect a change of the desorption energy when a phase transition takes place in the course of desorption.

Angular and Kinetic Energy Dependence. Ideally under the condition of thermal equilibrium, the differential flux of desorbing molecule is given by

$$\mathrm{d}r_{\mathrm{des}} = n \left(\frac{m}{2\pi k_{\mathrm{B}} T} \right)^{3/2} \cdot \exp\left(-\frac{mv^2}{2k_{\mathrm{B}}T} \right) v^3 \cos\vartheta \, \mathrm{d}v \, . \tag{12.19}$$

Thus, the ideal angular distribution of the desorption rate is a cosine law and the velocity distribution is Maxwellian. However, in reality, the deviations from (12.19) might be substantial. In the case of activated adsorption, the desorbing flux is more peaked in the normal direction as compared to the cosine distribution and the mean energy of the desorbing molecules is higher than one would expect from (12.19) for the substrate temperature.

Desorption of D_2 molecules from a Cu(100) surface furnishes a vivid example of such behavior. Figure 12.13 shows that the angular distribution of the D_2 flux follows $\sim \cos^8\vartheta$. Note that, in general, non-cosine distributions can be described phenomenologically by $\cos^n\vartheta$ with n commonly in the range from 1 to 10. Figure 12.14 demonstrates that the energy distribution of desorbing D_2 molecules corresponds to a much higher mean energy and is much narrower than the Maxwell distribution.

12.2.2 Thermal Desorption Spectroscopy

For the determination of the order of the kinetics and desorption energies two main experimental methods are employed, namely,

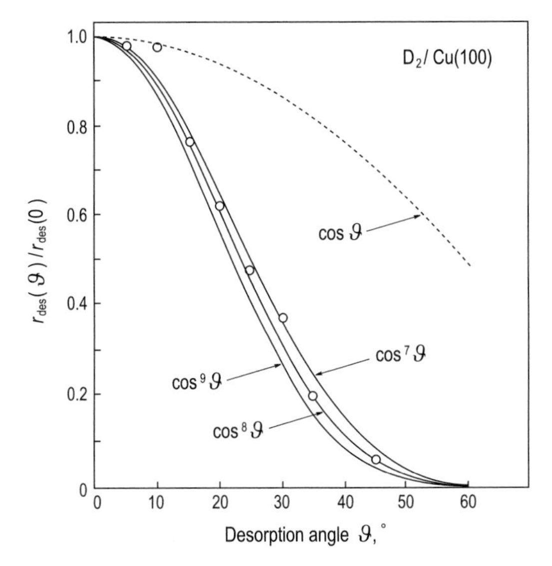

Fig. 12.13. Angular distribution of the flux of D_2 molecules desorbing from a Cu(100) surface heated to 1000 K. The experimental values follow the $\sim \cos^8 \vartheta$ law, but not the $\cos \vartheta$ of the Maxwell distribution (dashed line) (after Comsa and David [12.7])

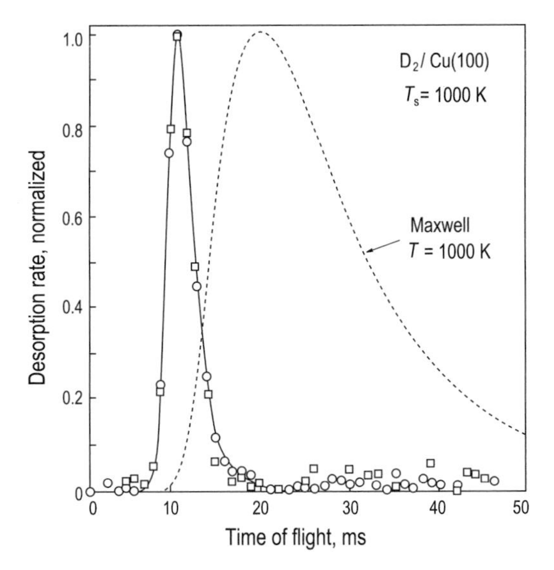

Fig. 12.14. Time-of-flight (TOF) spectra of D_2 molecules desorbed from a Cu(100) surface heated to 1000 K. The dashed curve represents the calculated Maxwell spectrum for 1000 K. In the TOF spectra, a shorter flight time corresponds to higher energy (after Comsa and David [12.8])

- the isothermal method;
- the temperature-programmed method.

Isothermal Method. In the isothermal method (which is also called *isothermal desorption spectroscopy (ITDS)*), the sample temperature is raised rapidly and then maintained at a fixed value T_s throughout the experiment, i.e., the desorption proceeds under *isothermal* conditions. In the experiment, the desorption rate or adsorbate coverage is measured as a function of time. The shape of the recorded dependences allows the evaluation of the kinetics order n and the value of the rate constant k_n (Fig. 12.12). Having the set of dependences measured at different temperatures, one can obtain the value of the desorption energy E_{des} from the slope of the Arrhenius plot, $\ln k_n$ versus $1/T_s$.

As an example, the results of the experimental study of thermal desorption in the Ag/Si(111) system are presented in Fig. 12.15 and Fig. 12.16. Figure 12.15 shows the kinetic dependences of Θ_{Ag} versus time, which are approximated by a set of zero-order kinetics. The original surface comprises the Si(111)$\sqrt{3}\times\sqrt{3}$-Ag surface phase of 1 ML Ag coverage with 3D Ag islands on it. Desorption takes place first from Ag islands and, when they are exhausted, from the $\sqrt{3}\times\sqrt{3}$-Ag phase. Decreasing the Ag coverage results in the appearance of domains of the Si(111)3×1-Ag surface phase of 1/3 ML Ag coverage. Upon further annealing, desorption from 3×1-Ag takes place, until the Ag-free Si(111)7×7 surface is obtained. Zero-order kinetics indicate that Ag desorption takes place from the 2D gas of Ag adatoms detached from the edges of the "solid" phase (i.e., Ag islands or a certain Ag/Si(111) surface phase). Each of the cases, Ag islands, $\sqrt{3}\times\sqrt{3}$-Ag phase, and 3×1-Ag phase, is characterized by the individual desorption rate constant. The Arrhenius plot of these rate constants (Fig. 12.16) reveals the following desorption energies: 1.47 eV for Ag 3D islands, 2.78 eV for $\sqrt{3}\times\sqrt{3}$-Ag, and 2.99 eV for 3×1-Ag.

Temperature-Programmed Method. In this technique, the temperature of the adsorbate-covered sample is monotonically increased according to the desired program and the increase of pressure induced by adsorbate desorption is simultaneously recorded as a function of temperature. The general relationship between the adsorbate desorption rate and its partial pressure in the UHV system is described by the pumping equation (3.11):

$$-A\frac{d\Theta}{dt} = \frac{V}{k_B T}\left(\frac{dp}{dt} + \frac{S}{V}p\right) , \qquad (12.20)$$

where A is the area of the sample surface, V the volume of the UHV chamber, p the partial pressure of the adsorbate, and S the pumping speed.

Depending on the relative values of the desorption rate and pumping speed, there are two possible regimes of data acquisition:

- If the pumping speed is extremely low, the rate of pressure increase is proportional to the desorption rate ($dp/dt \propto d\Theta/dt$) as seen from (12.20).

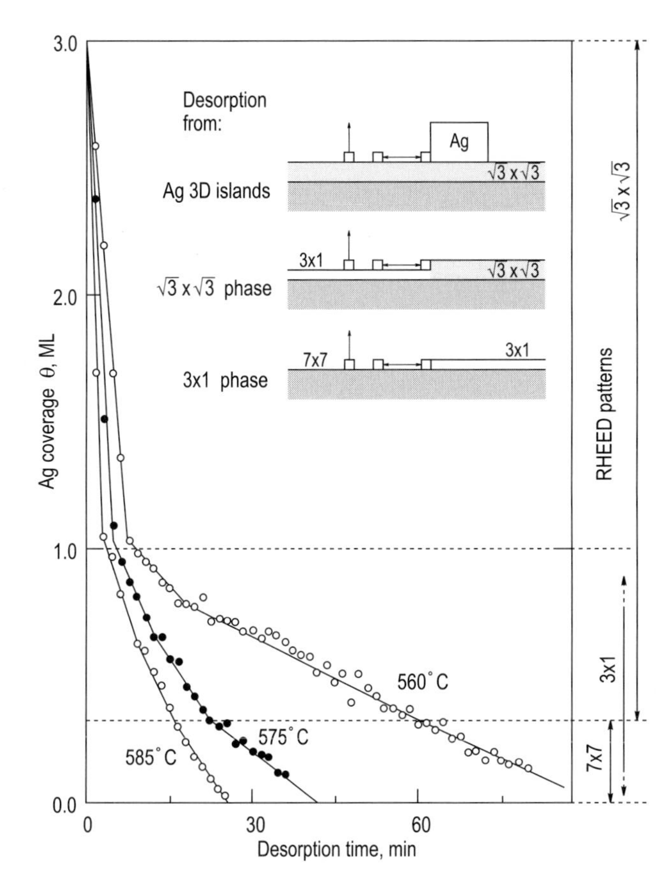

Fig. 12.15. Variation of the Ag coverage on Si(111) during isothermal desorption conducted at 560°C, 575°C, and 585°C. The desorption is represented as a set of zero-order kinetics with rate constants depending on the surface structure. The RHEED patterns that are observed simultaneously are shown in the right panel. The inset illustrates the desorption mechanism which accounts for zero-order kinetics (after Hasegawa et al. [12.9])

This regime is referred to as *flash desorption* and is rarely employed in practice. A possible application is the estimation of the overall amount of adsorbate.

- If the pumping speed is extremely high, the pressure is proportional to the desorption rate ($p \propto \mathrm{d}\Theta/\mathrm{d}t$). This is the most commonly used regime and the terms *temperature programmed desorption (TPD)* and *thermal desorption spectroscopy (TDS)* conventionally refer to this particular technique. The following discussion will concern exclusively this TPD mode.

In most TPD experimental set-ups, a linear temperature ramp is used, i.e.,

$$T(t) = T_0 + \beta t \,, \tag{12.21}$$

where t is time and β is the heating rate, typically on the order of 1–10 K/s.

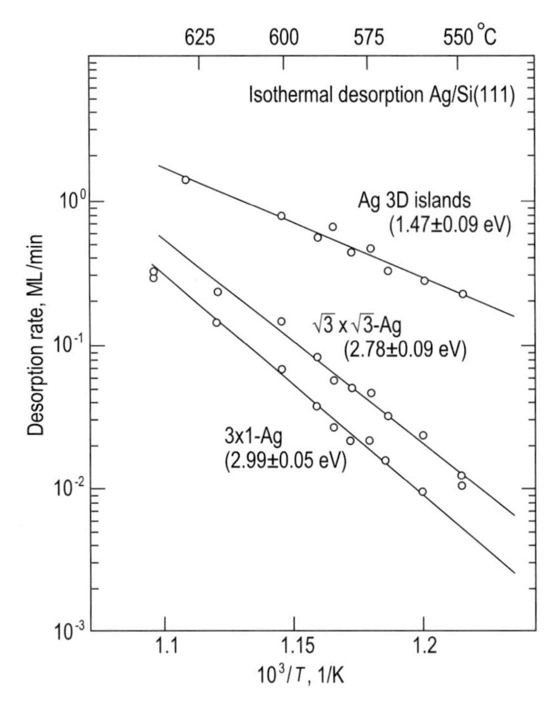

Fig. 12.16. Arrhenius plots of the desorption rates of Ag atoms from each phase in the Ag/Si(111) system. The desorption rates are obtained from the slope of the corresponding segment in the ITDS curves in Fig. 12.15 (after Hasegawa et al. [12.9])

From (12.18) and (12.21), one has

$$p \propto -\frac{\mathrm{d}\Theta}{\mathrm{d}t} = \frac{k_n^0 \Theta^n}{\beta} \exp(-E_{\mathrm{des}}/k_{\mathrm{B}}T) . \tag{12.22}$$

The general shape of the $p(T)$ dependence described by (12.22) can be visualized as follows. At low temperatures, the exponential term is negligible, hence the negligible desorption rate. At sufficiently high temperatures, the desorption rate increases rapidly following the growth of the exponential term. However, the inevitable decrease in adsorbate coverage in the course of desorption slows down the desorption rate until it reaches zero, when all adsorbate is exhausted. As a result, the $p(T)$ dependence displays a peak at a certain characteristic temperature T_{m}.

In general, the peak temperature, T_{m}, is related to the desorption energy, kinetics order, initial adsorbate coverage, etc. For the case of first-order kinetics and assuming E_{des} and ν_1 to be coverage independent, Redhead [12.10] established an approximate relationship between E_{des} and T_{m}:

$$E_{\mathrm{des}} = k_{\mathrm{B}} T_{\mathrm{m}} \left(\ln \frac{\nu_1 T_{\mathrm{m}}}{\beta} - 3.64 \right) . \qquad (12.23)$$

The plot of (12.23) for $\nu_1 = 10^{13}\ \mathrm{s}^{-1}$ and β =1, 10, 100, and 1000 K/s is shown in Fig. 12.17. It gives a simple method to estimate E_{des} from TPD data. However, one should bear in mind that the result depends on a guess at ν_1.

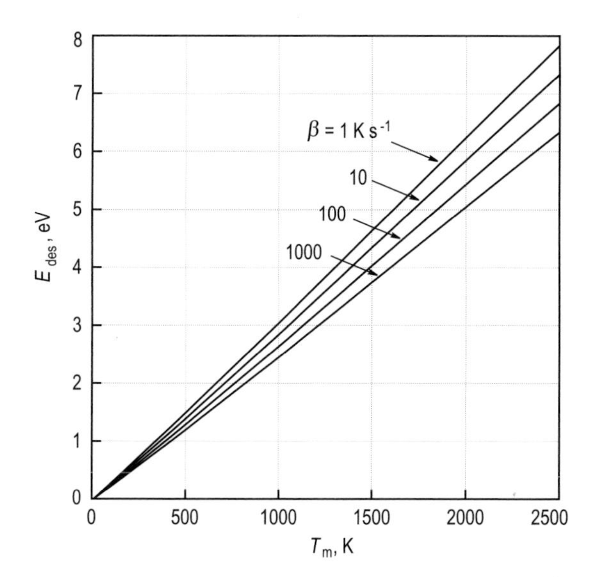

Fig. 12.17. Desorption energy E_{des} as a function of the TPD peak temperature T_{m} for first-order desorption kinetics and a linear temperature ramp, $T(t) = T_0 + \beta t$, taking $\nu_1 = 10^{13}\ \mathrm{s}^{-1}$ (after Redhead [12.10])

The shape of the TPD curves as a function of the initial coverage of adsorbate contains information on the order of the desorption kinetics. As an example, Fig. 12.18 shows a set of ideal TPD spectra calculated for zero-, first-, and second-order desorption kinetics. One can see the following features characteristic of a given order of kinetics:

- For *zero-order* kinetics ($n = 0$), the curves for all initial coverages have a common leading edge and rapid drop beyond T_{m}; the peak temperature T_{m} moves to higher temperatures with increasing initial coverage Θ_0.
- For *first-order* kinetics ($n = 1$), the peak has a characteristic asymmetric shape; the peak temperature T_{m} remains constant with increasing Θ_0.
- For *second-order* kinetics ($n = 2$), the peak is of nearly symmetric shape; the peak temperature T_{m} moves to lower temperatures with increasing Θ_0.

Experimental TPD spectra of Au from W(110) are shown in Fig. 12.19. With initial Au coverages below ~1 ML, a single peak at about 1430 K is

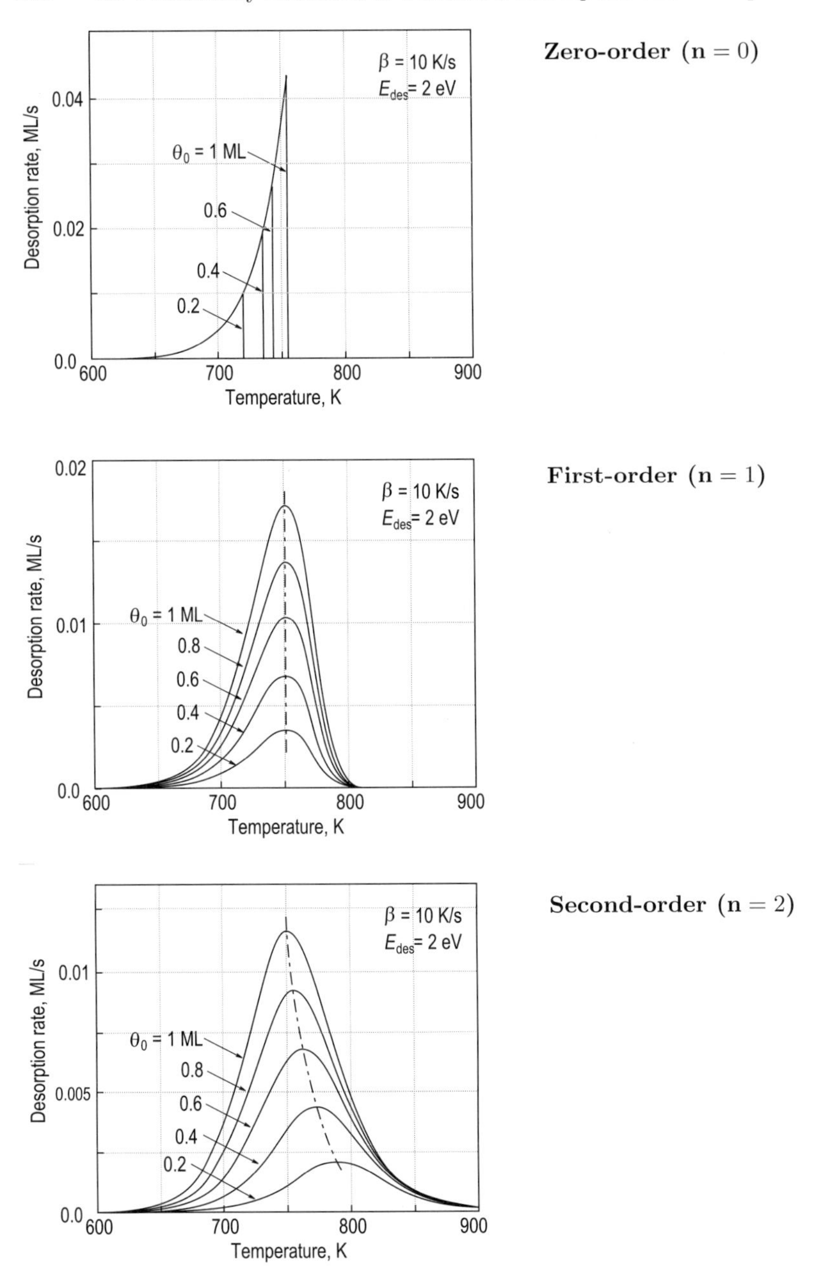

Fig. 12.18. TPD spectra calculated for zero-order, first-order, and second-order desorption kinetics and various initial coverages Θ_0. Both the pre-exponential factor and activation energy for desorption are assumed to be independent of the surface coverage Θ. The heating rate is assumed to be linear and of magnitude β. Note the different scales for different desorption orders. Only the integrated area under the peaks is proportional to the initial coverage Θ_0

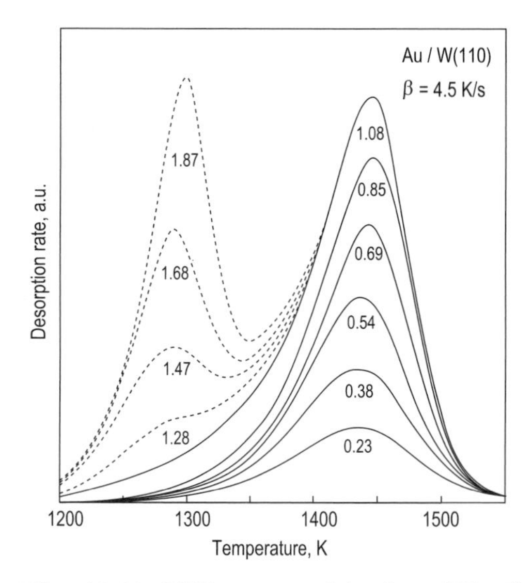

Fig. 12.19. TPD spectra of Au from W(110). The high-temperature peak (solid line) is due to the first Au monolayer, the low-temperature peak (dashed line) is due to the second Au monolayer. Parameters indicate the initial Au coverage in monolayers (after Bauer et al. [12.11])

observed (shown as a solid line in Fig. 12.19). When the Au coverage exceeds ~1 ML, the second peak appears at about 1300 K (shown as a dashed line in Fig. 12.19). Thus, the high-temperature peak is due to the first-monolayer Au atoms bonded to the W(110) substrate, while the low-temperature peak is due to the second-monolayer Au atoms residing above the first-monolayer Au. The second-monolayer peak indicates zero-order kinetics, while the first-monolayer peak is consistent with first-order kinetics with constant E_{des} and ν_1 in the coverage range from 0.2 ML to 0.8 ML with some deviations both at lower and higher coverages [12.11].

12.3 Adsorption Isotherms

Under conditions of thermodynamic equilibrium between an adsorbate layer and a gas phase, the rate of adsorption is equal to the desorption rate,

$$r_{\mathrm{ads}} = r_{\mathrm{des}} \,. \tag{12.24}$$

From (12.24) and taking into account (12.1), (12.3) and (12.17), one can write

$$p = \frac{1}{K} \frac{f^*(\Theta)}{f(\Theta)} \,, \tag{12.25}$$

where

$$K = \frac{\sigma}{\sigma^*} \frac{\exp(E_{\text{ads}}/k_B T)}{\sqrt{2\pi m k_B T}} , \tag{12.26}$$

which is essentially the ratio of the adsorption and desorption rate constants, $k_{\text{ads}}/k_{\text{des}}$, as $k_{\text{ads}} \propto 1/\sqrt{2\pi m k_B T}$ and $k_{\text{des}} \propto \exp(-E_{\text{ads}}/k_B T)$.

From (12.25) with an appropriate guess about the character of the $f^*(\Theta)$ and $f(\Theta)$ dependencies, one can establish a relationship between the equilibrium adsorbate coverage and gas pressure at constant temperature, i.e., evaluate the *adsorption isotherm*.

Henry's Law. In the simplest case, when all atoms adsorb independently of one another and all adsorption sites are equivalent, $f(\Theta) = 1$ and $f^*(\Theta) = \Theta$, and one readily has an isotherm of the form

$$\Theta(p) = Kp , \tag{12.27}$$

which is known as *Henry's law*, i.e., the equilibrium coverage of adsorbate is proportional to the gas pressure. Henry's law holds for many real systems at low pressures or in the initial stages of adsorption (i.e., at low coverages).

Langmuir Isotherm. For first-order Langmuir adsorption (see Sect. 12.1.1), $f(\Theta) = 1 - \Theta$ and $f^*(\Theta) = \Theta$ and thus one obtains the *Langmuir adsorption isotherm* as follows

$$\Theta(p) = \frac{Kp}{1 + Kp} . \tag{12.28}$$

The shape of the Langmuir-type isotherms for various values of K is illustrated by Fig. 12.20. As one can see, the greater the K value, the steeper the $\Theta(p)$ dependence. For the same adsorbate/substrate system, a greater K means a lower substrate temperature. For comparison of the different systems, a greater K means a greater adsorption energy E_{ads}, i.e., stronger adsorption. At large pressures, $\Theta \to 1$, i.e., a completely filled monolayer. At low pressures, $\Theta \to Kp$, i.e., a Langmuir-type isotherm can be approximated by Henry's law.

2D Condensation. To improve the Langmuir model one can take into account the interaction between adsorbed particles. The adsorption isotherm for the 2D gas of adsorbate with an attractive interaction is described by the relation known as the *Hill–DeBoer equation*

$$p(\Theta) = \frac{1}{K}\left(\frac{\Theta}{1-\Theta}\right) \exp\left[\left(\frac{\Theta}{1-\Theta}\right) - \left(\frac{2a\Theta}{k_B T b}\right)\right] , \tag{12.29}$$

where a characterizes the pair interaction energy between nearest-neighbour adsorbate particles and b is the minimal area occupied by a particle. As one can see, in addition to the Langmuir-type relation, (12.29) includes two new exponential terms. The first is responsible for the effects of the excluded area and the second for the energetics of lateral interaction.

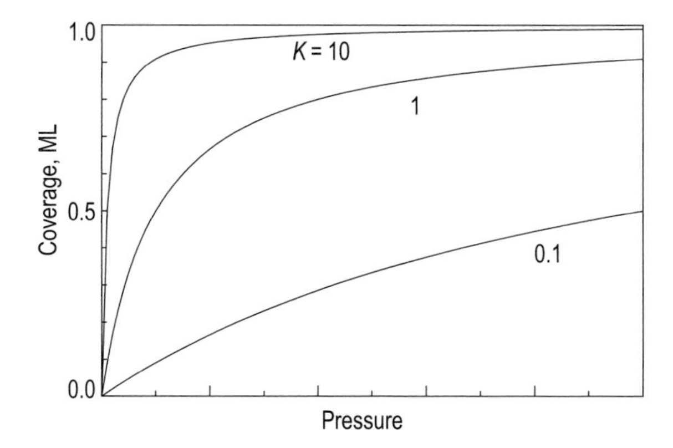

Fig. 12.20. Langmuir-type isotherms calculated for various K

Plots of the Hill–DeBoer isotherms calculated for a model system with $E_{\mathrm{ads}} = 200\,\mathrm{meV}$ and $2a/b = 60\,\mathrm{meV}$ are shown in Fig. 12.21. At high temperatures and low coverages, the isotherms are essentially of Langmuir type. However, below a critical temperature, the curves become double-valued, which means system instability, leading to a first-order transition from the 2D gas to the 2D condensed phase, i.e., *2D condensation.* The 2D condensed phase might be an ordered 2D crystal or a 2D liquid (the latter is a dense 2D phase without internal ordering). The region of phase coexistence, where the 2D condensed phase is in equilibrium with the 2D gas, is shown by the shaded area in the $\Theta - p$ diagram in Fig. 12.21. At higher coverages, only the 2D condensed phase is present. For temperatures above the critical point, there is no real phase separation between the 2D condensed phase and the 2D gas. Note that the phenomenon of 2D condensation is much like the condensation of the van der Waals gas in the 3D case.

Multilayer Adsorption. If the sticking probability remains non-zero after completion of the first monolayer, multilayer adsorption occurs. The case of multilayer adsorption was considered explicitly by Brunauer, Emmett, and Teller (BET), who derived the isotherm as follows:

$$\Theta(p) = \frac{K_{\mathrm{BET}}\, p\, p_0}{(p - p_0)[p_0 + K_{\mathrm{BET}}\,(p - p_0)]}\,, \tag{12.30}$$

where p_0 is the equilibrium vapor pressure of the bulk phase of adsorbate and K_{BET} characterizes the ratio of the mean lifetimes of the adsorbate species in the first and other layers. The BET model assumes that all layers beyond the first are characterized by a constant adsorption energy. The interaction between atoms of the same layer is not taken into account. The shapes of the BET isotherms for various K_{BET} is illustrated by Fig. 12.22.

Under the assumption of similar adsorption energetics for all layers beyond the first, one would expect growth of the next layer before the completion of the previous one, i.e., the formation of a rough surface. In practice,

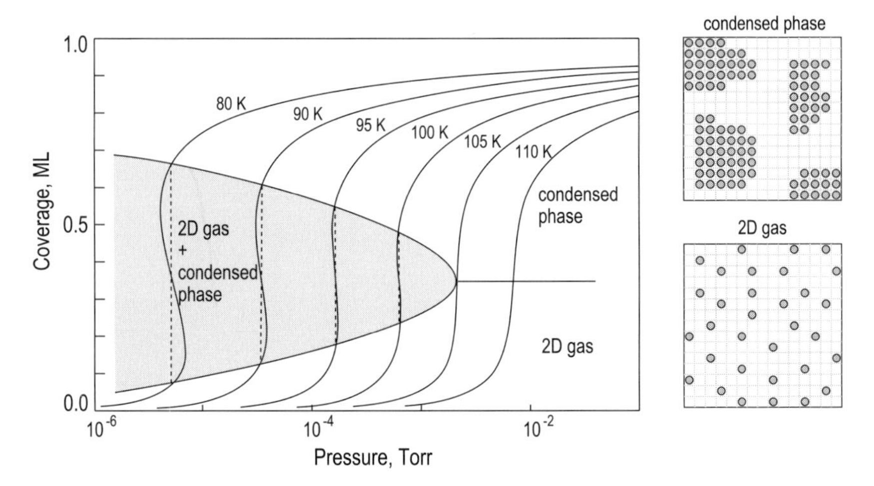

Fig. 12.21. Hill-DeBoer isotherms calculated from (12.29) for a model system with $E_{\mathrm{ads}} = 200\,\mathrm{meV}$ and $2a/b = 60\,\mathrm{meV}$, showing 2D condensation. The sketches in the right panel illustrate schematically the condensed 2D phase and diluted 2D gas

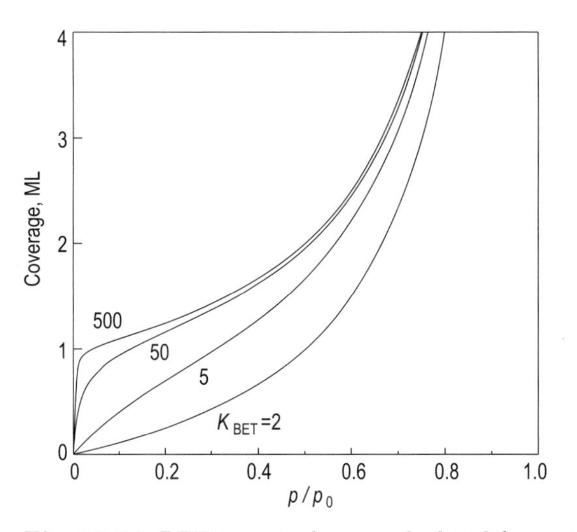

Fig. 12.22. BET-type isotherms calculated for various K_{BET}

multilayer adsorption often proceeds in a layer-by-layer fashion as illustrated by the experimental results for Kr adsorption on graphite (see Fig. 12.23). The layer-by-layer growth implies that the adsorption energy varies from layer to layer. An adequate description of the respective isotherm requires a more complicated formula than the relatively simple BET isotherm.

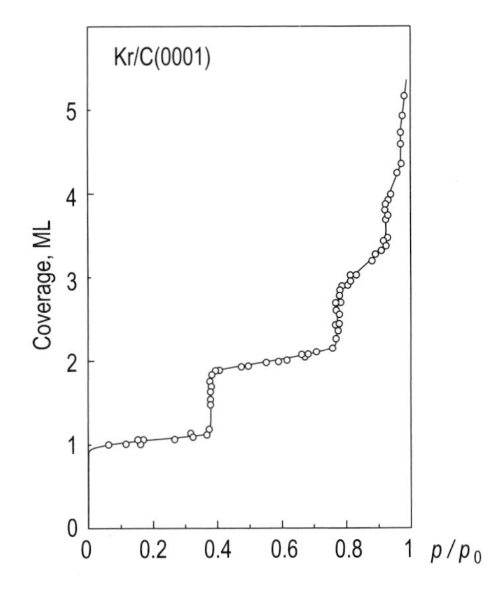

Fig. 12.23. Krypton adsorption isotherm at 77.3 K on graphite (after Thomy et al. [12.12])

12.4 Non-Thermal Desorption

There is set of desorption phenomena in which temperature is not of prime importance. The most important ones are as follows.

- electron-stimulated desorption;
- photodesorption;
- ion impact desorption;
- field desorption.

Electron-Stimulated Desorption (ESD). Typically the electron energy used in ESD is below ~500 eV. The kinetic energy transferred to the atom or molecule upon collision with an electron is rather small. For example, it is only ~0.2 eV even for light H atoms bombarded by 100 eV electrons (for an estimation, see (6.5)). Hence, desorption due to direct momentum transfer is possible for physisorbed species, but is unable to produce a noticeable effect on chemisorbed species. More effective are ESD processes involving excitation of the electronic system of the adsorbate. Two main mechanisms of ESD are known. According to the names of the founders, the first one is called the Menzel–Gomer–Redhead mechanism and the second one the Knotek–Feibelman mechanism.

The *Menzel–Gomer–Redhead mechanism* is illustrated schematically in Fig. 12.24 which shows the potential curves for the adsorbed species in its ground state (lower curve) and in the ionized (or excited) state (upper curve).

At large distances from the surface, the curves are separated by the ionization (excitation) energy of the free species. The electron-induced ionization (excitation) of the adspecies means a transition from the lower curve to the upper curve, as indicated by the vertical arrow in Fig. 12.24a. The transition is "vertical," since during a fast electronic transition the location of the adsorbate species remains essentially unchanged. As one can see, if the transition takes part into the repulsive part of the upper curve, the species can desorb as an ion or excited atom (or molecule) with kinetic energy from the range indicated in Fig. 12.24a.

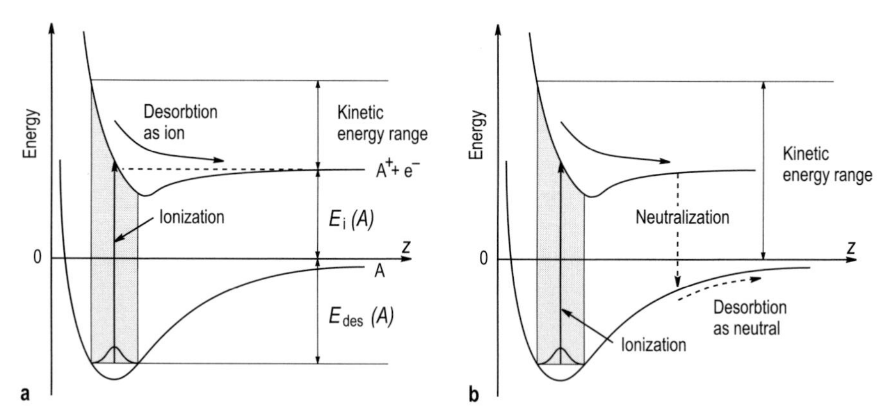

Fig. 12.24. Potential energy diagrams illustrating the processes involved in the electron-stimulated desorption of adspecies (**a**) in the form of an ion and (**b**) as a neutral species, in accordance with the Redhead–Menzel–Gomer model

However, on its way from the surface, the ion has finite probability of being neutralized again (consequently, the excited species has finite probability of being deexcited). For possible mechanisms of ion neutralization, see Sect. 6.1.6. As a result, the adspecies returns to the ground state energy curve but with the excess kinetic energy accumulated at the time the adspecies has been in the ionized (excited) state (see Fig. 12.24b). If the kinetic energy is sufficient, the adspecies desorbs as a non-excited neutral with the kinetic energy indicated in Fig. 12.24b.

The *Knotek–Feibelman mechanism* refers to the electron-induced decomposition of ionic crystals with preferential desorption of anions. Figure 12.25 illustrates this mechanism for TiO_2 oxide. As a first step, a hole is created by electron impact at the metal core level. In an ionic crystal, there are no valence electrons on the metal atoms and the relaxation of the metal core hole involves *interatomic Auger process*, i.e., one electron from the oxygen atom fills the hole and another is emitted as an Auger electron. Due to the loss of two electrons, the oxygen atom becomes neutral. In the double Auger process, the oxygen atom becomes a O^+ ion surrounded by positively charged metal atoms and will readily desorb from the surface.

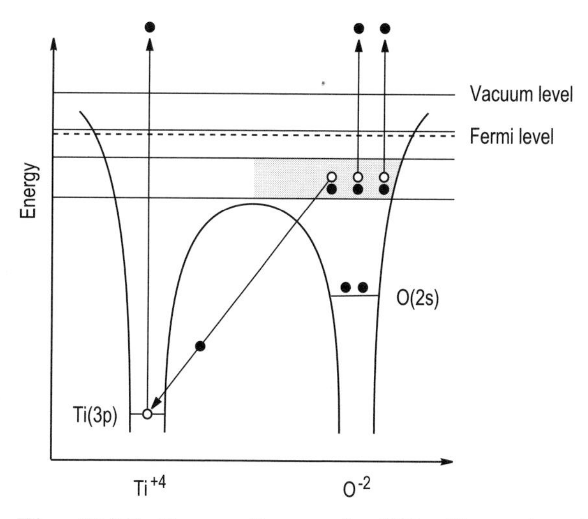

Fig. 12.25. Energy diagram for TiO_2 illustrating the Knotek–Feibelman ESD mechanism. A hole created by electron impact in the Ti(3*p*) level is filled via interatomic Auger decay from a neighboring oxygen site. This leaves the oxygen atom positively charged, causing it to be coulombically repelled from its lattice site. Electrons are shown by filled circles, holes by open circles (after Ramsier and Yates [12.13])

The angular distribution of the ions and neutrals produced in the ESD process reflects the bonding geometry of the adsorbed species, as they are desorbed approximately along the bonds ruptured by excitation. This peculiarity of ESD is utilized in an experimental technique called *electron-stimulated desorption ion angular distribution (ESDIAD).* An example of the application of ESDIAD application for studying the adsorption geometry of Cl on Si(100)2×1 [12.13] is shown in Fig. 12.26. ESDIAD observations indicate that upon low-temperature adsorption Cl atoms reside in the bridge positions producing a single-beam ESDIAD pattern. Upon annealing, the single-beam ESDIAD pattern transforms irreversibly to the four-beam pattern with a twofold azimuthal symmetry corresponding to Si dimer orientations in two domains. The change in the ESDIAD pattern reflects the formation of energetically stable Si-Cl bonds inclined $25°\pm4°$ from the surface normal.

Photodesorption (PD). Electronic excitations can be induced by photons of appropriate energy and the process of photodesorption is basically similar to ESD. However the cross-section of ionization is generally less for excitation by photons than by electrons. While using intense photon beams, one cannot avoid sample heating and it is often difficult to make a clear distinction between PD and conventional thermal desorption.

Ion Impact Desorption. Energetic ions (for example, 100 eV Ar^+) upon collision with a sample transfer their kinetic energy to the surf

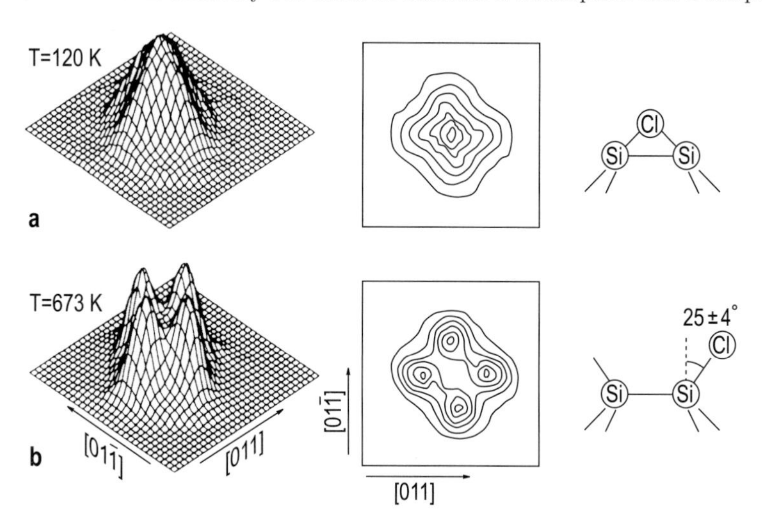

Fig. 12.26. ESDIAD patterns (perspective plots and contour plots) of Cl on Si(100) acquired (**a**) after dissociative chemisorption of Cl_2 at 120 K and (**b**) after subsequent annealing to 673 K for 60 s. The sketches show the corresponding bonding configurations of Cl on Si(100): (**a**) the metastable bridge-bonded structure and (**b**) the stable inclined bonding structure (after Cheng et al. [12.13])

causing their desorption. These processes referred to as *ion recoil* and *ion sputtering* are discussed in detail in Sects. 6.1.1 and 6.1.5, respectively.

Field Desorption. High electric fields on the order of 10^8 V/cm can also induced desorption of species from the sample surface. Such a process is used for surface imaging in the field-ion microscope (see Sect. 7.2) and for some atomic manipulations in STM (see Sect. 15.2).

Problems

12.1 Calculate the initial sticking probability of oxygen, if O_2 gas at 10^{-7} Torr is dissociatively adsorbed on a Ni(100) surface at an initial rate of 0.045 ML/s. The temperature is 300 K. Ni is a fcc crystal with a lattice constant of 3.52 Å.

12.2 A diatomic gas is chemisorbed dissociatively on a metal surface via a physisorbed precursor state. The activation energies for desorption from the chemisorbed and precursor physisorbed states are 1.3 eV and 0.2 eV, respectively. The barrier between the chemisorbed and physisorbed states is 0.1 eV. Estimate the equilibrium coverage in both adsorption states if $p = 10^{-4}$ Torr and $T = 400$ K. Assume that all rate constants are 10^{13} Hz and $n_0 = 10^{15}$ cm^{-2}.

12.3 The lifetime for which an aluminum atom remains adsorbed on a Si(111) surface is 30 s at 850°C and 1000 s at 755°C. Find the activation energy for Al desorption.

Further Reading

1. J.B. Hudson: *Surface Science: An Introduction* (John Wiley, New York 1994) Part II (gas–surface interactions, adsorption kinetics, gas scattering, physisorption, chemisorption)
2. S.J. Lombardo, A.T. Bell: *A Review of Theoretical Models of Adsorption, Diffusion, Desorption, and Reaction of Gases on Metal Surfaces.* Surf. Sci. Rep. **13**, 1–72 (1991)
3. D.A. King, M.G. Wells: *Reaction Mechanism in Chemisorption Kinetics: Nitrogen on the {100} Plane of Tungsten.* Proc. Roy. Soc. London **A339**, 245–269 (1974)
4. G. Comsa, R. David: *Dynamical Parameters of Desorbing Molecules.* Surf. Sci. Rep. **5**, 145–198 (1985)
5. R.D. Ramsier, J.T. Yates Jr.: *Electron-Stimulated Desorption: Principles and Applications.* Surf. Sci. Rep. **12**, 243–378 (1991)

13. Elementary Processes at Surfaces II. Surface Diffusion

Surface diffusion is the motion of adparticles, such as atoms or molecules, over the surface of a solid substrate. The diffusing particles might be the same chemical species as the substrate (the case referred to as a *self-diffusion*) or another one (the case of *heterodiffusion*). In most cases, the adparticle becomes mobile due to thermal activation and its motion is described as a random walk. In the presence of a concentration gradient (in the more general case, of the gradient of the chemical potential), the random walk motion of many particles results in their net diffusion motion in the direction opposite to the direction of the gradient. The diffusion process is affected by many factors, such as interaction between diffusing adspecies, formation of surface phases, presence of defects, etc.

13.1 Basic Equations

13.1.1 Random-Walk Motion

Consider the thermal motion of an adatom on an ideal crystal surface (Fig. 13.1). On the atomic scale, the surface comprises a periodic array of adsorption sites, which correspond to the positions of minimum energy (position 1 in Fig. 13.1a). Due to thermal excitations, an adatom can hop from one adsorption site to the next. The adatom motion along the surface can be visualized as a random site-to-site hopping process (random-walk motion), for which the mean-square displacement of the hopping atom in time t is given by

$$\langle \Delta r^2 \rangle = \nu a^2 t \, , \tag{13.1}$$

where a is the jump distance (i.e., the adsorption site spacing) and ν the frequency of hops. Note that νt gives the number of hops. For a single adatom, $\langle \Delta r^2 \rangle$ is averaged over many repeated observation periods of duration t.

The time-independent ratio of the mean-square displacement $\langle \Delta r^2 \rangle$ to time t is known as the *diffusion coefficient* (or *diffusivity*), D:

$$D = \frac{\langle \Delta r^2 \rangle}{zt} = \frac{\nu a^2}{z} \, , \tag{13.2}$$

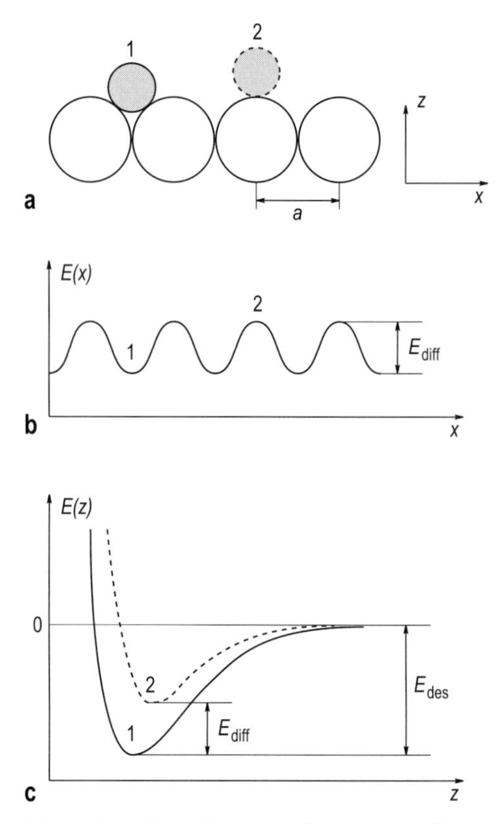

Fig. 13.1. One-dimensional schematic diagram showing (**a**) a substrate (open circles) and adatom (hatched circle) in an adsorption site (labeled 1) and in a transition (saddle point) state (labeled 2). z is the distance normal to the surface and x is the coordinate along the surface. (**b**) Schematic potential energy diagram for adatom motion along the surface. (**c**) Schematic diagram of the adatom potential energy as a function of z in positions 1 and 2 as in (**a**). The activation energy of surface diffusion E_{diff} equals the energy difference of the minima of the curves 1 and 2. The desorption energy E_{des} is shown for comparison (after Gomer [13.1])

where z is the number of neighboring sites which the atom can hop to. It is apparent that

$z = 2$ for one-dimensional diffusion (as shown in Fig. 13.1b, where the atom can hop either to the left or to the right neighboring sites),
$z = 4$ for surface diffusion on a square lattice, and
$z = 6$ for surface diffusion on a hexagonal lattice.

Atom hopping from site to site requires surmounting the potential barrier, i.e., this is a thermally activated process. If the oscillation frequency of the atom in the well (which is essentially an attempt frequency to overcome the barrier) is ν_0 and the barrier height is $E_{\rm diff}$, the hopping frequency can be expressed as

$$\nu = \nu_0 \exp(-E_{\rm diff}/k_{\rm B}T) , \tag{13.3}$$

where $k_{\rm B}$ is the Boltzmann constant and T the temperature. As one can see in Fig. 13.1, the activation energy of diffusion $E_{\rm diff}$ is the difference in potential energy of the adatom in the equilibrium adsorption site (position 1) and in the transition saddle point (position 2). $E_{\rm diff}$ is far less than the desorption energy $E_{\rm des}$ (typically, $E_{\rm diff} \sim$ 5–20% of $E_{\rm des}$).

For chemisorbed species, $E_{\rm diff} \gg k_{\rm B}T$ and the diffusion mechanism is referred to as *hopping* (or *jumping*) *diffusion.* If $E_{\rm diff}$ is less than $k_{\rm B}T$, the atoms transfer freely across the surface as a two-dimensional gas. This type of motion, called *mobile diffusion*, comprises a rather rare case detected only for a few physisorbed species and will be left out of the scope of further consideration.

13.1.2 Fick's Laws

In the presence of an atom concentration gradient, the random-walk motion of many atoms results in their net diffusion motion towards the region with lower concentration. The main regularities of such a diffusion process are described by Fick's two laws.

Fick's First Law. *Fick's first law* states that the diffusion flux $\boldsymbol{J}$ is proportional to the concentration gradient ∇c with diffusion coefficient D as a factor of proportionality. For one-dimensional diffusion, when the concentration varies only along a certain direction denoted by x, Fick's first law is written as:

$$J = -D\frac{\partial c}{\partial x} . \tag{13.4}$$

The negative sign in (13.4) indicates the opposite direction of the flux compared to the concentration gradient.

Fick's Second Law. *Fick's second law* describes the non-steady situation when the diffusion flux and concentration varies with time. For the one-dimensional case, it is written as:

$$\frac{\partial c}{\partial t} = \frac{\partial}{\partial x}\left(D\frac{\partial c}{\partial x}\right) . \tag{13.5}$$

If D is independent of concentration and, hence, of the coordinate x, (13.5) is reduced to

$$\frac{\partial c}{\partial t} = D\frac{\partial^2 c}{\partial x^2} . \tag{13.6}$$

Fick's second law is essentially a combination of (13.4) and the equation of continuity. It reflects the preservation of a substance during the diffusion process, i.e., it shows that if the number of atoms reaching a given local area differs from (say, exceeds) the number of atoms leaving it, the local concentration varies (increases) by the difference value.

Some simple analytical solutions of (13.6) for selected initial and boundary conditions are listed below. Note that the solutions comprise the *error function*

$$\mathrm{erf}(z) = \frac{2}{\sqrt{\pi}} \int_0^z e^{-\xi^2} \mathrm{d}\xi \tag{13.7}$$

and *complimentary error function*

$$\mathrm{erfc}(z) = 1 - \mathrm{erf}(z) \ . \tag{13.8}$$

Diffusion from a Source of Constant Concentration. If the initial concentration distribution has a step-like shape and the concentration at the boundary is maintained at a constant value c_0, i.e.,

$$\begin{aligned} c(0,t) &= c_0, \\ c(x,0) &= 0 \quad \text{for} \quad x > 0 \ , \end{aligned} \tag{13.9}$$

the solution of (13.6) is given by

$$c(x,t) = c_0 \, \mathrm{erfc}\left(\frac{x}{2\sqrt{Dt}}\right) \ . \tag{13.10}$$

Figure 13.2 shows a plot of (13.10) for four different values of $2\sqrt{Dt}$. The quantity $2\sqrt{Dt}$, often denoted as the *diffusion length*, is a scaling factor for diffusion and occurs in many diffusion problems.

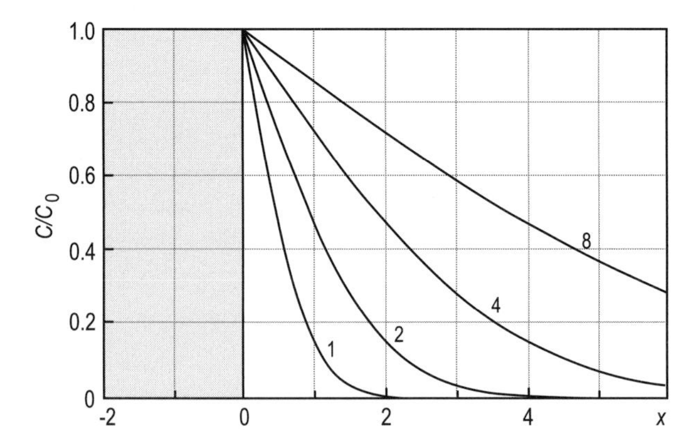

Fig. 13.2. Infinite source diffusion profiles. The concentration normalized to the constant initial concentration c_0 is plotted for four different values of the diffusion length $2\sqrt{Dt}$

In practice, this model situation simulates the case of surface diffusion from a stripe, which comprises a submonolayer film with 3D islands on it. Continuous supply of the mobile adatoms from the 3D islands might ensure the preservation of the adatom concentration at the stripe boundary at an approximately constant level.

Diffusion from a Semi-Infinite Source of Infinite Extent. If the initial distribution is given by

$$\begin{aligned} c(x,0) &= c_0 \quad \text{for} \quad x < 0, \\ c(x,0) &= 0 \quad \text{for} \quad x \geq 0\,, \end{aligned} \tag{13.11}$$

the solution is

$$c(x,t) = \frac{c_0}{2}\,\mathrm{erfc}\left(\frac{x}{2\sqrt{Dt}}\right)\,, \tag{13.12}$$

illustrated in Fig. 13.3

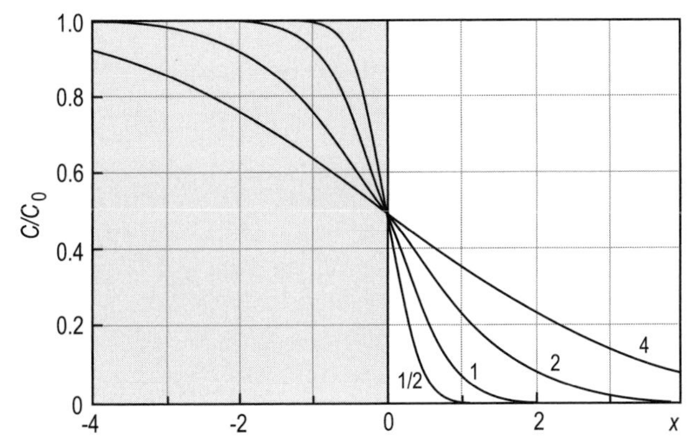

Fig. 13.3. Diffusion profiles for a semi-infinite source of infinite extent . The concentration normalized to the constant initial concentration c_0 is plotted for four different values of the diffusion length $2\sqrt{Dt}$. Note that $c = c_0/2$ at $x = 0$ for all $t > 0$

This model situation simulates surface diffusion in the case when half of a long sample is initially covered by a submonolayer film. The source may be considered as semi-infinite as long as $\sqrt{Dt}$ is much than less the source length in the diffusion direction.

Diffusion from a Finite Source of Limited Extent. If the source is a stripe of width $h \cong \sqrt{Dt}$,

$$\begin{aligned} c(x,0) &= c_0 \quad \text{for} \quad |\, x \,| < h \\ c(x,0) &= 0 \quad \text{for} \quad |\, x \,| \geq h \end{aligned} \tag{13.13}$$

the solution is

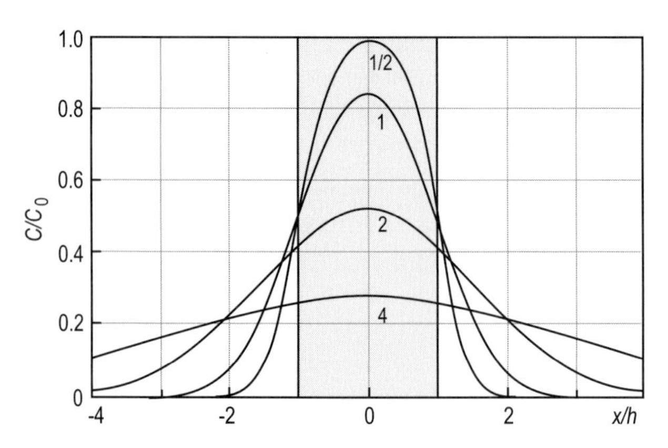

Fig. 13.4. Diffusion profiles for a finite source of limited extent. The concentration normalized to the constant initial concentration c_0 is plotted for four different values of the diffusion length $2\sqrt{Dt}/h$

$$c(x,t) = \frac{c_0}{2}\left[\operatorname{erf}\left(\frac{h-x}{2\sqrt{Dt}}\right) + \operatorname{erf}\left(\frac{h+x}{2\sqrt{Dt}}\right)\right] \tag{13.14}$$

(see Fig. 13.4).

This model situation simulates surface diffusion in the case when initially a submonolayer film is confined within a stripe of finite width. For example, such a stripe can be formed by adsorbate deposition through a mask.

Note that all solutions listed above are applicable only as long as $\sqrt{Dt}$ is much smaller than the sample size in the diffusion direction. Under this requirement, the sample can be treated as infinite or semi-infinite.

13.2 Tracer and Chemical Diffusion

Depending on the coverage of diffusable species, one distinguishes

- *tracer diffusion* (very low coverages, no interaction between adparticles);
- *chemical diffusion* (intermediate to high coverages, considerable interaction between adparticles).

Tracer Diffusion. In the ideal case, the tracer diffusion coefficient refers to the motion of a *single* adparticle, whose path or "trace" it describes. (It is worth noting that the term does not refer to the diffusion coefficient measured by the use of radioactive tracers!) The tracer diffusion coefficient is given by (13.2). For an ensemble of particles, it is still a good approximation as long as all particles migrate independently of one another. This holds for the case of very low coverages (typically, on the order of 0.01 ML) of non-interacting species. Note that dealing with an ensemble of many particles (ideally, of an infinite number of particles) one has no need to wait a very long time for a proper determination of D (as assumed in (13.2)), but rather one has to average the square displacement over all particles according to:

$$D = \frac{1}{zNt} \sum_{i=1}^{N} \langle \Delta r_i^2 \rangle , \tag{13.15}$$

where N is the number of particles and Δr_i^2 the square displacement of the i-th particle.

Chemical Diffusion. For larger coverages, where the interaction between diffusing species become important, Fick's first law (13.4) must be written in the more general form

$$J = -L \frac{\partial \mu}{\partial x} , \tag{13.16}$$

where μ is the chemical potential of the diffusing particles and the factor L is called the *transport coefficient*. The relation (13.16) can be rewritten as

$$J = -L \frac{\partial \mu}{\partial \Theta} \frac{\partial \Theta}{\partial x} = -D_{\mathrm{c}}(\Theta) \frac{\partial \Theta}{\partial x} \tag{13.17}$$

with the *chemical diffusion coefficient* D_{c} dependent on the adparticle coverage Θ and, hence, the coordinate x. It can be shown [13.1] that for a square lattice

$$D_{\mathrm{c}}(\Theta) = \frac{\nu(\Theta) a^2}{4 k_{\mathrm{B}} T} \Theta \left(\frac{\partial \mu}{\partial \Theta} \right)_T = \frac{1}{4} \nu(\Theta) a^2 \left(\frac{\partial (\mu / k_{\mathrm{B}} T)}{\partial (\ln \Theta)} \right)_T . \tag{13.18}$$

The term $\partial(\mu/k_{\mathrm{B}}T)/\partial(\ln \Theta)$ is called the *thermodynamic factor*. At low coverages ($\Theta \ll 1$) and in the absence of interaction, μ is simply given by $\mu = \mu_0 + k_{\mathrm{B}} T \ln \Theta$, where μ_0 is the standard chemical potential, and (13.18) is reduced to (13.2).

This semi-empirical equation (13.18) shows that both terms, the jump rate $\nu(\Theta)$ (which might depend on Θ) and thermodynamic factor, affect the coverage dependence of D_{c}. In particular, as μ depends on the magnitude and sign of the interaction between diffusing species, the diffusion coefficient exhibits a strong dependence on adsorbate coverage, especially when ordered surface phases are formed.

13.3 Intrinsic and Mass Transfer Diffusion

Depending on the peculiarities of the landscape, where diffusion takes place, one distinguishes

- intrinsic diffusion (i.e., diffusion in the absence of sources and traps for diffusing species);
- mass transfer diffusion (i.e., diffusion affected by the generation and/or trapping of diffusing species).

Intrinsic Diffusion. Ideally, the *intrinsic diffusion* coefficient describes the motion of particles across a surface of uniform potential (i.e., with equivalent adsorption sites). In practice, this means that to determine the intrinsic diffusion coefficient one has to monitor the adparticle motion within a single terrace, provided the measured area is defect-free. This sets a spatial limit for the diffusion length (typically on the order of 100 Å). Within the area under investigation any sources and traps are absent, hence, the number of mobile particles does not change in diffusion process. So, intrinsic diffusion is sometimes defined for situations in which the number of diffusable species remains constant as the temperature is varied [13.2]. As the experimental conditions for observing intrinsic and tracer diffusion are almost identical, often no strong distinction is made between these terms in the literature.

Mass Transfer Diffusion. Real surfaces (even well-ordered low-index single crystal surfaces) contain various defect sites, such as steps, kinks, adatoms, or vacancy clusters, etc. These defect structures constitute sites with binding energies different from that of the sites on a flat terrace. If the diffusion proceeds over a distance exceeding the average separation between these defect sites, the diffusion coefficient will be affected by trapping or generating mobile species in these sites and one speaks of *mass transfer diffusion*. In this case, the number of mobile particles is temperature dependent and this peculiarity might also be considered as a sign of mass transfer diffusion [13.2]. The mass transfer diffusion coefficient D_{M} is related formally to the intrinsic diffusion coefficient D_{I} by [13.3]:

$$D_{\mathrm{M}} = \frac{n}{n_0} D_{\mathrm{I}} \,, \tag{13.19}$$

where n is the actual number of mobile particles and n_0 their maximal number.

For self-diffusion, n_0 corresponds to the areal density of substrate atoms. Typically, $n/n_0 \ll 1$ and this ratio is strongly temperature dependent. Suppose that mass transfer surface self-diffusion occurs by the adatom mechanism. For equilibrium concentration of adatoms, one can write:

$$n_{\mathrm{A}} = n_0 \exp\left(-\frac{\Delta G_{\mathrm{A}}}{k_{\mathrm{B}} T}\right) \,, \tag{13.20}$$

where ΔG_{A} is the energy of adatom formation. Thus, from (13.2), (13.3), (13.19), and (13.20) one has

$$D_{\mathrm{M}} = \frac{\nu_0 a^2}{z} \exp\left(-\frac{\Delta G_{\mathrm{A}} + E_{\mathrm{diff}}}{k_{\mathrm{B}} T}\right) \,. \tag{13.21}$$

One can see that the activation barrier for mass transfer diffusion incorporates a barrier for migration and a barrier for the formation of the mobile particles, here adatoms. If the mass transfer diffusion is controlled by the formation of terrace vacancies or adatom–vacancy pairs, the appropriate formation energy should be substituted in (13.21).

For heterodiffusion, n_0 denotes the areal density of adsorbed foreign atoms. If all adsorbed particles are mobile, $n = n_0$ and, hence, $D_M = D_I$. However, adsorbate atoms might become immobile, being trapped at defect sites or involved in surface phase formation. In this case, n/n_0 may become small and D_M will deviate significantly from D_I.

13.4 Anisotropy of Surface Diffusion

Conventionally, two kinds of diffusion anisotropy are distinguished:

- orientational anisotropy and
- directional anisotropy.

Orientational Anisotropy. Surfaces of different crystallographic orientation have different atomic structures, hence exhibit different potential energy relief for surface diffusion. The dependence of the diffusion coefficient on the orientation of the surface is defined as the *orientational anisotropy.* A vivid example of orientational anisotropy is provided by the data of a comprehensive FIM investigation [13.4], in which the intrinsic self-diffusion of Rh adatoms on five different Rh planes was studied (see Fig. 13.5 and Table 13.1). As one can see, at a given temperature the difference in the diffusion coefficient for different planes can amount to several orders of magnitude.

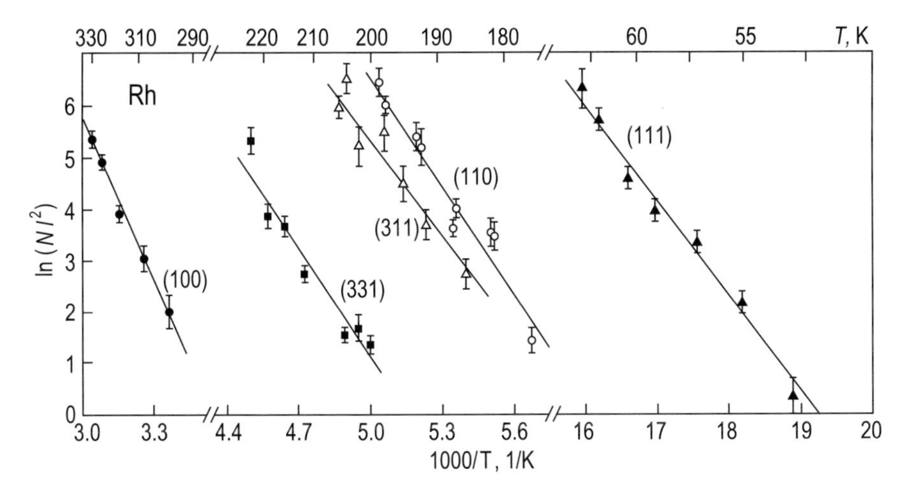

Fig. 13.5. Orientational dependence of the intrinsic surface self-diffusion of Rh. On the (110), (311), and (331) planes, diffusion is one-dimensional along [110], while diffusion perpendicular to [110] is too slow to be measured at these temperatures. On the (111) and (100) planes diffusion is two-dimensional (directionally isotropic). The diffusion interval is 3 min., N is the number of jumps per interval, and l is the jump distance. Diffusion parameters are given in Table 13.1 (after Ayrault and Ehrlich [13.4])

Table 13.1. Intrinsic self-diffusion parameters for single rhodium atoms on five different rhodium planes (after Ayrault and Ehrlich [13.4])

Plane	D_0, cm^2/s	E_{diff}, eV
(111)	$2 \cdot 10^{-4}$	0.16 ± 0.02
(311)	$2 \cdot 10^{-3}$	0.54 ± 0.05
(110)	$3 \cdot 10^{-1}$	0.60 ± 0.03
(331)	$1 \cdot 10^{-2}$	0.64 ± 0.04
(100)	$1 \cdot 10^{-3}$	0.88 ± 0.07

Directional Anisotropy. The anisotropy of surface diffusion for a given surface (i.e., when the diffusion coefficient depends on the direction at the surface) is referred to as *directional anisotropy.* There are two main causes of directional anisotropy of surface diffusion, namely,

- crystallographic anisotropy (related to the anisotropy of the surface atomic structure) and
- morphological anisotropy (related to the anisotropy of the step structure).

Consider first *crystallographic anisotropy.* While the surface diffusion is directionally isotropic on surfaces with square and hexagonal lattices, one would expect directional anisotropy for diffusion on a surface with a rectangular lattice, especially if the surface comprises atomic-scale channels along one of the principal directions. Indeed, diffusion along channels is generally faster than in the normal direction. If in the rectangular lattice the maximum diffusion coefficient D_x is along the x-axis and the minimum diffusion coefficient D_y is along the y-axis, then the diffusion coefficient in the direction defined by the polar angle φ is given by:

$$D(\varphi) = D_x \cos^2 \varphi + D_y \sin^2 \varphi \, . \tag{13.22}$$

The polar plots of this function for several values of the D_x/D_y ratio is shown in Fig. 13.6.

Diffusion of CO on Ni(110) furnishes an example of directional anisotropy of diffusion due to crystallographic anisotropy of the substrate surface. The diffusion is much faster in the $[1\bar{1}1]$ direction along the rows of the closed packed surface atoms than across these rows in the [100] direction. As one can see in Fig. 13.7, the difference in the diffusion coefficient in two directions is an order of magnitude and the angular dependence of the diffusion coefficient is well described by (13.22).

A clear demonstration of the influence of steps on the directional anisotropy of surface diffusion (i.e., *morphological anisotropy*) is shown in Fig. 13.8 for Ni mass transfer diffusion on a stepped W(110) surface. One can see that the

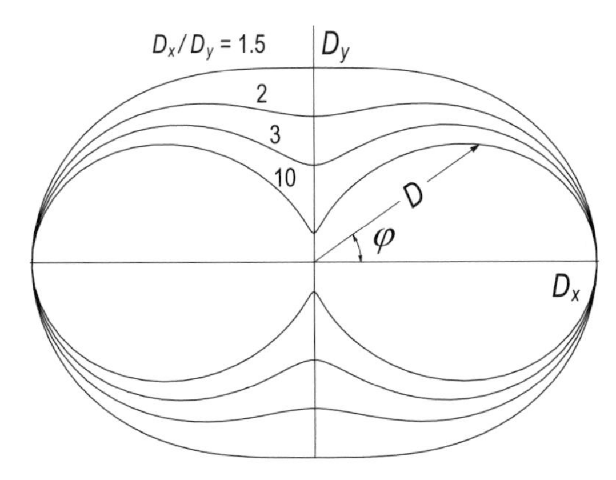

Fig. 13.6. Polar plot of (13.22) showing the anisotropy of surface diffusion with D_x and D_y being the maximum and minimum surface diffusion coefficients, respectively

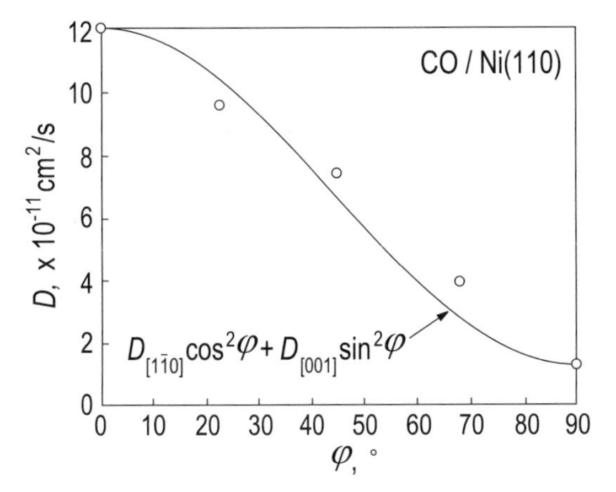

Fig. 13.7. Diffusion coefficient D for CO/Ni(110) as a function of the angle φ away from the $[1\bar{1}0]$ direction at $T \cong 110\,\mathrm{K}$ (after Xiao et al. [13.5])

diffusion is more rapid along the steps. There is also a certain difference in the diffusion coefficient in the "upstairs" and "downstairs" directions.

13.5 Atomistic Mechanisms of Surface Diffusion

Both experiment and theory show evidence that surface diffusion can occur via different atomistic mechanisms. Some of the main mechanisms are present below.

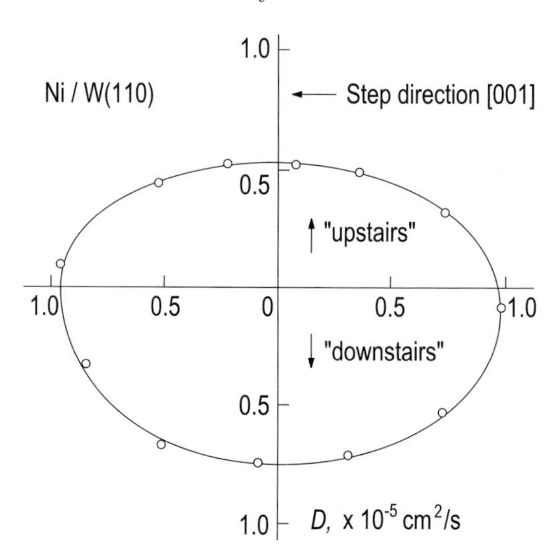

Fig. 13.8. Polar diagram of the mass transfer diffusion coefficient D for Ni on a stepped W(110) surface at $T = 1170\,\mathrm{K}$ (after Geguzin et al. [13.6])

13.5.1 Hopping Mechanism

The diffusion mechanism in which the fundamental step is a thermally activated jump from one equilibrium adsorption site to the next is known as a *hopping* (or *jumping*) mechanism. Though being the simplest conceivable mechanism, it appears to give a proper general description of self-diffusion and heterodiffusion in numerous real systems. As an example, Figs. 13.9 and 13.10 illustrate the hopping diffusion of individual nitrogen adatoms on an Fe(100) surface. In the experiment [13.7], STM movies of a selected area at the Fe(100) surface with adsorbed nitrogen adatoms were recorded at different temperatures from 299 to 325 K, with an image rate varying from 6 to 30 frames/min. Figure 13.9a shows a frame from an STM movie. Nitrogen atoms (seen as depressions) occupy the fourfold hollow sites on Fe(100). By monitoring the motion of individual nitrogen atoms in the STM movies, the hopping rate was established to follow an Arhenius law with the prefactor $\nu_0 = 4.3 \times 10^{12}\ \mathrm{s}^{-1}$ and diffusion barrier $E_{\mathrm{diff}} = 0.92 \pm 0.04\,\mathrm{eV}$ (Fig. 13.9b).

It should be noted that the physics behind hopping diffusion is not as simple as one might think at a glance. Theoretical calculations [13.7] reveal that nitrogen diffusion is strongly coupled to the distortion of the Fe(100) lattice. Another finding is the presence of a shallow metastable minimum at the bridge site, which corresponds to the saddle point for the diffusion jump (Fig. 13.10).

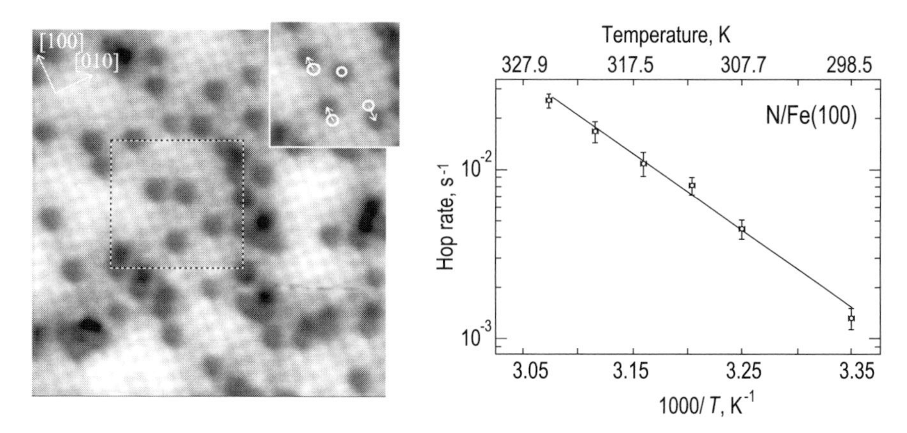

Fig. 13.9. Hopping diffusion of nitrogen adatoms on Fe(100). (**a**) 78×76 Å^2 STM image from an STM movie recorded at 300 K. The inset shows an area outlined by a dashed rectangular 62 sec later. Nitrogen adatoms are seen as dark depressions. Open circles in the inset indicate the original positions of four isolated nitrogen adatoms. One can see that, in the intervening time, adatoms have performed one or more fundamental jumps. (**b**) Arrhenius plot of the nitrogen adatom hopping rate. The straight line shows the least-squares best fit to the experimental points, yielding $E_{\mathrm{diff}} = 0.92 \pm 0.04$ eV and $\nu_0 = 4.3 \times 10^{12}$ s^{-1} (after Pedersen et al. [13.7])

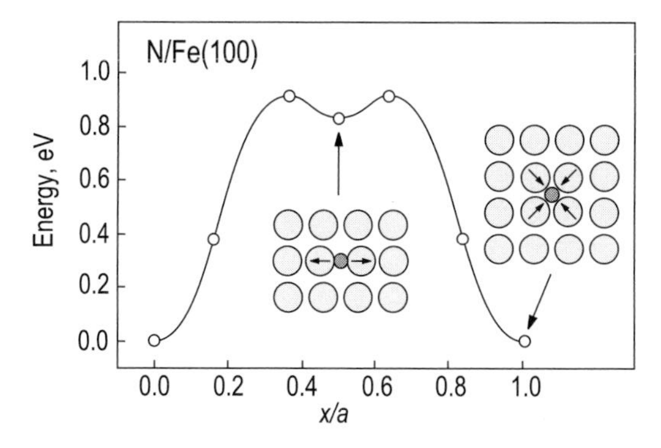

Fig. 13.10. Hopping diffusion of nitrogen adatoms on Fe(100): Calculated shape of the potential barrier between the two neighboring fourfold hollow adsorption sites. The energy of the maximum is 0.91 eV with a shallow metastable minimum at 0.83 eV. The insets illustrate the distortion in the Fe(100) top layer (indicated by arrows) induced by the nitrogen adatom in the stable fourfold hollow site and in the metastable bridge site (after Pedersen et al. [13.7])

13.5.2 Atomic Exchange Mechanism

The diffusion mechanism which involves an exchange between an adatom and a surface atom, is referred to as an *atomic exchange mechanism* (or a *place exchange mechanism*). In this mechanism (Fig. 13.11), the adatom replaces a substrate atom, while the replaced atom moves to the neighbor adatom site. During atom exchange, all the atoms involved preserve a high coordination number and in some cases the mechanism appears to be energetically more favorable than simple hopping of an adatom over a bridge site.

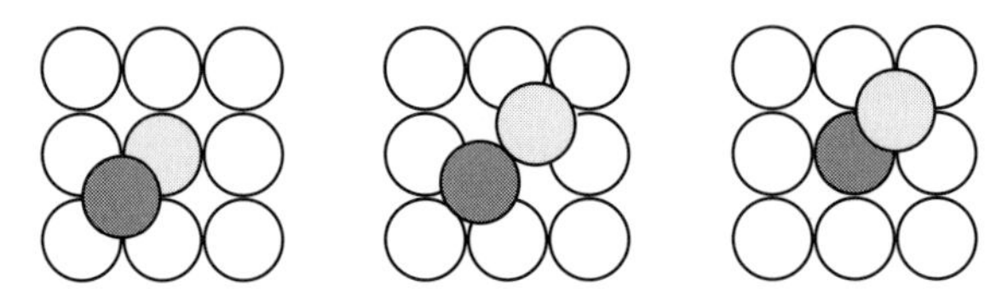

Fig. 13.11. Schematic representation of the atomic exchange mechanism

It has been confirmed that self-diffusion on non-reconstructed (110)1×1 and (100)1×1 surfaces of fcc metals, Pt and Ir, occurs by the atomic exchange mechanism. As an example, Fig. 13.12 illustrates confirmation of the atomic exchange mechanism for diffusion of a Pt adatom on Pt(100). One can see (Fig. 13.12a) that if the adatom migrates via the exchange mechanism, it can "jump" only along the [100] and [010] directions, hence visiting only half of the available sites, which form a $c(2\times2)$ sublattice. That is exactly what was observed in the FIM experiment [13.8] (Fig. 13.12b): the site-visitation map comprises a square $c(2\times2)$ lattice with its primitive vectors parallel to the ⟨100⟩ directions.

If an adatom is not the same chemical species as the substrate, an atomic exchange event would change the chemical identity of the adatom. In the atom-probe FIM experiment [13.9], it was shown that after a jump of a W adatom in the cross-channel direction on Ir(110), the new adatom is, in fact, an Ir adatom. The adatom–substrate atom exchange mechanism has been proved for a set of heterosystems (for example, Pt on Ni(110), Ir on Pt(100), Re on Ir(100)). It should be noted, however, that if atomic exchange occurs in the heterosystem, no long-range migration of the adsorbate atom is possible. Instead, atomic exchange produces single-atom alloying.

13.5.3 Tunneling Mechanism

If the diffusing particle has a small mass and the potential barrier against diffusion is low, the particle can tunnel across the barrier and its migration will occur via the *quantum tunneling mechanism*. At sufficiently low temperatures, the tunneling mechanism might dominate over the classical

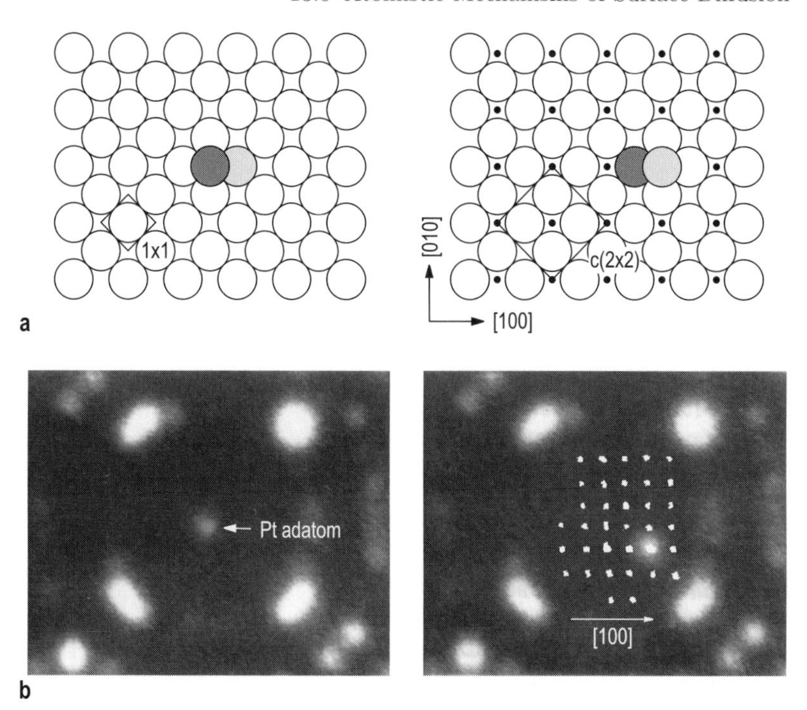

Fig. 13.12. (**a**) Schematic representation of the atomic exchange mechanism on an fcc(100) surface. Hops are possible only in the ⟨100⟩ directions, and accessible sites (indicated by solid dots) form a $c(2\times2)$ sublattice. (**b**) Field ion microscope images illustrating the atomic exchange mechanism for self-diffusion on a Pt(100) surface. The Pt adatom is seen as a bright spot (see left image). The right image shows the site-visitation map from 300 diffusion intervals superimposed on the image of the substrate plane. The $c(2\times2)$ orientation of the map confirms the exchange mechanism. The crystallographic directions are obtained from the symmetry of the overall field ion image (after Kellogg [13.8])

hopping mechanism. The obvious candidate for observing tunneling diffusion is hydrogen on a metal surface. Indeed, this was detected by FEM for H on W(110) [13.1] and by STM for H on Cu(100) [13.10]. In the latter work, diffusion of single hydrogen atoms was studied. The hopping rate of hydrogen atoms (seen in STM as small depressions) was recorded as a function of temperature and the obtained dependence is shown in Fig. 13.13. At temperatures above 60 K, the H atom readily acquires enough energy to surmount the barrier to diffusion, hence classical hopping prevails and the rate is characterized by the Arrhenius law with $\nu = 10^{12.9\pm0.3}\ \mathrm{s}^{-1}$ and $E_{\mathrm{diff}} = 0.197\pm0.04\,\mathrm{eV}$. At temperatures below 60 K, migration occurs via quantum tunneling at an almost temperature-independent rate.

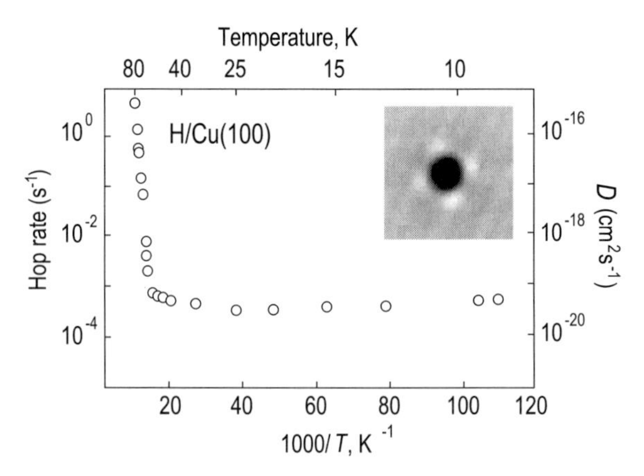

Fig. 13.13. Classical and quantum diffusion of H on Cu(100): Arrhenius plot of the hopping rate of the individual H atoms between 80 and 9 K. The right axis gives the equivalent single particle diffusion coefficient, D, which is related to the hop frequency ν by the expression $D = a^2\nu/4$, where the lattice constant is $a = 2.55$ Å. The inset shows a $15{\times}15$ Å^2 STM image of an isolated H atom on Cu(100) (after Lauhon and Ho [13.10])

13.5.4 Vacancy Mechanism

Migration of atoms within a completed surface atomic layer (i.e., in the case when the overwhelming majority of atomic sites are occupied) is often controlled by the formation and migration of vacancies or, in other words, proceeds according to a *vacancy mechanism.* In rare cases, vacancy motion can be observed directly [13.11], as illustrated in Fig. 13.14, which shows the creation and subsequent diffusion of a single atomic vacancy on the Ge(111)$c(2{\times}8)$ surface. (Recall that the atomic structure of Ge(111)$c(2{\times}8)$ comprises an ordered array of Ge adatoms occupying T_4 sites on the bulk-like terminated Ge(111)1×1 surface (Fig. 8.20).) The vacancy was created *artificially* by extracting a selected Ge adatom with the STM tip (Figs. 13.14a and b). The formed vacancy moves on the surface via thermally activated hopping of the neighboring atoms to the vacant site (Figs. 13.14c and d).

Heterodiffusion by a vacancy-exchange mechanism has been elucidated for diffusion of In and Pd atoms embedded within the first atomic layer of a Cu(100) surface [13.12, 13.13]. The atom hopping rate is determined by the concentration of the *natural* vacancies and their mobility. Note that individual vacancies could not be detected in those experiments directly, as the vacancy concentration was ultra-low and their mobility was, in contrast, very high. According to estimation [13.12], at room temperature the vacancy concentration in the Cu(100) surface is $\sim 10^{-10}$ and the vacancy hopping rate is $\sim 10^8$ Hz.

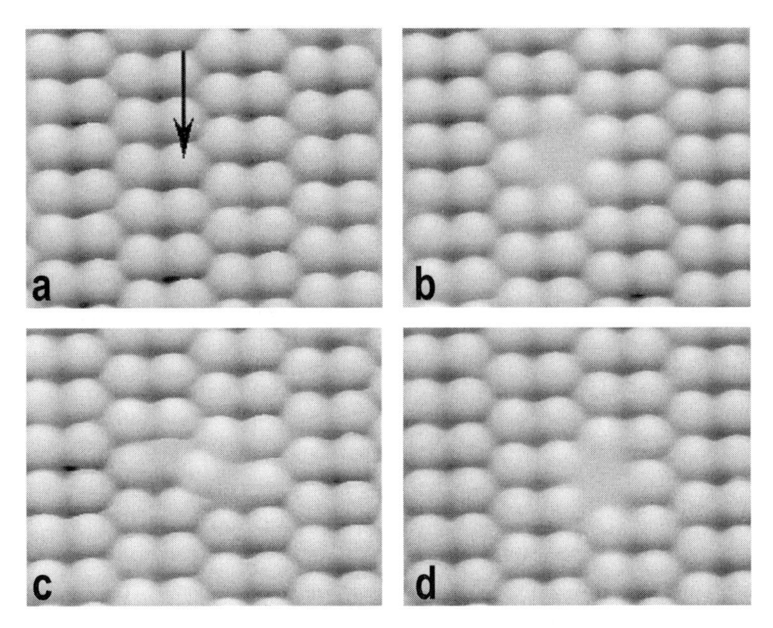

Fig. 13.14. STM topographic images showing the formation and subsequent diffusion of a single atomic vacancy. In (**a**) adatom to be selectively removed with the STM tip is indicated by the arrow. In (**b**) the adatom has been removed. The vacancy moves by the displacement of a neighboring adatom via a metastable site (**c**) to the vacant site (**d**) (after Mayne et al. [13.11])

13.6 Surface Diffusion of Clusters

In the course of migration, individual adatoms can occasionally approach each other and, in the presence of lateral attraction, can form a stable cluster. Depending on many circumstances, a cluster can contain from two atoms (a dimer) up to hundreds of atoms (an island). The cluster dynamics involves two aspects: the atom redistribution within a cluster and cluster diffusion. *Cluster surface diffusion* is characterized by a displacement of the center of mass of the cluster. In this resect, cluster migration can be treated in a manner analogous to single adatoms. Numerous experimental observations have revealed that, in general, the larger the cluster the lower its mobility. As an example, Fig. 13.15 shows the diffusivity of Pt clusters of various sizes on Pt(111). Figure 13.16 demonstrates that, in this case, the activation energy for cluster migration increases monotonically with cluster size.

It should be noted that the decrease in mobility with increase of cluster size is a general trend. However, noticeable deviations from it can take place in particular systems, as a size-dependent cluster shape also affects the cluster mobility. For example, oscillatory behavior of the size dependence of cluster diffusivity was detected for Rh on Rh(100) (Fig. 13.17). In this system, clus-

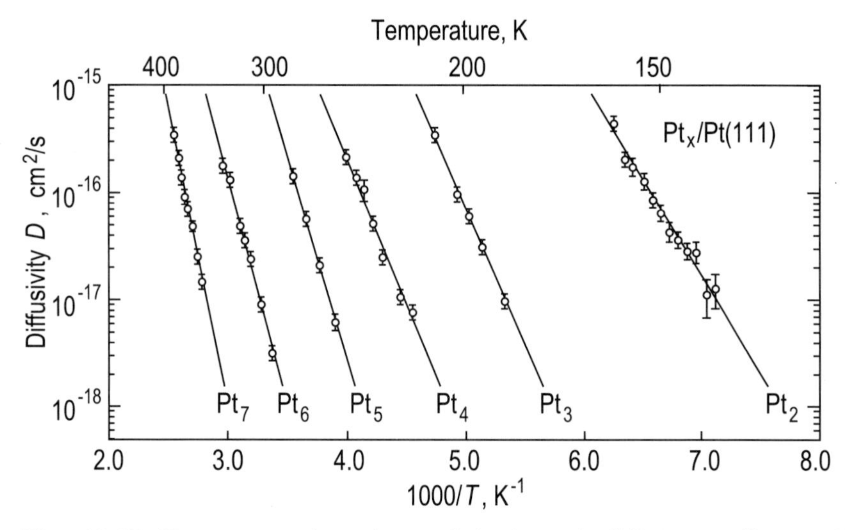

Fig. 13.15. Temperature dependence of the intrinsic diffusion coefficient of Pt clusters on Pt(111), as determined in FIM observations (after Kyuno and Ehrlich [13.14])

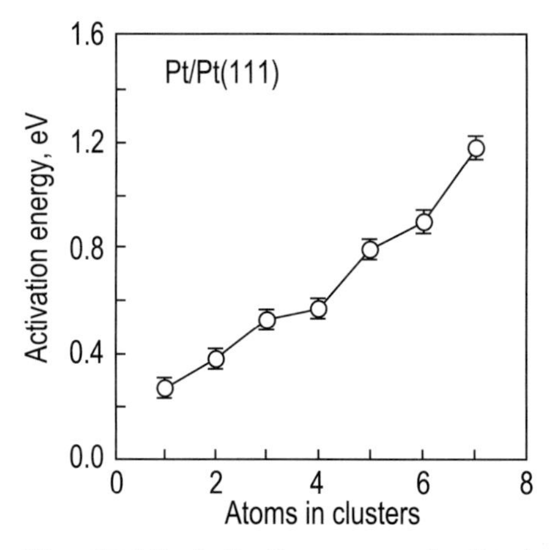

Fig. 13.16. Activation energy for Pt cluster diffusion on Pt(111) versus cluster size. The line is drawn to guide the eye (after Kyuno and Ehrlich [13.14])

ters with a compact configuration are characterized by a higher activation energy.

The variety and complexity of diffusion processes are naturally enhanced when the diffusion species is a cluster rather than a single adatom. Depending on particular conditions, the cluster motion can proceed in many ways, hence many different mechanism have been proposed. Some of them are listed below. All the proposed mechanisms can be subdivided into two main types:

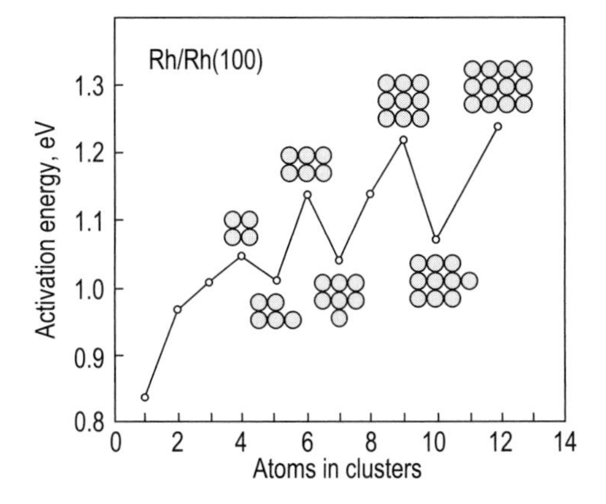

Fig. 13.17. Activation energy for diffusion of Rh clusters on Rh(100) versus cluster size. The stable shapes of the clusters are shown schematically. The low mobility of clusters with compact geometric shapes causes the activation energy to oscillate as a function of cluster size (after Kellogg [13.15])

- individual mechanisms (based on single atom motion) and
- concerted mechanisms (involving simultaneous motion of a group of atoms).

Individual Mechanisms. *Individual mechanisms* refer to the cases when the displacement of the whole cluster is a result of the independent motion of single atoms constituting the cluster. Examples of individual mechanisms are presented in Fig. 13.18. These are as follows:

- *Sequential displacement mechanism* (Fig. 13.18a), i.e., one-by-one motion of single atoms.
- *Edge* (or *periphery) diffusion mechanism* (Fig. 13.18b), in which the motion of edge adatom, vacancy, or kink along the cluster edge causes the displacement of the center of mass of the cluster.
- *Evaporation–condensation mechanism* (Fig. 13.18c), which describes cluster diffusion by exchange of atoms between the cluster and the 2D adatom gas.
- *Leapfrog mechanism* (Fig. 13.18d), in which one of the edge atoms is promoted on top of the cluster and becomes incorporated at the opposite side.

Concerted Mechanisms. *Concerted mechanisms* describe the situations when the cluster displacement is due to simultaneous correlated motion of, at least, several atoms of a cluster. As an example, Fig. 13.19 illustrates some of the proposed concerted mechanisms as follows:

- *Gliding mechanism* (Fig. 13.19a), which refers to the case when the cluster glides as whole.

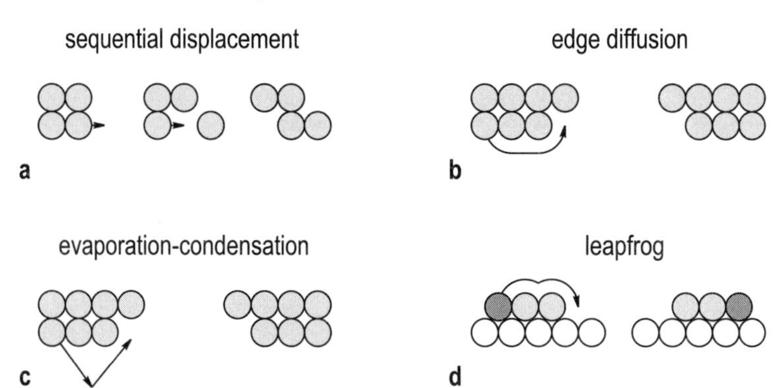

Fig. 13.18. *Individual mechanisms* of cluster motion, i.e., mechanisms in which the cluster moves as a whole due to uncorrelated motion of single adatoms constituting the cluster. (**a**) Sequential diffusion mechanism; (**b**) edge diffusion mechanism; (**c**) evaporation–condensation mechanism; (**d**) leapfrog mechanism. Each figure illustrates the initial and final stages of the elementary step. Figures (**a**), (**b**), and (**c**) display the top view, while figure (**d**) the side view

- *Shearing mechanism* (Fig. 13.19b), in which a group of atoms (for example, an atom row) in a cluster executes a concerted motion. In other words, this is a gliding, in which not the whole cluster, but a part of it is involved.
- *Reptation mechanism* (Fig. 13.19c), which includes successive shear translation of the adjacent subcluster regions, which results in a snake-like gliding motion.
- *Dislocation mechanism* (Fig. 13.19c), which refers to the case when two adjacent subcluster regions form a stacking fault and are separated by a misfit dislocation. Row-by-row motion of this misfit dislocation eventually eliminates the stacking fault and results in the shift of the center of mass of the cluster. The long-range cluster diffusion proceeds via nucleation and motion of misfit dislocations.

Concerning the validity of the proposed mechanisms for real diffusion of clusters, it should be noted that in most cases it cannot be established directly. Experimental techniques, FIM and STM, usually allow "snapshots" of a cluster before and after the diffusion event, but fail to fix the short-living transition states. Hence, the diffusion mechanism is typically inferred from indirect evidence such as the value of the activation barrier, the dependence of the diffusivity on the island size, or the evolution of the island shape in the course of diffusion. In this respect, the fascinating STM observation [13.16] of the transition state in the leapfrog diffusion of a Pt linear cluster along the missing-row trough on the Pt(110)2×1 surface (see Fig. 13.20) is especially valuable.

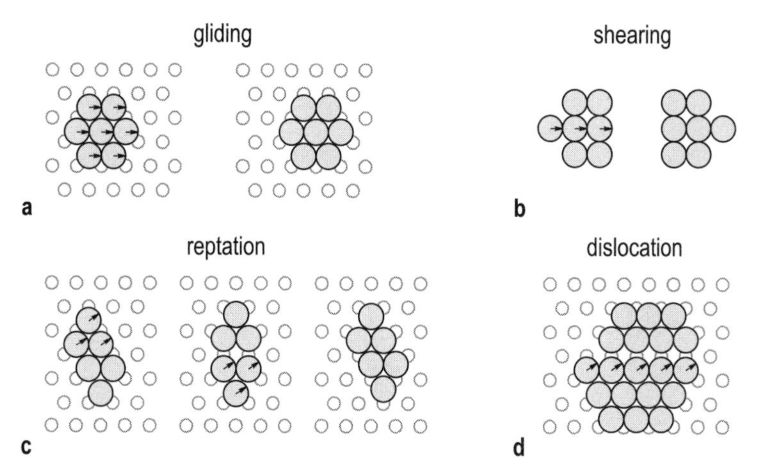

Fig. 13.19. *Concerted mechanisms* of cluster motion, i.e., mechanisms in which the cluster motion is due to correlated simultaneous motion of a group of atoms. (**a**) Gliding mechanism; (**b**) shearing mechanism; (**c**) reptation mechanism; (**d**) dislocation mechanism

13.7 Surface Diffusion and Phase Formation

If the adsorbate coverage is sufficient (typically, from $\Theta \gtrsim 0.1$ up to $\sim$1 ML) and the temperature is appropriate, the formation of surface phases takes place in accordance with the phase diagram. Under these conditions, the diffusion proceeds over surface phases and surface redistribution of the adsorbate atoms induces the phase transitions. Therefore, phase transitions and diffusion kinetics are closely interrelated and diffusion in the presence of phase transitions seemingly presents the most complicated case of chemical diffusion.

Generally, the main effect of surface phase formation on the diffusion is twofold:

- First, the mobile atoms after being incorporated into the surface phase become tightly bound to their residence sites, and hence become immobile and no longer contribute to the diffusion flux.
- Second, each adsorbate–substrate surface phase is characterized by an individual surface structure and, hence, by a specific potential relief, which differs from that of the substrate surface, as well as of the other surface phases. As a result, the diffusivity of adatoms varies from one phase to another (in some cases, over several orders of magnitude) and experiences a singularity in the transition point.

In simplified form, diffusion accompanied by phase formation can be visualized in terms of the so-called *unrolling carpet mechanism* (Fig. 13.21). According to this mechanism, the adsorbate atoms of the "first layer" are tightly bound to the substrate (i.e., chemisorbed) in contrast to the weakly

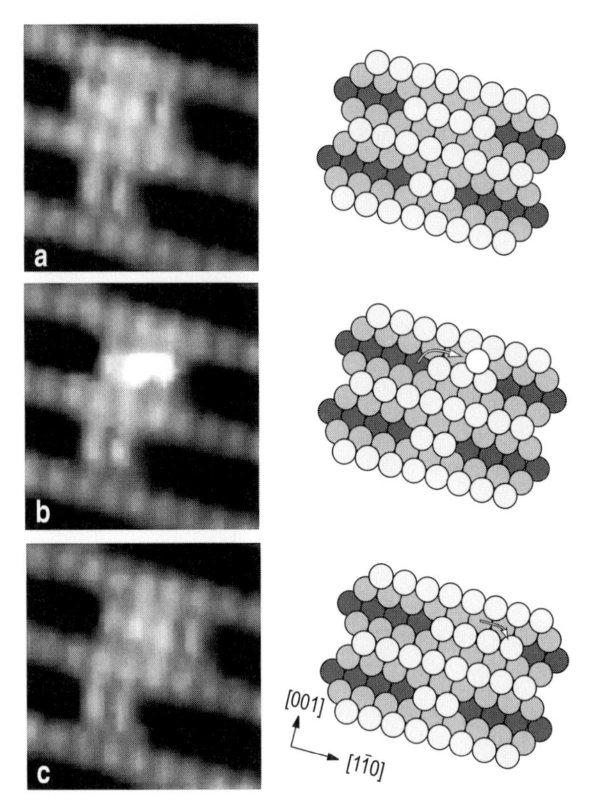

Fig. 13.20. Direct STM observation of the leapfrog diffusion of a four-atom-long Pt cluster along the missing-row trough on Pt(110)2×1. Three consecutive images, frames (**a**) to (**c**), show the movement of the cluster by one lattice spacing to the right. Note the frame (**b**), where the cluster length is reduced to three atoms, while an atom is clearly resolved on top of the remaining cluster. This provides direct evidence for the leapfrog diffusion mechanism illustrated by corresponding ball models of the stages observed (after Linderoth et al. [13.16])

bound (physisorbed) atoms of the "second layer." As a result, the first-layer atoms are essentially immobile, while the "second-layer" atoms are highly mobile. Under these conditions, the spreading of the adsorbate layer proceeds via the motion of the "second-layer" atoms towards the edge of the "first layer," where they eventually become incorporated as "first-layer" atoms. When dealing with surface phases, the term "first layer" actually refers to the adsorbate–substrate *surface phase* with a certain fixed coverage (not necessarily one monolayer, as one might infer from the simplified scheme of Fig. 13.21). The term "second-layer atoms" stands for the mobile *adsorbate adatoms on the surface phase*. Under appropriate modification, the unrolling carpet mechanism can be applied for qualitative consideration of diffusion

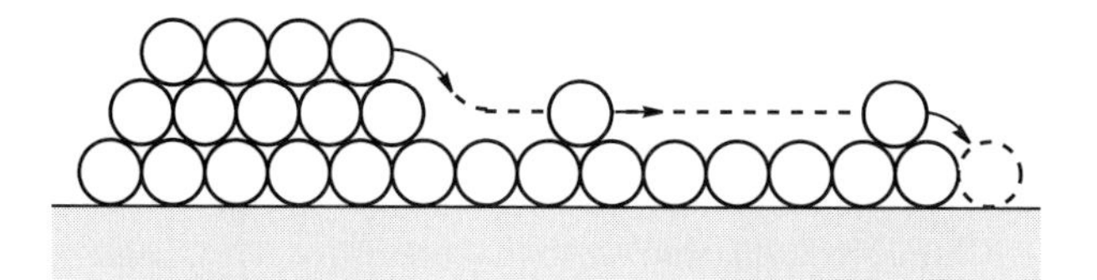

Fig. 13.21. *Unrolling carpet mechanism* illustrating the diffusion of adsorbate accompanied by formation of a surface phase. Mobile adatoms (represented by "second-layer atoms") move over the surface phase (represented by tightly bound "first-layer atoms") to become incorporated at its edge

in systems where two or more surface phases are formed successively with increasing adsorbate coverage.

The experimental example of surface diffusion accompanied by phase transitions is presented in Fig. 13.22, which shows the evolution of In concentration profiles in the course of In diffusion along the Si(111) surface. The fast diffusion of In adatoms over the ordered In/Si(111) surface phases, $\sqrt{7}\times\sqrt{3}$ and 4×1, manifests itself by the two plateaus in the concentration profiles. This behavior is in general agreement with what one would expect according to the unrolling carpet mechanism.

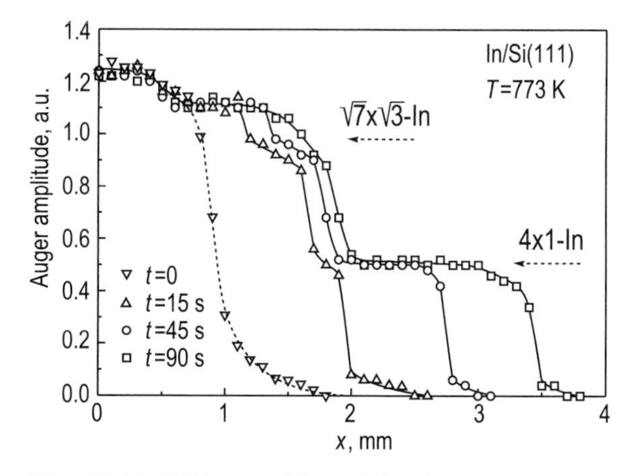

Fig. 13.22. Diffusion of In on Si(111). The terraces on the In concentration profiles correspond to the fast diffusion of In adatoms over the ordered In/Si(111) surface phases, $\sqrt{7}\times\sqrt{3}$ and 4×1 (after Bekhtereva et al. [13.17])

13.8 Surface Electromigration

The presence of an external force can greatly affect the diffusion process. A vivid example is *surface electromigration*, i.e., directional atomic motion on a surface in the case when electrical current flows through the sample. Surface electromigration has been found to occur both in homo- and heterosystems. In the case of self-electromigration, the change in the direction of current changes the surface morphology, in particular, the step structure (as an example, see Fig. 7.9). In hetero-electromigration, preferential mass transfer of the adsorbate occurs towards either cathode or anode. As an example, Fig. 13.23 illustrates the electromigration of Ag on a Si(111) sample. Upon annealing by passing a DC current through the sample, the patch of Ag thin film spreads mainly to the cathode side. The spread-out Ag layer has a constant thickness of 1 ML, which corresponds to the Si(111)$\sqrt{3}\times\sqrt{3}$-Ag surface phase.

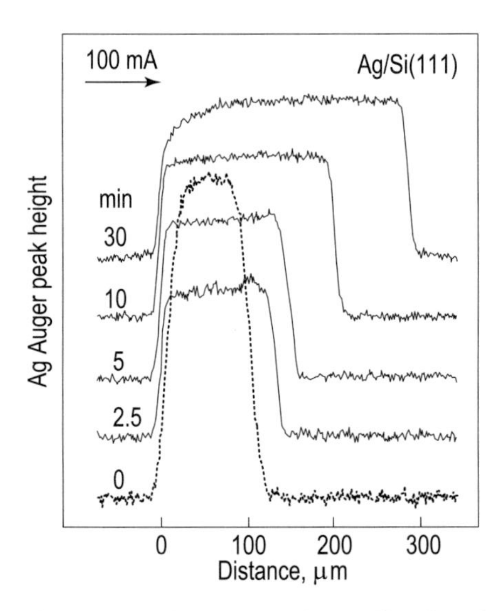

Fig. 13.23. Changes in the profile of an Ag layer on the Si(111)7×7 surface due to Ag electromigration. Ag spreads towards the cathode while the edge of the anode side makes no significant movement. Initially a 3.4 ML 100×100 μm^2 Ag patch was deposited on the surface and then annealed by a DC current of 100 mA. The sample temperature was 319°C (after Yasunaga et al. [13.18])

In the formal description of surface electromigration, the effective force $\boldsymbol{F}$ acting on a migrating adatom is proportional to the electric field $\boldsymbol{E}$ and is written as a sum of two components:

$$\boldsymbol{F} = \boldsymbol{F}_{\mathrm{d}} + \boldsymbol{F}_{\mathrm{w}} = e(Z_{\mathrm{d}} + Z_{\mathrm{w}})\boldsymbol{E} = eZ\boldsymbol{E} , \tag{13.23}$$

where

$Z = Z_d + Z_w$ is the total *effective charge* (or *effective valence*) of the atom (measured in units of the electron charge e).

$\boldsymbol{F}_d$ represents the *"direct" force*, acting on the atom (ion) through its electrostatic interaction with the applied field.

$\boldsymbol{F}_w$ is the *"wind" force*, which arises from momentum transfer from the current carriers (electrons) as they are scattered by the atom.

For adatom electromigration on metals, the "wind" force generally prevails, hence the sign of Z is always negative (i.e., the direction of adatom motion coincides with the direction of electron flow towards the anode). For electromigration on semiconductor surfaces, the sign of Z might be either negative (for example, for Ag or In on Si(111)) or positive (for example, for Au on Si(111)) depending on the particular case.

13.9 Experimental Study of Surface Diffusion

To characterize surface self- and heterodiffusion, a large variety of experimental methods have been developed. Some of them are considered below.

13.9.1 Direct Observation of Diffusing Atoms

In experiments of this type, the random-walk motion of the individual atoms is tracked directly and the *intrinsic* diffusion coefficient is readily obtained through the standard relation for the mean-square displacement (13.2). The method is suited for studying both *self-* and *heterodiffusion*. The experimental techniques, which allow single-atom imaging, are

- field ion microscopy and
- scanning tunneling microscopy.

Most *field ion microscopy* (FIM) experiments are conducted in the "image-anneal-image" manner as follows: After imaging of the adatom on a terrace, the field is removed and the tip is heated to a certain temperature (typically 200–500 K) to induce adatom diffusion. After annealing for a fixed time, the tip is quickly cooled and a new FIM image is recorded to trace the adatom displacement. A wealth of information on surface diffusion of individual atoms and clusters has been obtained with FIM (for example, see Figs. 13.5, 13.12, 13.15, 13.16, and 13.17). The limitation of FIM is that it can only be used to study metal atoms on refractory or noble metal surfaces.

Scanning tunneling microscopy is applicable for studying a great variety of adsorbates and substrates and has become the most common tool for imaging atoms at a surface. Besides the "image-anneal-image" regime, it also allows investigations in "image-while-hot" mode, if a variable-temperature scanning tunneling microscope is used. The set of images acquired in the course of annealing yields an STM movie of the atom motion. The time resolution of this procedure is determined by the time required to acquire the individual STM image (typically, from 1 to 100 s).

The time resolution is enhanced greatly if one uses the so-called *atom-tracking* technique [13.19]. While atom tracking, the STM tip is locked onto the selected adparticle (adatom or cluster) using two-dimensional lateral feedback. The feedback forces the tip to climb continually uphill, following the local surface gradient, and hence remaining at the top of the adparticle. When a diffusion event occurs, the tracking tip quickly relocates to the new position of the adparticle. Thus, the STM spends all of its time measuring the diffusion of the selected adparticle and the ability of STM to resolve individual dynamic events is increased by orders of magnitude. Figure 13.24 illustrates the atom tracking of a Si dimer migrating along the Si(100) surface.

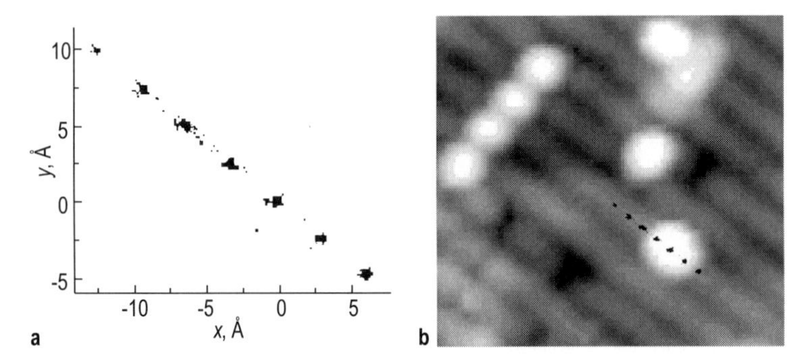

Fig. 13.24. STM atom tracking of a Si ad-dimer on Si(100). (**a**) X-Y scatter plot of lateral coordinates of the ad-dimer acquired at 65°C over 540 s. During the course of the measurements, the ad-dimer visited seven lattice sites along the top of the substrate dimer row, making 41 hops. (**b**) The atom-tracking data of (**a**) superposed on the STM image showing initial location of the ad-dimer (after Swartzentruber [13.19])

It should be noted that while imaging in STM the surface is subjected to very large electric fields and current densities. This circumstance raises the issue of the effect of the STM tip on surface diffusion, especially for adsorbates characterized by a large effective charge or polarizability. Therefore, quantitative STM measurements of the diffusivity requires preliminary tests for the tip effect in order to establish the tunneling parameters at which the tunneling process itself has little or no consequence on the acquired data.

13.9.2 Profile Evolution Method

The *profile evolution* method (also called the *concentration gradient* method) refers to the kind of experiments in which the smearing of the initially sharp concentration profile of adsorbate is monitored. To form a sharp initial profile, one conventionally employs adsorbate deposition through a mask. The evolution of the profile is characterized with techniques probing the surface concentration of adsorbate with spatial resolution (e.g, AES, SIMS, SEM, or local work-function measurements). The diffusion is treated by solving Fick's second equation (13.5) with the appropriate boundary conditions. As a result, the *mass transfer chemical* diffusion coefficient for *heterodiffusion* can be evaluated.

Dealing with chemical diffusion, one could expect the diffusion coefficient to be coverage-dependent. The *Boltzmann–Matano method* enables one to determine the diffusion coefficient as a function of coverage from the measured concentration profiles. This method exploits the variable transformation

$$\eta = \frac{x}{\sqrt{t}} , \tag{13.24}$$

which reduces Fick's second law (13.5) to the ordinary differential equation

$$-\frac{\eta}{2}\frac{\mathrm{d}\Theta(\eta)}{\mathrm{d}\eta} = \frac{\mathrm{d}}{\mathrm{d}\eta}\left(D(\Theta)\frac{\mathrm{d}\Theta}{\mathrm{d}\eta}\right) , \tag{13.25}$$

provided that the diffusion coefficient depends only on Θ.

For the smearing of the step-like profile (the case corresponding to the diffusion from a semi-infinite source of infinite extent (see (13.11) and Fig. 13.3), integrating (13.25) yields

$$D(\Theta') = -\frac{1}{2}\left.\frac{\mathrm{d}x}{\mathrm{d}\Theta}\right|_{\Theta=\Theta'} \int\limits_0^{\Theta'} x\,\mathrm{d}\Theta , \tag{13.26}$$

giving the diffusion coefficient for a particular coverage Θ'. The location of the initial step, $x = 0$, is crucial for the calculation of the integral (13.26). It is conventionally chosen at the measured profile so that

$$\int\limits_0^{\Theta_0} x\,\mathrm{d}\Theta = 0 , \tag{13.27}$$

which just describes the conservation of adsorbate particles in the coarse of diffusion.

From the set of $D(\Theta)$ dependencies acquired at different temperatures, on can formally extract the coverage-dependent pre-exponent $D_0(\Theta)$ and activation energy $E_\mathrm{a}(\Theta)$. However, one should bear in mind that these effective values result from some complex average of processes involved in chemical mass-transfer diffusion and does not refer to any microscopic process in particular.

An example of the experimental profiles evaluated by the Boltzmann–Matano method is presented in Fig. 13.25. In the experiment [13.20], the coverage profiles for Dy on Mo(112) were recorded using local work function measurements (Fig. 13.25a). The applicability of the Boltzmann–Matano method was checked through verifying that the profiles are invariant when

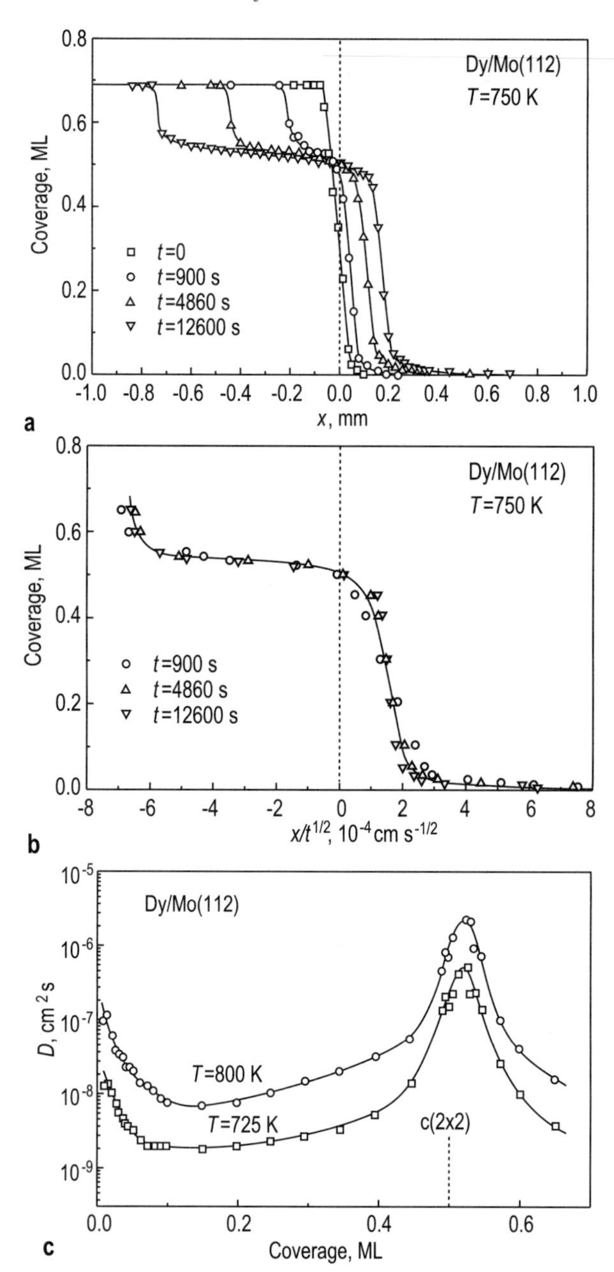

Fig. 13.25. Evaluation of the diffusivity of Dy on Mo(112) using the Boltzmann–Matano method. (**a**) Experimental coverage profiles recorded after annealing at 750 K for different times. (**b**) Coverage profiles of (**a**) replotted against $x/\sqrt{t}$. (**c**) Diffusivity of Dy on Mo(112) at 725 and 800 K as a function of coverage (after Loburets et al. [13.20])

plotted against $x/\sqrt{t}$ (Fig. 13.25b). The evaluated diffusivity as a function of Dy coverage is shown in Fig. 13.25c.

13.9.3 Capillarity Techniques

The diffusion constants for *mass transfer self-diffusion* can be evaluated in experiments treating the *capillarity-driven* redistribution of surface atoms. Capillarity-driven mass transport is a consequence of the dependence of the chemical potential μ on the local curvature K, which is described by the Gibbs–Thompson equation:

$$\mu(K) = \mu(0) + \gamma \Omega K \,, \tag{13.28}$$

where γ is the orientation-dependent specific surface energy and Ω the atomic volume. Therefore, spatial gradients in K give rise to gradients in μ, and hence induce diffusion (Fig. 13.26).

In the experiments, the surface is perturbed artificially from its lowest energy configuration (for example, by forming scratches, grooves, or roughnesses) and then permitted to relax via diffusion. From the measured rate of relaxation, the diffusion coefficient can be extracted using the appropriate solution of the diffusion equation. For example, the solution for the decay of a sinusoidal profile (Fig. 13.26) by surface self-diffusion is given by [13.21]:

$$A(t) = A(0)\exp(-Bq^4 t) \,, \tag{13.29}$$

$$B = \frac{\gamma D n_0 \Omega^2}{k_\mathrm{B} T}, \quad q = \frac{2\pi}{\lambda} \,, \tag{13.30}$$

where A is the amplitude and λ the wavelength of the sinusoidal profile, D the surface self-diffusion coefficient, n_0 the number of surface atoms per unit area, and t the diffusion time.

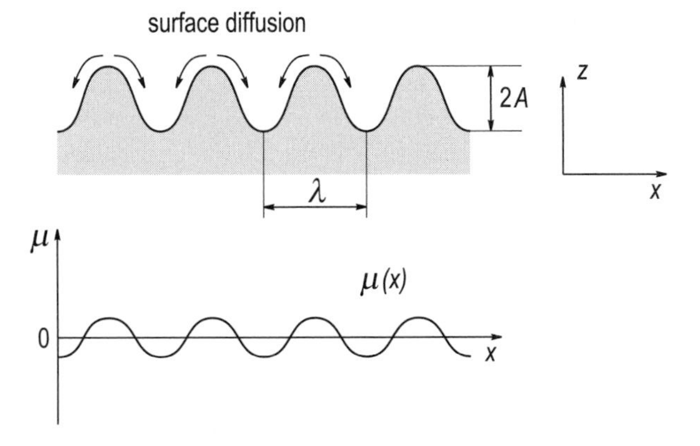

Fig. 13.26. Schematic diagram showing the sinusoidal profile of the surface and corresponding spatial variation of the chemical potential $\mu(x)$

Figure 13.27 illustrates the experimental results on Si(100) surface smoothing. In the experiment, the Si(100) surface was scratched to produce a one-dimensional stochastic relief. To monitor the smoothing of the profile, the surface was irradiated by monochromatic light from a He-Ne laser (0.6328 μm wavelength) and the angular distribution of the scattered light intensity was measured (Fig. 13.27a). The light scattering angle φ is tied through the diffraction law to the corresponding Fourier component of frequency q, while the intensity of the scattered light $I(\varphi)$ is proportional to the square amplitude $A^2(q)$ of this Fourier component. Taking into account (13.29) and (13.30), the time dependence of the intensity of the light scattered at angle φ is written as

$$I(\varphi, t) = I(\varphi, 0) \exp(-2Bq^4 t) \ . \tag{13.31}$$

Experimental $I(\varphi, t)$ plots for a set of scattering angles are shown in Fig. 13.27b. Evaluation of the time dependence of the intensity at a fixed temperature according to (13.29), (13.30), and (13.31) allows the determination of the diffusion constant; the same experiment conducted at several temperatures yields a data set for the determination of the pre-exponent and activation energy. For surface mass transfer self-diffusion on Si(100) in the temperature range from 1050 to 1200°C, the diffusion coefficient was found to be $9.5\times\exp(-2.2\,\text{eV}/k_B T)$ [cm^2/s].

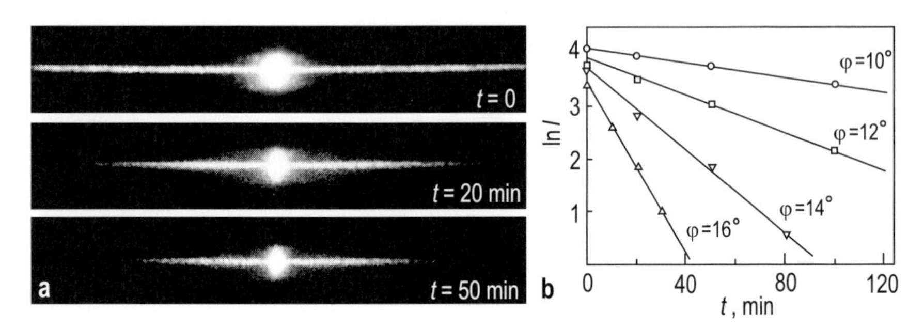

Fig. 13.27. Study of mass-transfer self-diffusion on Si(100) under capillarity-induced surface smoothing. (**a**) Light scattering patterns from the one-dimensional stochastic relief on Si(100) at various stages of annealing. (**b**) Time dependence of the scattered light intensity $I(\varphi, t)$ for selected scattering angles. Using (13.29), (13.30), and (13.31), the diffusion coefficient was extracted from the intensity decay rate. (after Gavrilyuk et al. [13.22])

13.9.4 Island Growth Technique

This technique is based on the analysis of the number density of stable islands formed after submonolayer deposition (see Sect. 14.2.1) and it allows the determination of the *intrinsic* diffusion coefficient of deposited adatoms. In the course of deposition, nucleation of the new islands and growth of the existing ones are competing processes; with increasing diffusion length of adatoms, the probability of adatom incorporation into an existing island dominates over nucleation. As a result, the number density of islands usually decreases with temperature. An accurate link between the island number density and the value of the surface diffusion coefficient is established within the framework of nucleation theory. In the general case, this relationship might be rather sophisticated. However, for some favorable physical situations it can be reduced to a simple power law. For example, if the re-evaporation is negligible, the critical island size is $i = 1$ (i.e., two atoms already form a stable island), the forming islands are two-dimensional, and the diffusion is isotropic, the number density of islands N is written as

$$N \sim \left(\frac{R\Theta}{\nu} \right)^{\chi} , \tag{13.32}$$

where R is the deposition rate, ν the adatom hopping rate, Θ the total coverage of the deposited adsorbate, and $\chi = i/(i+2) = 1/3$. An analysis based on the above assumptions was applied to island growth in the homoepitaxy of Fe on Fe(100) [13.23] and yielded a diffusion activation energy of $E_{\text{diff}} = 0.45 \pm 0.04$ eV and a pre-factor $D_0 = 7.2 \times 10^{-4}$ cm^{2}s^{-1} (Fig. 13.28).

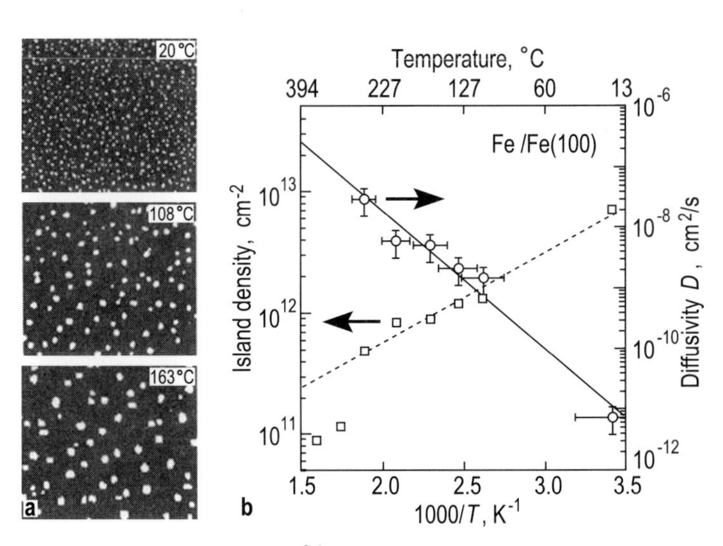

Fig. 13.28. (**a**) 1000×800 Å^2 STM images of single layer Fe islands on the Fe(100) surface formed upon deposition of 0.07 ML at a rate of 0.016 ML/s (1 ML = 1.214×10^{15} atoms/cm^2) at various temperatures. (**b**) The temperature dependence of the number density of Fe islands (squares) obtained from STM measurements as in (**a**) and the deduced diffusion coefficient (circles) using (13.32) (after Stroscio et al. [13.23])

Problems

13.1 Random-walk diffusion of Ag atoms occurs over the Si(111)$\sqrt{3}\times\sqrt{3}$-Ag surface. Estimate the mean displacement of the atom in a time of 1 s, 1 min, and 1 hour at 450 °C. $D_0 = 10^{-3}\,\mathrm{cm^2s^{-1}}$, E_{diff}=0.33 eV.

13.2 The hopping rate of a nitrogen atom on the Fe(100) surface is $10^{-3}\ \mathrm{s}^{-1}$ at 300 K and $3 \cdot 10^{-2}\ \mathrm{s}^{-1}$ at 330 K. Estimate the diffusion coefficient and calculate the activation energy. Take into account that Fe is a bcc crystal with lattice parameter 2.87 Å. Assume the vibration frequency is $\nu_0 = 4.3 \times 10^{12}\ \mathrm{s}^{-1}$.

13.3 After deposition of equal amounts of aluminum at the same deposition rate onto a Si surface, the number density of Al islands was found to be $10^{10}\,\mathrm{cm}^{-2}$ at 350 °C and $10^{12}\,\mathrm{cm}^{-2}$ at 80 °C. Estimate the activation energy for the surface diffusion of Al adatoms.

Further Reading

1. A.G. Naumovets, Yu.S. Vedula: *Surface Diffusion of Adsorbates.* Surf. Sci. Rep. **4**, 365–434 (1985)
2. R. Gomer: *Diffusion of Adsorbates on Metal Surfaces.* Rep. Prog. Phys. **53**, 917–1002 (1990)
3. G.L. Kellogg: *Field Ion Microscope Studies of Single-Atom Surface Diffusion and Cluster Nucleation on Metal Surfaces.* Surf. Sci. Rep. **21**, 1–88 (1994)
4. H.P. Bonzel: 'Surface Diffusion on Metals'. In: *Diffusion in Solid Metals and Alloys. Landolt Börnstein. Vol. III/26.* ed. by O. Madelung (Springer, Berlin, Heidelberg, New York 1990) Chapter 13 (basic definitions and comprehensive data set for diffusion in numerous particular systems)
5. H.P.E.G. Seebauer, M.Y.L. Jung: 'Surface Diffusion on Metals, Semiconductors, and Insulators'. In: *Physics of Covered Solid Surfaces. Landolt Börnstein. Vol. III/42A1.* ed. by H.P. Bonzel (Springer, Berlin, Heidelberg, New York 2001) Chapter 3.11 (basic definitions and comprehensive data set for diffusion in numerous particular systems)
6. H. Yasunaga, A. Natori: *Electromigration on Semiconductor Surfaces.* Surf. Sci. Rep. **15**, 205–280 (1992)

14. Growth of Thin Films

When the adsorbate coverage exceeds the monolayer range, one speaks about *thin film growth*. The oriented growth of a crystalline film on a single-crystal substrate is referred to as *epitaxy*, which, in turn, is subdivided into *homoepitaxy* (when both film and substrate are of the same material) and *heteroepitaxy* (when film and substrate are different). The film growth is controlled by the interplay of thermodynamics and kinetics. The general trends in film growth are understood within the thermodynamic approach in terms of the relative surface and interface energies. On the other hand, film growth is a non-equilibrium kinetic process, in which the rate-limiting steps affect the net growth mode. In this chapter, the surface phenomena involved in thin film growth and their effect on the growth mode, as well as on the structure and morphology of the grown films, are discussed.

14.1 Growth Modes

Three principal modes of film growth are generally distinguished (Fig. 14.1). These modes are named after their original investigators and are as follows:

- *Layer-by-layer*, or *Frank–van der Merve (FM)*, growth mode (Fig. 14.1a) refers to the case when the film atoms are more strongly bound to the substrate than to each other. As a result, each layer is fully completed before the next layer starts to grow, i.e., strictly two-dimensional growth takes place.
- *Island*, or *Vollmer–Weber (VW)*, growth mode (Fig. 14.1c) corresponds to the situation when film atoms are more strongly bound to each other than to the substrate. In this case, three-dimensional islands nucleate and grow directly on the substrate surface.
- *Layer-plus-island*, or *Stranski–Krastanov (SK)*, growth mode (Fig. 14.1b) represents the intermediate case between FM and VW growth. After the formation of a complete two-dimensional layer, the growth of three-dimensional islands takes place. The nature and thickness of the intermediate layer (often called the *Stranski–Krastanov layer*) depend on the particular case (for example, the layer might be a submonolayer surface phase or a strained film several monolayers thick).

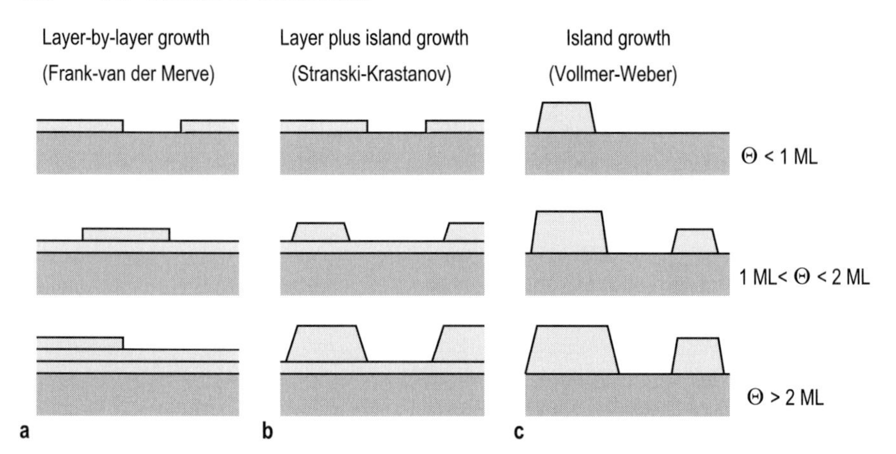

Fig. 14.1. Schematic representation of the three main growth modes: (**a**) layer-by-layer (Frank–van der Merve, FM) growth mode; (**b**) layer-plus-island (Stranski–Krastanov, SK) growth mode; (**c**) Island (Vollmer–Weber, VW) growth mode

The occurrence of different growth modes can be understood qualitatively in terms of the surface or interface tension, γ. Recall that the tension is defined as the work that must be performed to build a surface (or interface) of unit area. Bearing in mind that γ can also be interpreted as a force per unit length of boundary, consider the contact point of the film island and the substrate (Fig. 14.2). If the island wetting angle is φ, the force equilibrium can be written as

$$\gamma_{\mathrm{S}} = \gamma_{\mathrm{S/F}} + \gamma_{\mathrm{F}} \cos\varphi\ , \tag{14.1}$$

where γ_{S} is the surface tension of the substrate surface, γ_{F} the surface tension of the film surface, and $\gamma_{\mathrm{S/F}}$ the surface tension of the film/substrate interface. For the case of layer-by-layer (FM) growth, $\varphi = 0$, hence, from (14.1) one has the corresponding condition:

$$\gamma_{\mathrm{S}} \geq \gamma_{\mathrm{S/F}} + \gamma_{\mathrm{F}} \qquad \text{(layer-by-layer growth).} \tag{14.2}$$

For island growth (VW), $\varphi > 0$ and corresponding condition is

$$\gamma_{\mathrm{S}} < \gamma_{\mathrm{S/F}} + \gamma_{\mathrm{F}} \qquad \text{(island growth).} \tag{14.3}$$

For layer-plus-island (SK) growth, the condition (14.2) for layer growth is initially fulfilled, but the formation of the intermediate layer alters the values of γ_{S} and $\gamma_{\mathrm{S/F}}$, leading to the condition (14.3) for subsequent island growth.

The growth modes can be identified experimentally by monitoring the variation of the Auger signals from the film and substrate in the course of deposition. Typical plots are shown schematically in Fig. 14.3. Layer-by-layer growth manifests itself by segmented curves as in Fig. 14.3a. Island growth leads to a very slow increase and decrease of the film and substrate signals,

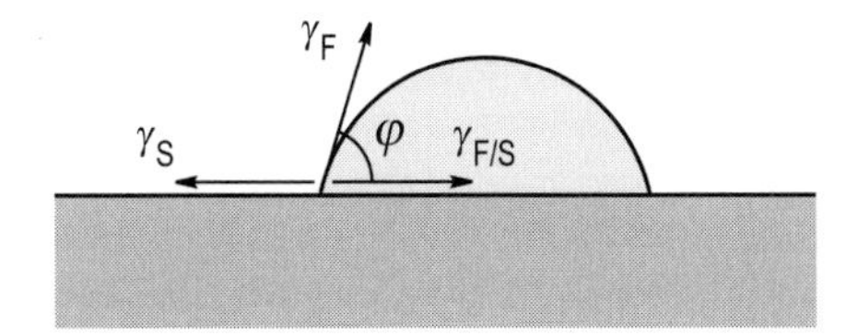

Fig. 14.2. Schematic of a film island on a substrate. γ_S is the surface tension of the substrate surface, γ_F the surface tension of the film surface, and $\gamma_{S/F}$ the surface tension of the film/substrate interface. The balance of forces acting along the substrate surface yields (14.1)

respectively (Fig. 14.3c). Stranski–Krastanov growth (Fig. 14.3b) is characterized by an initial linear segment, corresponding to an intermediate layer formation, followed by a sharp breakpoint, after which the Auger amplitude increases and decreases only slowly. The latter segment corresponds to island growth over the Stranski–Krastanov 2D layer.

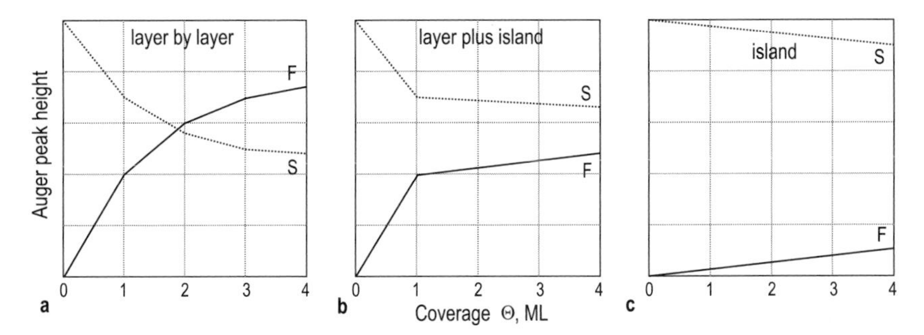

Fig. 14.3. Schematic plot of the Auger amplitude from the film (F) and substrate (S) versus the amount of deposited material for three main growth modes: (**a**) layer-by-layer (FM) growth; (**b**) layer-plus-island (SK) growth; and (**c**) island (VW) growth

14.2 Nucleation and Growth of Islands

14.2.1 Island Number Density

The main elementary processes involved in the formation and growth of islands are illustrated schematically in Fig. 14.4. Atoms arrive from the gaseous phase at a rate R and become accommodated at the surface as adatoms with a bound energy E_{ads}. This creates a population of single adatoms n_1 on the substrate with n_0 sites per unit area. Adatoms migrate over the surface with the diffusion coefficient $D = (\nu/4n_0)\exp(-E_{\text{diff}}/k_BT)$ until they are lost by

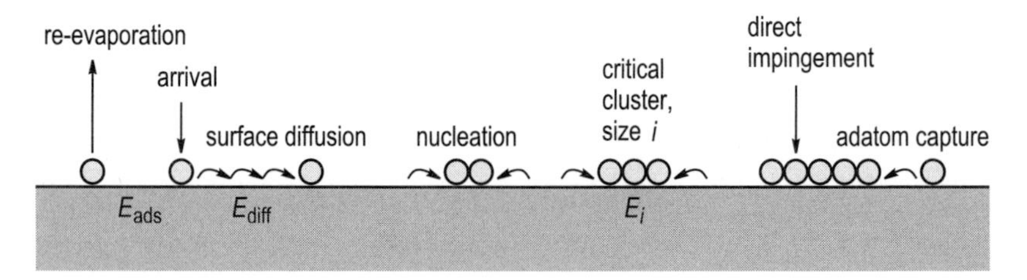

Fig. 14.4. Schematic diagram illustrating the atomic processes involved in nucleation and growth on surfaces

one of the following processes. First, they might be re-evaporated, if the substrate temperature is high enough. The re-evaporation is characterized by the residence time $\tau_{\mathrm{ads}} = \nu^{-1}\exp(E_{\mathrm{ads}}/k_{\mathrm{B}}T)$. Second, adatoms might become captured by existing clusters or at defect sites such as steps. Third, adatoms might combine with one another to form a cluster. The small clusters are metastable and often decay back into individual atoms. However, when the cluster grows in size, it becomes more stable and the probability of its growth is greater than the probability of decay. The *critical island size i* is defined as the minimal size when the addition of just one more atom makes the island stable.

The atomistic processes can be described quantitatively in terms of the rate equations. The consideration is based on the assessment of the formation and decay rates of the clusters. As an example, Fig. 14.5 illustrates the fluxes controlling the population n_j of the metastable clusters of size $j < i$, where i is the critical cluster size.

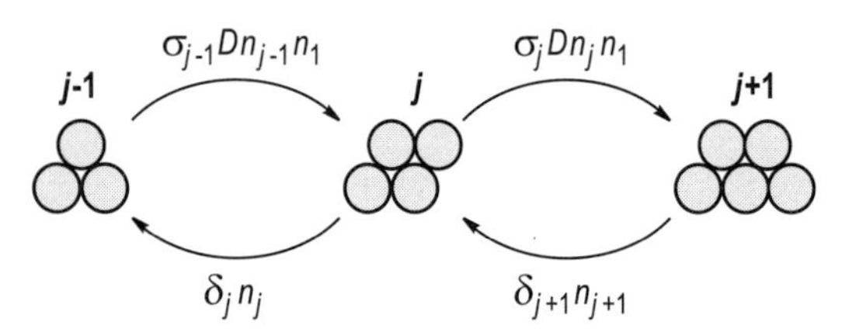

Fig. 14.5. Schematic diagram showing the fluxes which control the number density of clusters of size j

Four processes affect the number density of j-clusters, n_j.

- Two processes contribute to the increase of n_j.
  - First, an additional cluster of size j is formed, when an adatom is attached to the cluster of size $j - 1$. The net flux due to this process is expressed as $\sigma_{j-1}Dn_{j-1}n_1$.

  - Second, the detachment of the atom from the cluster of size $j+1$ (i.e., decay of the $(j+1)$-cluster) produces a cluster of size j and an adatom. The net flux of the decay is $\delta_{j+1} n_{j+1}$.
- Two processes decrease n_j.
  - First, attachment of atoms to j-clusters transforms them to $(j+1)$-clusters with a net rate $\sigma_j D n_j n_1$.
  - Second, decay of j-clusters produces $(j-1)$-clusters with a net rate $\delta_j n_j$.

Here n_1 is the number density of adatoms, D the diffusion coefficient, σ the capture number, which describes the capability of islands to capture diffusing adatoms, and $\delta_{j+1} \sim D \exp(-\Delta E_j^{j+1}/k_B T)$ the decay rate with ΔE_j^{j+1} being the energy difference between the $(j+1)$-cluster and the j-cluster.

Using an essentially similar approach for the evaluation of the number density of adatoms, n_1, and that of stable clusters with $j > i$, denoted n_x, one has the following set of rate equations:

$$\frac{dn_1}{dt} = R - \frac{n_1}{\tau_{ads}} + \left(2\delta_2 n_2 + \sum_{j=3}^{i} \delta_j n_j - 2\sigma_1 D n_1^2 - n_1 \sum_{j=2}^{i} \sigma_j D n_j \right) - n_1 \sigma_x D n_x \tag{14.4a}$$

$$\frac{dn_j}{dt} = n_1 \sigma_{j-1} D n_{j-1} - \delta_j n_j + \delta_{j+1} n_{j+1} - n_1 \sigma_j D n_j \tag{14.4b}$$

$$\frac{dn_x}{dt} = n_1 \sigma_i D n_i. \tag{14.4c}$$

Equation (14.4a) describes the time variation of the adatom density n_1. It denotes an increase in n_1 due to deposition with a flux R and decrease due to desorption at a rate n_1/τ_{ads}. The terms bracketed together represent the supply and consumption rates due to formation and decay of subcritical clusters. $2\delta_2 n_2$ and $2\sigma_1 D n_1^2$ stand for the decay and formation of dimers (the factor 2 indicates that in each of these processes adatoms are supplied or consumed as pairs). The σ terms are for the decay and formation of clusters of size from 3 to i. The last term is the net capture rate of stable clusters larger than i.

Equation (14.4b) is for the density of metastable clusters of size j. The terms in the right part are illustrated in Fig. 14.5 and have been discussed above. Equation (14.4c) describes the growth of the stable cluster density n_x due to attachment of adatoms to critical size clusters. Note that in these equations, some processes (for example, coalescence of clusters or direct impingement onto clusters and adatoms) are neglected, but can in principle be taken into account by adding appropriate terms.

Integration of (14.4a)–(14.4c) gives the time evolution of island and adatom densities. As an example, Fig. 14.6 shows the results of numerical calculations for the case of $i = 1$ (i.e., when a dimer is already a stable cluster) and sufficiently low temperatures, when one might neglect adatom re-evaporation and dimer mobility. One can see that the dynamic behavior of the density of adatoms (n_1) and islands (n_x) can be divided into four coverage regimes:

- low-coverage nucleation regime (labeled L in Fig. 14.6),
- intermediate-coverage regime (labeled I),
- aggregation regime (labeled A) and
- coalescence and percolation regime (labeled C).

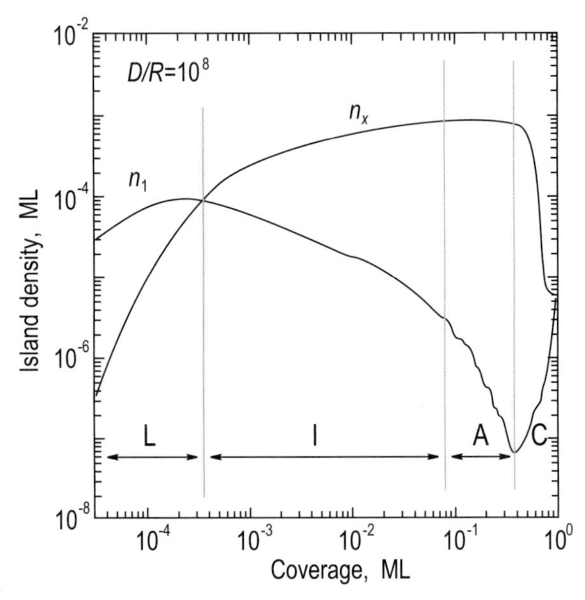

Fig. 14.6. Number density of adatoms (n_1) and islands (n_x) versus coverage in the case $i = 1$ and $D/R = 10^8$, showing four regimes: a low-coverage nucleation regime (L); an intermediate-coverage regime (I); aggregation regime (A); and a coalescence and percolation regime (C) (after Amar, Family and Lam [14.1])

In the initial, *low-coverage nucleation regime*, the adatom density is much higher than the island density, so the probability of island nucleation far exceeds the probability of an adatom to become incorporated into an existing island. In this regime, $n_1 \propto \Theta$ and $n_x \propto \Theta^3$. In the course of deposition, the number density of islands increases until it becomes comparable to the density of adatoms. The latter means that the probabilities of an adatom to encounter another adatom or an island become also comparable. This point corresponds to the onset of the *intermediate-coverage regime*. In this regime, the adatom density peaks and begins to decrease ($n_1 \propto \Theta^{-1/3}$), while the island density still increases but much more slowly ($n_x \propto \Theta^{1/3}$). When the density of islands increases so that the mean island separation is equal to the mean free path of migrating adatoms, any further deposition exclusively results in island growth and the island density attains a *saturation* value. This is defined as the *aggregation regime* and usually occurs at a coverage between 0.1 and 0.4 ML. Finally, in the *coalescence and percolation regime*, the islands join together (coalesce) and percolate, which leads to a decrease of the island number density. Eventually second-layer growth starts and the adatom density again increases.

For the case of *complete condensation* (i.e., with negligible re-evaporation), the *saturation density* of 2D islands, normalized to the density of adsorption sites n_0, is given by

$$\frac{n_x}{n_0} = \eta(\Theta, i) \left(\frac{R}{D n_0^2} \right)^{\chi} \exp\left(\frac{E_i}{(i+2) k_{\mathrm{B}} T} \right) , \tag{14.5}$$

where the scaling exponent is $\chi = i/(i+2)$ with i being the critical cluster size. E_i denotes the binding energy of the critical cluster, given approximately by the number of nearest-neighbor atom bonds. $\eta(\Theta, i)$ is a pre-exponential numerical factor, which can generally vary from 10^{-2} to 10, depending on the regime of condensation, the coverage, and the critical cluster size. In the coverage range of interest, 0.1 to 0.4 ML, $\eta(\Theta, i)$ is a weak function of Θ with typical values in the range from 0.1 to 1 (for example, for $i = 1$–5 $\eta \simeq 0.2$–0.3) [14.2].

Taking into account the temperature dependence of the diffusion coefficient, (14.5) can be rewritten for a square lattice as

$$\frac{n_x}{n_0} = \eta(\Theta, i) \left(\frac{4R}{\nu_0 n_0} \right)^{i/(i+2)} \exp\left(\frac{i E_{\mathrm{diff}} + E_i}{(i+2) k_{\mathrm{B}} T} \right) . \tag{14.6}$$

The application of the rate theory analysis for evaluation of the experimental data is illustrated by Figs. 14.7 and 14.8, demonstrating the results of the STM study of the initial stages of Cu epitaxy on Ni(100) [14.3]. Figure 14.7 shows the temperature dependence of the saturation island density as an Arrhenius plot. Three regions with different slopes are clearly distinguished. The temperature-independent regime (labeled post-nucleation) was attributed to the incorporation of adatoms into existing islands after termination of deposition. The two regimes entered above 160 and 320 K, respectively, were found to correspond to the different size of the critical cluster. In order to establish these sizes, the rate dependence of island density was measured at 215 and 345 K, located in the center of corresponding regions in the Arrhenius plot (Fig. 14.8). At 215 K the exponent corresponds to 0.32±0.01, which yields $i = 1$ according to (14.6). At 345 K exponent is 0.58±0.02 and, consequently, $i = 3$. With knowledge of the sizes of critical clusters, the diffusion activation energy E_{diff}, dimer bond energy E_{b}, and attempt frequency ν_0 can be evaluated by analyzing the Arrhenius plot of saturation island density in Fig. 14.7. They appear to be $E_{\mathrm{diff}} = 0.351 \pm 0.017$ eV, $E_{\mathrm{b}} = 0.46 \pm 0.19$ eV (in evaluation, taking into account that adsorption sites form a square lattice, it was assumed $E_3 \cong 2E_b$) and $\nu_0 = 4 \times 10^{11 \pm 0.3}$ Hz for $i = 1$ data and $5 \times 10^{12 \pm 2}$ Hz for $i = 3$ data.

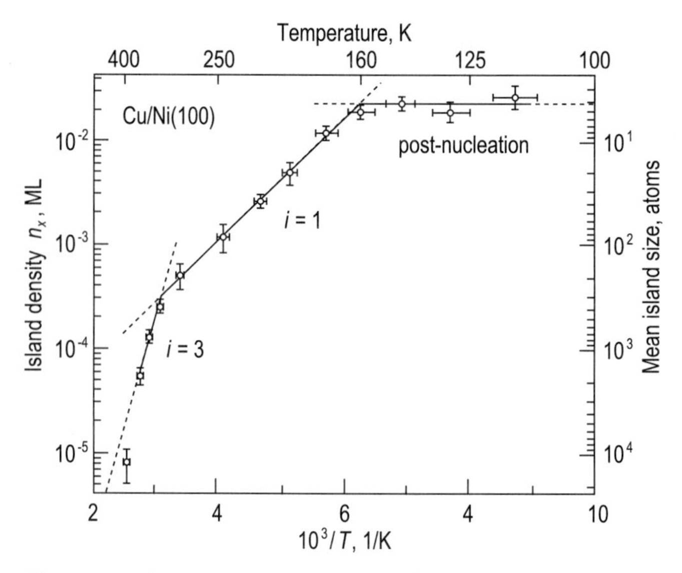

Fig. 14.7. Arrhenius plot of measured saturation island density of Cu on Ni(100) at a coverage of 0.1 ML and a flux of 1.34×10^{-3} ML/s (after Müller et al. [14.3])

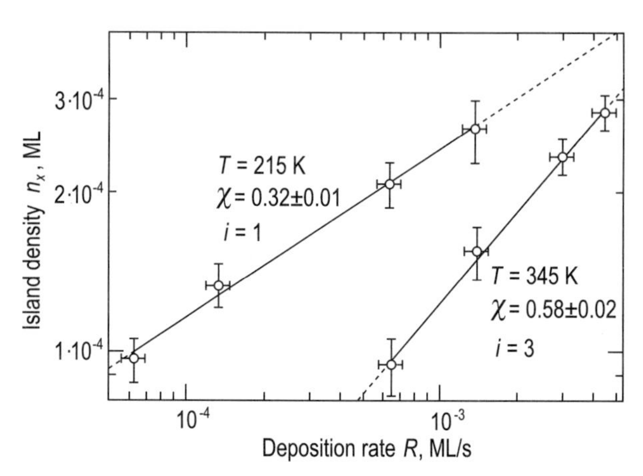

Fig. 14.8. Double-logarithmic plot of saturation island density versus deposition flux at different growth temperatures. The slope of the plot is described by the scaling exponent $\chi = i/(i+2)$, which allows the determination of the critical cluster size i (after Müller et al. [14.3])

14.2.2 Island Shape

Depending on growth conditions, the shapes of islands might be quit different. According to the compactness of their shape, islands can be subdivided into two general classes:

- ramified islands (i.e., fractal-like, dendrite shape, having rough island edges); and
- Compact islands (i.e., square, rectangular, triangular or hexagonal shape with relatively straight and equiaxial island edges).

The compactness of the island is largely controlled by the ability of the captured adatom to diffuse along the island edges and to cross corners, where two edges meet.

Typically, the formation of the *ramified islands* takes place at relatively low temperatures, when the edge diffusion is slow. In the ultimate case (so-called, *hit-and-stick* regime), an adatom sticks to an island and stays immobile at the impact site. This case is described by the classic *diffusion-limited-aggregation (DLA) model* [14.4], which predicts that under these conditions fractal islands with average branch thickness of about one atom width will be formed, irrespective of lattice geometry (Fig. 14.9). In real growth, the classic DLA mechanism does not occur, as adatoms reaching an island always walk a certain path to find an energetically more favorable site. The higher the rate of edge diffusion, the greater the branch thickness. The fractal islands observed in STM experiments all have branch thickness exceeding one atom width predicted by the classic DLA model. Even if the adatom can diffuse along the island edge, it might be unable to cross the corner of the island, since in passing across the corner the adatom has to lower its coordination. Without corner crossing, growth also leads to the formation of ramified fractal-like islands.

Low-temperature Pt growth on Pt(111) [14.5] furnishes an example of ramified island formation (Fig. 14.10). The developed fractal islands have a branch thickness of about four atoms. They display a clear trigonal symmetry. To account for the growth anisotropy, an asymmetric probability of a onefold-coordinated adatom at the corner to jump to either side was suggested. In Fig. 14.10c, for each of the onefold-coordinated atoms (shown as lightly shaded circles) two different paths to adjacent twofold coordinated sites are indicated: An atom moving along a path indicated by a solid arrow is able to remain at a higher coordination with respect to the substrate (passing via a bridge position) than an atom moving along the dashed arrow (via an atop position). Thus, the paths along the solid arrows are preferable. Monte Carlo simulation with the above assumptions (Fig. 14.10c) reproduces well the shape of Pt islands observed in experiment.

When adatoms can easily cross the island corners, the growth results in the formation of *compact islands*. The change from fractal-like to compact island growth takes place with increasing temperature. For the Pt/Pt(111)

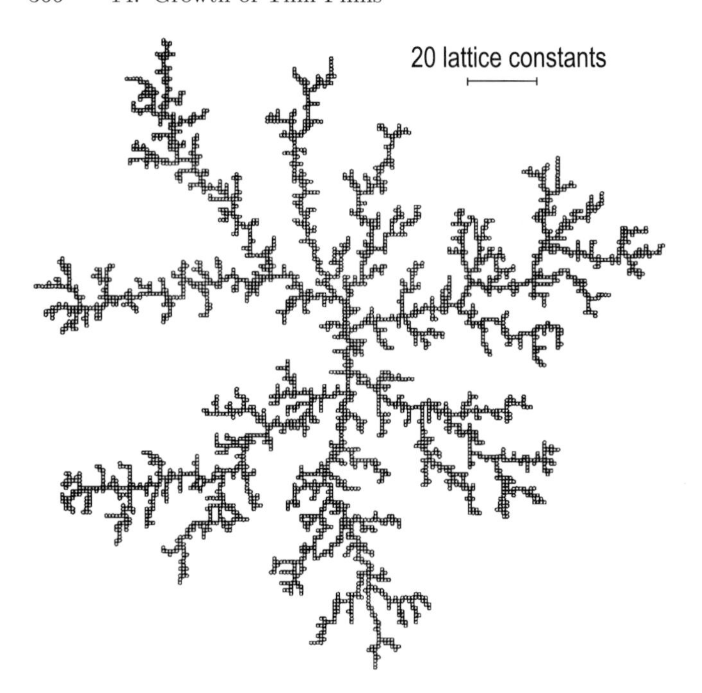

Fig. 14.9. Classical DLA aggregate of 3600 particles on a square lattice (after Witten and Sander [14.4])

system, the transition occurs between 300 and 400 K, leading to the growth of triangular Pt islands (Fig. 14.11b). In general, the compact island shape is controlled by the competition between steps of different orientations in accommodating arriving adatoms. Many rate processes are involved in the growth and the shape of the growing island might differ from the equilibrium shape due to kinetic limitations. For example, the equilibrium shape of a monolayer Pt island on Pt(111) is not a triangle (as formed in the course of growth) (Fig. 14.11b), but rather a hexagon (as obtained after subsequent annealing) (Fig. 14.12a).

For 3D crystallites, the equilibrium shape reflects the anisotropy of the specific surface free energy (see Sect. 10.1.2). For 2D islands, the equilibrium shape reflects the anisotropy of the specific *step free energy*. The 2D Wulff theorem reads: *In a 2D crystal at equilibrium, the distances of the borders from the crystal center are proportional to their free energy per unit length* (Fig. 14.12b). Hence, the ratio of the specific free energies of steps, $\gamma_{\mathrm{A}}/\gamma_{\mathrm{B}}$ can be readily evaluated from the aspect ratio, $r_{\mathrm{A}}/r_{\mathrm{B}}$. For Pt(111), the free energy ratio of B- and A-steps is determined to be 0.87 ± 0.02 [14.7].

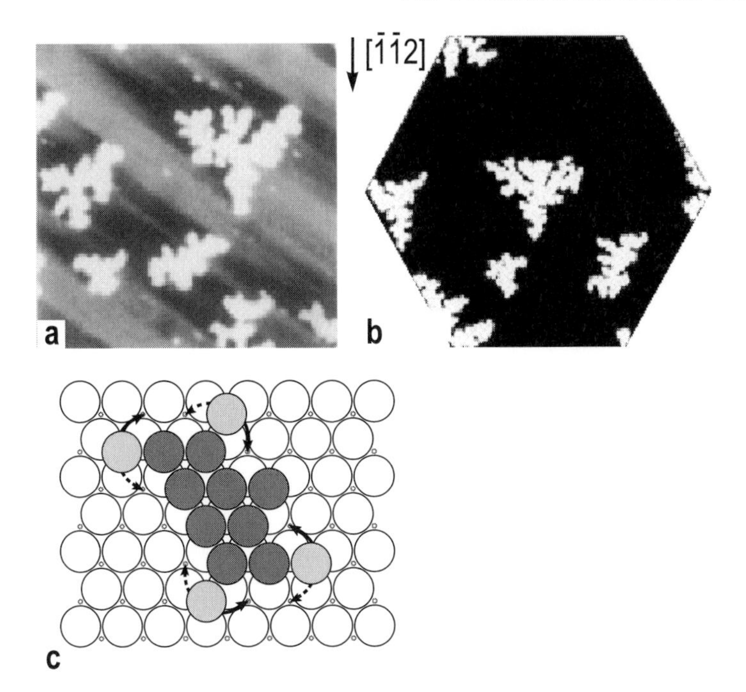

Fig. 14.10. (**a**) Experimental (500×500 Å^2 STM image) and (**b**) simulated island shapes in the Pt/Pt(111) system at 245 K. (**c**) Ball model illustrating the mechanism of anisotropic island growth. The lightly shaded atoms at the corners are onefold coordinated. Dashed and full arrows indicate diffusional jumps into twofold coordinated sites via atop positions (non-preferential path) and via bridge positions (preferential path), respectively (after Hohage et al. [14.5])

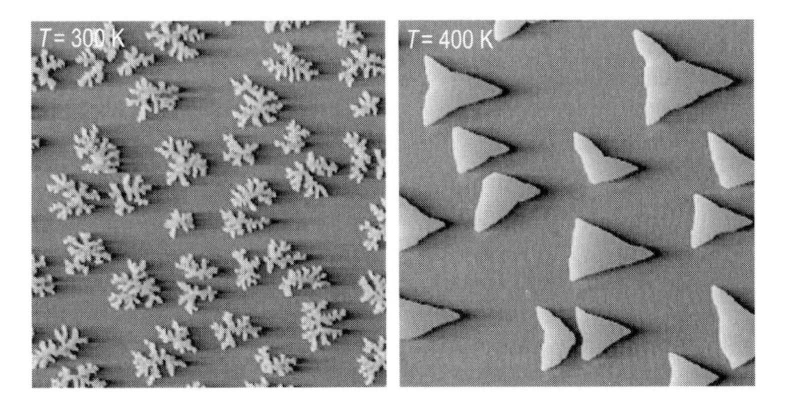

Fig. 14.11. Effect of growth temperature on island shapes in the homoepitaxial growth of Pt on Pt(111). (**a**) Growth at 300 K results in the formation of ramified islands. (**b**) At 400 K compact islands of triangular shape are formed (after Bott et al. [14.6])

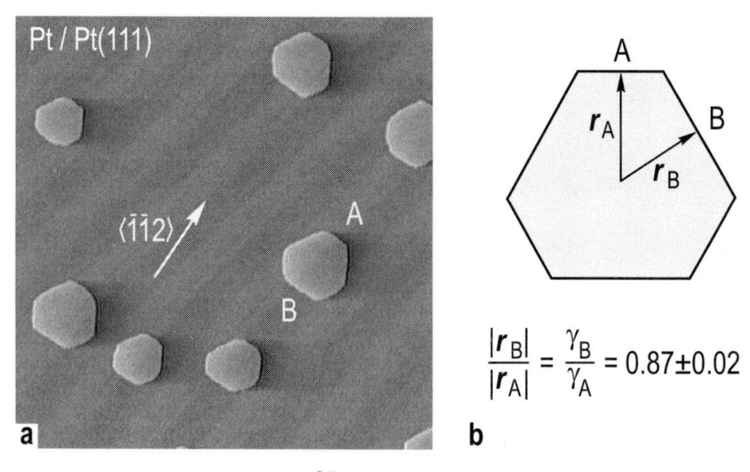

Fig. 14.12. (**a**) 2000×2000 Å^2 STM image of monatomic Pt islands on Pt(111) formed by deposition of 0.1 ML at 425 K followed by annealing to 700 K for 60 s. The islands reveal their equilibrium shape with A-step segments being considerably smaller than B-step segments (measured length ratio 0.66±0.05). (**b**) Schematic diagram illustrating the application of the 2D Wulff theorem for the determination of the ratio of the step free energies. For Pt(111) the ratio of the free energies of the B-step and A-step is found to be 0.87±0.02 [14.7] (after Michely [14.8])

14.2.3 Island Size Distribution

The size distribution of nucleated islands is affected by a set of parameters, including critical island size, coverage, and substrate structure. At relatively high coverages, it is strongly influenced by coarsening phenomena, such as ripening and coalescence of islands.

Effect of the Critical Island Size. The *island size distribution function* N_s gives the density of islands of size s (where s is the number of atoms in the island). Hence, the total number of stable islands N and coverage Θ can be expressed as

$$N = \sum_{s>i} N_s \quad \text{and} \quad \Theta = \sum_{s\geq 1} sN_s \,. \tag{14.7}$$

In these terms, the average island size $\langle s \rangle$ is defined as

$$\langle s \rangle = \frac{\sum\limits_{s>i} sN_s}{\sum\limits_{s>i} N_s} = \frac{\Theta - \sum\limits_{s\leq i} SN_s}{N} \,. \tag{14.8}$$

For a small critical size, such as $i = 1$,

$$\langle s \rangle = \frac{\Theta - N_1}{N} \simeq \frac{\Theta}{N} \,, \tag{14.9}$$

as the relative number of single adatoms N_1 is small compared to the total number of islands N. Under these assumptions, *scaling theory* [14.1] yields the scaling relation as

$$N_s = \Theta \langle s \rangle^{-2} f_i(s/\langle s \rangle) \ , \tag{14.10}$$

where $f_i(s/\langle s \rangle)$ is the *scaling function* for the island size distribution for a critical island size of i. Equation (14.10) holds when the Dn_0^2/R ratio is high. A plot of the simulated scaling function for i from 1 to 4 [14.1] is shown in Fig. 14.13. As one can see, the peak of the island size distribution increases and becomes more sharply peaked with increasing i.

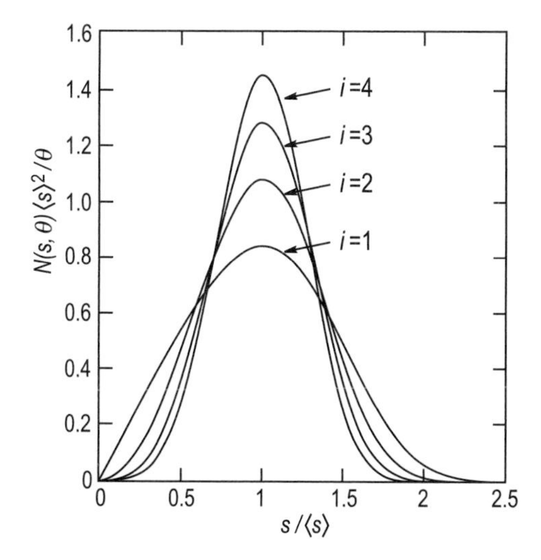

Fig. 14.13. Simulated scaling function f_i for the island size size distribution for critical island size of $i = 1$–4 (after Amar et al. [14.1])

Comparing the scaling function obtained in the experiment with the simulated ones (such as those shown in Fig. 14.13), one can determine the size of the critical island. To obtain the experimental scaling function, one has to scale the measured island size distribution $N(s)$ into $N_s(s/\langle s \rangle)$, and to plot $N_s(s/\langle s \rangle)\langle s \rangle^2/\Theta$ versus $s/\langle s \rangle$. From (14.10), it is clear that the plot will yield the scaling function f. The usefulness of such determination procedure is illustrated in Fig. 14.14, which presents the results of the evaluation for the case of Fe island growth on Fe(100). Figure 14.14a shows the island size distribution obtained from STM images for growth at relatively low temperatures, 20, 132, and 207°C. Figure 14.14b demonstrates the island size distribution at relatively high growth temperature of 356°C. Figure 14.14c shows the *scaled* island size distribution (i.e., the experimental scaling function f) for low-temperature growth at 20, 132, and 207°C. One can see that after scaling all three distributions collapse into a single curve, which resembles the simulated f plot for a critical size $i = 1$, taken from Fig. 14.13. For the higher growth temperature of 356°C, the scaled island size distribution fits the simulated f plot for $i = 3$ (Fig. 14.14d). Thus, the obtained results indicate that in the temperature range from 20 to 207°C the critical island size is one atom ($i = 1$), but it becomes three atoms ($i = 3$) at the higher temperature of 356°C.

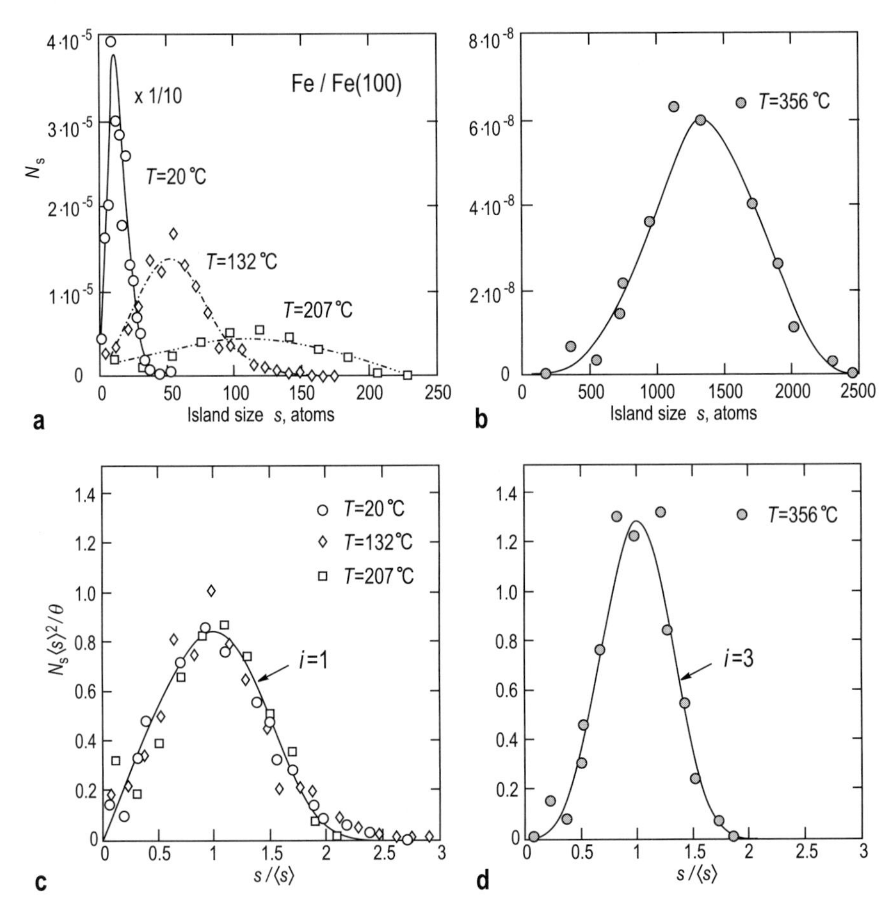

Fig. 14.14. Evaluation of the critical island size for Fe on Fe(100) growth from the scaled island size distribution. (**a**) Experimental Fe island size distribution at low growth temperatures, 20, 132, and 207°C. The island density per lattice site (2.8×2.8 Å^2) N_s is plotted against the number of atoms (size) s. (**b**) The same as (**a**) but for higher growth temperature of 356°C. (**c**) Scaled island size distributions for low growth temperatures, 20, 132, and 207°C, superposed on the simulated scaling function f for critical island size $i = 1$ taken from Fig. 14.13. (**d**) Scaled island size distributions for high growth temperature, 356°C, superposed on the simulated scaling function f for critical island size $i = 3$ (after Stroscio et al. [14.9] and Amar et al. [14.1])

Coarsening Phenomena. The process of increasing the mean island size at the expense of decreasing the number of islands is referred to as *coarsening*. There are two main mechanisms of coarsening:

- coalescence (merging of islands upon contact) and
- ripening (growth of larger islands due to the diffusion flux of adatoms detached from smaller islands).

In *coalescence*, two initially separated islands come into direct contact and transform to a single island (Fig. 14.15). For monatomic islands, the area of the new, larger island is the sum of the areas of the initial islands. The shape of the forming island depends on the edge mobility of atoms. In the case of low edge mobility, the islands attach to each other without any reshaping. However, often the mobility is sufficient to allow the forming island to recover the equilibrium shape, as illustrated in the schematic example in Fig. 14.15.

If the islands are mobile, they can encounter each other and coalesce into larger islands. This process is called *dynamic coalescence* or *Smoluchowski ripening* (named after Smoluchowski, who formulated in 1916 a kinetic theory of coarsening of colloid particles via their dynamic coalescence). *Static coalescence* corresponds to the case when neighboring immobile islands coalesce as their size increases in the coarse of deposition. Dynamic coalescence can take place at low coverages around 0.1 ML, while static coalescence requires higher coverages, typically of about 0.4–0.5 ML. As coverage increases further, the coalescing system reaches a threshold for *percolation* growth, i.e., for the formation of interconnected structures. The onset of percolation manifests itself by an abrupt change in several physical parameters (for example, in the conductivity of a metallic film).

Fig. 14.15. Schematic diagram showing sequential stages of coalescence

Ripening is often referred to in the literature as *Ostwald ripening* (named after Ostwald, who in 1900 described the change of size of granules imbedded in a solid matrix by the diffusion flow of particles from one grain to another). In the case of 2D monatomic islands, ripening is a thermodynamically driven effort to reduce the free energy cost associated with island edges. According to the Gibbs–Thompson relation, the chemical potential of a circular island of radius r is given by

$$\mu(r) = \Omega \frac{\gamma}{r} , \tag{14.11}$$

where γ is the step line tension and Ω is the area occupied by one atom. Hence, smaller islands have a higher pressure of adatoms, which results in a net flow of material from smaller to larger islands through the 2D adatom gas (Fig. 14.16). Note that ripening occurs only if the net supersaturation of adatoms is very small and, in most experimental studies of island ripening, the post-deposition behavior of islands is monitored.

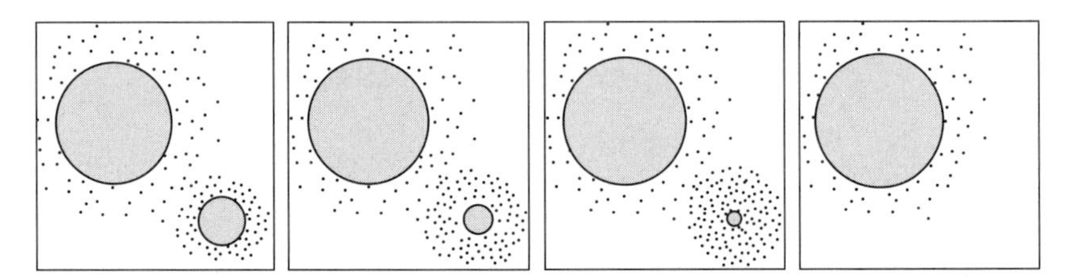

Fig. 14.16. Schematic diagram showing sequential stages of island ripening

Magic Islands. When due to their specific structure islands of selected sizes demonstrate enhanced stability, one speaks of *magic islands* (or *magic clusters*). The two examples shown below refer to triangular islands, for which the lateral growth of rows is hindered by the barrier for the formation of a new row. This leads to a kinetic stabilization of islands of magic sizes, which display pronounced peaks in the island size distribution.

The first example concerns the formation of atomic-scale 2D Ga islands upon deposition of ~0.15 ML of Ga onto the Si(111)$\sqrt{3}\times\sqrt{3}$-Ga surface followed by annealing to 200–500°C [14.10]. STM observations reveal that most of the forming islands are of triangular shape with apparent size preference (Fig. 14.17). The island size distribution clearly shows the existence of magic cluster sizes of $N(N+1)/2$, where N (2, 3, 4 or 5) is the number of atoms on each side of the triangular islands. Among others, decamer ($N = 4$) is the most abundant and stable species.

The second example is provided by the growth of epitaxial 2D Si islands upon deposition of 0.14 ML onto the Si(111)7×7 surface held at 450°C [14.11]. As in the above example, the Si islands also display a triangular shape, but they are much greater in size. Due to the 7×7 reconstruction, the island "building brick" is one-half of the 7×7 unit cell (HUC), containing 51 atoms. The island size distribution plotted in HUC units is shown Fig. 14.18. Several peaks corresponding to the magic island sizes N^2 ($N = 2, 3, 4, 5, 6$ or 7) HUCs are apparent.

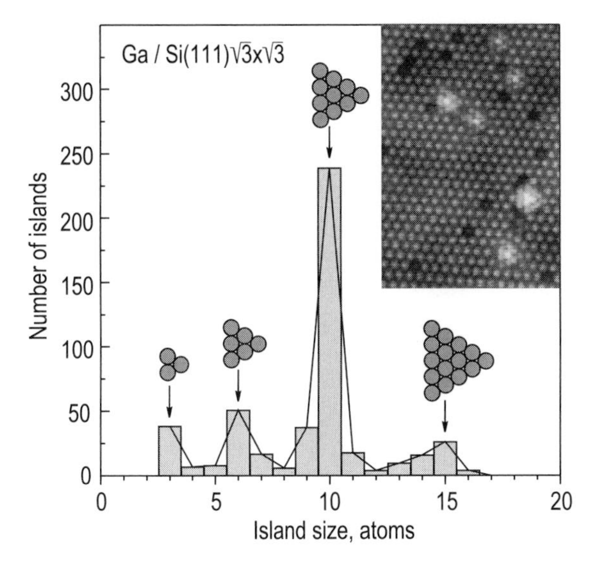

Fig. 14.17. Island size distribution of 2D Ga islands on Si(111)$\sqrt{3}\times\sqrt{3}$-Ga, indicating the existence of *magic sizes*. The inset shows an STM image (110×160 Å^2) of triangular magic islands (after Lai and Wang [14.10])

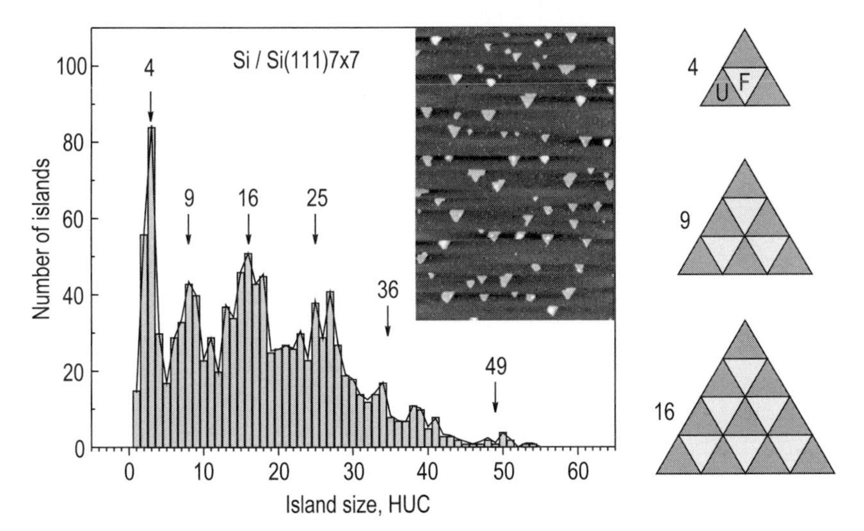

Fig. 14.18. Island size distribution of 2D islands epitaxially grown on Si(111) at 450°C. The distribution displays several peaks at *magic sizes*. The island size is expressed in units of one-half of the 7×7 unit cell (HUC). An STM image (2000×3000 Å^2) of epitaxial triangular islands is shown in the inset. As an example, the structure of magic islands of size 4, 9, and 16 HUCs is shown schematically in the right panel (after Voigtländer et al. [14.11])

14.2.4 Vacancy Islands

By analogy with the formation of adatom islands in the course of deposition, the agglomeration of surface vacancies produced by ion bombardment into pits can be visualized as the formation of *vacancy islands*. In this description, deep pit formation corresponds to the growth of three-dimensional vacancy islands and layer-by-layer sputtering to layer-by-layer growth. The regularities of the behavior of vacancy and adatom islands are almost common. For example, the equilibrium shape of both type of islands is controlled by the minimization of the free energy of the step edge bordering the island. As a result, vacancy and adatom islands display a similar equilibrium shape with the only difference being that they are rotated by 180° with respect to each other (Fig. 14.19). One can see that this is a natural consequence of the fact that step-down orientations are reversed in the vacancy island, as compared to the adatom island.

Fig. 14.19. STM image (2400×2400 Å^2) of equilibrium-shape adatom and vacancy islands on the Pt(111) surface (after Michely and Comsa [14.7])

Like that of adatom islands, the evolution of vacancy islands is also governed by processes such as island migration, island ripening, and coalescence. As an example, Fig. 14.20 illustrates the coalescence of two vacancy islands on Ag(111), as observed by STM.

14.3 Kinetic Effects in Homoepitaxial Growth

For the case of *homoepitaxy* (where the substrate and film are of the same chemical species), thermodynamic considerations, based on the balancing of the free energies of the film surface, substrate surface, and the substrate/film

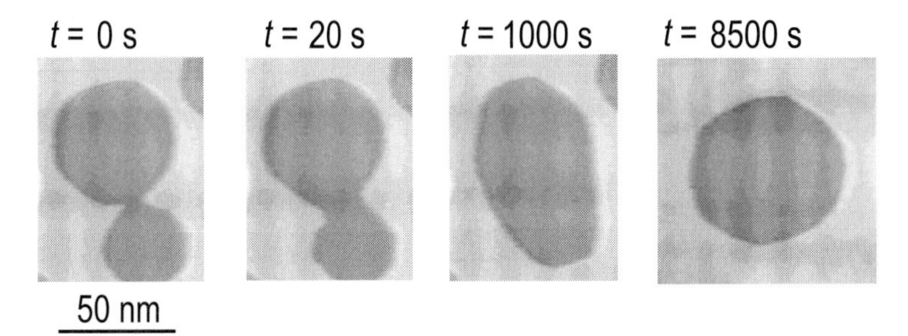

Fig. 14.20. STM snapshots showing sequential stages of the coalescence of two vacancy islands on Ag(111) at room temperature (after Rosenfeld et al. [14.12])

interface, predict a layer-by-layer, Frank–van der Merve, growth mode (see Sect. 14.1). However, growth occurs at conditions which are often far from equilibrium, and kinetic limitations associated with the finite rates of mass transport processes can greatly affect the actual growth mode.

There are two diffusion processes of prime interest:

- *intralayer* mass transport (i.e., diffusion of atoms on a flat terrace); and
- *interlayer* mass transport (i.e., diffusion of atoms across a step edge).

To characterize interlayer mass transport, consider the shape of the potential energy curve for an adatom near a step (Fig. 14.21). As one can see, an adatom encountering the step at the lower side would stick to the step, as the lower-terrace adsorption site adjacent to the step edge has higher coordination than that for an adatom on a terrace and, hence, a higher binding energy. Therefore, upward diffusion is usually neglected. An adatom reaching the step edge at the upper side encounters a barrier, which can be higher than the terrace diffusion barrier E_{diff}. The additional barrier E_{ES}, known as the *Ehrlich–Schwoebel barrier* [14.13, 14.14], appears as an adatom reduces its coordination while crossing the step edge. The efficiency of interlayer mass transport can be expressed by the transmission factor

$$s = \exp\left(-\frac{E_{\text{ES}}}{k_{\text{B}}T}\right) , \tag{14.12}$$

which gives the probability for an adatom to cross the step (relative to terrace diffusion and under the assumption of equal prefactors for terrace and step-down diffusion) [14.15]. If the Ehrlich–Schwoebel barrier is negligible, $s \simeq 1$, while for an insurmountable barrier, $s \simeq 0$.

Depending on the relative rate of intralayer and interlayer mass transport, homoepitaxial growth proceeds according to one of three possible growth modes shown in Fig. 14.22. These modes are

- step-flow growth;
- layer-by-layer growth;
- multilayer growth.

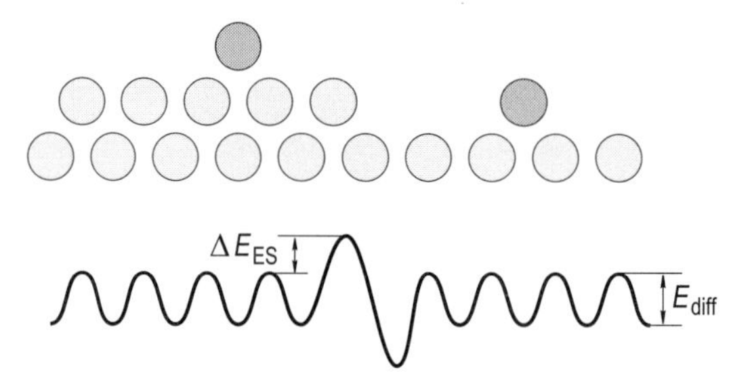

Fig. 14.21. Schematic representation of the potential energy variations associated with a monatomic step. E_{ES} denotes the Ehrlich–Schwoebel barrier, which an adatom (dark circle) has to surmount in addition to the terrace diffusion barrier E_{diff} in order to cross the step edge in the downward direction (after Schwoebel and Shipsey [14.14])

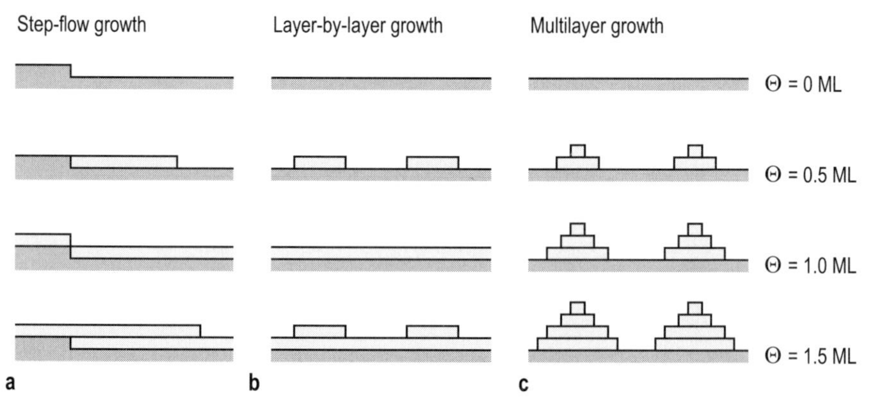

Fig. 14.22. Modes of homoepitaxial growth: (**a**) step-flow growth; (**b**) layer-by-layer growth; (**c**) multilayer growth (after Rosenfeld et al. [14.15])

Step-flow mode (Fig. 14.22a) takes place when the growth proceeds under conditions close to equilibrium. In this case, the adatom supersaturation is low and the adatom intralayer mobility is high, hence all adatoms reach the step before nucleation on the terrace sets in. As one can see, interlayer mass transport does not play an essential role in this growth mode.

If the terrace width exceeds the adatom migration length, the growth proceeds via nucleation and growth of adatom islands on terraces. In this case, depending on the rate of interlayer mass transport, *layer-by-layer* growth or *multilayer* growth take place. Layer-by-layer growth (Fig. 14.22b) requires a sufficient interlayer mass transport to ensure that all atoms, which are deposited onto the top of the growing island, would reach the island edge and jump to the lower layer. Ideally, $s = 1$ and a new layer starts to grow only after the previous layer has been entirely completed. If the interlayer mass transport is suppressed (i.e., $s \simeq 0$), the adatoms cannot escape from the top of the island and promote there the earlier nucleation of the new layer. As a result, multilayer 3D growth (Fig. 14.22c) takes place.

The growth mode can be altered by changing the deposition rate and/or substrate temperature. The dependence of the growth mode on these deposition parameters can be summarized in the form of a "growth mode diagram" shown in Fig. 14.23. The theoretical consideration of island nucleation assisted by interlayer diffusion shows that the dividing line between layer-by-layer and multilayer growth is defined by the condition [14.15]:

$$s{\cdot}\lambda = 1 \Leftrightarrow R = \lambda_0^{2(i+2)/i} \exp\left(-\frac{E_i/i + E_{\text{diff}} + E_{\text{ES}} \cdot 2(i+2)/i}{k_{\text{B}}T}\right), \quad (14.13)$$

while the transition to step-flow growth, which depends on the terrace width L, is given by the condition:

$$\lambda \cong L \Leftrightarrow R = \left(\frac{\lambda_0}{L}\right)^{2(i+2)/i} \exp\left(-\frac{E_i/i + E_{\text{diff}}}{k_{\text{B}}T}\right), \quad (14.14)$$

where $\lambda = N^{-1/2}$ is the average island separation, R the deposition rate, and i the critical island size. In (14.13) and (14.14), the lengths and densities are normalized to the lattice constant and the density of lattice sites, respectively. One can see that both transitions obey Arrhenius type relations with the difference in slope given by the Ehrlich–Schwoebel barrier and the critical island size.

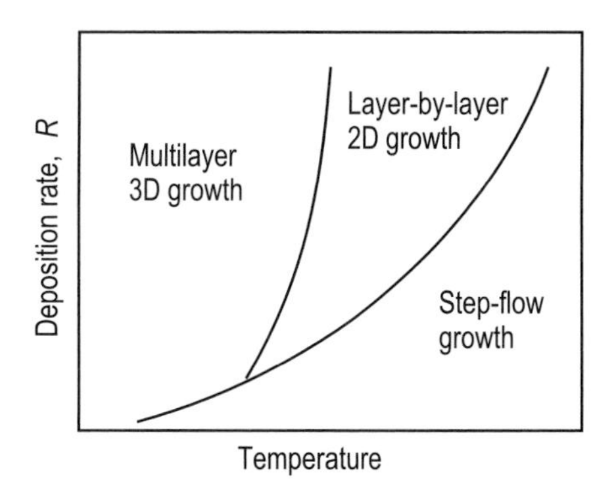

Fig. 14.23. Schematic growth mode diagram for the case of a non-zero Ehrlich–Schwoebel barrier (after Rosenfeld et al. [14.15])

14.4 Strain Effects in Heteroepitaxy

In *heteroepitaxy*, a crystalline film of one material is grown on the crystalline substrate of another material. The alignment of film lattice with respect to the substrate lattice is usually described in terms of the parallelism of crystal planes and directions: for example, Al(110)||Si(100), Al⟨001⟩ ||Si⟨011⟩ means that an epitaxial Al film is oriented so that its (110) plane is parallel to the Si(100) substrate surface, and the ⟨001⟩ azimuthal direction of the film coincides with the ⟨011⟩ orientation on the Si(100) surface.

As the substrate and the film are of different materials, it is a very rare case that they have the same lattice constant and ideal *lattice-matched, commensurate* growth takes place (Fig. 14.24a). Most often the crystal structure of the film and substrate is different. The usual measure to quantify this difference is the *misfit*, defined as the relative difference of lattice constants, i.e., $\varepsilon = (b - a)/a$. Relatively low misfits can be accommodated by elastic strain, i.e., deformation of the lattice of the epitaxial film in such a way that the strained film adopts the periodicity of the substrate in the interfacial plane, but can be distorted in the perpendicular direction in order to preserve the volume of the unit cell. This kind of growth, called *pseudomorphic* growth, is shown in Fig. 14.24b. At higher misfits, the strain is relieved by the formation of *misfit dislocations* at the film/substrate interface as shown in Fig. 14.24c. One can see that the distance between dislocations is given by

$$d = \frac{ab}{|b - a|} \ . \tag{14.15}$$

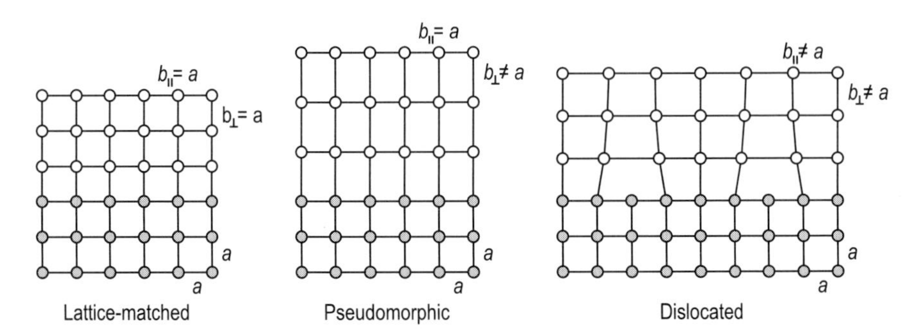

Fig. 14.24. Schematic of heteroepitaxial growth modes: (**a**) lattice-matched growth; (**b**) strained pseudomorphic growth; (**c**) relaxed dislocated growth

Which growth mode actually occurs in each particular case depends on the relationship between the free energy density associated with only strain (E_ε) and that with dislocations (E_D). The general energetics behind the transition from pseudomorphic to relaxed growth is illustrated qualitatively in Fig. 14.25. Figure 14.25a shows that the energy-versus-misfit curves for strained and dislocated films intersect at a certain critical misfit ε_c. Below the critical misfit, a purely strained film is energetically more favorable than a dislocated film, while above the critical misfit, the formation of dislocations becomes energetically more favorable. Figure 14.25b illustrates the effect of film

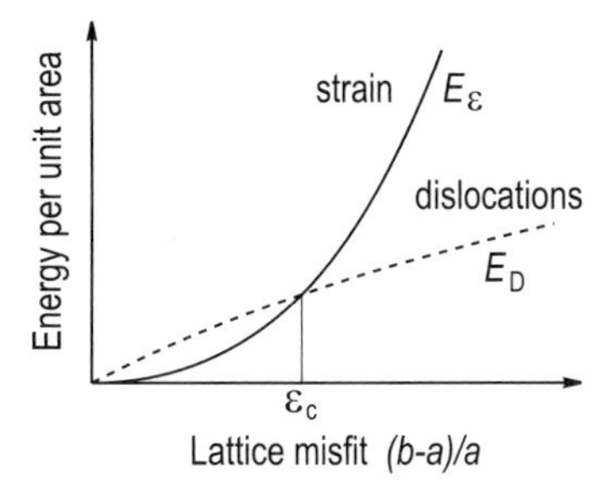

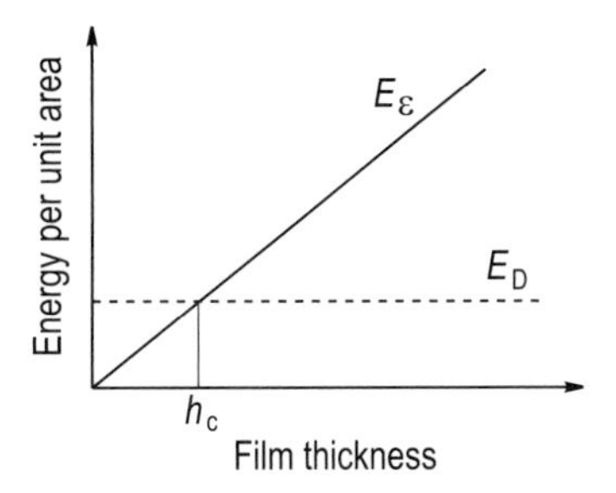

Fig. 14.25. Schematic plots of the lattice energy stored at a film/substrate interface per unit area (**a**) as a function of misfit and (**b**) as a function of film thickness. Below a critical misfit ε_c and below a critical film thickness h_c, strained pseudomorphous growth is more energetically favorable than relaxed dislocated growth

thickness. The strain energy increases with film thickness, while the energy of dislocations remains essentially constant. The intersect of the plots yields the critical thickness h_c, which indicates the transition from strained pseudomorphic growth to dislocated relaxed growth. As one can see in Fig. 14.26, the value of the critical thickness changes over several orders of magnitude with variation of misfit from a fraction of a percent to a few percent. The example shown refers to the growth of Ge_xSi_{1-x} alloy film on a Si(100) substrate, for which case the misfit is controlled by the Ge fraction x.

14.5 Thin Film Growth Techniques

There is a great variety of techniques used for the growth of thin films. The following concerns only the main techniques compatible with ultra-high vacuum.

14.5.1 Molecular Beam Epitaxy

In *molecular beam epitaxy (MBE)*, the material for the growing film is delivered to the sample surface by beams of atoms or molecules, i.e., via deposition. During growth, the substrate is conventionally kept at a moderately elevated temperature, which is, on the one hand, sufficient to ensure the arriving atoms migrate over the surface to the lattice sites, but, on the other hand, is not too high to induce diffusion intermixing between the layers in the bulk of the already grown film.

Figure 14.27 shows typical facilities for MBE growth. The substrate is mounted on a heater block and is faced to the deposition sources. Knudsen-type crucibles are usually used for deposition. Each source is supplied by a shutter driven from outside the UHV chamber, which allows the beam

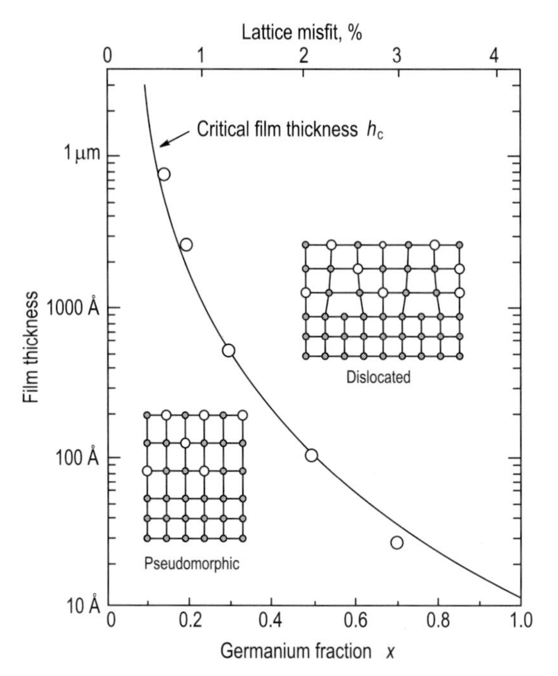

Fig. 14.26. Transition between strained pseudomorphous growth and relaxed dislocated growth for Ge_xSi_{1-x} alloy film grown on a Si(100) substrate. The boundary yields the plot of the critical film thickness versus lattice misfit, controlled by the Ge content x (after People and Bean [14.16])

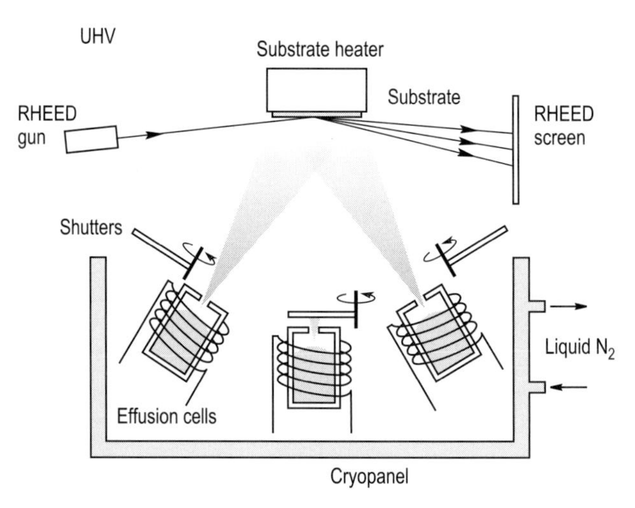

Fig. 14.27. Schematic sketch of the main facilities used in MBE growth, including deposition sources, shutters, sample at the heating block, RHEED system for monitoring sample surface structure in the course of deposition, all being installed in a UHV chamber

to be switched on and off almost instantly. The time required to change the composition of the flux is much shorter than that needed to grow a single atomic layer. Therefore, one has the important ability to control the in-depth composition of the growing film with atomic resolution. For in situ monitoring of the film crystal structure in the course of growth, RHEED is typically employed. Moreover, in case of layer-by-layer growth the intensity of the RHEED specular beam exhibits oscillations with a period corresponding to the time to complete a monolayer, which allows precise control of the film thickness (see Fig. 4.16).

Molecular beam epitaxy is a versatile technique for the formation of well-defined crystalline surfaces, films, and multilayer structures. It is used both for research applications and for semiconductor device fabrication.

14.5.2 Solid Phase Epitaxy

Solid phase epitaxy (SPE) is a specific regime of MBE growth, in which an amorphous film is first deposited at lower temperature and is then crystallized upon heating to higher temperature. For example, in the case of SPE of Si the deposition is conducted at room temperature, while typical temperatures for annealing are in the 500–600°C range. The crystallization proceeds by the motion of the amorphous/crystalline interface from the substrate to the outer surface of the film. Crystallization is a thermally activated process and the velocity of amorphous/crystalline interface motion (SPE rate) is described by an Arrhenius-type expression

$$V = V_0 \exp\left(-\frac{E_{\mathrm{SPE}}}{k_{\mathrm{B}}T}\right) , \tag{14.16}$$

where the activation energy E_{SPE} for Si-SPE growth is $\sim 3.0\,\mathrm{eV}$ (Fig. 14.28). One can see in Fig. 14.28 that the activation energy (given by the slope of the Arrhenius plots) is the same for different Si substrate orientation, while the pre-exponents vary significantly: the SPE rate is highest for Si(100) samples and is about 2 and 20 times that for Si(110) and Si(111) samples, respectively.

Generally, the crystallinity of films grown by SPE is slightly lower than that of MBE-grown films. However, SPE might be advantageous over MBE in achieving abrupt doping profiles in epitaxial semiconductor films. For example, many of the dopants, Sb, Ga, In, present difficulties during growth by conventional Si-MBE due to pronounced *surface segregation* incompatible with abrupt profiles. Surface segregation, i.e the capture of the doping impurities by the growing film surface, is controlled by the near-surface diffusion. In case of SPE growth, the smearing of the abrupt doping profile is controlled by bulk diffusion. For typical temperatures and durations of annealing used in SPE, the diffusion length does not exceed one interatomic distance and extremely abrupt concentration dopant profiles can be created. A vivid example is the fabrication of so-called *delta-doping layers*, in which dopants are confined to a few atomic layers within an intrinsic semiconductor material.

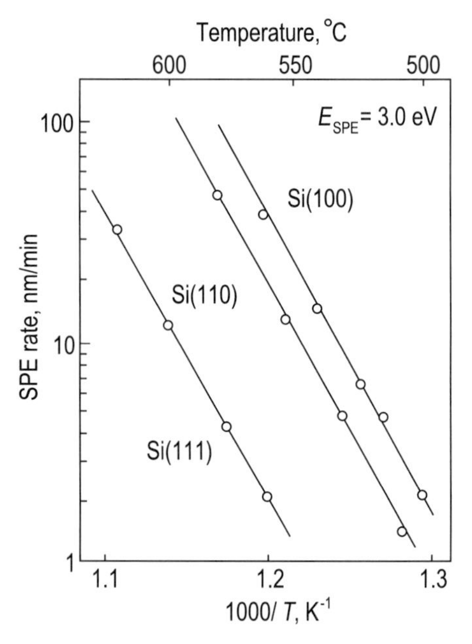

Fig. 14.28. Arrhenius plot of the SPE rate for epitaxial crystallization of amorphous Si films deposited in UHV onto Si(100), Si(110), and Si(111) substrates (after Zotov and Korobtsov [14.17])

14.5.3 Chemical Beam Epitaxy

Thin film growth performed by means of surface chemical reactions is defined by the general term *chemical vapor deposition (CVD).* In this technique, the source material is supplied to the sample surface in the form of its gaseous compounds. Precursor gas molecules decompose at the hot surface, leaving there the desired species, while the waste fragments of the molecules are desorbed from the surface. The group IV and group V components are supplied usually as hydrides, such as SiH_4, GeH_4, AsH_3, PH_3, etc., while group III components are supplied as metal-organic compounds, such as trimethyl gallium [$Ga(CH_3)_3$, TMGa], triethyl gallium [$Ga(C_2H_5)_3$, TEGa], triethyl indium [$In(C_2H_5)_3$, TEIn], etc. Conventionally the term metal-organic CVD (MOCVD) refers to growth conducted at relatively high pressures (~1–760 Torr). If the growth is performed under UHV conditions, the technique is called *metal-organic MBE (MOMBE)* or *chemical beam epitaxy (CBE).* Due to the large mean free path of the gas molecules at low pressure, CBE growth is not affected by reactions in the gas phase and only surface chemistry here determines the growth process.

For CBE growth, a UHV chamber is used, as in conventional MBE, but the reacting components are supplied by means of a gas inlet system. The latter is an important part of the CBE set-up. Material flux control is provided by adjusting the input pressure of the gas injection capillary. Sometimes it is required to crack (decompose) the gaseous compound already in the capillary, for which case a heated metal (for example, Ta) filament or foil is used. For example, as one can see in Fig. 14.29, the growth rate of GaAs film increases linearly with the TMGa beam pressure (i.e., with Ga supply), but saturates at a value which strongly depends on the cracking efficiency of the AsH_3 inlet capillary (i.e., on the elemental As supply).

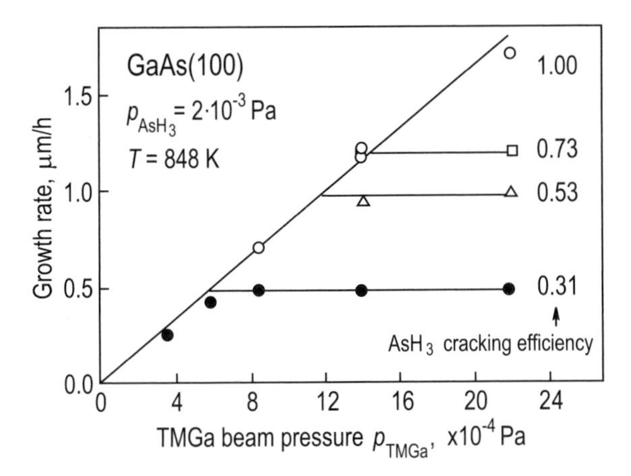

Fig. 14.29. Dependence of chemical beam epitaxy (CBE) growth rate of GaAs film on the beam pressure of trimethyl gallium [$Ga(CH_3)_3$, TMGa] for various cracking efficiencies of the AsH_3 inlet capillary (after Pütz et al. [14.18])

In comparison to MBE, CBE growth is far more complex and conventionally requires higher temperatures. However, it provides higher growth rates with conservation of high film crystallinity. An additional advantage of CBE lies in the ability to supply materials which are inconvenient for evaporation from effusion cells either due to the extremely low vapor pressure even at very high temperatures (for example, W, B, Nb) or due to extremely high vapor pressure already at low temperature (for example, P). Note that MBE and CBE techniques can be combined in the same growth process, when one component is supplied in the form of a gaseous compound and the other is deposited from the effusion cell.

14.6 Surfactant-Mediated Growth

By deliberately introducing a certain impurity, one can alter the growth mode in a desired direction, say, to change island growth to step-flow or layer-by-layer growth. This surface active impurity (typically in monolayer or submonolayer amounts) is referred to as a *surfactant*. The classical definition reads that a surfactant is "a substance that lowers surface tension, thereby increasing spreading and wetting properties" [14.19]. To be a surfactant, an impurity should satisfy the following requirements:

- It promotes 2D growth under conditions where normally 3D growth takes place.
- It is immiscible in the film, so that no or a negligible amount of surfactant atoms become incorporated in the film bulk.

The second requirement can be satisfied in two ways. First, the surfactant can be segregated to the film surface (i.e., due to continuous exchange with the adsorbing atoms, the surfactant atoms always remain on top of the growing film). Second, surfactant atoms can be trapped at the buried film/substrate interface. The latter kind of surfactant is sometimes referred to as *interfactant.*

An example of interfactant-mediated growth is given by the growth of Ag film on a hydrogen-terminated Si(111) surface as determined by TOF-ICISS [14.20] (Fig. 14.30). On the bare Si(111)7×7 surface and at 300°C, the growth of Ag film proceeds according to the Stranski–Krastanov mode (after completion of the $\sqrt{3}\times\sqrt{3}$ surface phase at 1 ML, rather thick Ag islands develop). The RT growth on bare Si(111)7×7 proceeds in layer-by-layer fashion, but the film contains rotational disorder (A- and B-type domains). In the case of Ag film growth on the H-terminated Si(111) surface held at 300°C, the growth proceeds according to the layer-by-layer mode and in a single domain orientation.

The classical example of growth with a segregating surfactant is Sb-mediated epitaxy of Ge on Si(111) (Fig. 14.31). Growth of Ge on the bare Si(111)7×7 proceeds in the Stranski–Krastanov growth mode, in which 3D islands are formed on top of a 3 ML thick pseudomorphic 5×5-reconstructed Ge-Si layer. With ~1 ML of Sb as a surfactant, the formation of 3D islands is suppressed and continuous Ge film is grown in layer-by-layer fashion. The surface of the film displays a well-ordered Ge(111)2×1-Sb (1 ML) reconstruction.

Depending on the particular case, various atomic mechanisms might be responsible for the surfactant effect. The surfactant-induced enhancement of adatom diffusion along the terrace leads to the earlier achievement of step-flow growth. However, in the case of layer-by-layer growth, the reduction of the adatom mobility might also lead to improving the 2D growth for the following reasons: First, lower mobility results in an increase in the island density. (Note that the process of site exchange between adatoms and surfactant atoms also leads to a reduction in the adatom migration length.) Second, the increase of the surface diffusion barrier leads to an effective decrease of the Ehrlich–Schwoebel barrier (Fig. 14.32) and, hence, to enhancement of the interlayer mass transport. Another possibility is a direct decrease of the Ehrlich–Schwoebel barrier by surfactant atoms incorporated at the step edge (Fig. 14.32b). The reduction of the edge atom mobility, inducing the ram-

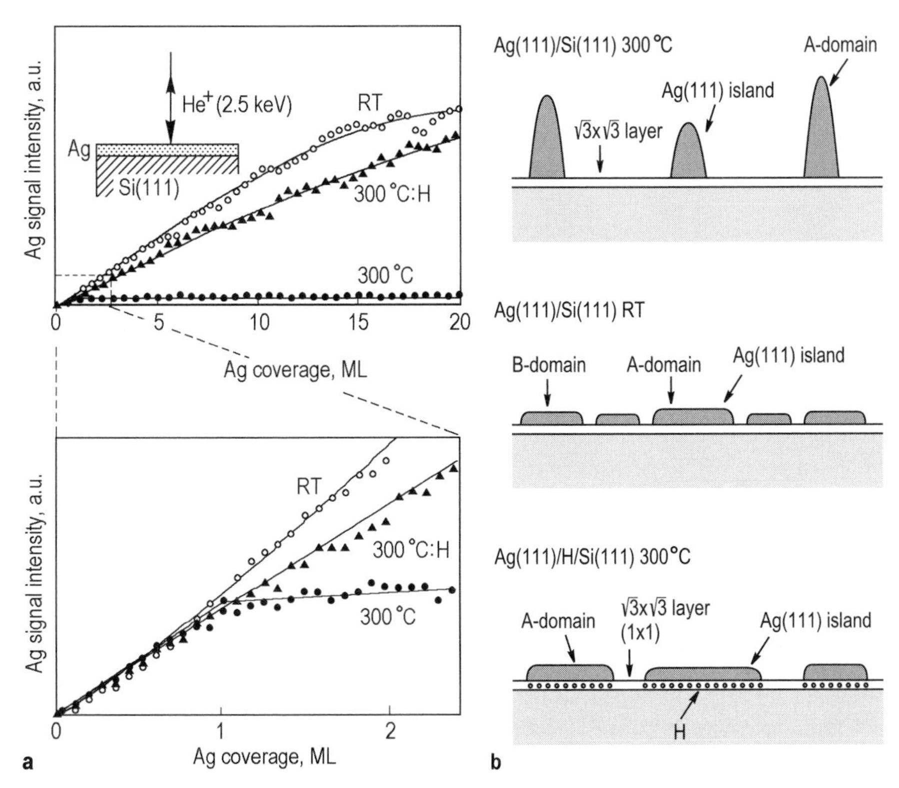

Fig. 14.30. (**a**) Variation of TOF-ICISS Ag signal intensities versus Ag coverage for three deposition conditions. Data points labeled RT and 300°C are for Ag deposition onto the bare Si(111)7×7 surface at RT and 300°C, respectively; those labeled 300°C:H are for deposition onto a H-terminated Si(111) surface at 300°C. Results for the initial stage of film growth below ~2.5 ML are enlarged in the bottom. (**b**) Schematic illustration of the growth modes derived for deposition of Ag onto the bare Si(111)7×7 surface at 300°C (top), at RT (middle), and onto H-terminated Si(111) surface at 300°C (bottom) (after Sumitomo et al. [14.20])

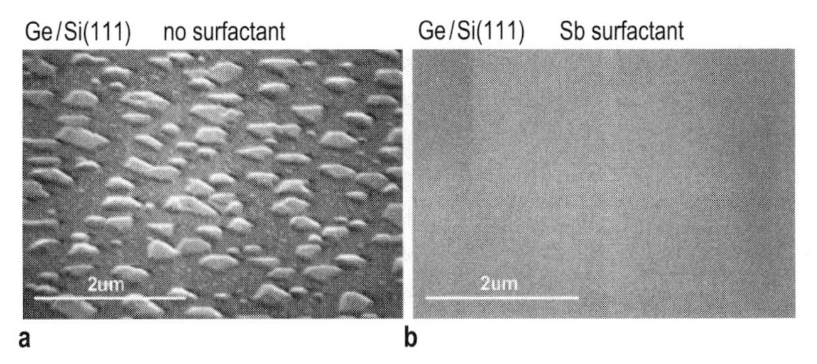

Fig. 14.31. SEM images of 50 ML thick Ge films grown on Si(111) (**a**) without a surfactant and (**b**) with Sb as a surfactant (after Zahl et al. [14.21])

ified island shape, also favors smoother growth. In conclusion, it is worth noting that none of the above mechanisms is related to the classical meaning of surfactant as a species that reduces the surface free energy. Actually, the surfactant affects film growth by modifying its kinetics.

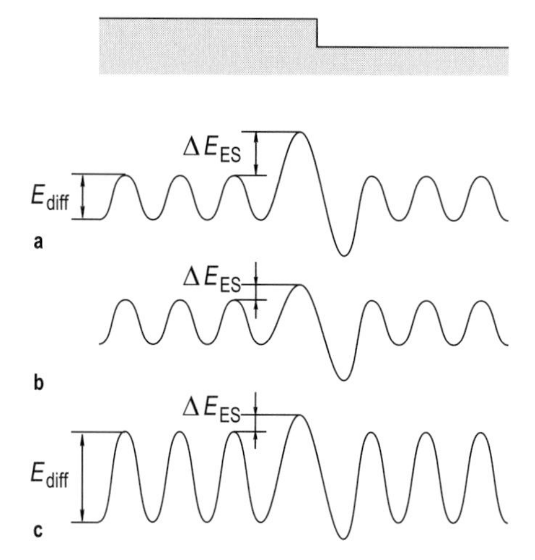

Fig. 14.32. Possible surfactant-induced modification of the surface potential at a step edge with the Ehrlich–Schwoebel barrier. The Ehrlich–Schwoebel barrier, ΔE_{ES}, on the clean surface (**a**) can be decreased in two ways: (**b**) by a local decrease of the total barrier at the step edge or (**c**) by a global increase of the surface diffusion barrier at a terrace, E_{diff} (after van der Vegt et al. [14.22])

Problems

14.1 If material A is grown on substrate B (a) in Frank–van der Merwe mode, (b) in Vollmer–Weber mode, what will be the growth mode of film B on substrate A in each case? Consider the various relations between surface and interface energies.

14.2 The activation energy of surface self-diffusion for a clean Ag(111) surface, 100 meV, is doubled by the presence of a Sb surfactant. Estimate how the saturation Ag island density will be changed at room temperature. Assume the critical island size is $i = 1$.

14.3 An Arrhenius plot of saturation island density, measured for Ag growth on a Pt(111) surface, shows two regimes with slopes of 56 and 122 meV for critical islands $i = 1$ and $i = 2$, respectively [14.23]. Calculate the Ag-Ag dimer bond energy and migration energy for Ag on Pt(111).

Further Reading

1. B. Lewis, J.C. Anderson: *Nucleation and Growth of Thin Films* (Academic Press, New York 1978) (solution of nucleation rate equations for various cases)
2. J.A. Venables, G.D.T. Spiller, M. Hanbücken: *Nucleation and Growth of Thin Films*. Rep. Prog. Phys. **47**, 399–459 (1984) (nucleation rate theory in great detail)
3. H. Brune: *Microscopic View of Epitaxial Metal Growth: Nucleation and Aggregation*. Surf. Sci. Rep. **31**, 121–229 (1998) (applications of the nucleation rate theory and Monte Carlo simulations for island growth)
4. J.G. Amar, F. Family: *Kinetics of Submonolayer and Multilayer Epitaxial Growth*. Thin Solid Films **272**, 208–222 (1996) (introduction to scaling theory of island growth)
5. M. Giesen: *Step and Island Dynamics at Solid/Vacuum and Solid/Liqued Interfaces*. Prog. Surf. Sci. **68**, 1–153 (2001) (equilibrium island shape, island ripening, and coalescence in detail)
6. M.A. Herman, H. Sitter: *Molecular Beam Epitaxy: Fundamentals and Current Status*, 2nd ed. (Springer, Berlin 1996)
7. J.R. Arthur: *Molecular Beam Epitaxy*. Surf. Sci. **500**, 189–217 (2002)
8. G.L. Olson, J.A. Roth: *Kinetics of Solid Phase Crystallization in Amorphous Silicon*. Mater. Sci. Rep. **3**, 1–78 (1988)
9. A.V. Zotov, V.V. Korobtsov: *Present Status of Solid Phase Epitaxy of Vacuum-Deposited Silicon*. J. Cryst. Growth **98**, 519–530 (1988)
10. H. Lüth: *Chemical Beam Epitaxy – A Child of Surface Science*. Surf. Sci. **299/300**, 867–877 (1994)

15. Atomic Manipulations and Nanostructure Formation

Recent progress in material science (and, in particular, in surface science) provides an opportunity for the fabrication of various artificial structures of nanometer size. The main approaches used for such fabrications are atomic manipulations (i.e., building up the structure atom by atom) and self-organization (i.e., spontaneous formation of many structures at once, as a result of certain processes). The growth process and the grown nanostructures themselves present great interest both for science and technology.

15.1 Nano-Size and Low-Dimensional Objects

The term *nanostructures* refer to the solid crystalline structures that have a minimum dimension on the nanometer scale (typically, in the range from 1 to 10 nm). The most peculiar properties of nanostructures arise as a consequence of two major effects:

- First, the number of surface atoms constitutes a large fraction of the total, which leads to *surface-dominated effects.*
- Second, in nanostructures the electrons are confined within a limited region of space, which leads to *quantum size effects* and *low-dimensionality effects.*

Surface-Dominated Effects. A vivid illustration of the surface-dominated effect is a drastic reduction in the melting temperature of nanocrystals (Fig. 15.1). An understanding of the melting point depression can be obtained if one recalls that the surface atoms have a lower coordination number than atoms in the bulk, and hence are more weakly bound and less constrained in their thermal motion. The smaller the nanocrystal, the higher the fraction of surface atoms, and the lower the melting temperature. Typically, the reduction in melting point, ΔT_{m}, is a linear function of the inverse nanocrystal size, i.e., $\Delta T_{\mathrm{m}} \propto 1/L$. For very small clusters of only a few atoms, the $\Delta T_{\mathrm{m}}(L)$ dependence might become non-monotonic, reflecting the existence of magic clusters with enhanced stability.

Quantum Size Effects. Quantum size effects arise when the region of space where an electron is confined, becomes comparable to or smaller than the de

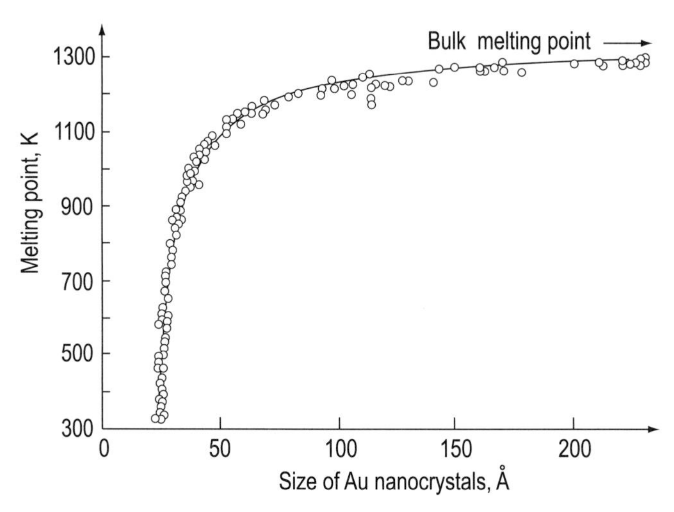

Fig. 15.1. Melting temperature of gold nanocrystals as a function of their size (after Buffat and Borel [15.1])

Broglie wavelength of the electron. In this case, the allowed energy states become discrete (i.e., quantized), rather than continuous. Note that the discrete energy levels can be resolved only in low-temperature experiments (i.e., when the energy separation between levels exceeds $k_{\mathrm{B}}T$). In general, electrons can be confined in one, two, or three spatial dimensions.

- Electron confinement in one spatial dimension corresponds to two-dimensional (2D) objects, called *quantum films.*
- Electron confinement in two dimensions corresponds to one-dimensional (1D) objects, called *quantum wires.*
- Electron confinement in three dimensions corresponds to zero-dimensional (0D) objects, called *quantum dots.*

To illustrate the effect of dimensionality on the electronic properties of nanostructures, consider the density of electronic states for the ideal 3D, 2D, 1D, and 0D cases. Recall that the solution of the Schrödinder equation for a free electron gas confined to a cube with edge L (under the assumption of an infinitely deep barrier function) yields the wave functions in the form of standing waves

$$\psi_n(\boldsymbol{r}) = \exp(i\boldsymbol{k}\cdot\boldsymbol{r}) \tag{15.1}$$

with the components of the wave vector $\boldsymbol{k}$ satisfying the conditions

$$k_x = \frac{2\pi n}{L}\,,\ k_y = \frac{2\pi n}{L}\,,\ k_z = \frac{2\pi n}{L}\,, \tag{15.2}$$

where n is a positive or negative integer. The energy E_n is given by

$$E_n = \frac{\hbar^2\boldsymbol{k}^2}{2m} = \frac{\hbar^2}{2m}\left(k_x^2 + k_y^2 + k_z^2\right)\,. \tag{15.3}$$

Density of States in the 3D Case. In 3D $\boldsymbol{k}$-space, each allowed state occupies a volume of $(2\pi/L)^3$. The volume of a spherical shell with radius k and thickness $\mathrm{d}k$ is $4\pi k^2\,\mathrm{d}k$. Thus, the number of states $\mathrm{d}N$ is given simply by dividing this volume by the volume of a single energy state and taking into account two allowed values of electron spin

$$\mathrm{d}N = 2 \cdot \frac{4\pi k^2\,\mathrm{d}k}{(2\pi/L)^3} = \frac{k^2 L^3\,\mathrm{d}k}{\pi^2}\,. \tag{15.4}$$

Since

$$k = \frac{(2m^* E)^{1/2}}{\hbar}\,, \tag{15.5}$$

the density of states per unit volume at an energy E is

$$D_{\mathrm{3D}}(E) = \frac{\mathrm{d}N}{L^3\,\mathrm{d}E} = \frac{1}{2\pi^2}\left(\frac{2m^*}{\hbar^2}\right)^{3/2} E^{1/2}\,. \tag{15.6}$$

This is a well-known result for the free electron gas approximation, giving the density of states proportional to the square root of the energy (see Fig. 15.2a).

Density of States in the 2D Case. In this case the procedure is much the same, but this time one of the $\boldsymbol{k}$-space components is fixed and the problem is to calculate the number of states lying in an annulus of radius k to $k + \mathrm{d}k$. Each allowed state occupies an area of $(2\pi/L)^2$, and the area of the annulus is given by $2\pi k\,\mathrm{d}k$. Dividing the area of the annulus by the area of the $\boldsymbol{k}$-state and remembering to multiply by 2 to account for the electron spin states one gets

$$\mathrm{d}N = 2 \cdot \frac{2\pi k\,\mathrm{d}k}{(2\pi/L)^2} = \frac{k L^2\,\mathrm{d}k}{\pi}\,. \tag{15.7}$$

Consequently, the density of states $D_{\mathrm{2D}}(E)$ per unit area is

$$D_{\mathrm{2D}}(E) = \frac{\mathrm{d}N}{L^2\,\mathrm{d}E} = \frac{m^*}{\pi\hbar^2}\,. \tag{15.8}$$

One can see that the 2D density of states $D_{\mathrm{2D}}(E)$ is independent of the energy. Taking into account the other energy levels E_n, the density of states takes on a staircase-like function (see Fig. 15.2b) given by

$$D_{\mathrm{2D}}(E) = \frac{m^*}{\pi\hbar^2}\sum_n H(E - E_n)\,, \tag{15.9}$$

where $H(E - E_n)$ is the Heaviside function:

$$H(E - E_n) = \begin{cases} 0, & E < E_n \\ 1, & E \geq E_n. \end{cases} \tag{15.10}$$

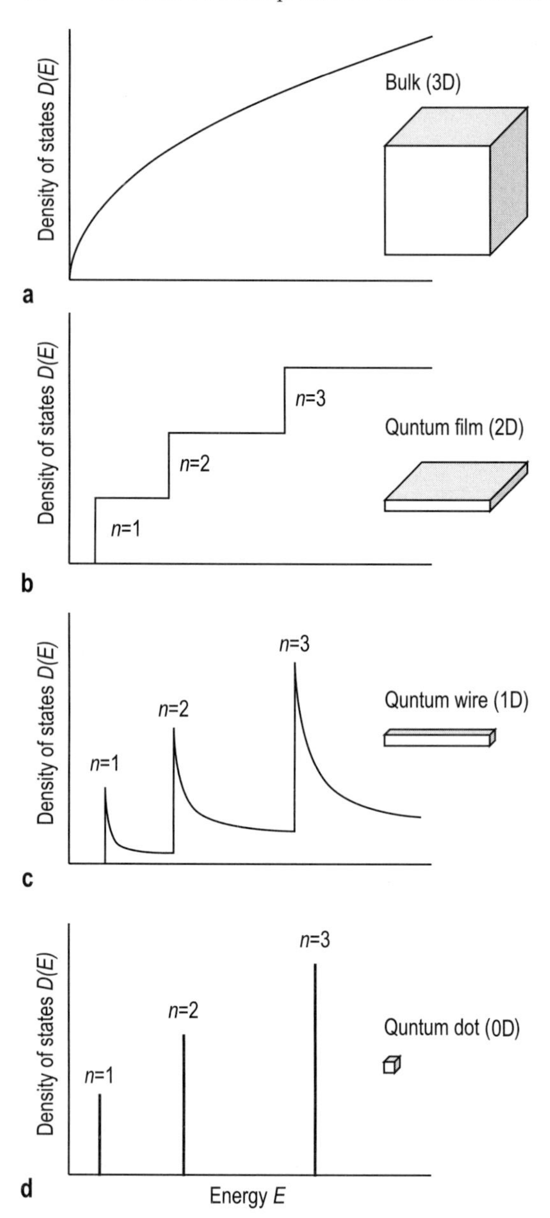

Fig. 15.2. Idealized density of states for (**a**) 3D bulk, (**b**) 2D quantum films, (**c**) 1D quantum wires, and (**d**) 0D quantum dots

Density of States in the 1D Case. In one dimension two of the $\boldsymbol{k}$ components are fixed, thus, the area of $\boldsymbol{k}$-space becomes a length and the area of the annulus becomes a line segment $2\,\mathrm{d}k$. The number of states in this length in 1D is given by

$$\mathrm{d}N = 2 \cdot \frac{\mathrm{d}k}{(2\pi/L)} = \frac{L\,\mathrm{d}k}{\pi} \,. \tag{15.11}$$

For one dimension, the density of states per unit length is

$$D_{1\mathrm{D}}(E) = \frac{\mathrm{d}N}{L\,\mathrm{d}E} = \frac{(2m^*)^{1/2}}{\pi\hbar}\frac{1}{E^{1/2}} \,. \tag{15.12}$$

Using more than the first energy level, the density of states function (see Fig. 15.2c) becomes

$$D_{1\mathrm{D}}(E) = \frac{(2m^*)^{1/2}}{\pi\hbar}\sum_n \frac{g_n H(E-E_n)}{(E-E_n)^{1/2}} \,, \tag{15.13}$$

where $H(E-E_n)$ is again the Heaviside function and g_n is the degeneracy factor.

Density of States in the 0D Case. In a 0D structure, the values of $\boldsymbol{k}$ are quantized in all directions. All the available states exist only at discrete energies and can be represented by delta functions (see Fig. 15.2d). In this respect, the ideal zero-dimensional structures can be considered as artificial atoms. Like natural atoms, these small electronic systems contain a finite number of electrons and have a discrete spectrum of energy levels.

15.2 Atomic Manipulation with STM

An ideal process for the fabrication of nanostructures is just to "arrange the atoms one by one the way we want," as Richard Feynman wrote in 1960 in his prophetic article on miniaturization [15.2]. With the development of scanning tunneling microscopy (STM) (see Sect. 7.7), this fascinating prospect has become a reality. Nowadays, of all the different approaches, STM has proved to offer the simplest and most general approach to atomic manipulation. An additional important advantage of STM is that atomic manipulation and the imaging of the surface with atomic resolution are performed using the same set-up. Thus, one has the possibility of imaging the surface, to choose an atomic feature of interest, to conduct the desired manipulations, and then to test the result. A variety of different atomic manipulation have been demonstrated. The main atomic manipulations with STM are as follows:

- lateral displacement of an atom along a surface (Fig. 15.3a);
- extraction of an atom from a surface (Fig. 15.3b);
- deposition of an atom from the STM tip to a surface (Fig. 15.3c).

Consider each kind of manipulation in greater detail.

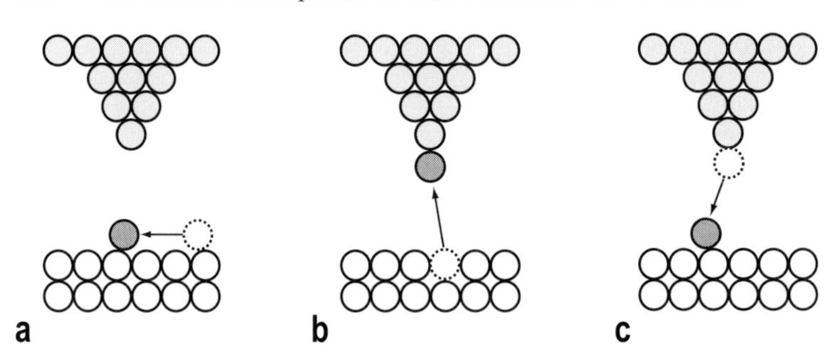

Fig. 15.3. Schematic diagram showing the main types of atomic manipulations with STM. (**a**) STM induces the lateral motion of an atom along the surface. (**b**) An atom is extracted from the surface and transferred to the STM tip. (**c**) An atom is deposited from the STM tip onto the surface

15.2.1 Lateral Atomic Displacement

Depending on the type of the utilized tip–sample interaction, the lateral atomic manipulations can be grouped into two modes:

- *Atom sliding*, induced by interatomic forces between the tip and a surface atom;
- *Field-assisted diffusion*, caused by the electrostatic field generated in the gap between the tip and the surface.

Atom Sliding. When the tip approaches an adatom on the surface, the van der Waals attractive interaction acts between the adatom and the tip. The interaction is increased with decreasing tip–adatom separation and eventually the adatom becomes trapped just below the tip and can be moved to the desired location. This is illustrated by Fig. 15.4 in terms of the potential energy curves. Figure 15.4b shows the energy of the adatom as a function of the distance normal to the surface at various tip–sample separations. The potential energy as a function of the lateral displacement in the absence (dotted line) and presence (solid line) of the tip is shown in Fig. 15.4c. When the tip is far away, the adatom experiences a periodic weakly corrugated potential. The close proximity of the tip results in developing a deep van der Waals minimum, which confines the adatom in the region under the tip. If the tip moves with the tip–sample separation being preserved constant, the adatom will follow the tip. When the adatom is moved to the desired destination, the tip is retracted, leaving the adatom at this new position. The successive stages of the sliding process are illustrated schematically in Fig. 15.5.

For reproducible and reliable utilization of atomic sliding, the system has to fit a number of requirements. First, it is necessary that the corrugation in the surface potential should be large enough that the adatoms could be imaged without inadvertently moving them, yet small enough that the atomic sliding could be conducted with the tip lowered towards the adatom. Second,

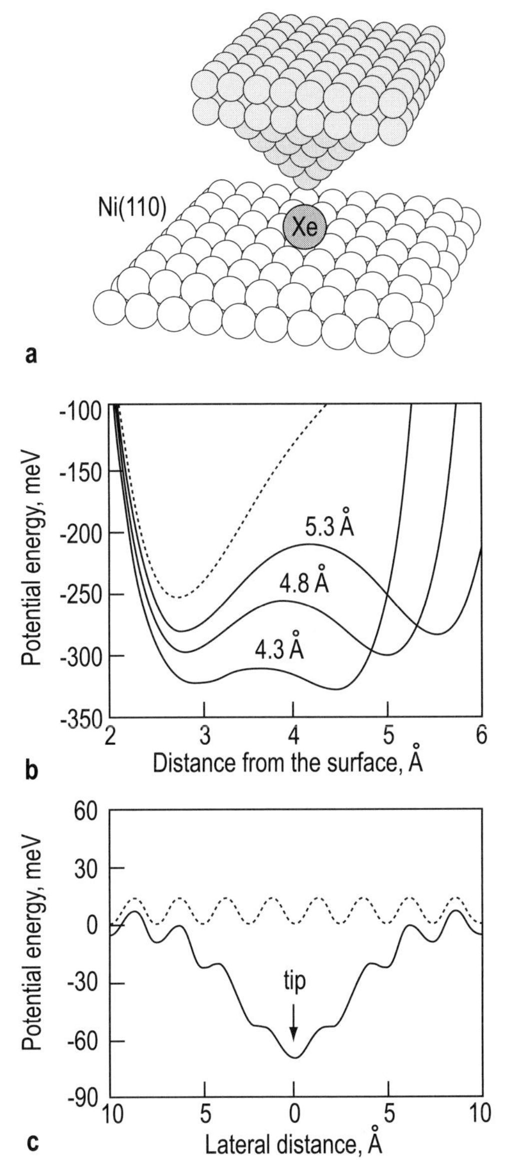

Fig. 15.4. Illustration of van der Waals interaction between the STM tip and a Xe adatom on a Ni(110) surface. (**a**) Atomic arrangement of the tip and sample used in the calculations. (**b**) Energy as a function of distance normal to the surface with a sequence of tip–sample separations. (**c**) Energy as a function of lateral displacement. The energy curves in the absence of the tip are shown as dotted lines; those in the presence of the tip as solid lines (after Avouris [15.3])

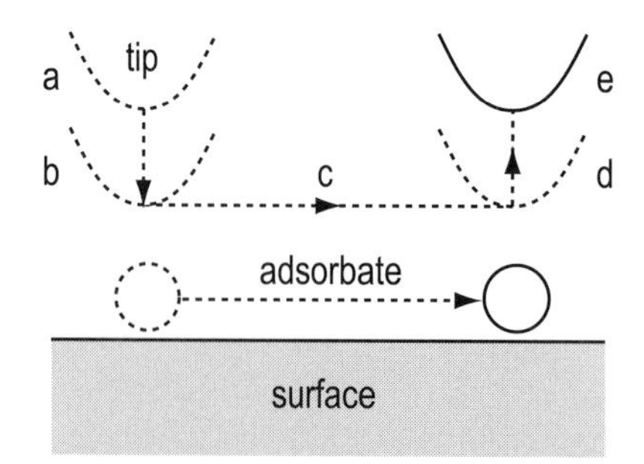

Fig. 15.5. Schematic illustration of the successive stages in the sliding process. (**a**) The atom is located and the tip is placed directly over it. The tip is lowered to position (**b**), where the atom–tip attractive force is sufficient to trap the atom beneath the tip, when the tip is moved across the surface (**c**) to the desired destination (**d**). Finally, the tip is retracted to position (**e**), where the atom–tip interaction is negligible, leaving the atom bound to the surface at the new location (after Eigler and Schweizer [15.4])

as the adatom should remain bound to the surface in the course of sliding, the diffusion barrier should be lower than the binding energy. Third, in order to maintain the desired arrangement of adatoms, thermal diffusion has to be suppressed by cooling the sample to low temperature. Metal surfaces with a weakly corrugated potential are good candidates. This capability to fabricate the desired rudimentary structures atom by atom was demonstrated first in 1990 by Eigler and Schweizer [15.4] with Xe atoms on a Ni(110) surface at 4 K (see Fig. 15.6). Xe atoms on a Ni(110) surface are characterized by a binding energy of ∼250 meV and an energy barrier for diffusion along the close-packed direction of ∼20 eV.

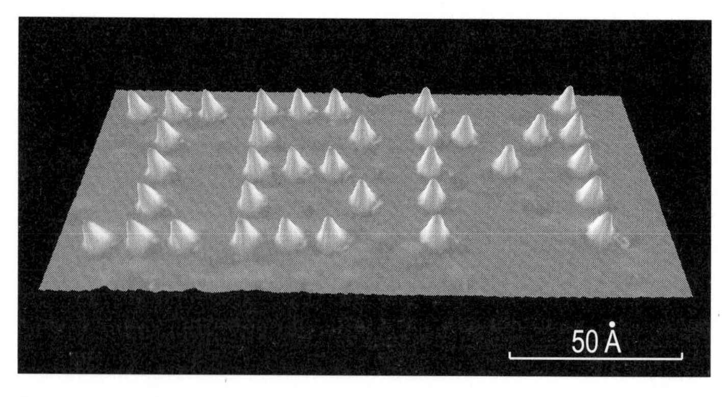

Fig. 15.6. STM image (quasi-3D representation) of Xe atoms (seen as 1.6 Å bumps) on Ni(110) arranged in the word "IBM" by using the atomic sliding process. Each letter is 50 Å from top to bottom (after Eigler and Schweizer [15.4])

Another fascinating example is the fabrication of the *quantum corral*, a circle of radius ∼71 Å built of 48 Fe atoms on a Cu(111) surface (Fig. 15.7). The quantum corral acts as a round two-dimensional box for surface-state electrons. The circular waves seen in the STM image in the interior of the corral are electron standing waves, predicted by a solution of the Schrödinger equation in that particular environment.

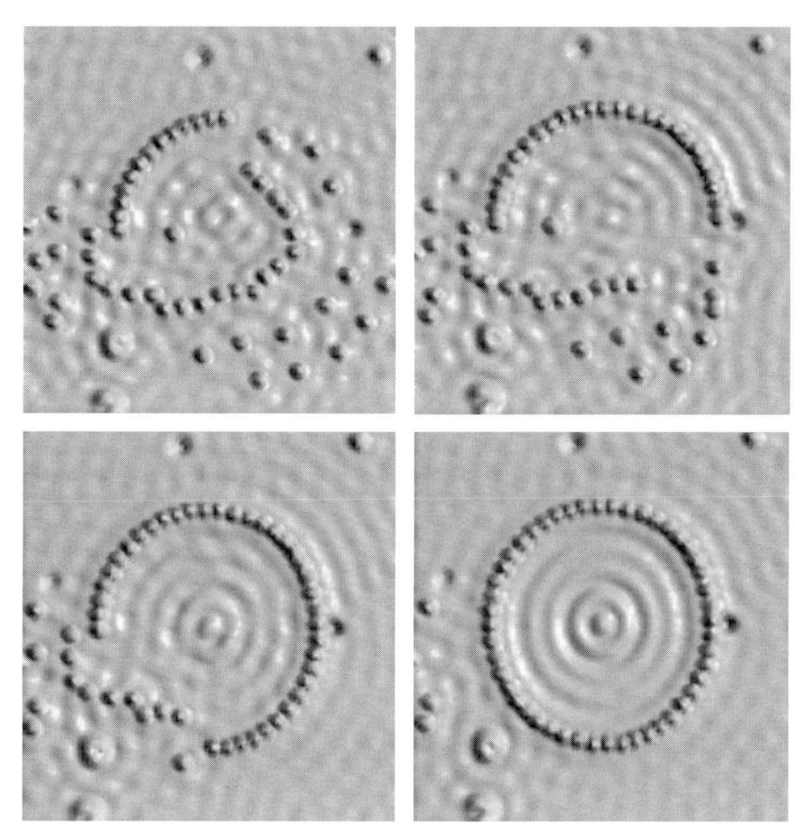

Fig. 15.7. A sequence of STM images taken during construction of the "quantum corral," which is built of 48 Fe atoms adsorbed on Cu(111) (after Crommie et al. [15.5])

For a given system, the characteristic parameter of the atomic manipulation with the sliding process is a threshold tip–surface separation. Above the threshold, the adsorbate–tip interaction is too weak to induce any manipulation. Below the threshold, the interaction is strong enough to allow the manipulation of the adsorbate. Since the absolute tip–surface separation is generally unknown, it is more convenient to express the separation in terms of the resistance of the tunnel junction: a greater resistance corresponds to a greater tip–sample separation, hence, to a weaker tip–sample interaction. The threshold resistance appears to be 5 MΩ for sliding Xe atoms along the rows of Ni(110), 200 kΩ for sliding Fe atoms along Cu(111) and CO molecules along Pt(111), and 20 kΩ for sliding Pt adatoms across Pt(111) [15.6]. The threshold resistance displays a clear tendency to decrease with increase of the adsorbate–substrate binding energy, i.e., a greater force has to be applied to move adsorbates, which are more strongly bound to the surface.

Field-Assisted Diffusion. The field generated in the tip–sample gap by applying a bias voltage can also be used for atomic manipulations. This field is inhomogeneous and concentrated underneath the apex of the tip. In the case of chemisorption, the adsorbate–substrate bonding involves charge transfer and formation of a dipole with static dipole moment $\boldsymbol{p}_0$. In addition, the electrostatic field $\boldsymbol{E}$ generates an induced dipole $\alpha\boldsymbol{E}$, where α is the polarizability of the adsorbate. Thus, an adsorbate atom in the electric field has a dipole moment given to first order in E as

$$\boldsymbol{p} = \boldsymbol{p}_0 + \alpha\boldsymbol{E} + \ldots \,. \tag{15.14}$$

The spatially dependent energy of the atom is then given by

$$U(\boldsymbol{r}) = -\boldsymbol{p}_0 \cdot \boldsymbol{E}(\boldsymbol{r}) - \frac{1}{2}\alpha E^2(\boldsymbol{r}) + \ldots \,. \tag{15.15}$$

As the field is non-uniform, the adsorbate would experience a potential gradient, i.e., a force. When the second (polarizability) term dominates, the adsorbate will always be attracted towards the region underneath the tip apex irrespective of the polarity of the applied bias voltage. This is a natural consequence of the fact that an induced dipole is oriented along the field and, therefore, moves towards the region of maximum field strength (see Fig. 15.8a). When the first (dipole) term dominates, the orientation of the dipole remains unchanged, hence the direction of adsorbate motion changes with the change of the bias polarity, as illustrated in Fig. 15.8.

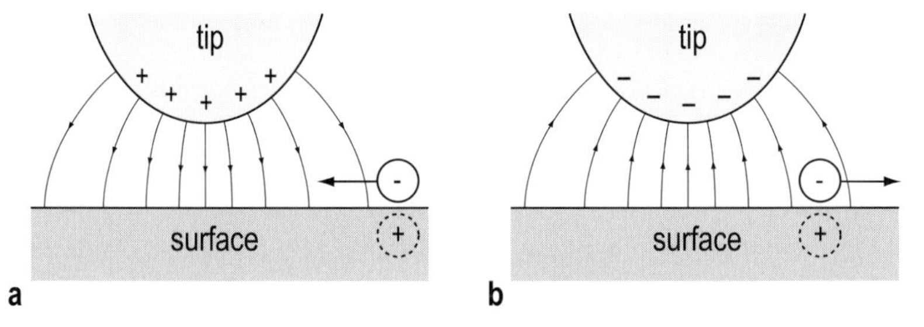

Fig. 15.8. (**a**) An adsorbate dipole oriented along the field is attracted towards the region underneath the tip (i.e., to the region of maximum field strength). (**b**) An adsorbate dipole oriented against the field is repelled away from the region underneath the tip

The field-assisted diffusion of In atoms on a Si(111) surface furnishes an example of the dipole-dominated case. Figure 15.9 illustrates the result of these atomic manipulations. In the experiment, In atoms were arranged in one of two phases, $\sqrt{3}\times\sqrt{3}$-In with 0.33 ML of In and 2×2-In with 0.75 ML of In. The less dense $\sqrt{3}\times\sqrt{3}$-In phase is seen in STM images as darker compared to the more dense and brighter 2×2-In phase. In Fig. 15.9a, the central area of the 400×400 Å^2 STM image is occupied by a $\sqrt{3}\times\sqrt{3}$-In phase, the bright region at the image bottom is an array of 2×2-In phase, and the dark region at the top is the next (lower) terrace. When the tip is lowered to the point indicated by a white dot and a negative bias, $V_t = -2.0$ V, is applied to the tip, In atoms are attracted to the region beneath the tip and form a patch of dense 2×2-In phase (Fig. 15.9b). Figures 15.9c and d illustrate the opposite process. When a positive bias, $V_t = +2.0$ V, is applied to the tip posed above an extended region of the dense 2×2-In phase (Fig. 15.9c), In atoms are repelled from the region underneath the tip and a "hole" occupied by less dense $\sqrt{3}\times\sqrt{3}$-In phase is formed (Fig. 15.9d).

It should be remarked that, compared to the sliding process, field-assisted diffusion provides a lower resolution due to the fact that the electrostatic field strength drops rather slowly with distance from the tip apex. For example, one can see in Figs. 15.9a and b that In atoms were attracted over a distance of about 200 Å, as indicated by dissolution of the "peninsular" of the 2×2-In phase near the bottom left corner of the image.

15.2.2 Atom Extraction

There are three main mechanisms that can be employed to remove a given atom from a surface using STM. These are

- interatomic interaction;
- field evaporation;
- electron-stimulated desorption.

Interatomic Interaction. When the tip approaches close to an atom on the surface, the adsorption wells on the tip and surface sides of the tip–sample junction overlap and the energy barrier separating the two wells diminishes greatly. This situation is illustrated in Fig. 15.4b. Hence, the adatom has a finite probability to hop to the tip well and, if the tip is then withdrawn, the atom is carried away with the tip. This technique is applicable mainly to weakly bonded adsorbates, for example, for extraction of Xe adatoms from Pt(111) or Ni(110) surfaces [15.8].

Field Evaporation. In the field evaporation process, the atom at the surface becomes ionized in the strong electric field and evaporates as an ion. The process takes place at both bias polarities: if an appropriate positive voltage is applied to the tip, the surface atoms are evaporated as negative ions and, vice versa, at negative tip voltage, atoms are evaporated as positive

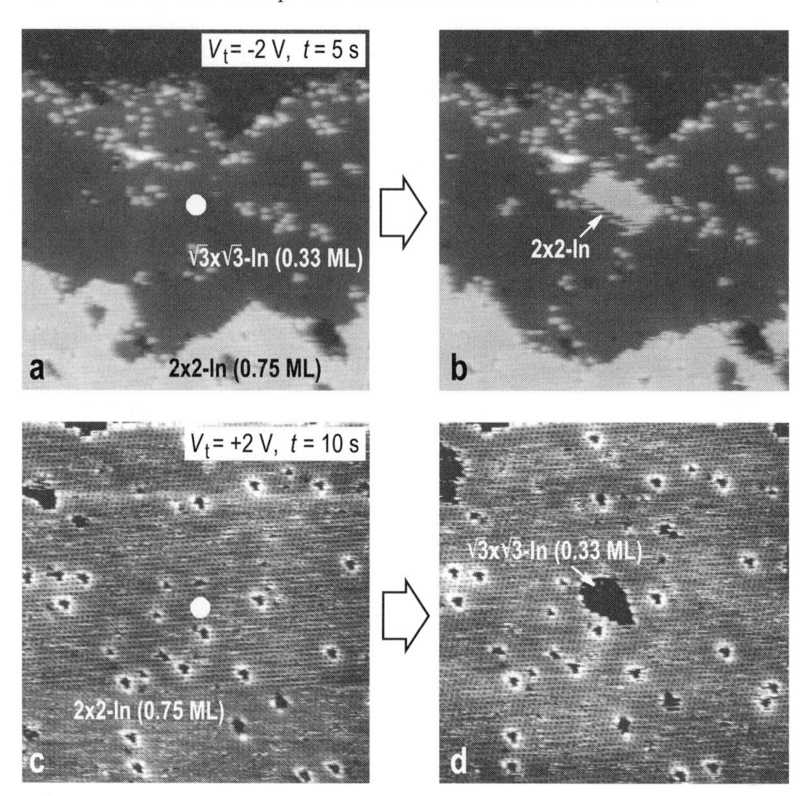

Fig. 15.9. Field-assisted diffusion of In on Si(111). (**a**) The tip is posed at the point indicated by a white dot above a region of the $\sqrt{3}\times\sqrt{3}$-In phase (0.33 ML of In) and a negative bias voltage, $V_t = -2.0$ V, is applied to the tip for 5 s. (**b**) Attraction of In atoms to the region underneath the tip results in a local increase in In coverage to 0.75 ML and formation of the patch of the 2×2-In phase. (**c**) The tip is posed at the point indicated by a white dot above a region of the 2×2-In phase (0.75 ML of In) and a positive bias voltage, $V_t = +2.0$ V, is applied to the tip for 10 s. (**d**) In atoms are repelled from the region beneath the tip and a "hole" occupied by the $\sqrt{3}\times\sqrt{3}$-In phase is formed. The 2×2-In phase is seen as brighter, the $\sqrt{3}\times\sqrt{3}$-In phase as darker. The scale of all images is 400×400 Å^2 (after Saranin et al. [15.7])

ions (Fig. 15.10). There is a threshold voltage, above which atom extraction via field evaporation takes place.

As an example, Fig. 15.11 shows the field-induced extraction of a single Si adatom from the Si(111)7×7 surface by applying a voltage pulse of +4 V to the tip. Figure 15.12 displays the relationship between the threshold voltage required to extract a Si atom and the logarithm of the tunneling current, which is roughly proportional to the tip–sample separation (see (7.2)). Tips made of Ag, W, Pt, and Au were used in the experiment. The general symmetry of the field evaporation process with respect to the polarity is clearly seen. The quantitative deviation from symmetry is due to the effect of the

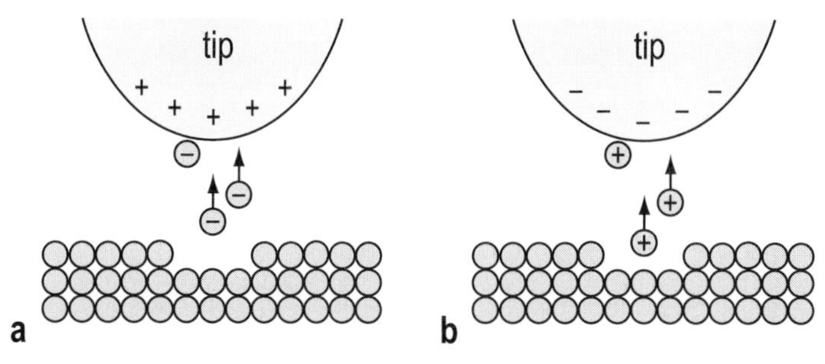

Fig. 15.10. Field evaporation is a symmetric process with respect to the polarity for the applied voltage bias. (**a**) When a positive bias voltage is applied to the tip, surface atoms evaporate as negative ions. (**b**) When a negative bias voltage is applied to the tip, surface atoms evaporate as positive ions

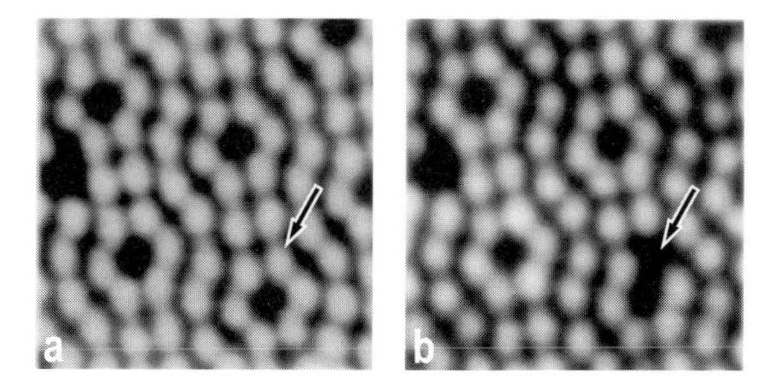

Fig. 15.11. STM images showing the field-induced extraction of a single Si atom from the Si(111)7×7 surface using a W tip. The Si atom indicated by the arrow in (**a**) is extracted by a voltage pulse, $V_t = +2.0$ V, applied to the tip for 10 ms. The formed adatom vacancy is indicated by the arrow in (**b**) (after Aono et al. [15.9])

work function of the tip, which enhances the field for negative tip biases, but weakens for positive biases (see Fig. 7.16).

Electron-Stimulated Desorption. When a negative bias voltage is applied to the tip, the flux of electrons will flow through the gap towards the surface. Due to the extremely small cross-section of the STM electron beam, very high current densities are conventionally present in the STM. One could expect local heating of the sample surface underneath the tip; however, estimation yields that for most crystalline materials the temperature rise is actually negligible ($\ll 1$ K) under ordinary STM conditions [15.3]. A more pronounced effect is the direct electronic excitation of the adsorbate–substrate, for example, as demonstrated for the case of the atoms adsorbed on Si surfaces.

The possibility of extracting hydrogen atoms one by one was demonstrated in the formation of "dangling-bond wires" on a monohydride Si(100) 2×1-H surface (see Fig. 15.13). In the original surface, all dangling bonds are

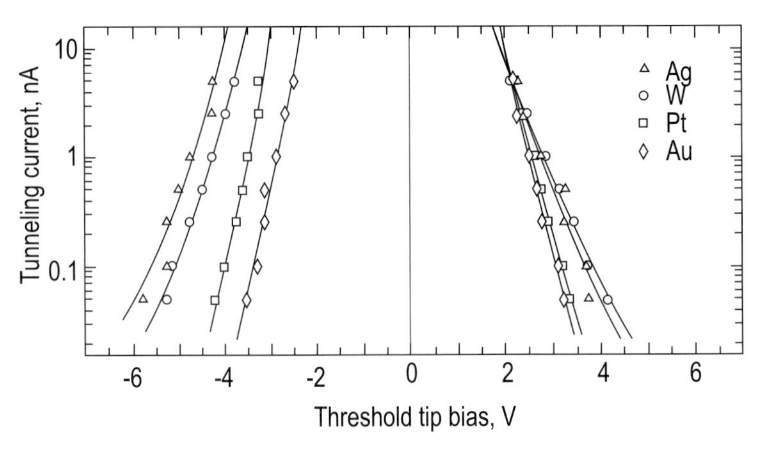

Fig. 15.12. Relation between the threshold tip voltage required to extract Si atoms from a Si(111)7×7 surface and the logarithm of the tunneling current, which is a measure of the tip–sample separation. The measurements were made with Ag, W, Pt, and Au tips, whose work functions increase in this order (after Aono et al. [15.9])

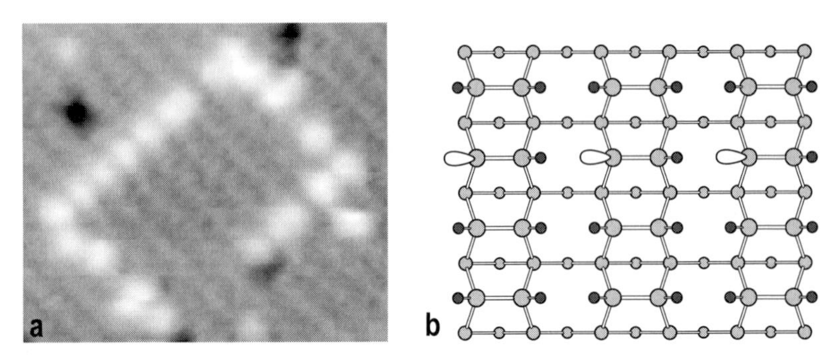

Fig. 15.13. (**a**) 100×100 Å^2 STM image showing a dangling-bond structure fabricated on the Si(100)2×1-H surface by extracting H atoms one by one. The structure contains dangling-bond wires both parallel and perpendicular to the dimer rows. (**b**) Schematic diagram showing an ideal dangling-bond wire perpendicular to the dimer-row direction made of single dangling bonds. Si atoms are shown as gray circles, H atoms are shown as small black circles, and dangling bonds formed after extraction of H atoms are shown by white ovals (after Hitosugi et al. [15.10])

saturated by H atoms, hence, removal of a single H atom produces a single dangling bond. This delicate procedure requires accurate adjustment of the parameters used: the tip bias voltage was $V_t = -2.9$ V, the tunneling current was 0.4 nA and the pulse duration was from 100 to 300 ms. For example, at slightly higher voltages (above -3.0 V) extraction of only a single H atom becomes difficult and extraction of several H atoms at once frequently occurs. In contrast, at slightly lower voltages (below -2.6 V), the extraction of even a single H atom rarely occurs.

15.2.3 Atom Deposition

When an atom is extracted from the surface, in principle it can be redeposited to a new place on the surface. This possibility is illustrated in Fig. 15.14. The Si adatom indicated by the arrow in Fig. 15.14a is extracted from the Si(111)7×7 surface and then redeposited to a site indicated by a cross in Fig. 15.14b. It should be noted that redeposition of the extracted atom is not such a well-reproduced process as single atom extraction. The main reason is that the extracted atom might migrate over the tip and its actual location on the tip surface is unknown. In experiments with a W tip and Si adatoms on a Si(111)7×7 surface discussed above, the probability of redeposition for every extracted atom was only ~20% at applied voltage $V_t = +6$ V and almost negligible at lower voltages. The exact place on the sample surface, where the atom will land, can hardly be controlled with high precision. The identity of the deposited atom with the extracted one, strictly speaking, might also cause some doubts.

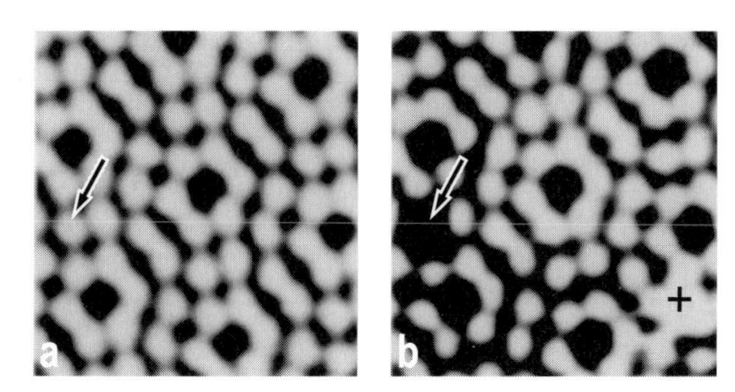

Fig. 15.14. Extraction and redeposition of a single Si adatom on the Si(111)7×7 surface using a W tip in an STM. (**a**) Surface before STM modification; the atom to be extracted is indicated by the arrow. (**b**) Surface after STM modification; the vacancy formed upon atom extraction is indicated by the arrow, and the redeposited atom is indicated by the cross (after Aono et al. [15.9])

A more reliable technique for nanostructure fabrication appears to be deposition of clusters rather than single atoms. Here nanometer-size hillocks are fabricated on the surface by transfer of the material from the tip. Two main methods are employed, namely,

- the z-pulse method and
- the voltage pulse method.

The *z-pulse method* is illustrated schematically in Fig. 15.15. When a voltage pulse is applied to the z-piezodrive of the tip, the tip is lowered until it comes into direct contact with the sample. As the tip is retracted back after the end of the pulse action, the connecting neck elongates and eventually breaks, leaving a hillock made of the tip material at the sample surface.

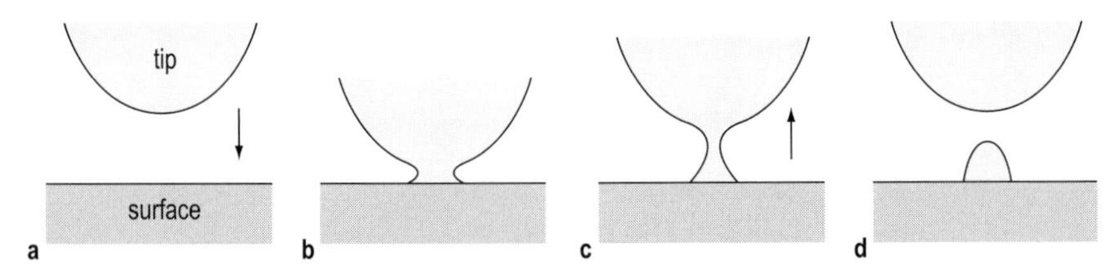

Fig. 15.15. Sequential stages of the formation of a nanometer-size hillock using the z-pulse method. (**a**) The tip approaches the surface. (**b**) A tip–sample contact is formed. (**c**) The tip is retracted and the connecting neck elongates. (**d**) The neck is broken and the hillock is left on the surface

The *voltage pulse method* allows fabrication of similar hillocks by applying an appropriate voltage pulse to the tip–sample junction. In this technique, the temporal contact between the tip and the sample is also plausibly established, since the mechanical stress due to the field can cause appreciable deformation of both tip and sample on the scale of the tunnel gap. Field evaporation cannot be excluded as an accompanying mechanism for hillock formation. It is worth noting that, depending on the tip and sample materials, under almost similar conditions pits instead of hillocks might form at the surface. The direction of the material transfer is controlled by the relative stiffness and elasticity of the tip and sample materials.

15.3 Self-Organization of Nanostructures

An alternative approach for nanostructure formation resides in the use of *self-organization* processes, i.e., to provide appropriate conditions for the system and to "let nature do it herself." In this case, the nanostructures are formed by spontaneous growth. Variety of surface phenomena are used for self-organized formation of different nanostructures. Some selected examples are considered below.

Ge Nanocrystal Growth on a Si(100) Surface. A typical example of self-organized nanostructure growth is island formation, which takes place in the Stranski–Krastanov (SK) growth mode in heteroepitaxy, for example, as in the case of the Ge/Si(100) system. According to SK growth, formation of uniform strained layer a few monolayers thick is followed by the growth of three-dimensional nanocrystals on top of the uniform layer. The shape, size, and number density of the forming nanocrystals have been found to depend on the growth conditions. Four distinct island shapes have been revealed for Ge nanocystals, called huts, pyramids, domes, and superdomes. *Huts* (or *hut-clusters*) are shown in Fig. 15.16a. They are rectangular based and bounded by {105} facets. The formation of hut-clusters dominates at the onset of 3D growth at relatively low temperature of 300–500°C. At higher temperatures (550–600°C), *pyramids*, which are a special case of a hut with a square base and four equal-area {105} facets, and *domes*, which are multifaceted structures bound by {311}, {518}, {105} and (001) facets, are formed. Pyramids and domes are shown in Fig. 15.16b (domes are larger in size). With further Ge deposition, *superdomes* grow. These are the largest islands, which are very similar in shape to domes, but have in addition {111} and other steep facets at the boundary with the substrate.

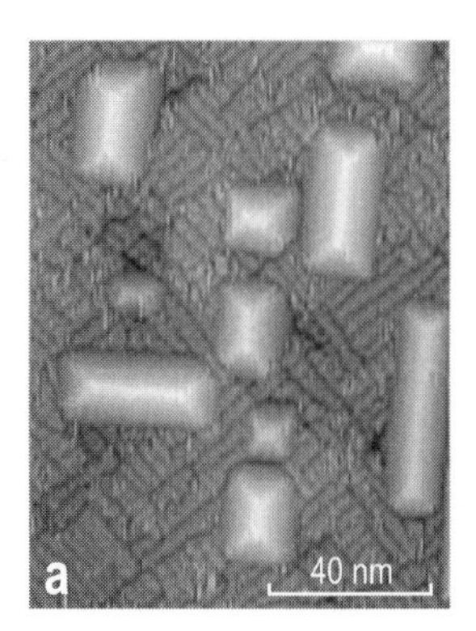

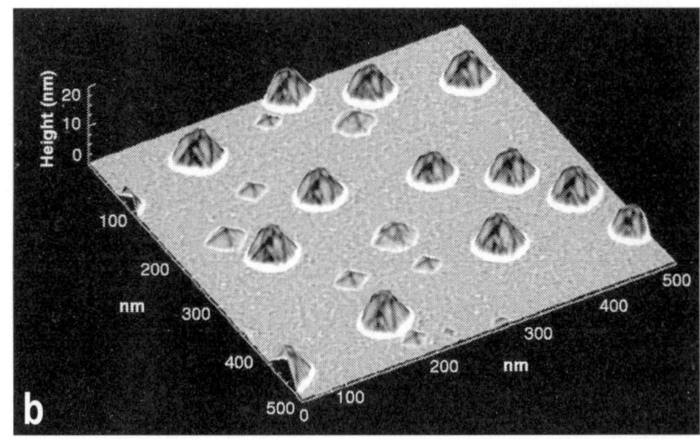

Fig. 15.16. The different shapes of strained Ge nanocrystals grown on a Si(100) surface: (**a**) huts, (**b**) pyramids and domes (after Voigtländer [15.11] and Williams et al. [15.12])

This example shows that the shape and the average size of Ge islands can be controlled by the coverage of deposited Ge and the growth temperature. However, the uniformity of island size and density can hardly be controlled in SK growth. An approach to improving the uniformity is illustrated in Fig. 15.17. The improvement is based on the growth of Ge/Si multilayers. When a spacer Si layer is deposited above Ge islands, the new surface is nearly flat, but its local strain is modulated by the embedded Ge islands. When Ge islands are grown again, they nucleate preferentially above embedded Ge islands. When two embedded islands are closely spaced, they produce a smoothed single minimum and a single island grows in the next layer. Conversely, above a region of low density of embedded Ge islands, new islands nucleate. Repetition of Ge and Si depositions improves the uniformity in island density and size, as one can see in Fig. 15.17b.

Hydrogen Interaction with Metal/Silicon Surface Phases. The interaction of atomic hydrogen with surface phases formed by metal adsorbates on silicon causes a drastic reordering of the surface structure. As illustrated schematically in Fig. 15.18a and b, the continuous adsorbate 2D layer agglomerates upon hydrogen adsorption into 3D metallic nanoclusters [15.14]. This effect has a common character and is typical for many metal adsorbates (for example, Ag, In, Al, Pb). The size, shape, and areal density of forming nanoclusters depend mainly on the structure and composition of the original

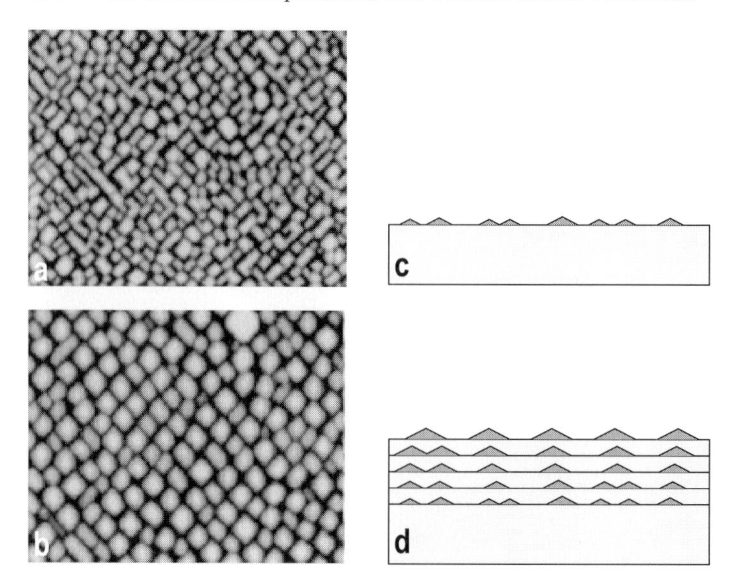

Fig. 15.17. Improvement of the uniformity of size and density of $Si_{0.25}Ge_{0.75}$ islands on Si(100) by the growth of Ge/Si multilayers: (**a**) AFM image (630×800 nm^2) of $Si_{0.25}Ge_{0.75}$ islands after deposition of the first alloy layer. (**b**) AFM image (1.250×960 nm^2) of $Si_{0.25}Ge_{0.75}$ islands after deposition of the 20th alloy layer. The Si spacer thickness is 10 nm and the average $Si_{0.25}Ge_{0.75}$ coverage is 2.5 nm (after Tersoff et al. [15.13])

metal/silicon phase and the temperature at which the sample is exposed to atomic hydrogen. It is remarkable that this transformation is reversible: upon H desorption, the original metal/silicon phase is restored.

Superlattice of Nanoclusters on Si(111)7×7. A stable crystal surface with a large unit cell can be used as a template for nanofabrication. The Si(111)7×7 surface (see Sect. 8.5.2) fits these requirements. Indeed, when adsorption of metal atoms is conducted at moderate temperatures, the preserved basic 7×7 structure modulates the adsorption process. As an example, Fig. 15.19a shows a superlattice of identical-size Al nanoclusters (magic clusters) formed by deposition of ∼0.35 ML of Al onto the Si(111)7×7 surface held at 575 °C. Each cluster is built of exactly six Al atoms linked by three Si atoms, as shown in Fig. 15.19b. They occupy both faulted and unfaulted 7×7 unit cell halves. This is in contrast to the 7×7 superlattice of thallium (Tl) nanoclusters, in which clusters consisting of ∼9 Tl atoms each occupy exclusively the faulted halves of the 7×7 unit cell (see Fig. 15.20)

Silicide Nanowires on a Si(100) Surface. If the lattice mismatch between an epitaxial layer and a substrate is small along one crystal axis and large along the perpendicular axis, highly anisotropic growth takes place: the growth of an epitaxial crystal is unrestricted in the first direction, but limited in the other. As a result, self-organized growth of nanowires might take place.

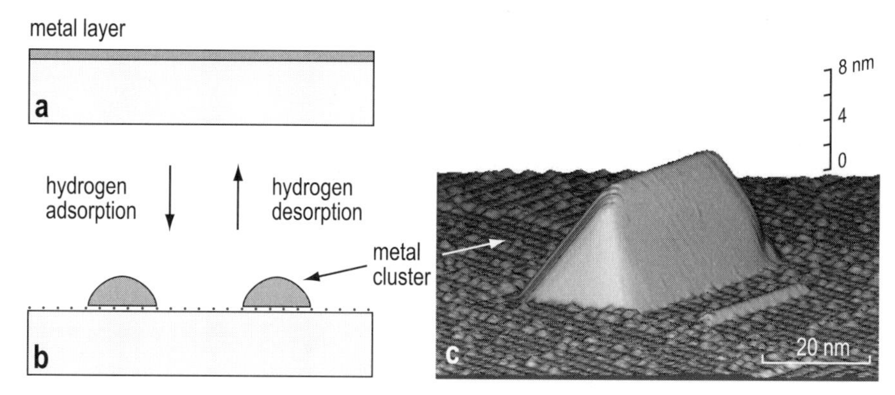

Fig. 15.18. Formation of nanoclusters by hydrogen interaction with the metal/silicon surface phase. Schematic diagram showing the surface structure (**a**) before and (**b**) after H interaction. (**c**) STM image (quasi-3D representation) of the Al nanocluster formed upon H interaction with the Si(100)$c(4\times12)$-Al surface phase (after Oura et al. [15.14])

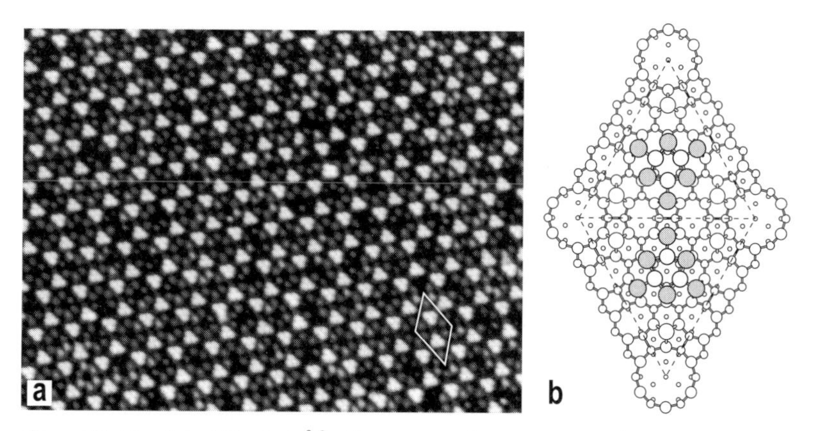

Fig. 15.19. (**a**) $465\times350\,\text{Å}^2$ filled state ($V_t = +2.0$ V) STM image of an ordered array of Al identical-size nanoclusters (magic clusters) formed by depositing ~0.35 ML onto the Si(111)7×7 surface held at 575°C. The 7×7 unit cell is outlined. (**b**) Schematic of the atomic structure of a magic cluster. The magic cluster is built of six metal atoms (gray circles) linked through three top Si atoms (large white circles) (after Kotlyar et al. [15.15])

The rare earth disilicides, $ErSi_2$, $DySi_2$, and $GdSi_2$, fit the above requirement, as they have lattice mismatches of 6.3%, 7.6%, and 8.9%, respectively, along one of the Si⟨011⟩ directions and mismatches of −1.6%, −0.1%, and 0.8%, respectively along the perpendicular Si⟨011⟩ direction. These silicides grow on Si(100) in the form of narrow straight nanowires, which have (depending on a particular case) widths in the range of 3–11 nm, heights in the range of 0.2–3 nm, and an average length in the range of 150–450 nm [15.17]. As an example, Fig. 15.21 shows $ErSi_2$ nanowires on Si(100).

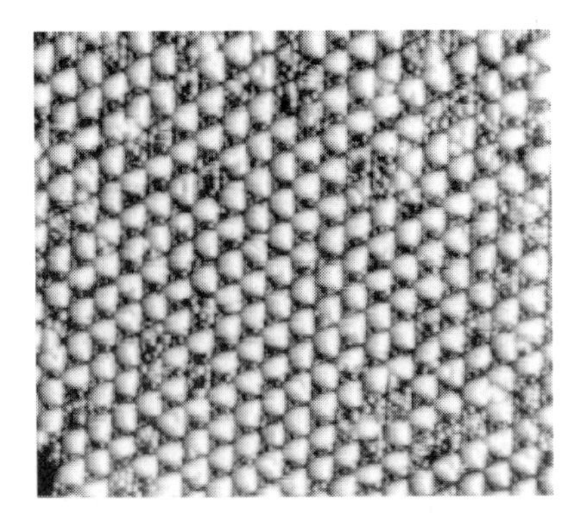

Fig. 15.20. (**a**) 500×425 Å^2 filled state ($V_t = +2.0$ V) STM image of an ordered array of Tl nanoclusters formed by depositing ~0.2 ML of Tl onto the Si(111)7×7 surface held at room temperature. The nanoclusters, containing ~ 9 Tl atoms each, occupy the faulted 7×7 unit cell halves (after Vitali et al. [15.16])

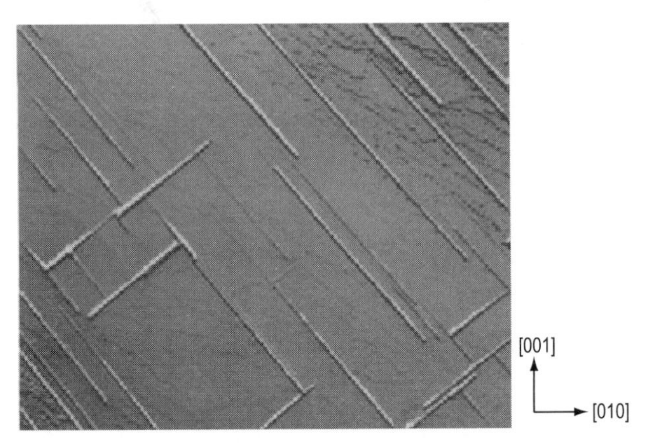

Fig. 15.21. 1000×800 nm^2 STM image showing $ErSi_2$ nanowires grown on a Si(100) surface (after Chen et al. [15.18])

15.4 Fullerenes and Carbon Nanotubes

Bulk carbon is generally known to exist in two forms: diamond and graphite. Recent progress in material science has brought to light new carbon modifications existing on the nanometer size scale. These are carbon spheroidal molecules of nanometer diameter, called *fullerenes*, and carbon hollow fibers of nanometer thickness, called *carbon nanotubes*.

Fullerenes. The C_{60} molecule was discovered in 1985 by Harold Kroto, Robert Curl, Richard Smalley, and co-workers [15.19] and, for this discovery, they were awarded the Noble prize in chemistry in 1996. In the original experiment, graphite was vaporized by laser irradiation and it was found that under appropriate conditions the cluster size distribution is dominated by remarkably stable clusters consisting of exactly 60 carbon atoms (see Fig. 15.22). The structure of the stable cluster was recognized as a truncated icosahedron, a polygon with 60 vertices and 32 faces, 12 of which are pentagonal and 20 hexagonal. This object is commonly encountered as a soccer ball. When a carbon atom is placed in each vertex of this structure, one obtains a model of the C_{60} molecule.

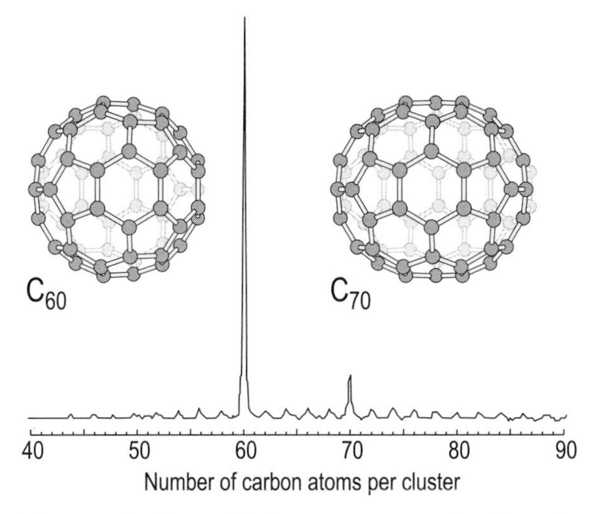

Fig. 15.22. Time-of-flight mass spectra of carbon clusters prepared by laser vaporization of graphite. The dominant peak corresponds to C_{60} fullerenes, the smaller peak to C_{70} fullerenes. The atomic arrangement is shown for both fullerene types (after Kroto et al. [15.19])

C_{60} appears to be a member of a set of spheroidal carbon molecules (for example, C_{70}, C_{74}, C_{84}, etc.), which are all built of hexagonal and pentagonal atomic rings. The number of pentagons is always twelve, while the number of hexagons grows with size. They are called *fullerenes*, after the American architect Richard Buckminster Fuller, who was renowned for his geodesic domes, which are based on hexagons and pentagons.

Each carbon atom in the fullerenes forms three bonds with the neighboring atoms (sp^2 hybridization), while the remaining valence electron forms a π-bond. The π-bond is delocalized, i.e., it is shared over the entire molecule, thus the inner and outer surfaces are covered with a sea of π electrons. The C_{60} molecules condense to form a weakly bound fcc crystal solid, called *fullerite*. After diamond and graphite, fullerite comprises a third form of pure carbon. It is an electrically insulating material with a band gap of about 2.3 eV. However, if alkali atoms (A) are added to solid C_{60}, new compounds like A_3C_{60} *alkali-doped fullerides* can be formed, which are superconducting when A is potassium (K) or rubidium (Rb). In the fulleride, the alkali atoms occupy the hollow sites between the C_{60} molecules. A_3C_{60} materials have a quite large superconducting transition temperatures in the range from 20 to 40 K.

The diameter of the C_{60} fullerene is ∼0.7 nm and up to ∼1.5 nm for the higher fullerenes. Hence, there is enough room inside the fullerene to incorporate a few atoms. Fullerenes with enclosed atoms are called *endohedral* fullerenes. The accepted notation for endohedral material is to use the symbol @ to show that the first material is inside the second, for example, $Sc_2@C_{84}$ means the C_{84} fullerene with two Sc atoms incorporated inside. It should be noted that most endohedral fullerene materials are made of C_{82}, C_{84}, or even higher fullerenes, which have more space inside.

Fullerenes are stable up to temperatures of over 1000°C, with the exact value depending on the particular fullerene. At temperatures of a few hundred °C, fullerenes can be sublimed directly from the solid fullerite. This property is used to study fullerene adsorption on different surfaces. The adsorbed fullerenes are free to diffuse on many surfaces and, as a result, can form ordered overlayers. The fullerene ordering is generally hexagonal with intermolecular separation close to that in the bulk fullerite, though some departures are possible due to competition between molecular–molecular and molecular–surface interactions. As an illustration of the ordering of the fullerene adsorbate, Fig. 15.23 shows an island of $La@C_{82}$ fullerenes grown on a Si(111)$\sqrt{3}\times\sqrt{3}$-Ag surface.

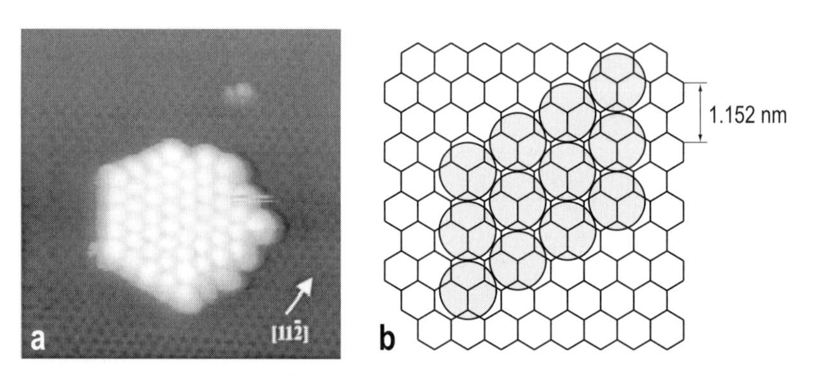

Fig. 15.23. (**a**) 170×170 Å^2 STM image showing a $La@C_{82}$ island nucleated on the Si(111)$\sqrt{3}\times\sqrt{3}$-Ag surface. (**b**) Simplified schematic showing the registry of the ordered array of $La@C_{82}$ molecules with respect to the underlying Si(111)$\sqrt{3}\times\sqrt{3}$-Ag reconstruction: the $La@C_{82}$ overlayer has a Si(111)3×3 periodicity (after Butcher et al. [15.20])

Carbon Nanotubes. Carbon nanotubes, which are tubules with diameters in the nanometer range and properties similar to an ideal graphite fiber, were discovered by Sumio Iijima in 1991 [15.21]. Two main types of carbon nanotubes are distinguished, namely,

- *single-wall nanotubes (SWNTs)* that consist of singular graphene cylindrical walls with diameter in the range of 1–2 nm (Fig. 15.24a).
- *multiwall nanotubes (MWNTs)* that contain several coaxial graphene cylinders separated by a spacing of 0.34–0.39 nm and having an outer diameter of 2–25 nm and inner hollow of 1–8 nm (Fig. 15.24b, c and d).

Chronologically, the MWNTs were the first to be discovered [15.21], while SWNTs were discovered about 2 years later in 1993 [15.22, 15.23].

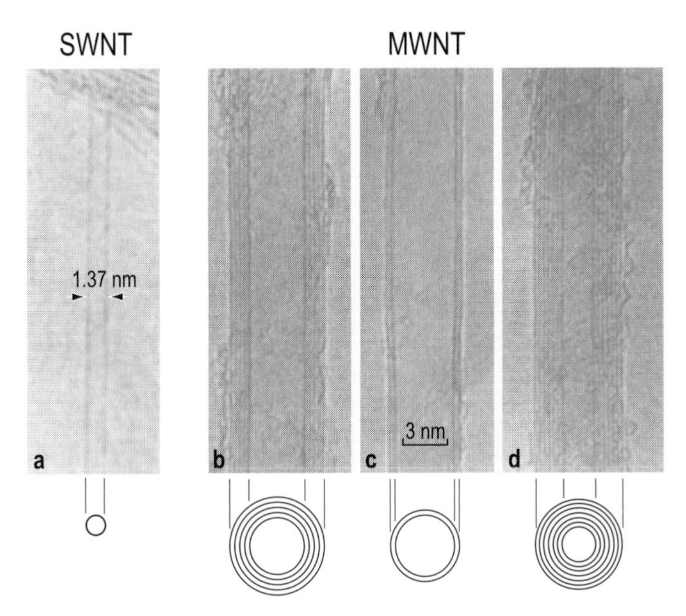

Fig. 15.24. TEM micrographs of (**a**) a single-wall carbon nanotube with diameter of 1.37 nm and **b**, **c**, (**d**) multiwall carbon nanotubes (MWNTs). The cross-section of each nanotube is illustrated. (**b**) A MWNT consisting of five graphitic shells with outer diameter of 6.7 nm; (**c**) a two-shell MWNT with outer diameter of 5.5 nm; and (**d**) a seven-shell MWNT with outer diameter of 6.5 nm and inner diameter of 2.2 nm (after Iijima [15.21] and Iijima and Ichihashi [15.22])

A single-wall nanotube may be thought of as a rolled up sheet of graphene. As illustrated in Fig. 15.25, the graphene sheet can be rolled up in different ways to form a variety of nanotube structures with different orientations of carbon hexagons relative to the nanotube axis. The two limiting cases are referred to as *zigzag* SWNTs and *armchair* SWNTs, while all the other cases are referred to as *chiral* SWNTs.

The structure of SWNTs is uniquely specified by the vector chosen on the plane graphene sheet so that by bending the sheet to superpose the head and tail of the vector one obtains a given nanotube (Fig. 15.26). This vector, called the *chiral vector*, can be expressed by the unit vectors, $\boldsymbol{a}_1$ and $\boldsymbol{a}_2$, of the hexagonal lattice as $m\boldsymbol{a}_1 + n\boldsymbol{a}_1$ and the components of the chiral vector (m, n)

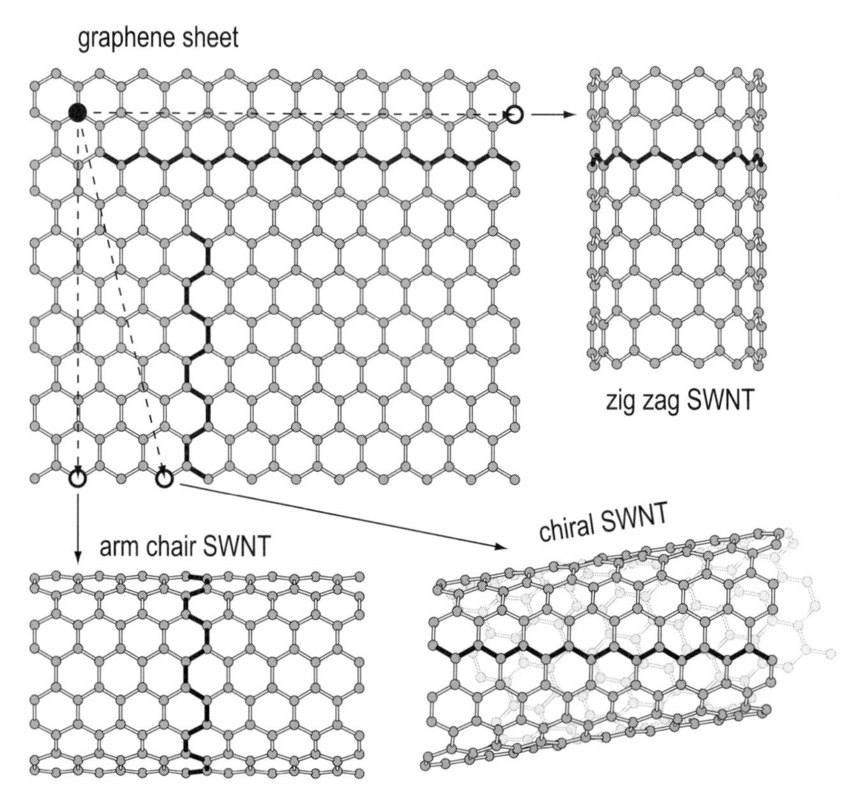

Fig. 15.25. Diagram illustrating how the manner in which the graphene sheet is rolled up produces nanotubes of various helical structures. The open circles in the graphene sheet denote the points that should be superposed at the origin in the rolling-up construction of the nanotube

comprise the index of the nanotube. The armchair nanotubes correspond to the case $n = m$, i.e., (n, n). Zigzag nanotubes correspond to the case $n = 0$, i.e., $(m, 0)$. All other (m, n) indices correspond to chiral nanotubes.

The length of the chiral vector is essentially the circumferential length, L, of the nanotube, hence the diameter of the nanotube, d, can be expressed in terms of its indices as:

$$d = \frac{L}{\pi} = \frac{a\sqrt{m^2 + n^2 + mn}}{\pi}, \tag{15.16}$$

where $a = 0.144\,\mathrm{nm}\times\sqrt{3} = 0.249\,\mathrm{nm}$ is the lattice constant of the honeycomb lattice (in the case of carbon nanotubes, the C-C bond length is 0.144 nm, which is slightly larger than that in graphite, 0.142 nm).

The angle between the chiral vector and the $\boldsymbol{a}_1$ direction is called the *chiral angle* (labeled α in Fig. 15.26). Its values are in the range $0 \leq |\alpha| \leq 30°$, because of the hexagonal symmetry of the honeycomb lattice, and it denotes

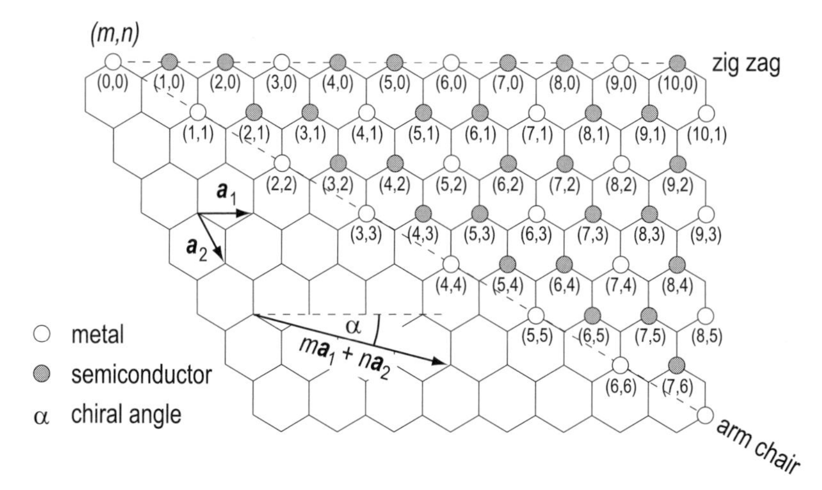

Fig. 15.26. Diagram illustrating the indexing scheme for nanotubes. When any index shown in the lattice is mapped onto the origin, a nanotube of that particular index is produced. Thus, the components of the chiral vector $\boldsymbol{a}_1$ and $\boldsymbol{a}_2$ comprise the index (m, n) of the nanotube. For example, all the (n, n) nanotubes are armchair nanotubes and all the $(m, 0)$ nanotubes are zigzag nanotubes. The chiral angle α and unit vectors, $\boldsymbol{a}_1$ and $\boldsymbol{a}_2$, of the hexagonal lattice are indicated. The indices (m, n) of the nanotubes, which are metallic and semiconducting, are denoted by open and solid circles, respectively

the tilt angle of the hexagons with respect to the direction of the nanotube axis. The chiral angle is given by

$$\alpha = \tan^{-1} \frac{\sqrt{3}\,n}{2m + n} \,. \tag{15.17}$$

In particular, zigzag nanotubes correspond to $\alpha = 0°$ and armchair nanotubes to $\alpha = 30°$.

The structure of carbon nanotubes defines their electronic properties. Depending on their helicity and diameter, nanotubes can be metallic or semiconducting. Figure 15.26 shows which carbon nanotubes are metallic (denoted by open circles) and which are semiconducting (denoted by closed circles). One can see that armchair nanotubes are always metallic, while zigzag and chiral nanotubes might be either metallic or semiconducting. The general rule states that if $m - n$ is divisible by 3, i.e.,

$$m - n = 3i \quad (i = \text{integer}) \,, \tag{15.18}$$

the (m, n) nanotubes are metallic; otherwise they are semiconducting. The band gap of semiconducting nanotubes depends inversely on their diameter, varying from ~0.8 eV to ~0.4 eV in the diameter range from 1 nm to 2 nm.

Since the diameter of a nanotube is smaller than the electron de Broglie wavelength, carbon nanotubes are predicted to be ideal 1D quantum wires. As a result, the calculated density of electronic states of both metallic and semiconducting nanotubes shows a series of spikes, called *Van Hove singularities* (see Fig. 15.27). At the Fermi level, the density of states is finite (albeit small) for metallic nanotubes and zero for semiconducting nanotubes. One can also notice the much larger separation between spikes around the Fermi level for metallic nanotubes, as compared to semiconducting nanotubes. The main predictions of the theory have found experimental confirmation in scanning tunneling spectroscopy (STS) studies of carbon nanotubes. As an example, Fig. 15.28 demonstrates a clear correspondence of the calculated density of states for (13,7) nanotubes to that determined in the experiment.

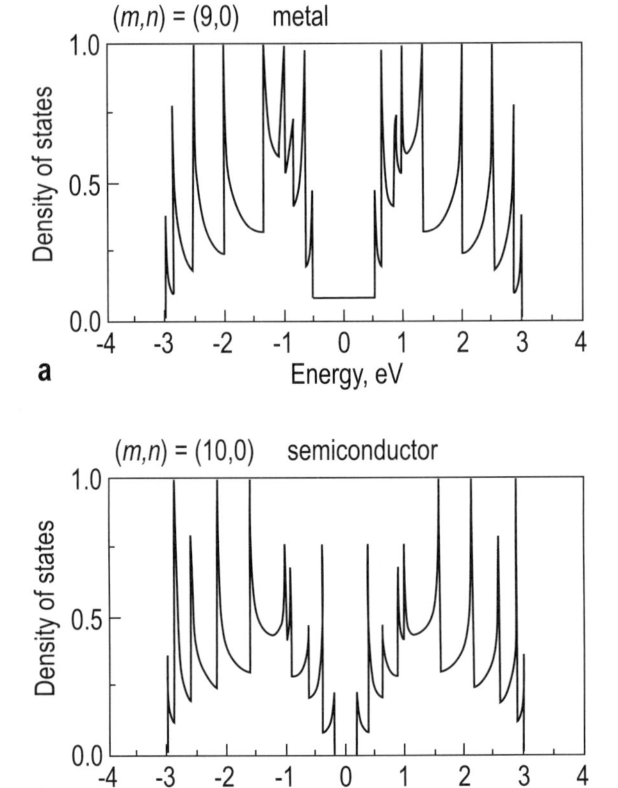

Fig. 15.27. Calculated 1D electronic density of states for (**a**) a metallic (9,0) nanotube and (**b**) a semiconducting (10,0) nanotube (after Dresselhaus [15.24])

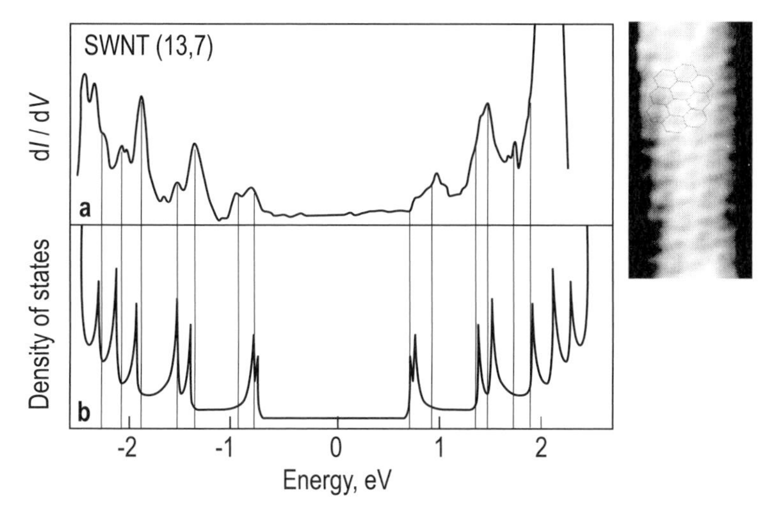

Fig. 15.28. Comparison of (**a**) the density of states obtained from an STS experiment at 77 K and (**b**) the tight-binding calculation for the (13,7) SWNT, which takes into account only π-electrons. In the right panel, an atomic resolution STM image of the carbon nanotube under investigation is shown. A portion of a hexagonal lattice is overlaid to guide the eye (after Kim et al. [15.25])

Further Reading

1. J.A. Stroscio, D.M. Eigler: *Atomic and Molecular Manipulation with Scanning Tunneling Microscope*, Science **254**, 1319–1326 (1991)
2. P. Avouris: *Manipulation of Matter at the Atomic and Molecular Levels*, Acc. Chem. Res. **28**, 95–102 (1995)
3. T. Ogino, H. Hibino, Y. Homma, Y. Kobayashi, K. Prabhakaran, K. Sumitomo, H. Omi: *Fabrication and Integration of Nanostructures on Si Surfaces*, Acc. Chem. Res. **32**, 447–454 (1999)
4. H.S. Nalwa (Ed.): *Nanostructured Materials and Nanotechnology* (Academic Press, New York 2002)
5. R. Saito, G. Dresselhaus, M.S. Dresselhaus: *Physical Properties of Carbon Nanotubes* (Imperial College Press, London 1998)
6. S.S. Sinnott, R. Andrews: *Carbon Nanotubes: Synthesis, Properties, and Applications*. Crit. Rev. Solid State Mater. Sci. **26**, 145–249 (2001)

References

Chapter 1

1.1 A. Zangwill: *Physics at Surfaces* (Cambridge University Press, Cambridge 1988) historical sketch, pp. 1–4
1.2 M.C. Desjonquères, D. Spanjaard: *Concepts in Surface Physics*, 2nd edn. (Springer, Berlin, Heidelberg, New York 1996) Introduction, pp. 1–3
1.3 C.B. Duke (Ed.): *Surf. Sci.: The First Thirty Years.* (North-Holland, Amsterdam 1994) pp. 1–1054. Reprinted from Surf. Sci. **299/300** (1994)
1.4 C.B. Duke, E.W. Plummer (Ed.): *Frontiers in Surface and Interface Science.* (North-Holland, Amsterdam 2002) pp. 1–1053. Reprinted from Surf. Sci. **500** (2002)

Chapter 2

2.1 B. Lang, R.W. Joyner, G.A. Somorjai: *Low Energy Diffraction Study of High Index Crystal Surfaces.* Surf. Sci. **30**, 440 (1972)
2.2 R.L. Park, H.H. Madden: *Annealing Changes on the (100) Surface of Palladium and their Effect on CO Adsorption.* Surf. Sci. **11**, 188 (1968)
2.3 E.A. Wood: *Vocabulary of Surface Crystallography.* J. Appl. Phys. **35**, 1306 (1964)
2.4 L.P. Bouckaert, R. Smoluchowski, E. Wigner: *Theory of Brillouin Zones and Symmetry Properties of Wave Functions in Crystals.* Phys. Rev. **50**, 58 (1936)

Chapter 3

3.1 W. Kern, D.A. Puotinen: *Cleaning Solutions Based on Hydrogen Peroxide for Use in Silicon Semiconductor Technology.* RCA Rev. **31**, 187 (1970)
3.2 J. Margrave (Ed.): *The Characterization of High Temperature Vapours.* (John Wiley, New York 1967)

Chapter 4

4.1 M. Henzler: *LEED Studies of Surface Imperfections.* Appl. Surf. Sci. **11/12**, 450 (1982)

4.2 H. Sakama, K. Murakami, K. Nishikata, A. Kawazu: *Structure of a Si(100)2×2-Ga surface.* Phys. Rev. B **50**, 14977 (1994)
4.3 E. Zanazzi, F. Jona: *A Reliability Factor for Surface Structure Determinations by Low-Energy Electron Diffraction.* Surf. Sci. **62**, 61 (1977)
4.4 J.B. Pendry: *Reliability Factors for LEED Calculations.* J. Phys. C **13**, 937 (1980)
4.5 S. Ino: *Some New Techniques in Reflection High Energy Electron Diffraction (RHEED) Application to Surface Structure Studies.* Japan J. Appl. Phys. **16**, 891 (1977)
4.6 J.H. Neave, B.A. Joyce, P.J. Dobson, N. Norton: *Dynamics of Film Growth of GaAs by MBE from RHEED Observations.* Appl. Phys. A **31**, 1 (1983)
4.7 C. Kittel: *Introduction to Solid State Physics*, 7th edn. (John Wiley, Chichester 1996)
4.8 R. Feidenhans'l: *Surface Structure Determination by X-ray Diffraction.* Surf. Sci. Rep. **10**, 105 (1989)
4.9 K. Takayanagi, Y. Tanishiro, M. Takahashi, S. Takahashi: *Structural Analysis of Si(111)−7×7 by UHV-Transmission Electron Diffraction and Microscopy.* J. Vac. Sci. Technol. A **3**, 1502 (1985)
4.10 K. Takayanagi, Y. Tanishiro, S. Takahashi, M. Takahashi: *Structure Analysis of Si(111)−7×7 Reconstructed Surface by Transmission Electron Diffraction.* Surf. Sci. **164**, 367 (1985)

Chapter 5

5.1 M.P. Seah, W.A. Dench: *Quantitative Auger Analysis.* Surf. Interface Anal. **1**, 2 (1979)
5.2 C.C. Chang: *Auger Electron Spectroscopy.* Surf. Sci. **25**, 53 (1971)
5.3 L.E. Davis, N.C. MacDonald, P.W. Palmberg, G.E. Piach, R.E. Weber: *Handbook of Auger Electron Spectroscopy*, 2nd edn. (Physical Electronic Industries, Eden Prairie, Minnesota 1976)
5.4 C.J. Powell, M.P. Seah: *Precision, Accuracy, and Uncertainty in Quantitative Auger Analysis.* J. Vac. Sci. Technol. A **8**, 735 (1990)
5.5 S. Tanuma, C.J. Powell, D.R. Penn: *Calculations of Electron Inelastic Mean Free Paths. II. Data for 27 Elements over 50–2000 eV Range.* Surf. Interface Anal. **17**, 911 (1991)
5.6 R. Shimizu: *Quantitative Analysis by Auger Electron Spectroscopy.* Japan J. Appl. Phys. **22**, 1631 (1983)
5.7 R.L. Gerlach: 'Applications of Ionization Spectroscopy'. In: *Proc. of Internat. Conf. Electron Spectroscopy at London, England 1972* ed. by D.A. Shirlley (North Holland, Amsterdam 1972) pp. 885–894
5.8 H. Ibach, J.E. Rowe: *Electron Transition of Oxygen Adsorbed on Clean Silicon (111) and (100) Surfaces.* Phys. Rev. B **9**, 1951 (1974)
5.9 C. Kittel: *Introduction to Solid State Physics*, 4th edn. (John Wiley, New York, London 1978)
5.10 C.J. Powell, J.B. Swan: *Origin of the Characteristic Electron Energy Losses in Aluminum.* Phys. Rev. B **115**, 869 (1959)
5.11 H. Ibach: *Electron Energy Loss Spectrometers.* (Springer, Berlin, Heidelberg, New York 1991)
5.12 F. Stietz, A Pantförder, J.A. Schaefer, G. Meister, A. Goldman: *High-Resolution Study of Dipole-Active Vibrations at the Ag(110)(n×1)O Surface.* Surf. Sci. **318**, L1201 (1994)

5.13 L. Vattuone, U. Valbusa, M. Rocca: *Coverage Dependence of the O-Ag(110) Vibration.* Surf. Sci. **318**, L1120 (1994)
5.14 H. Lüth: *Surfaces and Interfaces of Solid Materials*, 3rd edn. (Springer, Berlin, Heidelberg, New York 1995)
5.15 K.H. Frank, U. Karlson: 'sp-metals'. In: *Electronic Structure of Solids: Photoemission Spectra and Related Data, Landolt Börnstein III/23a.* ed. by A. Goldmann, E.-E. Koch (Springer, Berlin, Heidelberg 1989)
5.16 G.K. Wertheim: *Electron and Ion Spectroscopy of Solids* (Plenum Press, New York 1989)
5.17 T.C. Chiang, F.G. Himpsel: 'Band Structure of Tetrahedrally-Bonded Semiconductors'. In: *Electronic Structure of Solids: Photoemission Spectra and Related Data, Landolt Börnstein III/23a.* ed. by A. Goldmann, E.-E. Koch (Springer, Berlin, Heidelberg 1989)
5.18 F.J. Himpsel, F.R. McFeely, A. Taleb-Ibrahimi, J.A. Yarmoff, G. Hollinger: *Microscopic Structure of the SiO_2/Si Interface.* Phys. Rev. B **38**, 6084 (1988)
5.19 S.D. Kevan: *Evidence for a new Broadening Mechanism in Angle-Resolved Photoemission from Cu(111).* Appl. Phys. A **31**, 1 (1983)

Chapter 6

6.1 C. Linsmeier, H. Knözinger, E. Taglauer: *Ion Scattering and Auger Electron Spectroscopy Analysis of Alumina-Supported Rhodium Model Catalyst.* Surf. Sci. **275**, 101 (1992)
6.2 H. Niehus: 'Ion Scattering Spectroscopic Techniques'. In: *Practical Surface Analysis. Vol. 2. Ion and Neutral Spectroscopy* 2nd edn. ed. by D. Briggs, M.P. Seah (John Wiley, Chichester 1992) pp. 507–576.
6.3 R.S. Williams: 'Quantitative Intensity Analysis of Low-Energy Scattering and Recoiling from Crystal Surfaces'. In: *Low Energy Ion-Surface Interactions* ed. by J.W. Rabalais (John Wiley, Chichester 1994) pp. 1–54
6.4 J.A. Cookson: 'Analytical Techniques'. In: *Principles and Applications of High-Energy Microbeams* ed. by F. Watt, G.W. Grime (IOP Publishing, England 1987) pp. 21–78
6.5 C.S. Barrett, R.M. Müller, W. White: *Proton Blocking in Cubic Crystals.* J. Appl. Phys. **39**, 4695 (1968)
6.6 L.C. Feldman, J.W. Mayer, S.T. Picraux: *Materials Analysis by Ion Channeling. Submicron Crystallography* (Academic Press, New York 1982)
6.7 J.P. Ziegler, J.P. Biersack, U. Littmark: *The Stopping Power and Range of Ions in Solids*, Vol. 1 (Pergamon Press, New York 1985)
6.8 I.H. Wilson, S.S. Todorov, D.S. Karpuzov: *Profile Evolution During Ion Beam Etching of Clean Germanium Targets.* Nuclear Instr. Methods Phys. Res. **209-210**, 549 (1983)
6.9 H.D. Hagstrum: *Theory of Auger Injection of Electrons from Metals by Ions.* Phys. Rev. **96**, 336 (1954)
6.10 R.L. Erickson, D.P. Smith: *Oscillatory Cross Section in Low-Energy Ion Scattering from Surfaces.* Phys. Rev. Lett. **34**, 297 (1975)
6.11 D.P. Woodruff, T.A. Delchar: *Modern Techniques of Surf. Sci.*, 2nd edn. (Cambridge University Press, Cambridge 1994)
6.12 H. Niehus, E. Preuss: *Low Energy Alkali Backscattering at Pt(111).* Surf. Sci. **119**, 349 (1982)
6.13 H. Niehus: *Enhancement of the ICISS by Simultaneous Detection of Ions and Neutrals.* Surf. Sci. **166**, L107 (1986)

6.14 M. Aono, C. Oshima, S. Zaima, S. Otani, Y. Ishizawa: *Quantitative Surface Atomic Geometry and Two-Dimensional Surface Electron Distribution Analysis by a New Technique in Low-Energy Ion Scattering.* Japan J. Appl. Phys. **20**, L829 (1981)
6.15 M. Katayama, E. Nomura, N. Kanekama, H. Soejima, M. Aono: *Coaxial Impact-Collision Ion Scattering Spectroscopy a Novel Method for Surface Structure Analysis.* Nuclear Instr. Methods Phys. Res. B **33**, 857 (1988)
6.16 M. Aono, M. Katayama, E. Nomura, T. Chassè, M. Kato: *Recent Development in Low-Energy Ion Scattering Spectroscopy (ISS) for Surface Structural Analysis.* Nuclear Instr. Methods Phys. Res. B **37**, 264 (1989)
6.17 M. Aono, Y. Hou, C. Oshima, Y. Ishizawa: *Low-Energy Ion Scattering from the Si(100) surface.* Phys. Rev. Lett. **49**, 567 (1982)
6.18 I. Stensgaard, L.C. Feldman, P.J. Silverman: *Calculation of the Backscattering-Channeling Surface Peak.* Surf. Sci. **77**, 513 (1978)
6.19 L.C. Feldman, R.L. Kauffman, P.J. Silverman, R.A. Zuhr, J.H. Barrett: *Surface Scattering from W Single Crystal by MeV He^+ Ions.* Phys. Rev. Lett. **39**, 38 (1977)
6.20 W.C. Turkenburg, W. Soszka, F.W. Saris, H.H. Kersten, B.G. Colenbrander: *Surface Structure Analysis by Means of Rutherford Scattering: Methods to Study Surface Relaxation.* Nuclear Instr. Methods Phys. Res. **132**, 587 (1976)
6.21 J.W.M. Frenken, F. Huussen, J.F. van der Veen: *Evidence for Anomalous Thermal Expansion at a Crystal Surface.* Phys. Rev. Lett. **58**, 401 (1987)
6.22 F. Shoji, K. Kusumura, K. Oura: *A Si(100)-2×1:H Monohydride Surface Studied by Low-energy Recoil-Ion Spectroscopy.* Surf. Sci. **280**, L247 (1993)
6.23 A. Benninghoven: *Comparative Study of Si(111), Silicon Oxide, SiC and Si_3N_4 Surfaces by Secondary Ion Mass Spectroscopy (SIMS).* Thin Solid Films **28**, 59 (1975)
6.24 A. Benninghoven: *Developments in Secondary Ion Mass Spectroscopy and Applications to Surface Studies.* Surf. Sci. **53**, 596 (1975)
6.25 A. Casel, H. Jorke, M. Pawlik, R. Groves, E. Frenzel: *A Comparison of Electrical and Chemical Profiling of Dopant Superlattices in Silicon.* J. Appl. Phys. **67**, 1740 (1990)

Chapter 7

7.1 E.W. Müller: *Work Function of Tungsten Single Crystal Planes Measured by the Field Emission Microscope.* J. Appl. Phys. **26**, 732 (1955)
7.2 T.T. Tsong, C. Chen: 'Dynamics and Diffusion of Atoms at Stepped Surfaces'. In: *The Chemical Physics of Solid Surfaces. Vol. 8. Growth and Properties of Ultrathin Epitaxial Layers.* ed. by D.A. King, D.P. Woodruff (Elsevier, Amsterdam 1997) pp. 102–148
7.3 T.T. Tsong, J. Sweeney: *Direct Observation of the Atomic Structure of W(100) Surface.* Solid State Comm. **30**, 767 (1979)
7.4 G.A. Bassett, J.W. Menter, D.W. Pashley: *Electron Optical Studies of Imperfect Crystals and Their Surfaces.* Discussions Faraday Soc. **28**, 7 (1959)
7.5 C.H. Luo, F.R. Chen, L.J. Chen: *Atomic Structure of Si/$TbSi_2$ Double-Heterostructure Interfaces.* J. Appl. Phys. **76**, 5744 (1994)
7.6 K. Yagi: *Reflection Electron Microscopy: Studies of Surface Structures and Surface Dynamic Processes.* Surf. Sci. Rep. **17**, 305 (1993)
7.7 R.M. Tromp: *Low Energy Electron Microscopy.* IBM J. Res. Dev. **44**, 503 (2000)

7.8 R.M. Tromp: `http://www.research.ibm.com/leem/`

7.9 F. Cosandey, L. Zhang, T.E. Madey: *Effect of Substrate Temperature on the Epitaxial Growth of Au on TiO_2(110).* Surf. Sci. **474**, 1 (2001)

7.10 I.H. Wilson, S.S. Todorov, D.S. Karpuzov: *Profile Evolution During Ion Beam Etching of Clean Germanium Targets.* Nuclear Instr. Methods Phys. Res. **209-210**, 549 (1983)

7.11 R.J. Hamers, R.M. Tromp, J.E. Demuth: *Surface Electronic Structure of Si(111)–(7×7) Resolved in Real Space.* Phys. Rev. Lett. **56**, 1972 (1986)

7.12 R.M. Tromp: *Spectroscopy with the Scanning Tunneling Microscope: a Critical Review.* J. Phys.: Condensed Matter **1**, 10211 (1989)

7.13 E. Meyer, H. Heinzelmann: 'Scanning Force Microscopy'. In: *Scanning Tunneling Microscopy II. Vol. 28. Springer Series in Surf. Sci.* ed. by R. Wiesendanger, H.-J. Güntherodt (Springer Berlin, Heidelberg, New York 1992) pp. 99–149

7.14 G. Binnig, C.F. Quate, Ch. Gerber: *Atomic Force Microscope.* Phys. Rev. Lett. **56**, 930 (1986)

7.15 E. Meyer, H. Heinzelmann, P. Grütter, Th. Jung, H.-R. Hidber, H. Rudin, H.-J. Güntherodt: *Atomic Force Microscopy for the Study of Tribology and Adhesion.* Thin Solid Films **181**, 527 (1989)

7.16 K. Yokoyama, T. Ochi, A. Yoshimoto, Y. Sugawara, S. Morita: *Atomic Resolution Imaging on Si(100)2×1 and Si(100)2×1:H Surfaces with Noncontact Atomic Force Microscopy.* Japan J. Appl. Phys. **39**, L113 (2000)

Chapter 8

8.1 C.B. Duke: *Surface Structures of Tetrahedrally Coordinated Semiconductors: Principles, Practice, and Universality.* Appl. Surf. Sci. **65/66**, 543 (1993)

8.2 J.N. Andersen, H.B. Nielsen, L. Petersen, D.L. Adams: *Oscillatory relaxation of the Al(110) Surface.* J. Phys. C: Solid State Phys. **17**, 173 (1984)

8.3 F. Jona, P.M. Marcus: 'Surface Structures from LEED: Metal Surfaces and Metastable Phases'. In: *The Structure of Surfaces II.* ed. by J.F. Van Der Veen, M.A. Van Hove (Springer, Berlin, Heidelberg, 1988) pp. 90–99

8.4 Z.Q. Wang, Y.S. Li, F. Jona, P.M. Marcus: *Epitaxial Growth of Body-Centered-Cubic Nickel on Iron.* Solid State Comm. **61**, 623 (1987)

8.5 H.D. Shih, F. Jona, U. Bardi, P.M. Marcus: *The Atomic Structure of Fe{110}.* J. Phys. C: Solid State Phys. **13**, 3801 (1980)

8.6 J. Sokolov, F. Jona, P.M. Marcus: *Multilayer Relaxation of a Clean bcc Fe{111} Surface.* Phys. Rev. B **33**, 1397 (1986)

8.7 J. Sokolov, F. Jona, P.M. Marcus: *Multilayer Relaxation of the Fe{210} Surface.* Phys. Rev. B **31**, 1929 (1985)

8.8 M.M. Nielsen, J. Burchhardt, D. L. Adams: *Structure of Ni(100)-c(2×2)-Na: A LEED Analysis.* Phys. Rev. B **50**, 7851 (1994)

8.9 W. Reimer, V. Penka, M. Skottke, R.J. Behm, G. Ertl, W. Moritz: *A LEED Analysis of the (2×2)H-Ni(110) Structure.* Surf. Sci. **186**, 45 (1987)

8.10 W.R.A. Huff, Y. Chen, S.A. Kellar, E.J. Moler, Z. Hussain, Z.Q. Huang, Y. Zheng, D.A. Shirley: *Angle-Resolved Photoemission Extended Fine Structure of the Ni 3p, Cu 3s, and Cu 3p Core Levels of the Respective Clean (111) Surfaces.* Phys. Rev. B **56**, 1540 (1997)

8.11 D.L. Adams, W.T. Moore, K.A.R. Mitchel: *Multilayer Reconstruction of the Ni(311): A New LEED Analysis.* Surf. Sci. **149**, 407 (1985)

8.12 J. Sokolov, H.D. Shih, U. Bardi, F. Jona, P.M. Marcus: *Multilayer Relaxation of Body-Centred-Cubic Fe(211).* J. Phys. C: Solid State Phys. **17**, 371 (1984)
8.13 G. Ritz, M. Schmid, P. Varga, A. Borg, M. Rønning: *Pt(100) Quasihexagonal Reconstruction: A Comparison Between Scanning Tunneling Microscopy Data and Effective Medium Theory Simulation Calculations.* Phys. Rev. B **56**, 10518 (1997)
8.14 T.R. Linderoth, S. Horch, E. Lægsgaard, I. Stensgaard, F. Besenbacher: *Surface Diffusion of Pt on Pt(110): Arrhenius Behavior of Long Jumps.* Phys. Rev. Lett. **78**, 4978 (1997)
8.15 N.J. Wu, A. Ignatiev: *Low-Energy-Electron-Diffraction Structural Determination of the Graphite (0001) Surface.* Phys. Rev. B **25**, 2983 (1982)
8.16 R.E. Schlier, H.E. Farnsworth: *Structure and Adsorption Characteristics of Clean Surfaces of Germanium and Silicon.* J. Chem. Phys. **30**, 917 (1959)
8.17 T. Yokoyama, K. Takayanagi: *Anomalous Flipping Motions of Buckled Dimers on the Si(001) Surface at 5 K.* Phys. Rev. B **61**, R5078 (2000)
8.18 T. Abukawa, C.M. Wei, K. Yoshimura, S. Kono: *Direct Method of Surface Structure Determination by Patterson Analysis of Correlated Thermal Diffuse Scattering for Si(001)2×1.* Phys. Rev. B **62**, 16069 (2000)
8.19 S. Ferrer, X. Torrelles, V.H. Etgens, H.A. Vandervegt, P. Fajardo: *Atomic-Structure of the c(4×2) Surface Reconstruction of Ge(001) as Determined by X-Ray-Diffraction.* Phys. Rev. Lett. **75**, 1771 (1995)
8.20 K.C. Pandey: *New π-bonded Chain Model for Si(111)–(2 × 1) Surface.* Phys. Rev. Lett. **47**, 1913 (1981)
8.21 J.E. Northrup, M.L. Cohen: *Atomic Geometry and Surface-State Spectrum for Ge(111)–(2×1).* Phys. Rev. B **27**, 6553 (1983)
8.22 K. Takayanagi, Y. Tanishiro, S. Takahashi, M. Takahashi: *Structure Analysis of Si(111)–7×7 Reconstructed Surface by Transmission Electron Diffraction.* Surf. Sci. **164**, 367 (1985)
8.23 W.A. Harrison: *Surface Reconstruction on Semiconductors.* Surf. Sci. **55**, 1 (1976)
8.24 P.A. Bennett, L.C. Feldman, Y. Kuk, E.G. McRae, J.E. Rowe: *Stacking-Fault Model for the Si(111)–(7×7) Surface.* Phys. Rev. B **28**, 3656 (1983)
8.25 R.J. Culbertson, L.C. Feldman, P.J. Silverman: *Atomic Displacements in the Si(111)–(7×7) Surface.* Phys. Rev. Lett. **45**, 2043 (1980)
8.26 R.M. Tromp, E.J. Van Loenen, M. Iwami, F.W. Saris: *On the Structure of the Laser Irradiated Si(111)–(1×1) Surface.* Solid State Comm. **44**, 971 (1982)
8.27 G. Binning, H. Rohrer, Ch. Gerber, E. Weibel: *(7×7) Reconstruction on Si(111) Resolved in Real Space.* Phys. Rev. Lett. **50**, 120 (1983)
8.28 F.J. Himpsel: *Structural Model for Si(111)–(7×7).* Phys. Rev. B **27**, 7782 (1983)
8.29 E.G. McRae: *Surface Stacking Sequence and (7×7) Reconstruction at Si(111) Surfaces.* Phys. Rev. B **28**, 2305 (1983)
8.30 R.M. Feenstra, J.A. Stroscio, J. Tersoff, A.P. Fein: *Atom-Selective Imaging of the GaAs(110) Surface.* Phys. Rev. Lett. **58**, 1192 (1987)
8.31 C.B. Duke: *Structural Chemistry of the Cleavage Faces of Compound Semiconductors.* J. Vac. Sci. Technol. B **1**, 732 (1983)
8.32 K.W. Haberern, M.D. Pashley: *GaAs(111)A-(2×2) Reconstruction Studied by Scanning Tunneling Microscopy.* Phys. Rev. B **41**, 3226 (1990)
8.33 R.M. Feenstra, A.J. Slavin: *Scanning Tunneling Microscopy and Spectroscopy of Cleaved and Annealed Ge(111) Surfaces.* Surf. Sci. **251/252**, 401 (1991)

Chapter 9

9.1 J.M. Carpinelli, H.H. Weitering, E.W. Plummer: *Charge Rearrangement in the Ge_xPb_{1-x}/Ge(111) Interface.* Surf. Sci. **401**, L457 (1998)
9.2 A. Shibata, Y. Kimura, K. Takayanagi: *On the Restructed Layer of the Si(111)$\sqrt{3}\times\sqrt{3}$-Ag Structure Studies by Scanning Tunneling Microscopy.* Surf. Sci. **275**, L697 (1992)
9.3 D.W. McComb, R.A. Wolkow, P.A. Hackett: *Defects on the Ag/Si(111)-($\sqrt{3}\times\sqrt{3}$) Surface.* Phys. Rev. B **50**, 18268 (1994)
9.4 A.A. Saranin, A.V. Zotov, V.G. Lifshits, J.-T. Ryu, O. Kubo, H. Tani, T. Harada, M. Katayama, K. Oura: *Ag-Induced Structural Transformations on Si(111): Quantitative Investigation of the Si Mass Transport.* Surf. Sci. **429**, 127 (1999)
9.5 Y. Tanishiro, K. Takayanagi, K. Yagi: *Density of Silicon Atoms in the Si(111)$\sqrt{3}\times\sqrt{3}-$Ag Structure Studied by* in situ *UHV Reflection Electron Microscopy.* Surf. Sci. **258**, L687 (1991)
9.6 R.M. Tromp, T. Michely: *Atomic-Layer Titration of Surface Reaction.* Nature **373**, 499 (1995)
9.7 M. Nielsen, J.P. McTague, W. Ellenson: *Adsorbed Layers of D_2, H_2, O_2, and ^{3}He on Graphite Studied by Neutron Scattering.* J. Physique **38**, 10 (1977)
9.8 R. Imbihl, R.J. Behm, K. Christmann, G. Ertl, T. Matsushima: *Phase Transitions of a Two-Dimensional System: H on Fe(110).* Surf. Sci. **117**, 257 (1982)
9.9 H. Hirayama, S. Baba, A. Kinbara: *Electron Energy Loss Measurements of In/Si(111) Superstructures: Correlation of the Spectra with Surface Superstructures.* Appl. Surf. Sci. **33/34**, 193 (1988)
9.10 A.A. Saranin, A.V. Zotov, V.G. Lifshits, J.-T. Ryu, O. Kubo, H. Tani, T. Harada, M. Katayama, K. Oura: *Analysis of Surface Structures through Determination of their Composition Using STM: Si(100)4×3-In and Si(111)4×1-In Reconstructions.* Phys. Rev. B **60**, 14372 (1999)
9.11 A.A. Saranin, A.V. Zotov, A.N. Tovpik, M.A. Cherevik, E.N. Chukurov, V.G. Lifshits, M. Katayama, K. Oura: *Composition and Atomic Structure of the Si(111)$\sqrt{31}\times\sqrt{31}$-In Surface.* Surf. Sci. **450**, 34 (2000)
9.12 J. Kraft, M.G. Ramsey, F.P. Netzer: *Surface Reconstructions of In on Si(111).* Phys. Rev. B **55**, 5384 (1997)
9.13 P. Sprunger, F. Besenbacher, I. Stengaard: *STM Study of the Ni(110)-2×1-2CO System: Structure and Bonding-Site Determination.* Surf. Sci. **324**, L321 (1995)
9.14 C. Nagl, M. Pinczolits, M. Schmid, P. Varga: *p(n×1) Superstructures of Pb on Cu(110).* Phys. Rev. B **52**, 16796 (1995)
9.15 L. Lottermoser, T. Buslaps, R.L. Johnson, R. Feidenhans'l, M. Nielsen, D. Smilgies, E. Landemark, H.L. Meyerheim: *Bismuth on Copper (110): Analysis of the c(2×2) and p(4×1) Structures by Surface X-ray Diffraction.* Surf. Sci. **373**, 11 (1997)
9.16 H.Tochihara, S. Mizuno: *Composite Surface Structures Formed by Restructuring-Type Adsorption of Alkali-Metals on fcc Netals.* Progress Surf. Sci. **58**, 1 (1998)
9.17 M. Foss, R. Feidenhans'l, M. Nielsen, E. Findeisen, T. Buslaps, R.L. Johnson, F. Besenbacher, I. Stengaard: *X-ray Diffraction Investigation of the Sulphur Induced 4×1 Reconstruction of Ni(110).* Surf. Sci. **296**, 283 (1993)

9.18 J. Zegenhagen, J.R. Patel, P.E. Freeland, D.M. Chen, J.A. Golovchenko, P. Bedrossian, J.E. Northrup: *X-ray Standing-Wave and Tunneling-Microscope Location of Gallium Atoms on a Silicon Surface.* Phys. Rev. B **39**, 1298 (1989)
9.19 P. Mårtensson, G. Meyer, N.M. Amer, E. Kaxiras, K.C. Pandey: *Evidence for Trimer Reconstruction of Si(111)$\sqrt{3} \times \sqrt{3}$–Sb: Scanning Tunneling Microscopy and First-Principles Theory.* Phys. Rev. B **42**, 7230 (1990)
9.20 K. Spiegel: *Untersuchungen zum Schichtwachstum von Silver auf der Silizium (111)-Oberflache Durch Beugung Langsamer Electronen.* Surf. Sci. **7**, 125 (1967)
9.21 M. Katayama, R.S. Williams, M. Kato, E. Nomura, M. Aono: *Structure Analysis of the Si(111)$\sqrt{3}\times\sqrt{3}R30$–Ag Surface.* Phys. Rev. Lett. **66**, 2762 (1991)
9.22 T. Takahashi, S. Nakatani, N. Okamoto, T. Ishikawa: *A Study of the Si(111)$\sqrt{3}\times\sqrt{3}$–Ag Surface by Transmission X-ray Diffraction and X-ray Diffraction Topography.* Surf. Sci. **242**, 54 (1991)
9.23 K. Oura, J. Yamane, K. Umezawa, M. Naitoh, F. Shoji, T. Hanawa: *Hydrogen Adsorption on Si(100)–2×1 Surfaces Studied by Elastic Recoil Detection Analysis.* Phys. Rev. B **41**, 1200 (1990)
9.24 J.J. Boland: *Role of Bond-Strain in the Chemistry of Hydrogen on the Si(100) Surface.* Surf. Sci. **261**, 17 (1992)

Chapter 10

10.1 A.A. Chernov: *Modern Crystallography III. Crystal Growth* (Springer, Berlin, Heidelberg 1984)
10.2 I.V. Markov: *Crystal Growth for Beginners* (World Scientific, Singapore 1995)
10.3 C. Herring: *Some Theorems on the Free Energies of Crystal Surfaces.* Phys. Rev. **82**, 87 (1951)
10.4 R.M. Tromp, M. Mankos: *Thermal Adatoms on Si(001).* Phys. Rev. Lett. **81**, 1050 (1998)
10.5 G. Brocks, P.J. Kelly: *Dynamics and Nucleation of Si Ad-Dimers on the Si(100) Surface.* Phys. Rev. Lett. **76**, 2362 (1996)
10.6 N. Sato, T. Nagao, S. Hasegawa: *Two-Dimensional Adatom Gas Phase on the Si(111)-$\sqrt{3}\times\sqrt{3}$-Ag Surface Directly Observed by Scanning Tunneling Microscopy.* Phys. Rev. B **60**, 16083 (1999)
10.7 R.J. Hamers, U.K. Köhler: *Determination of the Local Electronic Structure of Atomic-Sized Defects on Si(001) by Tunneling Spectroscopy.* J. Vac. Sci. Technol. A **7**, 2854 (1989)
10.8 Ph. Ebert: *Nano-Scale Properties of Defects in Compound Semiconductor Surfaces.* Surf. Sci. Rep. **33**, 121 (1999)
10.9 Ph. Ebert, P. Quadbeck, K. Urban: *Identification of Surface Anion Antisite Defects in (110) Surfaces of III-V Semiconductors.* Appl. Phys. Lett. **79**, 2877 (2001)
10.10 R.J. Hamers, J.E. Demuth: *Electronic Structure of Localized Si Dangling-Bond Defects by Tunneling Spectroscopy.* Phys. Rev. Lett. **60**, 2527 (1988)
10.11 H. Hibino, T. Ogino: *Exchange Between Group-III (B,Al,Ga,In) and Si Atoms on Si(111)-$\sqrt{3}\times\sqrt{3}$ Surfaces.* Phys. Rev. B **54**, 5763 (1996)
10.12 M. Krohn, H. Bethge: 'Step Structures of Real Surfaces'. In: *1976 Crystal Growth and Materials. Vol. 2.* ed. by E. Kaldis, H.J. Scheel (North Holland, Amsterdam 1977) pp. 142–164

10.13 O. Rodríguez de la Fuente, M.A. González, J.M. Rojo: *Ion Bombardment of Reconstructed Metal Surfaces: From Two-Dimensional Dislocation Dipoles to Vacancy Pits*. Phys. Rev. B **63**, 085420 (2001)
10.14 C. Nagl, M. Schmid, P. Varga: *Inverse Corrugation and Corrugation Enhancement of Pb Superstructures on Cu(111) and (110)*. Surf. Sci. **369**, 159 (1996)
10.15 J. Zegenhagen, P.F. Lyman, M. Böhringer, M.J. Bedzyk: *Discommensurate Reconstructions of (111) Si and Ge Induced by Surface Alloying with Cu, Ga and In*. Physica Status Solidi (b) **204**, 587 (1997)
10.16 D.J. Chadi: *Stabilities of Single-Layer and Bilayer Steps on Si(001) Surfaces*. Phys. Rev. Lett. **59**, 1691 (1987)
10.17 B.S. Swartzentruber, Y.-W. Mo, R. Kariotis, M.G. Lagally, M.B. Webb: *Direct Determination of Step and Kink Energies on Vicinal Si(100)*. Phys. Rev. Lett. **65**, 1913 (1990)
10.18 D.R. Bowler, M.G. Bowler: *Step Structures and Kinking on Si(001)*. Phys. Rev. B **57**, 15385 (1998)
10.19 V. Zielasek, F. Liu, Y. Zhao, J.B. Maxon, M.G. Lagally: *Surface Stress-Induced Island Shape Transition in Si(001) Homoepitaxy*. Phys. Rev. B **64**, 201320 (2001)
10.20 F.K. Men, W.E. Packard, M.B. Webb: *Si(100) Surface Under an Externally Applied Stress*. Phys. Rev. Lett. **61**, 2469 (1988)
10.21 A.V. Latyshev, A.L. Aseev, A.B. Krasilnikov, S.I. Stenin: *Reflection Electron Microscopy Study of Structural Transformations on a Clean Silicon Surface in Sublimation and Homoepitaxy*. Surf. Sci. **227**, 24 (1990)
10.22 H. Hibino, T. Fukuda, M. Suzuki, Y. Homma, T. Sato, M. Iwatsuki, K. Miki, H. Tokumoto: *High-Temperature Scanning-Tunneling-Microscopy Observation of Phase Transitions and Reconstruction on Vicinal Si(111) Surface*. Phys. Rev. B **47**, 13027 (1993)
10.23 H. Minoda, K. Yagi, F.-J. Meyer zu Heringdorf, A. Meier, D. Kähler, M. Horn-von Hoegen: *Gold-Induced Faceting on a Si(001) Vicinal Surface: Spot-Profile-Analyzing LEED and Reflection-Electron-Microscopy Study*. Phys. Rev. B **59**, 2363 (1999)
10.24 H. Minoda, T. Shimakura, K. Yagi, F.-J. Meyer zu Heringdorf, M. Horn-von Hoegen: *Formation of Hill and Valley Structures on Si(001) Vicinal Surfaces Studied by Spot-Profile-Analyzing LEED*. Phys. Rev. B **61**, 5672 (2000)

Chapter 11

11.1 P. Hohenberg, W. Kohn: *Inhomogeneous Electron Gas*. Phys. Rev. **136**, B864 (1964)
11.2 W. Kohn, L.J. Sham: *Self-Consistent Equations Including Exchange and Correlation Effects*. Phys. Rev. **140**, A1133 (1965)
11.3 N.D. Lang, W. Kohn: *Theory of Metal Surfaces: Charge Density and Surface Energy*. Phys. Rev. B **1**, 4555 (1970)
11.4 M.F. Crommie, C.P. Lutz, D.M. Eigler: *Imaging Standing Waves in a Two-Dimensional Electron Gas*. Nature **363**, 524 (1993)
11.5 N.D. Lang, W. Kohn: *Theory of Metal Surfaces: Work Function*. Phys. Rev. B **3**, 1215 (1971)
11.6 S.D. Kevan: *Evidence for a New Broadening Mechanism in Angle-Resolved Photoemission from Cu(111)*. Phys. Rev. Lett. **50**, 526 (1983)

11.7 S.L. Hulbert, P.D. Johnson, N.G. Stoffel, W.A. Royer, N.V. Smith: *Crystal-Induced and Image-Potential-Induced Empty Surface States on Cu(111) and Cu(001).* Phys. Rev. B **31**, 6815 (1985)

11.8 R.I.G. Uhrberg, G.V. Hansson, J.M. Nicholls, S.A. Flodström: *Experimental Evidence for One Highly Dispersive Dangling-Bond Band on Si(111)2×1.* Phys. Rev. Lett. **48**, 1032 (1982)

11.9 F.J. Himpsel, P. Heimann, D.E. Eastman: *Surface States on Si(111)−2×1.* Phys. Rev. B **24**, 2003 (1981)

11.10 P. Perfetti, J.M. Nicholls, B. Reihl: *Unoccupied Surface-State Band on Si(111)2×1.* Phys. Rev. B **36**, 6160 (1987)

11.11 K.C. Pandey: *Theory of Semiconductor Surface Reconstruction: Si(111)−7×7, Si(111)−2×1, and GaAs (110).* Physica B **117/118**, 761 (1983)

11.12 P. Mårtensson, W.-X. Ni, G.V. Hansson: *Surface Electronic Structure of Si(111)7×7−Ge and Si(111)5×5−Ge Studied with Photoemission and Inverse Photoemission.* Phys. Rev. B **36**, 5974 (1987)

11.13 R.I.G. Uhrberg, T. Kaurila, Y.-C. Chao: *Low-Temperature Photoemission Study of the Surface Electronic Structure of Si(111)7×7.* Phys. Rev. B **58**, R1730 (1998)

11.14 R.J. Hamers, R.M. Tromp, J.E. Demuth: *Surface Electronic Structure of Si(111)−(7×7) Resolved in Real Space.* Phys. Rev. Lett. **56**, 1972 (1986)

11.15 F.J. Himpsel: *Inverse Photoemission from Semiconductors.* Surf. Sci. Rep. **12**, 1 (1990)

11.16 R. Wolkow, Ph. Avouris: *Atom-Resolved Surface Chemistry Using Scanning Tunneling Microscopy.* Phys. Rev. Lett. **60**, 1049 (1988)

11.17 F.J. Himpsel: 'Experimental Probes of the Surface Electronic Structure'. In: *Handbook of Surf. Sci.. Vol. 2. Electronic Structure.* ed. by K. Horn, M. Scheffler (Elsevier, Amsterdam 2000)

11.18 R. Losio, K.N. Altmann, F.J. Himpsel: *Fermi Surface of Si(111)7×7.* Phys. Rev. B **61**, 10845 (2000)

11.19 R.I.G. Uhrberg, R.D. Bringans, M.A. Olmstead, R.Z. Bachrach, J.E. Northrup: *Electronic Structure, Atomic Structure and Passivated Nature of the Arsenic-Terminated Si(111) Surface.* Phys. Rev. B **35**, 3945 (1987)

11.20 M.S. Hybertsen, S.G. Louie: *Theory of Quasiparticle Surface States in Semiconductor Surfaces.* Phys. Rev. B **38**, 4033 (1988)

11.21 T. Kinoshita, S. Kono, T. Sagawa: *Angle-Resolved Photoelectron-Spectroscopy Study of the Si(111) $\sqrt{3}\times\sqrt{3}$−Sn Surface: Comparison with Si(111) $\sqrt{3}\times\sqrt{3}$−Al, -Ga, and -In Surfaces.* Phys. Rev. B **34**, 3011 (1986)

11.22 J.M. Nicholls, P. Martensson, G.V. Hansson, J.E. Northrup: *Surface States on Si(111)$\sqrt{3}\times\sqrt{3}$−In: Experiment and Theory.* Phys. Rev. B **32**, 1333 (1985)

11.23 J.M. Nicholls, B. Reihl, J.E. Northrup: *Unoccupied Surface States Revealing the Si(111) $\sqrt{3}\times\sqrt{3}$−Al, -Ga, and -In Adatom Geometries.* Phys. Rev. B **35**, 4137 (1987)

11.24 F. Bauerle, W. Monch, M. Henzler: *Correlation of Electronic Surface Properties and Surface Structure on Cleaved Silicon Surfaces.* J. Appl. Phys. **43**, 3917 (1972)

11.25 I. Shiraki, F. Tanabe, R. Hobara, T. Nagao, S. Hasegawa: *Independently Driven Four-Tip Probes for Conductivity Measurements in Ultrahigh Vacuum.* Surf. Sci. **493**, 633 (2001)

11.26 I. Shiraki, T. Nagao, S. Hasegawa, C.L. Petersen, P. Bøggild, T.M. Hansen, F. Grey: *Micro-Four-Point Probes in a UHV Scanning Tunneling Microscope for in situ Surface-Conductivity Measurements.* Surf. Rev. Lett. **7**, 533 (2000)

11.27 S. Hasegawa, F. Grey: *Electronic Transport at Semiconductor Surfaces – from Point-Contact Transistor to Micro-Four-Point Probes.* Surf. Sci. **500**, 84 (2002)
11.28 S. Hasegawa, I. Shiraki, Y. Tanigawa, C.L. Petersen, F. Grey: *Measurement of Surface Electron Conductivity Using Micro 4-Probe.* Kotai Butsuri (Solid State Phys.) **37**, 299 (2002)
11.29 K. Jakobi: 'Electronic Structure of Surfaces: Metals'. In: *Physics of Solid Surfaces. Landolt Börnstein III/24b.* ed. by G. Chiarotti (Springer, Berlin, Heidelberg, New York 1993) pp. 29–351
11.30 P.J. Goddard, R.M. Lambert: *Adsorption-Desorption Properties and Surface Structural Chemistry of Chlorine on Cu(111) and Ag(111).* Surf. Sci. **67**, 180 (1977)
11.31 S.Å. Lindgren, L. Walldén: *Electronic Structure of Clean and Oxygen-Exposed Na and Cs Monolayers on Cu(111).* Phys. Rev. B **22**, 5967 (1980)
11.32 E.W. Müller: *Work Function of Tungsten Single Crystal Planes Measured by the Field Emission Microscope.* J. Appl. Phys. **26**, 732 (1955)
11.33 G.F. Smith: *Thermionic and Surface Properties of Tungsten Crystal.* Phys. Rev. **94**, 295 (1954)
11.34 R.H. Fowler: *The Analysis of Photoelectric Sensitivity Curves for Clean Metals at Various Temperatures.* Phys. Rev. **38**, 45 (1931)
11.35 P.O. Gartland, S. Berge, B.J. Slagsvold: *Photoelectric Work Function of a Copper Single Crystal for the (100), (110), (111), and (112) Faces.* Phys. Rev. Lett. **28**, 738 (1972)

Chapter 12

12.1 P. Brault, H. Ranger, J.P. Toennies: *Molecular Beam Studies of Sticking of Oxygen on the Rh(111) surface.* J. Chem. Phys. **106**, 8876 (1997)
12.2 S.J. Lombardo, A.T. Bell: *A Review of Theoretical Models of Adsorption, Diffusion, Desorption, and Reaction of Gases on Metal Surfaces.* Surf. Sci. Rep. **13**, 1 (1991)
12.3 S.A. Barnett, H.F. Winters, J.E. Greene: *The Interaction of Sb_4 Molecular Beams with Si(100) Surfaces Modulated-Beam Mass Spectrometry and Thermally Stimulated Desorption Studies.* Surf. Sci. **165**, 303 (1986)
12.4 D.A. King, M.G. Wells: *Reaction Mechanizm in Chemisorption Kinetics: Nitrogen on the {100} Plane of Tungsten.* Proc. Roy. Soc. London **A339**, 245 (1974)
12.5 P. Bratu, W. Brenig, A. Groß, M. Hartmann, U. Höfer, P. Kratzer, R. Russ: *Reaction Dynamics of Molecular Hydrogen on Silicon Surfaces.* Phys. Rev. B **54**, 5978 (1996)
12.6 C.T. Rettner, L.A. DeLousie, D.J. Auerbach: *Effect of Incident Kinetic Energy and Surface Coverage on the Dissociative Chemisorption of Oxygen on W(110).* J. Chem. Phys. **85**, 1131 (1986)
12.7 G. Comsa, R. David: *The Purely "Fast" Distribution of H_2 and D_2 Molecules Desorbing from Cu(100) and Cu(111) Surfaces.* Surf. Sci. **117**, 77 (1982)
12.8 G. Comsa, R. David: *Dynamical Parameters of Desorbing Molecules.* Surf. Sci. Rep. **5**, 145 (1985)
12.9 S. Hasegawa, H. Daimon, S. Ino: *A Study of Adsorption and Desorption Processes of Ag on Si(111) Surface by Means of RHEED-TRAXS.* Surf. Sci. **186**, 138 (1987)
12.10 P.A. Redhead: *Thermal Desorption of Gases.* Vacuum **12**, 203 (1962)

12.11 E. Bauer, F. Bonczek: *Thermal Desorption of Metals from Tungsten Single Crystal Surfaces.* Surf. Sci. **53**, 87 (1975)
12.12 A. Thomy, X. Duval, J. Regnier: *Two-Dimensional Phase Transitions as Displayed by Adsorption Isotherms on Graphite and Other Lamellar Solids.* Surf. Sci. Rep. **1**, 1 (1981)
12.13 R.D. Ramsier, J.T. Yates Jr.: *Electron-Stimulated Desorption: Priciples and Applications.* Surf. Sci. Rep. **12**, 243 (1991)
12.13 C.C. Cheng, Q. Gao, W.J. Choyke, J.T. Yates: *Transformation of Cl bonding structures on Si(100)-(2×1).* Phys. Rev. B **46**, 12810 (1992)

Chapter 13

13.1 R. Gomer: *Diffusion of Adsorbates on Metal Surfaces.* Rep. Prog. Phys. **53**, 917 (1990)
13.2 E.G. Seebauer, M.Y.L. Jung: 'Surface Diffusion on Metals, Semiconductors and Insulators'. In: *Physics of Covered Solid Surfaces. Landolt Börnstein III/42.* ed. by H.P. Bonzel (Springer, Berlin, Heidelberg, New York 2001) pp. 455–530
13.3 H.P. Bonzel: 'Surface Diffusion on Metals'. In: *Diffusion in Solid Metals and Alloys. Landolt Börnstein III/26.* ed. by O. Madelung (Springer, Berlin, Heidelberg, New York 1990) pp. 717–748
13.4 G. Ayrault, G. Ehrlich: *Surface Self-Diffusion on an fcc Crystal: An Atomic View.* J. Chem. Phys. **60**, 281 (1974)
13.5 X.-D. Xiao, X.D. Zhu, W. Daum, Y.R. Shen: *Anisotropic Surface Diffusion of CO on Ni(110).* Phys. Rev. Lett. **66**, 2352 (1991)
13.6 Ya.E. Geguzin, Yu.S. Kaganovski, E.G. Mikhailov: *On the Structural Anisotropy of ^{63}Ni Diffusion Spread Over the W Surface.* Ukrainski Fizicheski Z. **27**, 1865 (1982)
13.7 M.Ø. Pedersen, L. Österlund, J.J. Mortensen, M. Mavrikakis, L.B. Hansen, I. Stensgaard, E. Lægsgaard, J.K. Nørskov, F. Besenbacher: *Diffusion of N Adatoms on the Fe(100) Surface.* Phys. Rev. Lett. **84**, 4898 (2000)
13.8 G.L. Kellogg: *Field Ion Microscope Studies of Single-Atom Surface Diffusion and Cluster Nucleation on Metal Surfaces.* Surf. Sci. Rep. **21**, 1 (1994)
13.9 J.D. Wrigley, G. Ehrlich: *Surface Diffusion by an Atomic Exchange Mechanism.* Phys. Rev. Lett. **44**, 661 (1980)
13.10 L.J. Lauhon, W. Ho: *Direct Observation of the Quantum Tunneling of Single Hydrogen Atoms with a Scanning Tunneling Microscope.* Phys. Rev. Lett. **85**, 4566 (2000)
13.11 A.J. Mayne, F. Rose, C. Bolis, G. Dujardin: *A Scanning Tunneling Microscopy Study of the Diffusion of a Single or a Pair of Atomic Vacancies.* Surf. Sci. **486**, 226 (2001)
13.12 R. Van Gastel, E. Somfai, S.B. Van Albada, W. Van Saarloos, J.W.M. Frenken: *Nothing Moves a Surface: Vacancy Mediated Surface Diffusion.* Phys. Rev. Lett. **86**, 1562 (2001)
13.13 M.L. Grant, B.S. Swartzentruber, N.C. Bartelt, J.B. Hannon: *Diffusion Kinetics in the Pd/Cu(001) Surface Alloy.* Phys. Rev. Lett. **86**, 4588 (2001)
13.14 K. Kyuno, G. Ehrlich: *Diffusion and Dissociation of Platinum Clusters on Pt(111).* Surf. Sci. **437**, 29 (1999)
13.15 G.L. Kellogg: *Atomic View of Cluster Diffusion on Metal Surfaces.* Prog. in Surf. Sci. **53**, 217 (1996)

13.16 T.R. Linderoth, S. Horch, L. Petersen, S. Helveg, E. Lægsgaard, I. Stensgaard, F. Besenbacher: *Novel Mechanism for Diffusion of One-Dimensional Clusters: Pt/Pt(110)-(1×2).* Phys. Rev. Lett. **82**, 1494 (1999)
13.17 O.V. Bekhtereva, Yu.L. Gavrilyuk, V.G. Lifshits, B.K. Churusov: *Indium Surface Phase Formation on Si(111) Surface and their Role in Diffusion and Desorption.* Poverkhnost'. Physika, Khimiia i Mekhanika **8**, 54 (1988)
13.18 H. Yasunaga, S. Sakomura, T. Asaoka, S. Kanayama, N. Okuyama, A. Natori: *Electromigration of Ag Ultrathin Films on Si(111)7×7.* Japan J. Appl. Phys. **27**, L1603 (1988)
13.19 B.S. Swartzentruber: *Direct Measurement of Surface Diffusion Using Atom-Tracking Scanning Tunneling Microscopy.* Phys. Rev. Lett. **76**, 459 (1996)
13.20 A.T. Loburets, A.G. Naumovets, Yu.S. Vedula: *Diffusion of Dysprosium on the (112) Surface of Molybdenum.* Surf. Sci. **399**, 297 (1998)
13.21 W.W. Mullins: *Flattening of a Nearly Plane Solid Surface due to Cappilarity.* J. Appl. Phys. **30**, 77 (1959)
13.22 Yu.L. Gavrilyuk, Yu.S. Kaganovskii, V.G. Lifshits: *Diffusin Mass Transfer on (111) and (100) Surfaces of Silicon Single Crystals.* Kristallografiya **26**, 561 (1981). Sov. Phys. Crystallography (English translation) **26**, 317 (1981)
13.23 J.A. Stroscio, D.T. Pierce, R.A. Dragoset: *Homoepitaxial Growth of Iron and a Real Space View of Reflection-High-Energy-Electron Diffraction.* Phys. Rev. Lett. **70**, 3615 (1993)

Chapter 14

14.1 J.G. Amar, F. Family, P.-M. Lam: *Dinamic Scaling of the Island -Size Distribution and Percolation in a Model of Submonolayer Molecular-Beam Epitaxy.* Phys. Rev. B **50**, 8781 (1994)
14.2 J.A. Venables, G.D.T. Spiller, M. Hanbücken: *Nucleation and Growth of Thin Films.* Rep. Prog. Phys. **47**, 399 (1984)
14.3 B. Müller, L. Nedelmann, B. Fischer, H. Brune, K. Kern: *Initial Stages of Cu Epitaxy on Ni(100): Postnucleation and a Well-Defined Transition in Critical Island Size.* Phys. Rev. B **54**, 17858 (1996)
14.4 T.A. Witten Jr., L.M. Sander: *Diffusion-Limited Agregation, a Kinetic Critical Phenomenon.* Phys. Rev. Lett. **47**, 1400 (1981)
14.5 M. Hohage, M. Bott, M. Morgenstern, Z. Zhang, T. Michely, G. Comsa: *Atomic Processes in Low Temperature Pt-Dendrite Growth on Pt(111).* Phys. Rev. Lett. **76**, 2366 (1996)
14.6 M. Bott, T. Michely, G. Comsa: *The Homoepitaxial Growth of Pt on Pt(111) Studied with STM.* Surf. Sci. **272**, 161 (1992)
14.7 T. Michely, G. Comsa: *Temperature Dependence of the Sputtering Morphology of Pt(111).* Surf. Sci. **256**, 217 (1991)
14.8 T. Michely: Atomare Prozesse bei der Pt-Abscheidung auf Pt(111). Habilitationsschrift, Bonn (1996)
14.9 J.A. Stroscio, D.T. Pierce, R.A. Dragoset: *Homoepitaxial Growth of Iron and a Real Space View of Reflection-High-Energy-Electron Diffraction.* Phys. Rev. Lett. **70**, 3615 (1993)
14.10 M.Y. Lai, Y.L. Wang: *Direct Observation of Two Dimensional Magic Clusters.* Phys. Rev. Lett. **81**, 164 (1998)
14.11 B. Voigtländer, M. Kästner, P. Šmilauer: *Magic Islands in Si/Si(111) Homoepitaxy.* Phys. Rev. Lett. **81**, 858 (1998)

14.12 G. Rosenfeld, K. Morgenstern, M. Esser, G. Comsa: *Dinamics and Stability of Nanostructures on Metal Surfaces.* Appl. Phys. A **69**, 489 (1999)
14.13 G. Ehrlich, F.G. Hudda: *Atomic View of Surface Self-Diffusion: Tungsten on Tungsten.* J. Chem. Phys. **44**, 1039 (1966)
14.14 R.L. Schwoebel, E.J. Shipsey: *Step Motion on Crystal Surfaces.* J. Appl. Phys. **37**, 3682 (1966)
14.15 G. Rosenfeld, B. Poelsema, G. Comsa: 'Epitaxial Growth Modes Far from Equilibrium'. In: *The Chemical Physics of Solid Surfaces. Vol. 8. Growth and Properties of Ultrathin Epitaxial Layers.* ed. by D.A. King, D.P. Woodruff (Elsevier, Amsterdam 1997) pp. 66–101
14.16 R. People, J.C. Bean: *Calculation of Critical Layer Thickness Versus Lattice Mismatch for Ge_xSi_{1-x}/Si Strained-Layer Heteroepitaxy.* Appl. Phys. Lett. **47**, 322 (1985)
14.17 A.V. Zotov, V.V. Korobtsov: *Present Status of Solid Phase Epitaxy of Vacuum-Deposited Silicon.* J. Cryst. Growth **98**, 519 (1989)
14.18 N. Pütz, E. Veuhoff, H. Heinecke, M. Heyen, H. Lüth, P. Balk: *GaAs Growth in Metal-Organic MBE.* J. Vac. Sci. Technol. B **3**, 671 (1983)
14.19 D.J. Eaglesham, F.C. Unterwald, D.C. Jacobson: *Growth Morphology and the Equilibrium Shape: The Role of "Surfactants" in Ge/Si Island Formation.* Phys. Rev. Lett. **70**, 966 (1993)
14.20 K. Sumitomo, T. Kobayashi, F. Shoji, K. Oura, I. Katayama: *Hydrogen-Mediated Epitaxy of Ag on Si(111) as Studied by Low-Energy Ion Scattering.* Phys. Rev. Lett. **66**, 1193 (1991)
14.21 P. Zahl, P. Kury, M. Horn-von Hoegen: *Interplay of Surface Morphology, Strain Relief, and Surface Stress During Surfactant Mediated Epitaxy of Ge on Si.* Appl. Phys. A **69**, 481 (1999)
14.22 H.A. Van Der Vegt, J. Vrijmoeth, R.J. Behm, E. Vlieg: *Sb-Enhanced Nucleation in the Homoepitaxial Growth of Ag(111).* Phys. Rev. B **57**, 4127 (1998)
14.23 H. Brune, G.S. Bales, J. Jacobsen, C. Boragno, K. Kern: *Measuring Surface Diffusion from Nucleation Island Density.* Phys. Rev. B **60**, 5991 (1999)

Chapter 15

15.1 Ph. Buffat, J-P. Borel: *Size Effect on the Melting Temperature of Gold Particles.* Phys. Rev. A **13**, 2287 (1976)
15.2 R.P. Feynman: *There's Plenty of Room at the Bottom.* Engineering and Science (California Institute of Technology) **23**, 22 (1960). Reprinted in: Journal of Microelectromechanical Systems **1**, 60 (1992) (see also `http://www.its.caltech.edu/~feynman/plenty.html`)
15.3 P. Avouris: *Manipulation of Matter at the Atomic and Molecular Levels.* Acc. Chem. Res. **28**, 95 (1995)
15.4 D.M. Eigler, E.K. Schweizer: *Positioning Single Atoms with a Scanning Tunnelling Microscope.* Nature **344**, 524 (1990)
15.5 M.F. Crommie, C.P. Lutz, D.M. Eigler: *Imaging Standing Waves in a Two-Dimensional Electron Gas.* Nature **363**, 524 (1993). (`http://www.almaden.ibm.com/vis/stm/corral.html#stm16`)
15.6 J.A. Stroscio, D.M. Eigler: *Atomic and Molecular Manipilation with Scanning Tunneling Microscope.* Science **254**, 1319 (1991)
15.7 A.A. Saranin, T. Numata, O. Kubo, H. Tani, M. Katayama, V.G. Lifshits, K. Oura: *STM Tip-Induced Diffusion of In Atoms on the Si(111)$\sqrt{3}\times\sqrt{3}$–In Surface.* Phys. Rev. B **56**, 7449 (1997)

15.8 D.M. Eigler, C.P. Lutz, W.E. Rudge: *An Atomic Switch Realized with the Scanning Tunneling Microscope.* Nature **352**, 600 (1991)
15.9 M. Aono, A. Kobayashi, F. Grey, H. Uchida, D.-H. Huang: *Tip-Sample Interactions in the Scanning Tunneling Microscope for Atomic-Scale Structure Fabrication.* Japan J. Appl. Phys. **32**, 1470 (1993)
15.10 T. Hitosugi, T. Hashizume, S. Heike, Y. Wada, S. Watanabe, T. Hasegawa, K. Kitazawa: *Scanning Tunneling Spectroscopy of Dangling-Bond Wires Fabricated on the Si(100)-2×1-H Surface.* Appl. Phys. A **66**, S695 (1998)
15.11 B. Voigtländer: *Fundamental Processes in Si/Si and Ge/Si Epitaxy Studied by Scanning Tunneling Microscopy During Growth.* Surf. Sci. Rep. **43**, 127 (2001)
15.12 R.S. Williams, G. Medeiros-Ribeiro, T.I. Kamins, D.A.A. Ohlberg: *Chemical Thermodynamics of the Size and Shape of Strained Ge Nanocrystals Grown on Si(001).* Acc. Chem. Res. **32**, 425 (1999)
15.13 J. Tersoff, C. Teichert, M.G. Lagally: *Self-Organization in Growth of Quantum Dot Superlattices.* Phys. Rev. Lett. **76**, 1675 (1996)
15.14 K. Oura, V.G. Lifshits, A.A. Saranin, A.V. Zotov, M. Katayama: *Hydrogen Interaction with Clean and Modified Silicon Surfaces.* Surf. Sci. Rep. **35**, 1 (1999)
15.15 V.G. Kotlyar, A.V. Zotov, A.A. Saranin, T.V. Kasyanova, M.A. Cherevik, I.V. Pisarenko, V.G. Lifshits: *Formation of the ordered array of Al magic clusters on Si(111)7×7.* Phys. Rev. B **66**, 165401 (2002)
15.16 L. Vitali, M.G. Ramsey, F.P. Netzer: *Nanodot Formation on the Si(111)-(7×7) Surface by Adatom Trapping.* Phys. Rev. Lett. **83**, 316 (1999)
15.17 Y. Chen, D.A.A. Ohlberg, R.S. Williams: *Nanowires of Four Epitaxial Hexagonal Silicides Grown on Si(001).* J. Appl. Phys. **91**, 3213 (2002)
15.18 Y. Chen, D.A.A. Ohlberg, G. Medeiros-Ribeiro, Y.A. Chang, R.S. Williams: *Self-Assembled Growth of Epitaxial Erbium Disilicide Nanowires on Silicon (001).* Appl. Phys. Lett. **76**, 4004 (2000)
15.19 H.W. Kroto, J.R. Heath, S.C. O'Brien, R.F. Curl, R.E. Smalley: *C_{60}: Buckminsterfullerene.* Nature **318**, 162 (1985)
15.20 M.J. Butcher, J.W. Nolan, M.R.C. Hunt, P.H. Beton, L. Dunsch, P. Kuran, P. Georgi, T.J.S. Dennis: *Orientaionally Ordered Island Growth of Higher Fullerenes on Ag/Si(111)-$(\sqrt{3}\times\sqrt{3})R30^\circ$.* Phys. Rev. B **64**, 195401 (2001)
15.21 S. Iijima: *Helical Microtubules of Graphitic Carbon.* Nature **354**, 56 (1991)
15.22 S. Iijima, T. Ichihashi: *Single-Shell Carbon Nanotubes of 1-nm Diameter.* Nature **363**, 603 (1993)
15.23 D.S. Bethune, C.H. Kiang, M.S. de Vries, G. Gorman, R. Savoy, J. Vazquez, R. Beyers: *Cobalt-Catalysed Growth of Carbon Nanotubes with Single-Atomic-Layer Walls.* Nature **363**, 605 (1993)
15.24 M.S. Dresselhaus: *New Trics with Nanotubes.* Nature **391**, 19 (1998)
15.25 J.-L. Huang P. Kim, T.W. Odom, C.M. Lieber: *Electronic Density of States of Atomically Resolved Single-Walled Carbon Nanotubes: Van Hove Singularities and End States.* Phys. Rev. Lett. **82**, 1225 (1999)

Index

γ-plot 234
π-bonded chain model 183
2D condensation 317
2D crystallography 3

Accumulation layer 279
Adatom 230
– external 237
– islands 374
– thermal 237
Adsorbate 195
Adsorption 195, 295
– angular dependence 305
– dissociative 297
– isotherm 315
– kinetic energy dependence 305
– multilayer 317
– non-dissociative 296
– precursor-mediated 298
– precursor-mediated for interacting adsorbates 299
– precursor-mediated for non-interacting adsorbates 299
AED *see* Auger electron diffraction
AES *see* Auger electron spectroscopy
AES spectrum *see* Auger spectrum
AFM *see* Atomic force microscopy
Ag(110) 97
Al 94
Al_2O_3 112
Al(110)1×1 173
Alkali ISS 124
Angle-integrated UPS 106
Angle-resolved UPS 106
ARUPS 106
Atom sliding 394
Atomic force microscopy 164
– contact mode 166
– – constant-force mode 167
– – constant-height mode 167
– non-contact mode 167
– tapping mode 168

Atomic form factor 68
Atomic manipulation 393
Au(100) 246
Auger
– de-excitation 122
– depth profiling 87
– electron diffraction 74
– electron energy 83
– electron spectroscopy 82
– – general equation 87
– – quantitative analysis 87
– electrons 82
– emission 82
– mapping 87
– neutralization 121
– process 82
– spectrum 85
– spectrum differentiation 85

Backscattering factor 87, 88
Bakeout 22
Band bending 278
Base pressure 22
BET isotherm 317
Blocking 116
– critical angle 116
Blocking cone 116
Bouckaert–Smoluchowski–Wigner notation 16
Bragg's law 47
Brillouin zone 16
– bcc lattice 16
– fcc lattice 16
– hcp lattice 17
BSW notation *see* Bouckaert–Smoluchowski–Wigner notation
Built-in potential 278
Bulk plasmon 93

C(0001) 179
CAICISS *see* Coaxial impact-collision ion scattering spectroscopy

Carbon nanotubes 410
CBE *see* Chemical beam epitaxy
CHA *see* Concentric hemispherical analyzer
Channeling 117
– critical angle 117
Chemical beam epitaxy 382
Chemical diffusion 331
– coefficient 331
Chemical shifts 104
Chemical treatment
– ex situ 37
– in situ 37
Chemical vapor deposition 382
Chemisorption 196
CITS *see* Current-imaging tunneling spectroscopy
Cleavage 35
CLEELS *see* Core level electron energy loss spectroscopy
Cluster surface diffusion 341
Cluster surface diffusion mechanism
– concerted 343
– – dislocation 344
– – gliding 343
– – reptation 344
– – shearing 344
– individual 343
– – edge diffusion 343
– – evaporation–condensation 343
– – leapfrog 343
– – sequential displacement 343
CMA *see* Cylindrical mirror analyzer
CNT *see* carbon nanotubes
Coaxial impact-collision ion scattering spectroscopy 127
Coherence length 56
compact islands 365
Concentric hemispherical analyzer 81
Condensation coefficient 296
Conductance 21
Conflat flanges 32
Contact potential difference 290
Core level electron energy loss spectroscopy 90
Coverage
– adsorbate 197
– substrate atoms 199
CPD *see* Contact potential difference
Critical island size 360
Crystal structure 3
Crystal truncation rods 72
Cu(110)$n\times$1-Pb 216
Cu(110)5$\times$1-Pb 216
Cu(111) 107, 251, 270
Current-imaging tunneling spectroscopy 163
CVD *see* Chemical vapor deposition
Cylindrical mirror analyzer 80

DAS model *see* Dimer–adatom–stacking fault model
Defect
– substitutional 244
Defects
– anti-site 243
– missing-dimer 239
– one-dimensional 229
– zero-dimensional 229
Deflection type analyzer 79
Degassing 22
Delta-doping layers 381
Density functional theory 261
Density of states
– 0D 393
– 1D 393
– 2D 391
– 3D 391
Depletion layer 278
Depth ion profiling 120
Desorption 305
– angular dependence 308
– energy 308
– kinetic energy dependence 308
– non-thermal 319
– thermal 305
DFT *see* Density functional theory
Diamond structure 180
Diffusion
– field-assisted 398
– interlayer 375
– intralayer 375
Diffusion length 328
Diffusion-limited-aggregation model 365
Diffusivity *see* Diffusion coefficient
Dimer–adatom–stacking fault model 184
– adatoms 186
– corner hole 186
– dimer rows 186
– rest atoms 186
Direct recoil spectroscopy 137
Dislocation
– misfit 250
– screw 230, 245
– surface 246

DLA model *see* Diffusion-limited-aggregation model
Domain boundary 247
– antiphase 247
Domain wall 247
– heavy 250
– light 250
Domains
– antiphase 247
– orientational 248
Domes 405
DRS *see* Direct recoil spectroscopy
Dynamic coalescence 371

EELS *see* Electron energy loss spectroscopy
Ehrlich–Schwoebel barrier 375
Elastic recoil detection analysis 137
Electromigration 348
Electron
– attenuation length 87, 88
– mean free path 87
Electron beam evaporator 42
Electron energy analyzer 78
Electron energy loss spectroscopy 89, 92
Electron spectroscopy 77
– for chemical analysis 99
Electron-stimulated desorption 319, 401
Electron-stimulated desorption ion angular distribution 321
Elemental sensitivity factor 88
ERDA *see* Elastic recoil detection analysis
ESCA *see* Electron spectroscopy for chemical analysis
ESD *see* Electron-stimulated desorption
ESDIAD *see* Electron-stimulated desorption ion angular distribution
Ewald construction 47

Facets 236
Facetting 257, 258
Fe(211)1×1 174
FEM *see* Field emission microscopy
Field desorption 322
Field emission 287
Field emission microscopy 145
Field evaporation 399
Field ion microscopy 147
FIM *see* Field ion microscopy
Flanges 32
Flash desorption 310
Focusing effect 114
Formation phase diagram 205, 209–211
Four grid analyzer 79
Fowler–Nordheim equation 287
Fractal islands 365
Frank–van der Merve growth mode 357
Friedel oscillations 264, 270
Fullerenes 408
– endohedral 410
Fulleride 409
Fullerite 409

GaAs($\bar{1}\bar{1}\bar{1}$)2×2 192
GaAs(100) 65
GaAs(110)1×1 189
GaAs(111)2×2 190
GaP(110) 243
Gauge 27
– Bayard–Alpert 28
– ion 28
– ionization 28
– Pirani 27
– thermocouple 27
Ge(100)c(4×2) 183
Ge(100)2×1 183
Ge(111)$\sqrt{3}\times\sqrt{3}$-Bi 218
Ge(111)$\sqrt{3}\times\sqrt{3}$-Pb 218, 222
Ge(111)$\sqrt{3}\times\sqrt{3}$-Sn 218
Ge(111)c(2×8) 187
Ge(111)-Ga γ-phase 251
Ge(111)2×1 183, 187
Gibbs–Thompson equation 371
GIXRD *see* Grazing incidence x-ray diffraction
GIXRD structural analysis 69
Grazing incidence x-ray diffraction 66
Growth mode
– Frank–van der Merve 357
– island 357
– lattice-matched 378
– layer-by-layer 357, 375
– layer-plus-island 357
– multilayer 375
– pseudomorphic 378
– relaxed dislocated 378
– step-flow 375
– strained pseudomorphic 378
– Stranski–Krastanov 357
– Vollmer–Weber 357

Half-crystal position 232
HCT model *see* Honeycomb-chained-trimer model
HEIS *see* High-energy ion scattering spectroscopy
Henry's law 316
Heterodiffusion 325
Heteroepitaxy 377
High-energy ion scattering spectroscopy 129
High-resolution electron energy loss spectroscopy 95
High-resolution TEM 151
Hill–DeBoer equation 316
"Hill-and-valley" structure 258
Hit-and-stick regime 365
Homoepitaxy 374
Honeycomb-chained-trimer model 222
HREELS *see* High-resolution electron energy loss spectroscopy
Hut-clusters 404

Impact-collision ion scattering spectroscopy (ICISS) 126
Inelastic mean free path 77
InP(110) 241
INS *see* Ion neutralization spectroscopy
InSb(111)2×2 71
Interfactant 384
Intrinsic diffusion 331
Inverse photoemission 269
Inversion layer 279
Ion impact desorption 321
Ion neutralization spectroscopy 122
Ion scattering 109
– energy 110
Ion scattering spectroscopy 123
Ion spectroscopy 109
– basics 109
Ion sputtering 37
Ionization cross-section 87
Island aggregation regime 362
Island coalescence 371
– regime 362
Island coarsening 371
Island growth mode 357
Island nucleation 360
– intermediate-coverage regime 362
– low-coverage regime 362
– rate equation 361
Island percolation 371
– regime 362
Island ripening 371
Island saturation density 363
Island shape 365
Island size distribution 368
Isothermal desorption spectroscopy 310
ISS *see* Ion scattering spectroscopy
ITDS *see* Isothermal desorption spectroscopy

Jellium model 263

$\boldsymbol{k}$-resolved inverse photoemission 269
Kelvin-probe 290
Kinematic approximation 68
Kink 230
Kinked surface 11
Knudsen cell 39
Kohn–Sham equations 262
KRIPS *see* $\boldsymbol{k}$-resolved inverse photoemission

Langmuir adsorption isotherm 316
Langmuir Adsorption Model 296
Langmuir formula 93
Lattice 3
– Bravais 4
– centered rectangular 5
– hexagonal 5
– oblique 5
– rectangular 5
– square 5
Lattice misfit 378
Layer-by-layer growth mode 357
Layer-plus-island growth mode 357
LDA *see* Local density approximation
Leak valve 44
Leaks
– real 22
– virtual 22
LEED *see* Low-energy electron diffraction
LEED spot profile 54
LEED-AES device 79
LEEM *see* Low-energy electron microscopy
LEIS *see* Low-energy ion scattering
Local density approximation 263
Low-dimensionality effects 389
Low-energy electron diffraction 47
– I–V curves 57
Low-energy electron microscopy 154
– bright-field image 155
– dark-field image 155

Low-energy ion scattering 123
Low-index planes
– bcc crystal 8
– diamond crystal 9
– fcc crystal 8
– hcp crystal 9

Magic clusters 372, 406
Magic islands 372
Mass transfer diffusion 331
MBE *see* Molecular beam epitaxy
MDS *see* Metastable de-excitation spectroscopy
Medium-energy ion scattering spectroscopy 129
MEIS *see* Medium-energy ion scattering spectroscopy
Metal-organic chemical vapor deposition 382
Metal-organic molecular beam epitaxy 382
Metallic surface 277
Metastable de-excitation spectroscopy 122
Miller indices 6
– four-index notation 7
– hcp lattice 7
– three-index notation 7
Misfit 249
– dislocation 378
– dislocation network 251
Misoriented surface 8
– tilt angle 8
– tilt azimuth 8
– tilt zone 8
ML *see* Monolayer
MOCVD *see* Metal-organic chemical vapor deposition
Molecular beam epitaxy 379
Molecular flow 22
MOMBE *see* Metal-organic molecular beam epitaxy
Monolayer 197
Multiwall nanotubes (MWNT) 411

Na 102
NaCl 151
NaCl(100) 245
Nanoclusters 406
Nanostructures 389
Nanowires 406
Ni(110)2×1-CO 213
Ni(110)4×1-S 217

Ostwald ripening 371

Particle-induced Auger electron spectroscopy 121
Particle-induced x-ray emission spectroscopy 121
Patterson function 69, 70
Pb110 135
PD *see* Photodesorption
PED *see* Photoelectron diffraction
PES *see* Photoelectron spectroscopy
– quantitative analysis 101
PES experimental set-up 99
Phase transition
– first-order 208
– order–disorder 208
– order–order 208
– second-order 208
Photodesorption 321
Photoelectric effect 98
Photoelectron diffraction 74
Photoelectron spectroscopy 98
– bremsstrahlen 270
– isochromate 269
– two-photon 270
Photoemission threshold 290
Physisorption 195
Plasmon 93
Point defects 229
Polanyi–Wigner equation 306
Polar plot of γ 234
Polar surface 189
Precursor state 298
Pressure units 20
Pt(100) quasi-hexagonal 176
Pt(110)2×1 178
Pt(111) 125, 126
Pump
– cryosorption 25
– ion 26
– rotary 24
– sublimation 26
– turbomolecular 26
Pumping equation 22
Pumping speed 21
Pumping system 30
Pumps 24
Pyramids 404

Quantum corral 397
Quantum dots 390
Quantum films 390
Quantum wires 390

Quartz crystal monitor 43
Quasi-resonant neutralization 121

Ramified islands 365
RBS *see* Rutherford backscattering spectroscopy
Reciprocal lattice rods 48
Recoil atom
– energy 110
Reconstruction 172
– conservative 199
reconstruction
– non-conservative 199
Reflection electron microscopy 152
Reflection high energy electron diffraction 59
Relaxation 171
– normal 171
– parallel 171
– tangential 171
REM *see* Reflection electron microscopy
Residual gas analyzer 29
Resonance ionization 121
Resonance neutralization 121
Retarding field analyzer 79
RFA *see* Retarding field analyzer
RHEED *see* Reflection high energy electron diffraction
RHEED oscillations 64
RHEED rocking curves 64
Richardson–Dushman equation 287
Rodscans 70
Rutherford backscattering spectroscopy 129
Rutherford cross-section 113

S_A step *see* Si(100) step
S_B step *see* Si(100) step
SARS *see* Scattering and recoiling spectroscopy
Scaling theory 369
Scanning electron microscopy 156
Scanning tunneling microscopy 159
– constant-current mode 161
– constant-height mode 162
Scanning tunneling spectroscopy 161, 162
Scattered intensity 68
Scattering amplitude 68
Scattering and recoiling spectroscopy 137
Scattering cross-section 113
Schottky effect 289
Screening function 113
Secondary ion mass spectroscopy 119, 138
– dynamic 141
– general equation 140
– ion yield 140
– quantitative analysis 140
– static 141
Segregation 381
Self-diffusion 325
Self-organization 404
SEM *see* Scanning electron microscopy
Semiconducting surface 278
Shadow cone 114
Shadowing 114
– critical angle 114
Si(100) 85, 105
Si(100) defect
– "split-of" dimer 241
– A-type 241
– B-type 241
– C-type 241
Si(100) step
– non-bonded 253
– rebonded 254
– type A 253
– type B 253
Si(100) terrace
– type A 253
– type B 253
Si(100)$c(4\times2)$ 183
Si(100)1×1-H 225
Si(100)2×1 130, 153, 155, 167, 181
– single-domain 256
Si(100)2×1-H 139, 224
Si(100)2×2-Ga 58
Si(100)3×1-H 225
Si(100)4×3-In 55
Si(111) 140
Si(111)$\sqrt{3}\times\sqrt{3}$-Ag 200, 222, 239
Si(111)$\sqrt{3}\times\sqrt{3}$-Al 218, 244
Si(111)$\sqrt{3}\times\sqrt{3}$-B 220
Si(111)$\sqrt{3}\times\sqrt{3}$-Bi 218, 220
Si(111)$\sqrt{3}\times\sqrt{3}$-Ga 218, 244
Si(111)$\sqrt{3}\times\sqrt{3}$-In 55, 218, 244, 247, 277
Si(111)$\sqrt{3}\times\sqrt{3}$-Pb 218, 222
Si(111)$\sqrt{3}\times\sqrt{3}$-Sb 220
Si(111)$\sqrt{3}\times\sqrt{3}$-Sn 218
Si(111)1×1 155